This book provides a state-of-the-art review of wind tunnel modeling for civil engineering applications. It contains over forty papers prepared by internationally recognized experts in the use of wind tunnel models for studying wind related engineering problems.

The first sessions helps set the stage for other sessions by presenting a discussion of problems that may require wind tunnel tests and methods for combining test results with other information in order to solve design problems.

The second session is devoted to the simulation of atmospheric winds and includes both a discussion of simulation requirements and methods that are being used to achieve the simulation. This session also contains a paper that probes questions concerning the consequences of distorting the flow simulation.

This is one of the first papers to specifically address effects of compromise in simulation for a variety of applications.

Sessions III, IV, and V discuss modeling and measurement techniques for a variety of problems. Emphases are placed on the measurement of surface pressures and on aeroelastic modeling.

Session VI provides an assessment of the validity of wind tunnel testing by comparing model and full-scale test results. This session provides a good overview of both the successes and failures of model studies.

Wind Tunnel Modeling for Civil Engineering Applications

Wind Tunnel Modeling for Civil Engineering Applications

Proceedings of the International Workshop on Wind Tunnel Modeling Criteria and Techniques in Civil Engineering Applications, Gaithersburg, Maryland, USA, April 1982

Edited by
Timothy A. Reinhold

National Bureau of Standards
U.S. Department of Commerce
United States of America

CAMBRIDGE UNIVERSITY PRESS

Cambridge

London • **New York** • **New Rochelle** • **Melbourne** • **Sydney**

Published by the Press Syndicate of the University of Cambridge

The Pitt Building, Trumpington Street, Cambridge, CB2 1RP

32 East 57th Street, New York, NY 10022, USA

296 Beaconsfield Parade, Middle Park, Melbourne 3206, Australia

First Published 1982

Printed in the United States of America

Library of Congress Cataloging in Publication Data

International Workshop on Wind Tunnel Modeling

    Criteria and Techniques in Civil Engineering

    Applications (1982: Gaithersburg, Md.)

    Wind tunnel modeling for civil engineering

    applications.

    1. Wind tunnel models--Congresses. 2. Civil

engineering--Congresses. 3. Engineering models--

Congresses. I. Reinhold, Timothy A., 1951-

II. Title

TA177.I57 1982   624.1'75'00287   82-14594

ISBN 0-521-25278-4

Foreword

Within the past twenty years the field of wind engineering has experienced tremendous growth in research and testing activities. As the field has developed, the wind tunnel (and more recently the boundary layer wind tunnel) has become the primary tool for use in studying wind engineering problems related to civil engineering applications. While there have been significant advances in mathematical modeling of wind effects for some special cases, it is expected that the wind tunnel will continue to be the primary means for investigating wind effects.

Some specialized wind tunnel facilities have been built for industrial aerodynamics applications, but in many cases research is being carried out in tunnels originally designed and built for aeronautical applications. The large differences in tunnel characteristics such as cross-sectional area, test section length, speed range, speed and temperature control, and tunnel layout have led to the development of many different approaches to and techniques for wind tunnel modeling. Consequently it is important to clearly define modeling criteria and acceptable limits of approximation. The need for defining modeling criteria and for developing some level of standardization in wind tunnel model tests is becoming more important as building codes and loading standards move toward recognition of wind tunnel model tests as an alternative to prescriptive wind loads.

This book contains the proceedings of an International Workshop on Wind Tunnel Modeling Criteria and Techniques for Civil Engineering Applications which was held at the U.S. National Bureau of Standards, Gaithersburg, Maryland, from April 14 to 16, 1982. The Workshop provided an international forum for the exchange of information, for the assessment of research accomplishments which help define modeling criteria and acceptable limits of approximation, for the discussion of different modeling techniques, and for defining needs and directions for future research.

More than forty papers, written by recognized experts, comprise the bulk of these proceedings and provide a state-of-the-art review of wind tunnel modeling for civil engineering problems. The honest portrayals of both successes and failures of modeling endeavors are an indication of the level of maturity that has been achieved in the field of wind engineering model studies. It is the hope of the organizers that these proceedings will serve as a unified resource document for engineers and architects who find a need for commissioning wind tunnel tests and for standards development bodies as they seek to define minimum requirements for acceptable wind tunnel tests.

The first session helps provide an introduction to the other sessions and to the need for wind tunnel model studies by presenting a discussion of problems which may require wind tunnel testing. Methods are also discussed for combining tests results with other information in order to solve design problems. The increased attention being given to assessing the reliability of designs which are based on test results are mirrored in the last three papers of this session.

The second session is devoted to the simulation of atmospheric winds and includes both a discussion of simulation requirements and methods which are being used to achieve the simulation. The first paper in the session provides a good description of the complexities and uncertainties involved in defining atmospheric conditions, especially when neutrally stable flow conditions do not exist. The fourth paper in this session probes questions concerning the consequences of distorting the flow simulation and is one of the first papers to specifically address effects of compromise in simulation for a number of different types of studies.

Sessions three, four, and five discuss modeling and measurement requirements and techniques for a variety of problems. Measurement of surface pressures and aeroelastic modeling are emphasized.

Session six provides an assessment of the validity of wind tunnel testing by comparing model and full-scale test results for high- and low-rise buildings, bridges, and chimneys. This session provides an overview of the successes and failures of model studies.

This material is based upon activities supported by the National Science Foundation under Agreement No. CEE-8103177 and by the National Bureau of Standards.

Any opinions, findings, and conclusions or recommendations expressed in this publication are those of the authors and do not necessarily reflect the views of the National Science Foundation or the National Bureau of Standards.

Editor

Contents

Acknowledgments

This publication represents the integration of contributions from many individuals and groups. The editor would like to begin by expressing a special word of appreciation to Dr. Richard D. Marshall of the National Bureau of Standards who originally brought the idea of holding such a workshop to the editor's attention and who provided guidance and assistance in securing the funding necessary to making the workshop a reality. Later, despite heavy commitments to other tasks, Dr. Marshall found time to provide encouragement, valuable advice on organizing the workshop and served as chairman of the fifth session. A second word of special appreciation is extended to Dr. Michael P. Gaus of the National Science Foundation for his encouragement and advice.

Many thanks are also due to the members of the steering committee for their valuable suggestions which helped shape the final form of the workshop and for their assistance in both selecting topics and potential authors. The steering committee members also provided organizational assistance by serving as session chairmen and many provided written contributions as well.

The editor is deeply indebted to all the authors for their many hours spent in preparing the written contributions which are contained in this book and for their participation in the workshop.

For assistance in handling extensive correspondence and administrative details before, during, and after the workshop, the efforts of other members of the organizing committee are gratefully acknowledged.

The National Science Foundation and the National Bureau of Standards provided necessary financial support for this workshop. The workshop was co-sponsored by the Wind Engineering Research Council, Inc. and by both the Aerodynamics and Wind Effects Committees of the American Society of Civil Engineers.

STEERING COMMITTEE

Frank Durgin, Wright Brothers Wind Tunnel, Mass. Inst. of Technology
Michael Gaus, Directorate for Engineering, National Science Foundation
Gary Hart, Englekirk and Hart Consulting Engineers, Los Angeles, California
Nicholas Isyumov, Boundary Layer Wind Tunnel Lab., Univ. of Western Ontario
Ahsan Kareem, Department of Civil Engineering, University of Houston
Jon Peterka, Department of Civil Engineering, Colorado State University
Robert Scanlan, Department of Civil Engineering, Princeton University

ORGANIZING COMMITTEE

Timothy Reinhold
Richard Marshall
Anita Sweigert
Ann Pararas

Session I

Interfacing Wind Tunnel Tests and Client's Needs

Chairman

Gary C. Hart

Englekirk and Hart, Inc.
Los Angeles, California
USA

WHEN A DESIGNER SHOULD CONSIDER HAVING WIND TUNNEL TESTS CONDUCTED TO ESTIMATE STRUCTURAL LOADS

Carlos M. Dobryn, Leslie E. Robertson, Saw-Teen See
Skilling, Helle, Christiansen, Robertson, P.C.

INTRODUCTION

The title of this presentation: "When a Designer Should Consider Having Wind Tunnel Tests Conducted to Estimate Structural Loads," can be said to be a subset of a more general problem confronting today's professional -- that is, when to draw in the rapidly developing field of professional expertise in highly specialized aspects of building technology.

As with the more general problem, the answer to our specific question defies the straightforward analytical treatment desired by design professionals. A particular design problem implies a solution within the real life constraints imposed by tight budgets, rigid fee structures and severe time constraints. And yet, these constraints notwithstanding, the need for special expertise -- particularly in the field of wind engineering -- is not only unmistakably clear but pressing and fast becoming inescapable.

Modern Engineering, by reason of its probing and ever-changing nature, is, and will continue to be, a balanced combination of art and science. A successful engineering solution implies a thorough understanding of the problem in terms of the expected structural behavior and the available means to accomplish it. Analytical and empirical approaches are undoubtedly of the essence, but, unless backed by sound judgment and experience, may lead to inadequate if not unsafe solutions. So, even though an algorithmic type of answer to our question defies present day knowledge, it is possible to provide some insight into the nature of the problem.

A basic component can be singled out; that is, the economics of the real life problem at hand. And by this is not meant just a cost-effectiveness analysis that would indicate the direct benefits of a wind-tunnel study, but perhaps a more subtle understanding of the costs involved as a kind of risk-insurance investment. It is obvious that, were wind-tunnel testing a cost-free endeavor, our question would disappear from engineering problematic. We will return to this later on in our presentation.

AVAILABLE WIND ENGINEERING STUDIES

New materials, lighter buildings, daring architectural forms and functions backed by sophisticated analysis techniques create a compelling need to obtain, in the final analysis, an accurate description of wind effects and a thorough understanding of wind-building interactions.

Analytical approaches to the problem do exist, but they fall often outside the boundaries of a specific design problem. It is at this point that laboratory techniques coupled with probablistic methodologies find their, so to say, "ecological niche" in the overall design process.

A growing number of wind laboratory studies are available. Among the most commonly used:

o   Cladding pressures: rapidly developing new building technologies and the number of inadequately performing or failed curtain wall systems make this a most important wind-tunnel study. For certain design situations, internal pressures may be as important as external ones. The influence of factors such as stack effects, air leakage and air handling equipment need further exploration and better understanding.

o   Aeroelastic studies: while collapse failure of complete buildings is not, historically, a cause of concern, serviceability limits associated with damaged architectural finishes and partitions, impaired mechanical systems, and the many problems associated with human perception of vibrations, make this a most important laboratory study.

In depth analysis techniques for tall buildings may include the interrogation of the ductile load-deformations characteristics of the structure, with due consideration given to second order effects and low cycle fatigue of the member joints. These type of analyses acquire meaning only when based on a complete description of the integrated mean and fluctuating pressures associated with given wind directions. A laboratory study of the static and dynamic components of the loading in the sway and torsional modes may, in some cases, be the only way of providing meaningful design information. Aeroelastic wind studies normally provide input information with respect to expected peak accelerations for various return periods. This information is of utmost importance in passing judgment as to the adequacy of tall buildings from the view point of occupants comfort.

o   Pedestrian wind environment: a new structure is in itself the cause of a disturbance in the existing wind environment. A proper understanding of the aerodynamic phenomena associated with the flow of air and the changes in flow patterns will permit timely design changes to restore and even improve the wind climate at the pedestrian level. Wind direction and velocity predictions associated with seasonal changes are the main output of this study.

o   Other significant wind studies include roof level windiness (vertical take-off aircraft), disposal of effluents and intake pollution (cooling towers, ambulance discharge zones, etc.), impact of the newly created flow patterns on surrounding structures, wind noise, snow drift and curtain wall mock-up testing.

FACTORS INFLUENCING THE DECISION

Of the several available laboratory wind studies, two are of the greatest concern for the professional architectural-engineer consultant: the cladding and aeroelastic series.

As indicated previously, economics lie at the root of the problem. In deciding whether or not to commission a wind tunnel or mock-up test, the architect-engineer is confronted with the task of evaluating the need for such studies and justifying this evaluation to the owner.

To set the problem in perspective, let us consider a nominal 120 feet square building, forty stories high (costs are indicative of a range and are purely for comparative purposes):

Building cost	$48,000,000
Structural cost	$12,000,000
Curtain wall cost	$ 4,000,000
Wind tunnel aeroelastic	$30,000
Wind tunnel cladding	$40,000
Mock-up test	$30,000

The wind tunnel aeroelastic test represents 0.25% of the cost of the structure and .06% of the cost of the building. The cladding test, 1.0% of the cladding cost and .08% of the cost of the building. The mock-up test cost -- routine for any major cladding system -- is comparable to the cladding study cost.

These figures, eloquent in themselves, will not provide any clue to an owner or developer as to their cost-effectiveness. To compound the problem, these are normally advance expenditures, the kind liked least by those in charge of financial management. Something more is needed. While wind tunnel design information cannot be forecast, it may not be improper to indicate that, in general, aeroelastic testing may predict lower loads than those of current codes. Cladding results may show a larger dispersion, with lower than code pressures in some areas and larger ones in others. In both studies, lower loads have a very clear and appealing economic implication, while higher ones can be understood within the context of the risk-insurance concept. Confidence in the wind tunnel predictions, of course, lies at the heart of economic justification.

In evaluating the need for wind laboratory studies, consideration of the factors influencing the decision will provide insight into the problem.

o Geographical location: wind loads grow very rapidly with wind velocities. For a given building and a given set of external actions -- dead, live, seismic, temperature, etc. -- the influence of wind loads may be very important or may be negligible. In the latter case, a tunnel study may not be justifiable in view of the small advantage or the low risk implied in adopting code loads.

o Topographical location: within a given geographical zone, consideration of the local topography and the pressence of surrounding structures will provide insight into the need for a laboratory study.

o Geometrical configuration: a quick examination of the building spacial shape and its relationship with the form and behavior of structural resisting systems will indicate very early whether the structure lies within the boundaries of the governing codes. Unusual building surface shapes and textures, the presence of crenellations and uncommon roof shapes, particularly if rising above surrounding buildings, may indicate that a wind study is necessary.

o Frequency of oscillation: for buildings with a high slenderness ratio, an estimate of the building density and its correlation with the stiffness criteria will permit a quick evaluation of magnification factors due to P-Delta effects. The need to avoid significant magnification factors brings up the question of how best to accomplish it. A wind tunnel study could provide insight into the effect of varying the design parameters, leading to an optimization of the final architectural-engineering solution. It is the function of the designer to provide model information, such as aerodynamic shape, mass matrix, sway and torsional stiffness or flexibility matrices and structural damping.

o Expected performance: beyond the standard safety and serviceability limit states, a building may have very special performance requirements dictated either by its function or by the owner's own performance criteria. Very strict performance requirements in terms of expected building behavior or human response may, by themselves, make the need for a wind laboratory study an unavoidable requirement.

o Any other unusual condition, particularly if not covered by the provisions of a modern code, should point out the need for a closer examination to determine the need for a wind laboratory study. Large open volume structures such as convention halls, hangars, stadiums, light cable hanging roof structures, shells, etc. deserve special consideration and an understanding of their internal and external pressures and their correlation.

CONCLUDING REMARKS

The recognition of the dynamic character of the wind and of its probabilistic nature has brought present knowledge closer to a true description of the wind-building interaction. This knowledge, however, is not yet sufficient to permit an algorithmic type of response to the main topic of this paper. This is particularly so for that narrow band of design problems for which not enough empirical data has been gathered to permit an analytical treatment. We are confident that further testing will move this boundary band up to encompass most of the real life problems facing the design professional. Meanwhile, a careful evaluation of the factors involved in a particular design problem will provide powerful insight into the need for wind laboratory testing.

As buildings increase in complexity, exploration in highly specialized areas grows in sophistication. However, further developments in wind research and the formulation of new and better methods of analysis will not prove fruitful unless accompanied by a parallel development in the area of acceptability criteria.

On the part of the wind laboratory consultant, a clear recognition of the needs of the design professional is needed. These needs can be stated in very simple terms: early release of sound preliminary predictions and clear and concise communication. It is a truism that early provision of model information by the design professional, leading to early release of laboratory predictions, constitutes the key to a successful wind engineering program.

As a final word, it should be noted that modern codes, such as the recently published ANSI A58.1-82, constitute a very important step forward in the transmission of established wind knowledge to the design profession. Within it, the upper boundaries of those problems amenable to analytical treatment can be found. Beyond that, the wind laboratory consultant is called upon to provide the answers, and hopefully with them -- the sure sign of uninterrupted scientific development --, further questioning.

REFERENCES

1. Building Code Requirements for Minimum Design Loads in Buildings and Other Structures, American National Standard A58.1-1982, American National Standards Institute, New York (1982).

2. Robertson, L.E., "Limitations on Swaying Motion of Tall Buildings Imposed by Human Response Factors," Proceedings of the Australian Conference on Planning and Design of Tall Buildings, University of Sydney (August 1972).

3. Robertson, L.E., "Le Phenomene P-Delta, Evalue sur la Base de la Periode Fundamentale D'Oscillation," Construction Metallique, No. 3 (1981).

4. Robertson, L.E., "The Relevance of Static and Dynamic Loading to Building Systems," Proceedings of the 4th International Conference on Wind Effects on Buildings and Structures, Heathrow (1975) pp. 777-779.

5. Isyumov, N. and Davenport, A.G., "Comparison of Full-Scale and Wind Tunnel Wind Speed Measurements in the Commerce Court Plaza," Journal of Industrial Aerodynamics, Vol. 1, No. 2 (1975) pp. 201-212.

6. Davenport, A.G. and Zilch, K., "Wind Loads and Safety of Tall Buildings," (Windlasten und Sicherheit von Hochlausern), Sicherheit von Betonbauten, Deutscher Beton-Verein, Wiesbaden, Germany.

7. Cermak, J.E., "Wind Loading and Wind Effects," Proceedings of 5th Regional Conference (Chicago, Ill.), Lehigh University, Bethlehem, Pennsylvania (1971) pp. 49-52.

8. Peterka, J.A., "Wind Pressures on Buildings - Probability Densities," Journal of the Structural Division, ASCE, Vol. 101, No. ST6, Proc. Paper 11373 (1975) pp. 1255-1267.

9. Melbourne, W.H., "Wind Tunnel Test Expectations," Proceedings of the International Conference on Planning and Design of Tall Buildings, Vol. 16, Lehigh University, Bethlehem, Pennsylvania (August 1972) pp. 441-444.

ON CONSIDERING PEDESTRIAN WINDS DURING BUILDING DESIGN

Ed Arens, Associate Professor, Department of Architecture, University of California, Berkeley

INTRODUCTION

Tall or exposed buildings adjacent to public open spaces may cause local winds at ground level that are much more intense than winds found elsewhere at ground level. These winds may affect the comfort and safety of pedestrians and thus reduce the usefulness of the outdoor open spaces. In recent years, wind problems have become more common, as more tall buildings are built and as cities and building owners place increasing emphasis on public plazas and open space. Since both the cost and economic benefits of such plazas and open spaces may be very high, significant financial losses may occur when such spaces are rendered unusable due to wind.

The designers of buildings and their sites would benefit from being able to anticipate, in the planning stage, the possibility of local wind flow zones that cause unacceptable discomfort to users of outdoor space. If such zones are found, appropriate design decisions can eliminate them or direct pedestrians away from them.

Designers concerned with acceptable outdoor environments will have to address the following issues during the design of a building and its surrounding open space:

1. How does the wind affect comfort and safety? What are the wind descriptors that measure comfort and safety? What are the limits of these descriptors at which the wind becomes uncomfortable or dangerous? These physiological and psychological questions can be answered by research in laboratories and in the field.

2. At any outdoor site, wind will probably exceed comfort or safety limits for a certain number of hours, minutes, or seconds during the year. The time during which the limits are exceeded, measured as a percentage of the total time that the project is occupied, indicate the users' probability of discomfort or danger. How large should these probabilities be? It is up to the designer and owner to decide acceptable probabilities for any given project, but certain levels have been suggested by researchers for use as design criteria.

3. How strong are the regional winds (as measured at the weather station) when local wind flows on site exceed comfort or safety limits? The relationship between the project's local wind and the wind of the surrounding region is a ratio based on the aerodynamic configuration of the project and its surroundings. The ratio varies with wind direction. Certain generalizations can be made, but complex configurations are best modeled in an appropriate wind tunnel.

Part of this paper appeared in the Transportation Engineering Journal, ASCE, Vol. 107, No. TE2, March 1981, and is reproduced here with permission of ASCE.

4. How often do these excessive regional winds occur? This information, extracted from climatological records, determines the probability of the local wind on the site exceeding the limits for comfort or safety.

5. If local winds exceed the comfort or safety limits an unacceptable percentage of time, what design measures will reduce their strength so that the velocity limits are exceeded less often? These measures are devised by a combination of common sense, experience, and wind tunnel testing.

The following five sections summarize the state-of-the-art in acceptability criteria and design procedures. An additional issue is when and to what level designers should deal with environmental winds during the normal process of designing buildings. What are the respective roles of the designer, wind consultant, wind tunnel modeling facilities, codes requirements, and general education? When is testing most effectively utilized, and at what level of detail and expense? Some suggestions are given in the sixth section.

WIND AND PEDESTRIANS

Wind influences comfort both mechanically, through its pressure effects and particle transport, and thermally, through wind chill. Pedestrian safety in the urban pedestrian environment is affected by mechanical pressure. Both mechanical and thermal effects are reviewed in the following to establish the wind speed limits above which comfort and safety are jeopardized. These limits form the basis for acceptability criteria used in design.

1.1 Mechanical Influences of Wind

Wind influences comfort mechanically through pressure effects and particle transport. Wind pressure causes disturbance of clothing and hair, resistance to walking, and buffeting of the body and carried objects such as umbrellas. Comfort also is affected when the wind lifts dust and grit particles to eye level, or drives rain laterally into the eyes or beneath clothing. At higher velocities, wind interferes with walking and endangers people by causing them to lose their balance. At such velocities, eye damage from dust or possibly from flapping hair is also a safety problem. Some of the effects described are caused by the wind on a continuing basis, while others are precipitated by sudden unexpected peak gusts: a summary of such effects is provided in Refs. 1 and 2.

Wind at pedestrian level is accompanied by turbulence, and perceived as varying velocity, gusts, or eddies. The intensity of turbulence for any given wind speed varies from place to place, tending to be greater in urban or built-up surroundings than in open countryside. The effects of turbulence on pedestrians may be described as an addition to mean velocity. An "equivalent steady wind" defined as giving the same comfort or safety effect as the turbulent wind was introduced by Hunt, Poulton, and Mumford (3):

$$u_s = \bar{u}(1 + a \cdot TI) \dots\dots\dots\dots\dots\dots\dots\dots\dots\dots\dots\dots\dots (1)$$

in which u_s = the equivalent steady wind; $\bar{u}$ = the mean wind speed; a = an empirically determined coefficient; and TI = the relative turbulence intensity (the root mean square of the instantaneous deviations from the mean velocity divided by $\bar{u}$).

Hunt et al. found a $\simeq 3$ in experimental observations of the performance of pedestrians in a wind tunnel with controllable turbulence characteristics. The steady wind, u_s is a value greater than the mean, reflecting turbulent fluctuations superposed on the mean speed.

Jackson (4) used this relationship in defining a "standard equivalent mean wind speed" $\bar{u}_{se}$, in which terms he assembled a wide range of previously observed wind effects into a table combining the effects caused by steady uniform winds, turbulent wind fluctuations, and infrequently occurring peak gusts (see Table 1).

Certain standardized conditions were necessary to compile Table 1:

1. The standardized equivalent wind speed, $\bar{u}_{se}$, is measured at a height of 2m over an averaging time of 5 min. Published observations for different averaging times are corrected to this time.

2. The horizontal relative turbulence intensity is 18%. Observations for which TI is unspecified are assumed to have occurred with a TI of 18%. Observations with specified TI values different from 18% are converted by the following relationship:

$$\bar{u}(1 + 3\ TI) = \bar{u}_{se}(1 + 3 \times 0.18) \dots\dots\dots\dots\dots\dots\dots\dots\dots\dots\dots\dots (2)$$

3. For effects caused by a maximum gust of some minimum required duration, a value of $\bar{u}_{se}$ is calculated that would be expected to produce one such gust in the 5-min averaging period under 18% TI.

Wind effects observations as summarized in Table 1 and Ref. 5 are the basis for making judgments about wind acceptability in buildings and open space. Recommended acceptability criteria will be described later.

1.2 Thermal Influences of Wind

The familiar concept of wind chill reflects the thermal influence of wind. In cool climates, wind increases the rate of cooling of the body by removing the insulating film of still air found next to skin and clothing in calm conditions. The increased rate of cooling may cause discomfort. In hot climates, the increased wind-induced convection and evaporation may be beneficial to comfort.

Thermal comfort is influenced by the following climatic variables: air, temperature, radiation (solar and terrestrial), humidity, and wind. The pedestrian's clothing and activity level are also important variables. Thermal comfort is a function of body and skin temperature, the rate of heat transfer, and in overheated conditions, of skin wettedness as well.

To date, attempts to develop thermal models of human comfort outdoors have been confined to steady-state thermal balance models (2) in which thermal equilibrium is assumed to assure comfort. Such models have not been useful in practice primarily because thermal equilibrium with the surroundings requires 1-2 h continuous exposure to the outside environment. This is a rare situation for pedestrians, although such exposures may be experienced at bus stops and sports stadia. The average exposure is much shorter.

TABLE 1 Wind Effects Versus Standard Equivalent Mean Wind Speed Under Standard Conditions

Standard Equivalent mean wind speed, in meters per second	Effects Observed or Deduced
0	Calm, no noticeable wind
2	Wind felt on face
	Clothing flaps (5)*
	Newspaper reading becomes difficult (Ref. 2)
4	
	Hair disarranged (5), dust and paper raised, rain and sleet driven (Ref. 2).
6	
8	
	Control of walking begins to be impaired
	Violent flapping of clothes (5), progress into wind slightly slowed
10	Umbrella used with diffuculty
12	Blown sideways (2), inconvenience felt walking into wind, hair blown straight
	Difficult to walk steadily, appreciably slowed into wind (10)
	Noise on ears unpleasant
14	
	Generally impedes progress
	Almost halted into wind, uncontrolled tottering downwind (10)
	Difficulty with balance in gusts (2)
16	
	Unbalanced, grabbing at supports (2)
18	People blown over in gusts (3)
20	
22	Cannot stand (3)

*The minimum gust duration required for each effect to be experienced is given, in seconds, in parentheses.

A computer program that iteratively calculates the thermal response of the body over a series of 1-min intervals shows promise for providing thermal design criteria for the wind environment (6). The effect of any period of exposure to outdoor climates on body temperature, heat transfer, thermal sensation, and comfort sensation can be predicted in this way. This model currently suffers from lack of experimental information on wind penetration or infiltration of clothing, and requires validation in outdoor conditions.

2. DESIGN CRITERIA: COMFORT, SAFETY AND PROJECT ACCEPTABILITY

With an understanding of wind effects on people, it is possible to suggest limits above which wind should not be permitted or is not desired. The Building Research Establishment, in England, recommended the following well-known values for mean wind at pedestrian height over an unspecified averaging period, in cool conditions (7): at 5 m/s, there is an onset of discomfort; at 10 m/s, it is definitely unpleasant; and at 20 m/s, it becomes dangerous.

Hunt et al. (3) refined these recommendations to the velocities given in Table 2. Murakami et al. (5) corroborated Hunt's limits, with some values somewhat more restrictive, finding control of walking difficult in the 10 m/s–15 m/s range. Nonuniformity of wind in time and space greatly affects walking and can cause the observed effect "walking difficult to control" in wind speeds as low as 3 m/s.

TABLE 2 Wind Velocity Limits

Wind and Effect	Criterion*
Steady uniform wind:	
For comfort and little effect on performance	$\bar{u} < 6$ m/s
For ease of walking	$\bar{u} < 13$ m/s–15 m/s
For safety of walking	$\bar{u} < 20$ m/s–30 m/s
Nonuniform winds ($\bar{u}$ varies by at least 70% over a distance less than 2 m):	
To avoid momentary loss of balance and to be able to walk straight	$\bar{u} < 9$ m/s
For safety (for elderly people this criterion may be too high)	$\bar{u} < 13$ m/s–20 m/s
Gusty Winds**:	
For comfort and little effect on performance	$u_s < 6$ m/s
Most performance unaffected	$u_s < 9$ m/s
Control of walking	$u_s < 15$ m/s
Safety of walking	$u_s < 20$ m/s

*$\bar{u}$ is averaged over short periods, on the order of seconds.
** u_s is defined in Eq. 1.

With a wind speed limit in hand, the designer must judge what percentage of the time (or with what frequency in a year or season) it may be exceeded. The percentage of time exceeded equals the probability of discomfort or danger. For comfort limits, 10%–20% may seem reasonable. For safety limits, lower percentages of time exceeded (say 0.1%) should be applied to the higher velocities. Various researchers have suggested limits and acceptable frequencies for the limits. These have been summarized and compared by Melbourne (8).

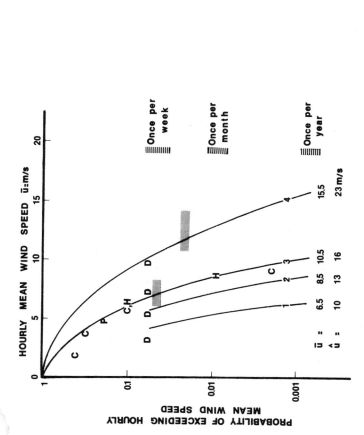

13

LEGEND FOR COMFORT CRITERIA GRAPH

CRITERIA:		SYMBOL:
MELBOURNE (9)		
Acceptable below for:		
Walking		1
Stationary, short exposure		2
Stationary, long exposure		3
Unacceptable, dangerous above:		4
DAVENPORT (10)		
Acceptable for:		
Walking fast	If $P(\bar{u} > 10) < 0.05$	D
Strolling	If $P(\bar{u} > 7.5) < 0.05$	D
Standing, Sitting		
Short Exposure	If $P(\bar{u} > 5.5) < 0.05$	D
Standing, Sitting		
Long Exposure	If $P(\bar{u} > 3.5) < 0.05$	D
PENWARDEN & WISE (19)		
Acceptable	If $P(\bar{u} > 5) < 0.2$	P
LAWSON (11)		
Acceptable	If $P(\bar{u} > 6 \text{ to } 8) < 0.04$	
Unacceptable	If $P(\bar{u} > 11 \text{ to } 14) < 0.02$	
HUNT ET AL. (3)		
Acceptable for:		
Strolling	If $P(\bar{u} > 6) < 0.1$	H
Acceptable for:		
Walking	If $P(\bar{u} > 9) < 0.01$	H
COHEN ET AL. (12)		
Data from Table 3.		C

FIG. 1 Comparison of Various Criteria for Environmental Wind Conditions for Daylight Hours, for Relative Turbulence Intensity of 15%.

14

Fig. 1 is a reproduction of a figure from Ref. 8 with some additions as noted later. The figure compares the various criteria using the probability of exceeding various hourly mean speed limits in any given year. The probabilities are adjusted to apply to half the hours in the year representing the daylight hours when most buildings are occupied. The curves imply a relative turbulence intensity of 15%.

Isyumov (13,14) presents a plot of comfort and safety versus hourly mean wind speed similar to Melbourne's, but expressed in terms of number of occurrences per year where the hourly mean wind speed is greater than the comfort limits indicated. The return periods are calculated using a method by Davenport (15).

Cohen et al. (12) recommends a number of acceptability criteria based on their extensive field observations (see Table 3). These findings have been incorporated into Fig. 1 correcting for daylight hours for consistency with Melbourne's method. The safety-related values show lower wind speed limits in the acceptability criteria than the other researchers, almost certainly because the peak gusts recorded per hour in this investigation tended to three times the hourly mean, whereas such gusts in Melbourne's

TABLE 3 Pedestrian Safety/Comfort Standards for Urban Winds

Activity Area	Hourly mean wind speed in meters per second (miles per hour)	Permitted occurrence frequency, as a percentage	Permitted frequency considering only daylight hours, as a percentage
All pedestrian areas limit for safety	9.1 (20)	0.1 (≈10 h/yr)	0.2
Major walkways, especially principal egress path for high-rise buildings	9.1	0.1	0.2
Other pedestrian walkways, including street and arcade shopping areas	6.4 (14)	5	10
Open plazas and park areas walking, strolling activities	3.6 (8)	15	30
Open plaza and park sitting areas, open-air restaurants	2.3 (5)	20	40

assumed turbulence would be only 1.5 times the hourly mean. Because the turbulence intensity was not reported with Cohen's data, it is not possible to adjust these values to make them fully comparable. The comparison does show the significance of assumed turbulence intensity in these acceptability criteria for safety purposes, where the influence of the peak gust predominates.

Penwarden (1) analyzed cases of shopping centers that had experienced wind complaints. He found that in centers where the single limit of 5 m/s hourly mean wind speed was exceeded 20% of the time or more, the owners invariably spend money to add protective screens or roofs. Centers with frequencies of 10-20% caused complaints but no remedial action was taken. Few complaints were registered at centers with frequencies below 10%. Penwarden's study gives the most concrete economic evidence in support of specific acceptability criteria to date.

Caution should be exercised in applying such criteria to wind data available from the United States National Oceanic and Atmospheric Administration (NOAA). In a nonsteady wind, the length of the averaging interval will affect the value of $\bar{u}$ associated with the maximum wind effect observed during that interval. This is because most wind effects are sensed by the pedestrian over very short intervals of time, 1 sec-10 sec, while wind is usually observed and recorded for longer intervals, of 1 min or more. Melbourne's and Cohen's effects on the chart are either observed against hourly means, or converted to be so. The actual effects, however, may refer to instances that may have occurred only once, during a gust, within that hour.

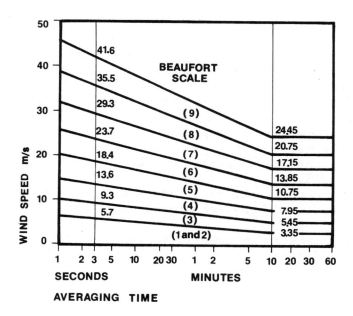

FIG. 2 Variation of Wind Speed with Averaging Time, for Velocity Ranges of Beaufort Scale

The NOAA wind records available in the United States are termed hourly mean wind speeds, but are actually 1-min means measured once an hour. This means that the distribution of windspeeds from NOAA data will show greater variability than would true hourly means. The pedestrian effects specified for gusts implied within Melbourne's hourly mean wind speeds would therefore occur at higher mean wind speeds when using NOAA data, speeds closer to those of the gust itself. Fig. 2, reproduced from Lawson (16), gives an example of the variation of wind speed with averaging time at a relative turbulence intensity of 0.28. One may estimate from this that the hourly mean criteria should be 25% higher when using NOAA data; that the 5-min mean wind speed limits described above should be 20% higher; and that criteria based on direct peak gust measurement (2 sec-3sec), should be 25% lower.

Finally, caution is needed when using any of the criteria based on mechanically caused wind discomfort, for thermal discomfort in cool environments usually begins at lower velocities than mechanical discomfort. Reliance on these may underestimate the actual discomfort, especially for prolonged or relatively inactive outdoor activities.

3. DETERMINING SURFACE WINDS IN PROPOSED BUILT ENVIRONMENT

The designer should consider outdoor comfort and safety early in the design process. Because the relationships between the physical form of a building and site, the climate, and resulting comfort and safety around it are complex, he or she may have to follow a climatic design process in order to find a satisfactory solution. The process, basically: (1) Determines the climatic characteristics of the site and the preliminary project, partly by model tests; (2) assesses its effects on the acceptability of the project; and (3) modifies the project design and tests the climate until a solution is reached.

Winds at pedestrian level often will be strongly affected by the building, planting, and grading configuration of the project. Briefly, there are basically three flow fields that cause enhanced wind at pedestrian level: (1) Vortices at the base of the windward face of a building caused by greater pressures on that face at higher elevations; (2) flows caused by pressure differences between low pressure regions at the sides and lee of a building and the relatively higher pressure regions at the windward side (open passageways between these regions will experience greatly enhanced wind); and (3) flows through constrictions between buildings. Two publications by Gandemer (17, 18) provide visualizations of wind flow around a wide variety of building configurations. They are very useful for obtaining an intuitive feel for wind behavior around buildings.

Penwarden and Wise (19) have provided generalized rooftop-to-ground-level velocity ratios for such configurations that have been widely quoted. However, note that these ratios do not apply when buildings of similar height are located in the upwind direction.

If the configuration of a yet-unbuilt project seems likely to cause high local winds, it should be tested in model form in a wind tunnel. This technique is also useful for defining winds on existing sites, since the flow strength and direction can be controlled during the tests. Physical modeling with limited field verification is most desirable.

Physical modeling requires the use of a specialized wind tunnel that reproduces the boundary-layer conditions above the actual site. Both the velocity and the turbulence intensity profiles should be modeled to scale. The most satisfactory means of achieving this at present is to generate the boundary layer with turbulence generators and long fetches of roughness similar to that of the terrain upstream of the project site. The testing of architectural models for environmental wind conditions may be carried out at low wind speeds (5 m/s-10 m/s) because turbulent flow patterns around bluff(sharp edged) objects do not vary over a wide range of velocities. This is fortunate, for it allows tests to be performed at costs comparable to other design consulting fees.

Velocities measured in the wind tunnel are nondimensionalized and are expressed as a percentage of a reference velocity. The reference velocity in the tunnel is measured at a reference height, often chosen as the height (at model scale) of the wind instrumentation of the weather station providing climatological wind data. By relating wind tunnel measurements to climatological data, wind speed frequency distributions are found for important locations within and around the proposed object.

Measurements are normally made at a network of grid points on the project model. A hot-wire anemometer is usually used to measure wind speed and turbulence at each grid point. Its small size allows it to measure speeds within distances of the order of millimeters above the surface, representing pedestrian height at model scale. In addition, when the wire is held vertically, the hot-wire anemometer is insensitive to the azimuth angle of approaching wind (within a sector of about 270°). Since the wind turbulence at pedestrian level near buildings often results in rapid lateral directional changes, it is important that the measuring instrument exhibit minimal directional sensitivity in the horizontal plane.

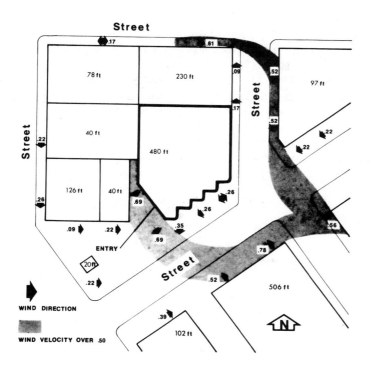

FIG. 3 Street-Level Wind Directions and Velocities during West Wind, High-Rise Project

It is also possible to mount small pressure-sensing devices on the model at pedestrian height that will give omnidirectional wind response. Although such devices are more complicated to install at the outset, they have the advantage of not requiring repositioning as the model is rotated for different wind directions.

The entire network of points on the model is tested separately for each wind direction, the number of directions normally corresponding to the number of points of the compass considered in the meteorological wind data base.

Fig. 3 is a representation of a strong air flow pattern occurring around a wind tunnel model of a San Francisco high-rise project during exposure to west winds. The building's entrance is located on the southwest side of the building. The wind directions at several grid points are marked and the velocities expressed as decimal fractions of the velocity at the reference height, which in this case is 132 feet, the height of the anemometer at the local weather station in the city.

4. DETERMINING FREQUENCIES OF UNCOMFORTABLE OR DANGEROUS WINDS IN PROPOSED BUILT ENVIRONMENT

Wind data are usually recorded at airports in open terrain. For most sites wind information must be extrapolated geographically from the recording station to the vicinity of the site. The effects of topography, vegetation, and structures must be carefully considered in this extrapolation.

The meteorological data base should provide information necessary to determine the amount of time that pedestrians will be uncomfortable or endangered on the site. This requires predictions of expected winds preferably by time of day, and if thermal comfort is being estimated, coincident data on temperature and sun. The most useful data format is a cumulative directional frequency distribution providing the percentage of time that each wind velocity is exceeded for each wind direction. The National Climatic Center of the NOAA may have this information in existing published summaries, although most locations have not been analyzed in detail. The Center will prepare such summaries from original hourly observations upon request. It also provides magnetic tapes with the hourly observations, usually with 10 yr/tape. The data from the tape is then commonly fitted to a model such as the Weibull distribution to provide a smooth frequency distribution.

The frequency of discomfort may then be determined for either thermal or mechanical wind effects by following the procedure in Fig. 4 taken from Ref. 2. The same approach may be used to predict dangerous wind velocity frequencies. The procedure for thermal comfort is carried out for selected hours of the day to incorporate sun/shade patterns and temperature data into the comfort determination.

Fig. 5 shows the probability of discomfort around the high-rise project considering all wind conditions expected in San Francisco in the early afternoon during the autumn season. The predominance of west winds over other winds during this period causes the discomfort values to closely follow the wind patterns shown in Fig. 3. The threshold velocity used here for determining discomfort is 5 m/s. The shadow patterns for this period are also shown, to give a suggestion of the thermal environment.

Figure 4   Process for Determining Probabilities of Discomfort or Danger

DISCOMFORT OR DANGER
BASED ON MECHANICAL
WIND EFFECTS

CLIMATOLOGICAL
AND MODEL INPUT

DISCOMFORT BASED ON
A THERMAL MODEL OF
CLIMATIC EFFECTS

For each point of interest
define discomfort or danger
threshold windspeed, $\bar{u}$
or equivalent steady wind,
U_s*

Average temperature,
sun-shade patterns,
comfort curves

For each point of interest
define discomfort
threshold windspeed
in sun and shade

For each wind direction
compute reference wind-
speed corresponding to
threshold

 Nondimensional wind-
speeds and turbulence
from wind tunnel

For each wind direction
compute reference wind-
speed corresponding to
threshold windspeed for
sun and shade

Record frequency of winds
greater than computed
reference windspeed
for each wind direction

 Cumulative windspeed
frequency distribution

Record frequency of wind
greater than computed
reference windspeed
for sun and shade for
each wind direction

Weight each direction
by directional frequency
and sum overall direc-
tions

 Wind direction fre-
quency distribution

Weight each direction
by directional frequency
and sum overall direc-
tions

Percent possible
sunshine

Weight by frequency of
sun and shade

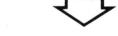

Probability of discomfort
or danger

Probability of discomfort

*See Table 2

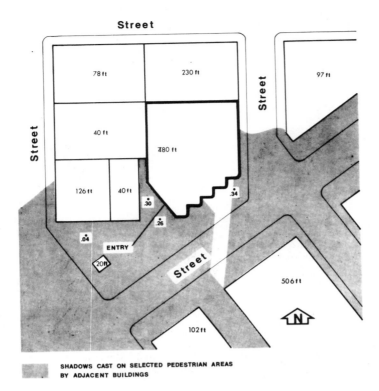

SHADOWS CAST ON SELECTED PEDESTRIAN AREAS
BY ADJACENT BUILDINGS

FIG. 5 Probability of Discomfort at 1:00 P.M. Autumn Season

5. AVOIDING AND MITIGATING WIND PROBLEMS

If the expected comfort levels are unacceptable, design modifications to improve conditions may have to be investigated. The following measures are appropriate for tall buildings (20,9):

1. Large slab buildings should not be oriented in a direction normal to the prevailing wind to avoid downwash on the windward face. Circular and polygonal towers tend to have advantageous wind climates at ground level because of reduced downwash.

2. Tall buildings benefit from significant horizontal projections to break up downward-directional winds. Protective awnings and canopies near ground level need to be large to influence wind over appreciable areas around buildings.

3. Important pedestrian thoroughfares and building entrances should not be planned at the windward corners of tall slab buildings as these are regions of accelerated corner flows.

4. Openings through buildings near the ground, especially with openings facing the prevailing wind, will experience strong winds unless revolving doors are used.

5. Vegetation may be used to absorb horizontal wind energy in pedestrian areas. Trees and shrubbery are not usually effective at protecting appreciable areas from downdraft winds.

Based on trials with wind tunnel models, design modifications within the site and budgetary limitations of the project may be selected. Modifications to reduce winds can range from changes in building height, bulk, or orientation, to the provision of vegetation or landscaping.

For any given problem there is a range of possible solutions varying in effectiveness and cost that should be optimized. For example, roofing over a shopping mall would surely reduce winds, but construction of wind deflectors or latticework or use of vegetation may yield the same results at far less cost. The influence of such devices on sunlight should also be considered.

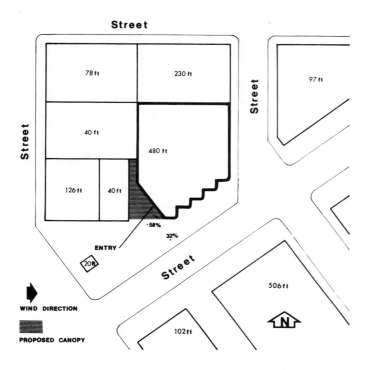

FIG. 6 Percentage Reductions in Wind Speeds Near Entry due to Proposed Canopy above Entry

A model of the modified design is tested in the wind tunnel and the analysis of comfort repeated. Further modifications are again suggested and the process is repeated until a satisfactory design is obtained.

Fig. 6 shows the reduction in the wind speeds at the sidewalk for the west wind case when a canopy is extended over the entrance. The reduction in expected discomfort on the sidewalk may be determined from this, allowing the designer to assess the feasibility of the canopy.

6. THE PROCESS OF SOLVING WIND PROBLEMS DURING DESIGN

The foregoing sections summarize the essential steps of wind design. This section comments on how these steps are carried out: by whom, when, and to what level of effort. It suggests that the process of wind design could become more widespread and successful if it were possible to inform the project architect(s) about wind earlier in the design process than is presently the case.

U.S. architects are generally not informed about the architectural causes of the pedestrian wind environment. At present, the proportion of designers sensitive to wind concerns is small, and the number that have expertise, or a systematic procedure for designing for those concerns, is even smaller. Wind consciousness may come from bad experience with particular buildings and climates, or may be promoted by the small number of cities that require wind assessments as part of the permit-granting procedure. In general, it has not come from training in architecture schools. After school, the main source of technical information to U.S. architects is the architectural professional press. The architectural journals have not provided the case histories or the types of qualitative instruction to be found in the illustrated publications of Gandemer (18, 17) or the Building Research Establishment (19). It has been the engineers, acquainted with aerodynamic effects through their concern with structural wind loading, that have initiated most environmental wind studies.

The early design decisions have the largest impact on the eventual wind environment. Such decisions include siting, building, massing, building orientation, and location of walkways, entries, and other uses of outdoor space. Although the structural engineer is a member of the design team from the outset of the project, he will not require structural wind analysis until the geometry of the building is largely fixed, and a detailed model of the building can be built for wind tunnel testing. (If the building is not tall, there is unlikely to be any wind analysis regardless of its expense and potential for causing environmental winds.) If environmental winds are evaluated at the same time as structural loads, it is usually too late to eliminate the causes of problem wind flows. The wind analysis can only identify the problems and propose relatively superficial remedial action. The wind information is applied at the wrong time, if at all, causing both the design process and the resulting built environment to be less effective than they might be.

Environmental wind design will be improved by a combination of developments:

1. Awareness. The first is increased awareness that wind is a design problem that can be solved. This awareness is helped along by the creation of environmental wind codes in cities with histories of problems (Toronto, San Francisco, Melbourne and Tokyo are examples). Codes also promote the creation of local wind consultants who become experienced with both the city's specific climatic characteristics and its design community. In pursuing their business, they distribute information on the wind problem and the process of dealing with it.

2. Initial design. Architects need help in making better design decisions in schematic design, since such decisions tend to become locked in by subsequent design work. This includes avoiding mistakes or issues such as siting, massing and positioning of passageways. The level of information needed here could be contained in systematic handbooks such as Gandemer's. It could also come from published analysis of specific problem situations such as the discussion of Croyden Center in Ref. 19, and from publications of wind criteria and wind design methods. There is a gap in the U.S. literature at

this level. Finally, the designer may rely on advice of a wind consultant, prior to any form of model testing.

3. <u>Model testing during schematic design</u>. Most design situations are too complex to be evaluated to any level of detail by handbook methods. If there is a potential wind problem, it is desirable to test models of various alternatives at the same time as the design is being developed. These exploratory texts should be quick, inexpensive, and involve the actual designer in the tunnel as much as possible. The precision of the simulation can be relaxed somewhat, particularly in the number of measurement points and wind direction tested; since the objective is to identify problems and solutions and to evaluate them comparatively. Ideally, the process continues during design as successive design developments are brought in for checking.

Here is an example of how such a test is usefully carried out. The designer has discussed the project with a wind consultant and they have agreed that there are potential wind problems that should be looked at in detail. The consultant has access to a wind tunnel facility. For the large majority of projects, there is a distinct advantage in having an inexpensive facility near the architect's office. Is is likely to be an old aeronautical or open jet tunnel in a nearby university, or possibly a low-speed boundary layer wind tunnel which can be built relatively inexpensively. In any case, it is desirable to have side access to the test section while the tunnel is running, in order to manipulate smoke probes and anemometers. The Building Research Establishment's enclosed open jet configuration is ideal for this purpose. The wind velocity during testing is usually between 5 and 10 meters per second.

The testing is carried out in the space of a day or two. Cardboard or foam preliminary massing models of the building and site are prepared by the architect and delivered to the tunnel. The wind tunnel technician sets up an appropriate profile in the tunnel and makes a rough simulation of the site surroundings using foam or wood blocks. The technician may then look at the windfield in the model in order to spot significant features. This allows him to better direct the investigation by selecting the problem wind directions and measurement points in advance. It helps if there are few predominant wind directions, as caused by the climate or by the influence of the surroundings.

The architect then joins the consultant or technician in the tunnel for up to 1/2 a day. They look at the problem flow areas and modify the model to test design solutions. The tools used are pieces of cardboard, polystyrene foam, a hot-wire foam cutter, a smoke probe, a preferably non-directional hot wire or hot film anemometer reading mean velocity and turbulence intensity output, floodlights and a camera. It is desirable to have the anemometer either loose or on a small tripod allowing it to be quickly repositioned to any point of interest. A velocity wind rose, preferably one for each season, and windspeed criteria should be at hand for immediate evaluation of certain conditions. If the test is conducted effectively, the architect very quickly gains a large number of insights about configuring the design, at a time when configuration is still open to change.

4. <u>Checking Environmental Winds While Testing Structural Loads</u>. A more limited number of projects, nearly all of them high-rise buildings, will be tested for wind-induced loads. These tests require a detailed model, tested for all significant wind directions. Wind engineers now commonly evaluate environmental winds at the same time as pressure tests are performed on the building itself. Although the information tends to come late in the architect's design, there may still be remedies to problems that are encountered. Analyses at this stage may also be used as documentation to prove code compliance.

7. CONCLUSIONS

Wind effects on comfort and safety have been described and limits for acceptable velocities proposed by various authors have been summarized. These limits are strongly influenced by the averaging interval selected, and by the turbulence component of the wind. The paper presents suggested design criteria for the amount of time that these velocity limits may be exceeded.

The designer of a project affecting pedestrians outdoors may estimate whether the project meets these criteria by synthesizing information on the aerodynamic characteristics of the project and climatological information on wind frequency distributions for the region. Some general prolem-causing building geometries can be identified, but many urban sites are sufficiently complex to justify model testing in a wind tunnel. A procedure for determining project acceptability through model testing is outlined.

Suggestions are made for staging the wind analysis in order to make it occur earlier in the design process than is now commonly the case. This would allow designers to avoid problem-causing configurations from the outset.

REFERENCES

1. Penwarden, A.D., Acceptable Wind Speeds in Towns, <u>Building Science</u>, 8 (1973) 259-267.

2. Arens, E.A. and Ballanti, D.B., Outdoor Comfort of Pedestrians in Cities, <u>Proceedings of the Conference on the Metropolitan Physical Environment</u>; also <u>General Technical Report NE-25</u>, United States Department of Agriculture Forest Service 1977, 115-129.

3. Hunt, J.C.R., Poulton, E.C., and Mumford, J.C., The Effects of Wind on People: New Criteria Based on Wind Tunnel Experiments, <u>Building and Environment</u>, 11 (1976) 15-28.

4. Jackson, P.S. The Evaluation of Windy Environments, <u>Building and Environment</u>, 13 (1978) 251-260.

5. Murakami, S., Uehara, K., and Deguchi, K., Wind Effects on Pedestrians: New Criteria Based on Outdoor Observation of Over 2000 Persons, *Proceedings of the Fifth International Conference on Wind Engineering*, Pergamon Press, New York, N.Y., 1 and 2 (1979 and 1980).

6. Gagge, A.P., Nishi, Y., and Nevins, R.G., The Role of Clothing in Meeting FEA Energy Conservation Guidelines, *Transactions*, American Society of Heating, Refrigerating, and Air-Conditioning Engineers, 82 (1977) 234-247.

7. Wise, A.F.E., Wind Effects Due to Groups of Buildings, *Current Paper 23/70*, Building Research Station, Watford, England, 1970.

8. Melbourne, W.H., Criteria for Environmental Wind Conditions, *Journal of Industrial Aerodynamics*, 3 (1978) 241-249.

9. Melbourne, W. H., and Joubert, P.N., Problems of Wind Flow at the Base of Tall Buildings, *Proceedings of the Third International Conference on Wind Effects on Buildings and Structures*, Saikon Co., Tokyo, Japan, 1971.

10. Davenport, A.G., An Approach to Human Comfort Criteria for Environmental Wind Conditions, CIB/WMO Colloquia, *Teaching the Teachers*, Swedish National Building Research Institute, Stockholm, Sweden, 1972.

11. Lawson, T.V., The Wind Environment of Buildings: A Logical Approach to the Establishment of Criteria, *Report No. TVL 7321*, Department of Aeronautical Engineering, University of Bristol, Bristol, England, 1973.

12. Cohen, H., McLaren, T., Moss, S., Petyk, R., and Zube, E., *Pedestrians and Wind in the Urban Environment*, UMASS/IME/R-77/13, University of Massachusetts, Amherst, Mass., 1977.

13. Isyumov, N., Studies of the Pedestrian Level Wind Environment at the Boundary Layer Wind Tunnel of the University of Western Ontario, *Journal of Industrial Aerodynamics*, 3 (1978) 187-200.

14. Isyumov, N., and Davenport, A.G., The Ground Level Wind Environment in Built-up Areas, *Proceedings Fourth International Conference on on Buildings and Structures*, Cambridge University Press, Cambridge, England, 1977.

15. Davenport, A.G., On the Statistical Prediction of Structural Performance in the Wind Environment, presented at the April 1971, ASCE National Structural Engineering Meeting held at Baltimore, Md. (Preprint 1420).

16. Lawson, T.V., The Wind Content of the Built Environment, *Journal of Industrial Aerodynamics*, 3 (1978) 93-105.

17. Gandemer, J., Discomfort Due to Wind Near Buildings: Aerodynamic Concepts, Report of the Centre Scientific et Technique du Batiment, Paris, France, translated and issued as *NBS Technical Note 710-9*, National Bureau of Standards, Washington, D.C., 1978.

18. Gandemer, J., and Guyot, A., Integration du Phenomene Vent dans la Conception du Milieu Bati, Report of the Secretariat General du Group Central des Villes Nouvelles, Paris, France, 1976.

19. Penwarden, A.D., and Wise, A.F.E., Wind Environment Around Buildings, Building Research Establishment Report, London, England, HMSO, 1975.

20. Aynsley, R.M., Effects of Airflow on Human Comfort, Building Science, 9 (1974) 91-94.

RELIABILITY - BASED DESIGN CONSIDERATIONS FOR RELATING LOADS MEASURED IN WIND TUNNEL TO STRUCTURAL RESISTANCE

Gary C. Hart, Principal, Englekirk & Hart Consulting Engineers, Inc. Professor of Engineering & Applied Science, University of California, Los Angeles, California.

Bruce Ellingwood, Center for Building Technology, National Engineering Laboratory, National Bureau of Standards, Washington, D.C.

1.   INTRODUCTION

Considerable effort has recently been devoted to the development of a reliability based design methodology. Load factors have been proposed which are appropriate for all types of building materials[1]. Resistance factors (ϕ factors) for use in the design of different materials that are consistent with these factored loads have or are being developed. The development of these load/strength relationships requires that we define the statistical nature of loads and resistances and that we specify a certain level of acceptable risk in setting performance criteria for design.

The specification of structural loads due to wind has become more comprehensive and complex over the past two decades[2]. At the same time, there has been an increasing use of the wind tunnel as a design tool, particularly on large projects[3]. There are several reasons for these trends. Complex structures have been introduced for which reference pressure coefficients in codes and standards may not be adequate. Light buildings or tall, slender buildings in which the dynamic response to wind is significant are becoming more common. The sensitivity of certain structures to wind direction or to the channeling effects caused by nearby structures and cross-wind response is becoming more widely appreciated. Structural engineers have made significant advances in the analytical procedures they use for analyzing structural frames and components. Post yield and inelastic analysis procedures are frequently used in the design verification phase on major projects. These advances and the move toward limit states design, with its focus on the behavior of the structure in resisting extreme and service loads, demand a more accurate characterization of structural loads.

The increasing reliance being placed on wind tunnel testing can have a positive effect on probability-based loading criteria. The wind tunnel study may improve the estimation of effects associated with site exposure and the distribution of pressure coefficients. The load criteria depend on the probability distributions, means and variabilities of the load variables[1]. The wind tunnel studies refine the statistical estimates of loads and, in most cases, reduce the level of uncertainty in the loads. Sources of uncertainty in the specification of wind load and the role of wind tunnel testing in reducing these uncertainties are described in the following sections.

If the structural engineer uses the wind loads in the A58 Standard[2], then the steps in the design procedure are clear. The implicit level of risk associated with the use of the new proposed A58 load factors has been quantified. The wind loads calculated using ANSI A58 were selected such that they have a known probability of being exceeded for certain idealized conditions. The wind loads obtained from a wind tunnel study of a more complex situation should, at the minimum, have a probability of being exceeded consistent with the ANSI A58 levels.

2. ANALYSIS OF WIND LOADS

The wind load is a function of basic wind climate, nearby terrain features, and aeroelastic and aerodynamic response of the structure. The wind pressure, W, may be given as [2,4],

$$W = C K_z GC_p V^2 \qquad (1)$$

in which C is a constant; K_z is a coefficient to account for variation of wind speed with height and for terrain roughness; G is a gust factor to account for turbulence and resonant amplification; C_p is a pressure coefficient that is dependent on structural shape; V is the reference wind speed at a height of 10 m (33 ft). The parameters K_z, GC_p and V in Eq. 1 are random variables; see Figure 1. The probability distribution of W is,

$$F_W(w) = \int\int\int f_V(v) \, f_{K_z}(u) \, f_{GC_p}(t) \, dt \, du \, dv \qquad (2)$$

in which $f_i(x)$ are probability density functions (PDF) and the integration is performed over the region where $CV^2 K_z GC_p \le W$. The product GC_p in Eqs. 1 and 2 has been treated as a single variable for reasons that will become apparent presently.

Equation 1 provides a useful definition of equivalent static wind load on a building or structure, regardless of whether the structure is designed by analysis, through wind tunnel tests, or a combination of these approaches. The probability density functions in Eq. 2 show greater dispersion if the load criteria are to apply to a wide class of buildings than if they are to apply to a small group of buildings or to a single building.

Appropriate nominal values of the variables in Eq. 1 must be specified for code purposes. The nominal wind load W_n according to ANSI Standard A58.1-1972[2] is calculated for most permanent structures using a 50-year mean recurrence interval wind speed. This nominal wind load can be expressed in equation form as:

$$W_n = C K_z G C_p V_{50}^2 \qquad (3)$$

In Eq. 3, V_{50} is the 50-year mean recurrence interval wind speed referenced to a height of 10 m (33 ft) determined from a statistical analysis of annual extreme fastest mile wind speeds[5]; K_z is based on one of three exposures (open country, suburban, urban); and G and C_p are selected for design purposes from dynamic analysis and from wind tunnel pressure measurements, most of which were obtained under conditions of relatively smooth flow.

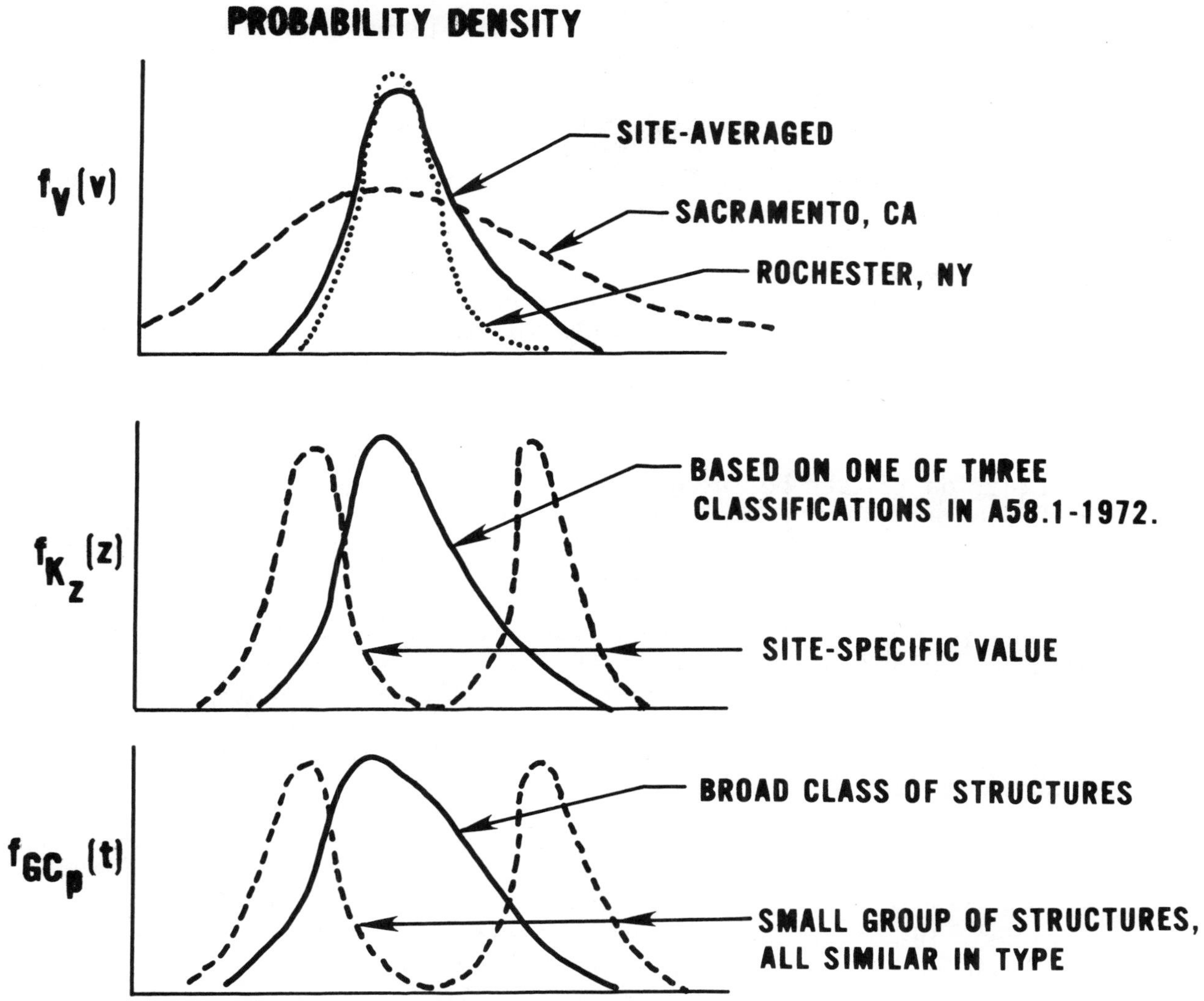

PROBABILITY DENSITY

Figure 1 – Dependence of Wind Load Probability
Densities on Scope of Load Criteria.

The 1982 edition of the ANSI A58 loading criteria presents, in addition, the factored load equations[1]:

$$U = 1.2 D_n + 0.5 L_n + 1.3 W_n \qquad (4a)$$
$$U = 0.9 D_n - 1.3 W_n \qquad (4b)$$

In most practical cases Equation (4a) is used for the wind safety check. The terms D_n, and L_n are the ANSI Standard-specified nominal dead and live loads.

The load factors in Eqs. (4a) and (4b) were developed using probabilistic methods to integrate available statistical data on wind and other structural loads[1]. In probability-based design, acceptable performance and risk may be quantified using the Reliability Index, β. The reliability index incorporates the mean and variability (and, in some methods, the probability distribution) of the load and strength variables. The load combinations in Eqs. (4a) and (4b) were intended to lead to a reliability index β of approximately 2.5 with reference to a 50-year period of time.

Thus, the wind load (W) given in Eq. (1) used in the reliability analysis leading to Eqs. (4a) and (4b) is a random variable and is defined to be the maximum 50-year wind load. The reference wind speed at a 10 meter height, V, is also a random variable defined as the maximum 50-year fastest mile wind speed at a height of 10 meters. The distribution of the 50-year maximum wind speed can be related to the available site-dependent distribution of the annual extreme wind speed using order statistics[4]. Based on our review of historical data, the coefficient of variation in the annual extreme wind speed averages about 0.15, considering a number of sites, so that when sampling and observation errors are included the coefficient of variation in the 50-year maximum averages about 0.12[1,4]. The variability in GC_p is on the order of 0.12 - 0.20, depending on the component loaded and its tributary area, while the variability in K_z is about 0.15[4].

Equation (2) can be integrated numerically to yield $F_W(w)$, assuming appropriate probability distributions for V, GC_p, and K_z. The Type I Extreme Value distribution fitted to the upper percentiles of $F_W(w)$, as described in Ref. 4, has a mean equal to 78% of the nominal ANSI wind load W_n determined from Eq. (3) and a coefficient of variation equal to 37%. These statistics include a crude correction for wind directionality effects. Figure 2 shows a plot of the PDF of W, the maximum 50 year wind pressure. From this figure, the nominal ANSI wind load (W_n) has a 19% chance of being exceeded in 50 years. If we multiply W_n by the load factor 1.3, then the probability that this factored load will be exceeded in the next 50 years is 5.4%.

It should be emphasized that these means, coefficients of variation and probability distributions, are intended to apply to a rather broad range of structures falling within the scope of the A58 wind load provisions. The distributions might be expected to change if the scope of application is restricted to a small group of buildings or to one site. Figure 1 shows, qualitatively, how the densities of probability in Eq. (2) might change as the scope of application narrows. While the dispersion of wind speed may become either more or less favorable to reliability, the variability in GC_p and K_z invariably decreases. The mean values of GC_p and K_z may either increase or decrease. However, experience has shown that these means usually decrease, particularly when the orientation of the building with respect to the prevailing direction of wind is taken into account.

3.   WIND TUNNEL TESTING

Wind tunnel test results can be used to improve code provisions and to refine the assessment of wind effects beyond what would be possible using a code. Here, we shall be concerned particularly with its uses in refining the estimated probability distributions in Eq. 2 and Figure 1.

A wind tunnel test strives to reproduce wind climate, terrain features, and aeroelastic and aerodynamic response in miniature. Care must be taken to insure proper similation of the variation of mean wind speed with height, turbulence intensity and longitudinal scale of turbulence. The selection of the structural model (rigid or aeroelastic) and the response characteristics of the wind tunnel instrumentation will also affect the usefulness of the test results.

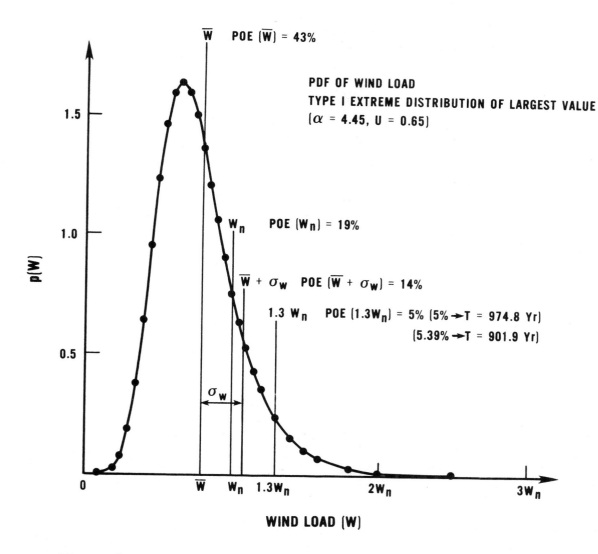

Figure 2 – PDF of Maximum 50 Year Wind Load Based on ANSI A58.

The process of conducting a wind tunnel test on a building involves three basic steps. First, the design wind speed(s) must be established. Second, wind tunnel test results are obtained from simulation of the atmospheric boundary layer and appropriately scaled structural models. Third, the results from the first two steps must be combined to produce the design wind forces. In order to satisfy the reliability requirements corresponding to the ANSI wind standard and also to produce a cost effective design, we must quantify the mean values and uncertainties of the random variables involved in each step.

Some of the sources of uncertainties in a wind tunnel study are:

(1) Statistical scatter in NWS data and other historical data.

(2) NWS site exposure characterization.

(3) Building site exposure characterization.

(4) Wind tunnel model of building site mean velocity profile.

(5) Wind tunnel model of building site turbulance.

(6) Structural model (rigid or aeroelastic).

(7) Wind tunnel instrumentation and response character-istics.

Wind tunnel studies are made for a specific building and site. One might expect that they would be of considerable value in improving estimates for GC_p and reducing their variability. To the extent that the surrounding terrain features and adjacent buildings can be modeled, the wind tunnel usually helps in establishing K_z and reducing its variability as well.

Pressure coefficients C_p determined in wind tunnel studies are obtained from the measured pressures or structural response of the model through the relation,

$$C_p = W/(1/2 \rho V_r^2) \tag{5}$$

in which V_r is a reference wind speed. Depending on the manner in which these coefficients are analyzed and reported, the values thus obtained may actually be equivalent to GC_p rather than C_p. This makes a separate determination of G and C_p unnecessary and even artificial. A typical sample function of GC_p, showing its mean and fluctuating components as a function of time, is shown in Figure 3.

Probability densities of the response peaks and of the maximum response may be of interest in structural reliability studies, and are shown in Figure 3. Many wind tunnel studies report the peak distributions. The probability distributions of the individual peaks appear to be Weibull for suction pressures and Normal[6] for positive pressures. The peak distributions may be of significance for fatigue studies or for service-ability checking. However, calculations of failure probabilities for safety-related limit states usually require the distribution of the maximum reponse during period T_0 (see Fig. 3).

The specification of GC_p depends on the averaging time for the mean wind speed V. If V in Eq. 1 is mean hourly (or 10 minute average) wind speed, GC_p will be higher than if V is the fastest mile wind speed, which corresponds to an averaging period of 3600/V seconds. To be useful in reliability studies, the wind speed with which the distribution of GC_p is compatible must be specified. The maximum wind pressure computed from Eq. 1 or 3 corresponds to a gust with an averaging time of from 1 to 10 seconds. Although this characterization of peak response is adequate for analyzing main wind force-resisting frames, cladding and local areas of the structure may respond to gusts of shorter duration. Accordingly, the probability distribution of GC_p may depend on influence area as well.

The statistics of GC_p determined from Eq. 5 also include the uncertainties in modeling of the atmospheric boundary layer, which would otherwise be reflected by the mean and coefficient of variation in K_z. If the wind tunnel test could simulate perfectly the characteristics of the atmospheric boundary layer at the building site for all storms, the dispersion in K_z would be negligible. In practice, this is not possible; thus, while the wind tunnel test would reduce the uncertainty in the estimate of K_z, it would not eliminate it. Thus, the variability in K_z might still be significant, particularly in areas of highly irregular terrain that would be difficult to model in a wind tunnel.

Much of the variability in W is due to variability in the wind speed V, which is squared in Eq. (1). The component of variability in W due to V is not reduced by a wind tunnel study. However, a reduction in design load for a prescribed reliability at a specific site may be realized if the site is one at which the inherent variability in the wind speed is small (e.g., at Rochester, NY, the coefficient of variation in the annual extreme is 0.10, as opposed to the more typical value of 0.15). Moreover, one particular value of wind tunnel testing is in showing the dependence of structural response on building orientation with respect to the mean wind direction. It has become apparent[7] that the use of maximum pressure coefficients in combination with maximum wind speeds, both of which are reported without regard to direction, may lead to overly conservative results.

Validation of wind tunnel studies through the use of in-situ pressure measurements on buildings is essential in establishing the uncertainties associated with wind tunnel modeling. The limited in-situ measurements that exist indicate that the wind tunnel can capture the essential features of building response, if properly executed. However, there are some differences. Validations are made difficult by the complex

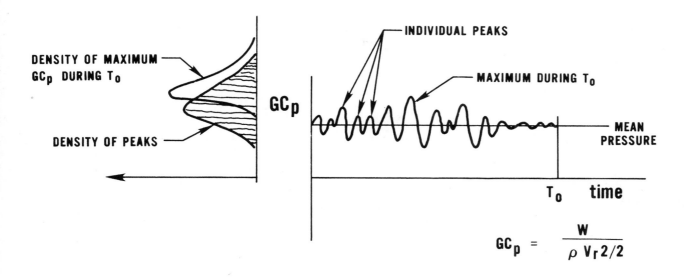

Figure 3 - Pressure Coefficients.

interaction between wind climate and structural response parameters and the fact that full scale structural properties such as frequency and damping are uncertain quantities. Therefore, in reliability analysis studies, we incorporate our uncertainty in our ability to model the real world with a modeling confidence random variable.

4.   IMPLICATIONS FOR PROBABILITY BASED DESIGN

The load factor γ_w on nominal wind load W_n may be computed, approximately, as[1]

$$\gamma_w = (W/W_n)(1 + \alpha \beta \rho_w) \qquad (6)$$

in which α is a sensitivity factor approximately equal to 0.72, ρ_w is the coefficient of variation of the wind load, and W is the mean 50 year maximum wind speed. If the variability in GC_p and K_z are reduced by a properly designed and executed wind tunnel test to about 0.08 and 0.10, respectively, the variability in W decreases to about 0.30 from the previously determined value of 0.37. The load factor on nominal wind load that is consistent with a reliability index β of 2.5 would then be about

$$\gamma_w = (0.78)(1 + 0.72 \times 2.5 \times 0.3) = 1.2 \qquad (7)$$

rather than 1.3. Thus, using the wind tunnel results, one could design the structure for a wind load that is about 8 percent less than would otherwise be required and still expect a comparable level of reliability.

The pressure coefficients in A58.1-1972 used to design main wind force-resisting frames tend to be conservative for the leeward side of buildings. If wind tunnel studies are used as part of a reliability assessment of a structure, the mean pressure coefficient on the leeward side would be more like -0.3 than -0.5. This would reduce the total mean drag coefficient to about 1.1 from 1.3, and the mean and coefficient of variation of W would be about 0.66 W_n and 0.30, respectively. Thus, the estimate of the probability that the wind load exceeds 1.3 W_n in 50 years, where W_n is determined from Eq. 3 and Ref. 2, could be revised downward from 0.054 to about 0.009.

5.   CONCLUSIONS

Reliability analysis and probability-based design require a knowledge of the probability distributions of the load variables. In this context, wind tunnel testing leads to improved estimates of these probability distributions for specific classes of structures. As an analysis tool, these improvements permit the probabilities of unsatisfactory performance to be calculated with greater confidence. As an aid in design, prescribed reliability objectives frequently may be achieved at some savings, because of decreases in the dispersions in the load variables to which the load factors and load combinations are directly related.

APPENDIX A

REFERENCES

1. Ellingwood, B., Galambos, T.V., MacGregor, J.G., and Cornell, C.A., "Development of a Probability-Based Load Criterion for American National Standard A58," National Bureau of Standards Special Publication 577, June 1980, 222 pp.

2. "American National Standard Building Code Requirements for Minimum Design Loads in Buildings and Other Structures (ANSI A58.1-1972)," American National Standards Institute, New York, 1972; see also ANSI A58.1-1982, the recent version.

3. Davenport, A., Isyumov, N., and Surry, D., "The Role of Wind Tunnel Studies in Design Against Wind," presented at the American Society of Civil Engineers Spring Convention, Boston, MA, April 1979.

4. Ellingwood, B., "Wind and Snow Load Statistics for Probabilistic Design," Journal of the Structural Division, ASCE, Vol. 107, No. ST7, July 1981, pp. 1345 - 1350.

5. Simiu, E., Filliben, J., "Weibull Distributions and Extreme Wind Speeds," Journal of the Structural Division, ASCE, Vol. 106, No. ST12, December 1980, pp. 2365 - 2374.

6. Peterka, J., Cermak, J., "Wind Pressures on Buildings - Probability Densities," Journal of the Structural Division, ASCE, Vol. 101, No. ST6, June 1974, pp. 1255 - 2267.

7. Simiu, E., Filliben, J., "Wind Direction Effects on Cladding and Structural Loads," Engineering Structures, Vol. 3, July 1981, pp. 181 - 186.

8. Simiu, E., Changery, J.J., Filliben, J., "Extreme Wind Speeds at 129 Stations in the Contiguous United States," NBS Building Science Series 118.

9. Hart, G.C., "Loads and Safety in Structural Engineering," Prentice-Hall, Inc., 1982.

APPENDIX B

MAXIMUM 50 YEAR WIND SPEED

The wind speed in Eq. (1) is the maximum 50 year wind speed. Therefore, we are using the maximum wind speed for a design life of 50 years. Confusion often exists because this "maximum 50 year wind speed" is a random variable and as such does not have a unique value.

The National Weather Service (NWS) wind data we most frequently see referenced is the "maximum annual wind speed". The maximum annual wind speed is a random variable and is denoted V_a. The PDF of V_a is usually assumed to be an Extreme Type I. The 50 and 100 year mean return period wind speeds, denoted V_{50} and V_{100}, are deterministic variables. V_{50} has a 2% probability of being exceeded any year and V_{100} has a 1% chance of being exceeded in any single year. If we denote the mean of the maximum annual wind speed data as $\overline{V}_a$ and its coefficient of variation as ρ_{V_a}, then we can write

$$V_{50} = \overline{V}_a [1 + 2.59 \ \rho_{V_a}] \tag{B-1}$$

and

$$V_{100} = \overline{V}_a [1 + 3.14 \ \rho_{V_a}] \tag{B-2}$$

The mean and coefficient of variation of the maximum 50 year wind speed, V and ρ_V, can be obtained from the mean and coefficient of variation of the maximum annual wind speed, V_a. It follows that

$$\overline{V} = \overline{V}_a [1 + 3.05 \ \rho_{V_a}] \tag{B-3}$$

and

$$\rho_V = \rho_{V_a} / [1 + 3.05 \ \rho_{V_a}] \tag{B-4}$$

The PDF of V is usually assumed to be an Extreme Type I.

Representative values for the mean and coefficient of variation of the maximum annual winds are 50 mph and 18%, respectively[e]. Therefore, using Equations (B-1) and (B-2) it follows that the 50 and 100 year mean return period winds are 1.47 and 1.56 times the mean maximum annual wind speed, respectively. The mean of the 50 year maximum wind speed is 1.55 times the maximum annual wind speed and the corresponding coefficient of variation is 11.6%. The PDF of V_a and V is shown in Figure B-1. The 50 and 100 year mean return period wind speeds are also shown in Figure B-1.

The nominal wind pressure is calculated using Eq. (3) and the 50 year mean return period wind speed (V_{50}). It is clear from Figure B-1 that V_{50} is less than V and it can be shown that it has approximately a 64% chance of being exceeded in 50 years. V_{100} is approximately equal to V and it has approximately a 40% chance of being exceeded in 50 years.

The above example illustrates the relationship between the different wind speeds. It is clear from Eqs. (B-1) to (B-4) that the mean and coefficient of variation of the maximum annual wind speed are both important parameters. Therefore, it is clear that the incorporation of a description of wind speed that is a function of direction, as is usually done in a wind tunnel study, is very important.

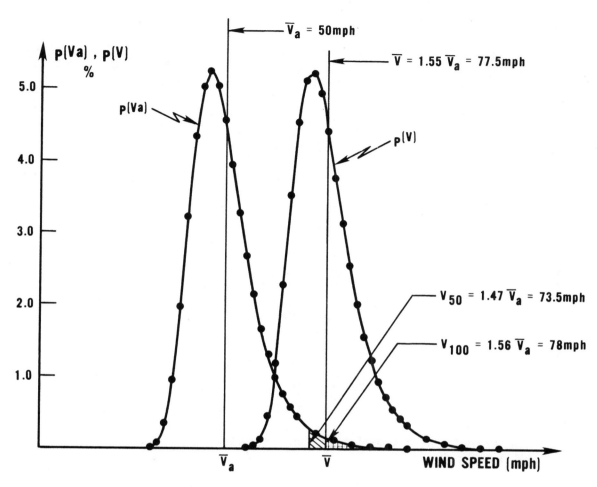

Figure B-1 – PDF of Maximum Annual Wind Speed (ρ_{v_a} = 18%) and Maximum 50 Year Wind Speed (ρ_v = 11.62%).

APPENDIX C

DUCTILE FRAME DESIGN

The safe and economical design of high rise buildings for extreme loads usually requires the structural frame to be ductile. This especially applies to concrete and steel building frames. The earthquake design of ductile frames uses an earthquake response spectra provided by a geotechnical engineer. This response spectra describes the earthquake loading to the structural frame. Like the wind loading, W, the earthquake loading is a random variable and is the maximum 50 year load.

A two load level design is sometimes used in earthquake engineering and it can also be used for wind engineering. The two load levels are deterministic values of load that have been assigned so as to have a specific probability of being exceeded in the next 50 years. The probabilities are:

Load Level I   The probability of the earthquake load or wind load being exceeded in 50 years is 39 percent. This is often referred to as the Damage or Yield Level Design Load.

Load Level II   The probability of the earthquake load or wind load being exceeded in 50 years is 10 percent. This is often referred to as the Ultimate Level Design Load.

The structural frame is designed such that its performance at these two load levels is:

Level I   An elastic frame analysis is performed and all component moments, forces, and deformations are less than their yield values.

Level II   An inelastic frame analysis is performed and all component moments, forces, and deformations are less than their ultimate values but may exceed their yield values.

The selection of the 10% and 39% probability of exceedence values is based upon the earthquake experience and also the fact that the ANSI 50 year maximum value would be about 5 to 6%. Therefore, it seems reasonable to retain the above noted values for both earthquake and wind design of ductile frames.

If a wind tunnel study is performed then the results obtained should be compatible with this ductile frame design approach. In order to illustrate how this might be achieved, consider the following example.

Example:

The wind loading acting on a building component or system can be written

$$W = P_T (XV)^2 \qquad (C-1)$$

where

P_T = wind tunnel load coefficient random variables.
X = NWS and building site exposure random variables.
V = maximum 50-year fastest mile wind speed NWS site.
W = the maximum 50-year wind loading.

W is a random variable and, for this example, it will be assumed to have an Extreme Type I PDF.

We will consider the following three cases in this illustrative example:

Case 1: Only the NWS wind data is considered to have uncertainty.

Case 2: The NWS data and the quantification of NWS and building site exposure has uncertainty.

Case 3: The NWS data, site exposure, and measured wind tunnel pressure coefficients have uncertainty.

We will assume that the NWS open country maximum annual fastest mile wind speed at 10 meters has a mean of 50 mph and a coefficient of variation of 18%. These values we selected to be representative of the values obtained for 129 U.S. National Weather Service stations[8]. We selected the power law formula for transferring NWS site data to building site data[9]. The mean power law parameter values used were the ANSI A58 values. The corresponding coefficient of variations were 10 percent. The pressure coefficient measured in the wind tunnel was assumed to have a mean value corresponding to the "best estimate" and a coefficient of variation equal to 10%.

If we view the wind problem in a deterministic context, then the nominal design wind speed corresponding to a 50-year mean return period wind speed would be

$$W_{50} = \overline{P}_T (\overline{X} V_{50})^2 \qquad (C-2)$$

where

$\overline{P}_T$ = mean of P_T (best estimate if a normal PDF).
$\overline{X}$ = mean of X (best estimate if a normal PDF).
V_{50} = 50-year mean return period of the annual fastest mile wind speed.
W_{50} = wind loading corresponding to a nominal 50-year return period wind speed.

If we consider W to be a random variable, then for Cases 1 to 3 the mean and coefficient of variations are as given in Table C-1. For this example, we selected P_T such that W for Case 1 was approximately 20 psf. If we write

Table C-1   Statistics of Maximum 50 Year Wind Load

	Case 1	Case 2	Case 3
Mean ($\overline{W}$)	20.2 psf	20.7	21.1
Coefficient of Variation (ρ_W)	25.6%	29.6	31.9
Damage Level Wind Load (W_D)*	20.7 psf	21.3	21.8
Ultimate Level Wind Load (W_C)**	26.9 psf	28.7	29.9
Damage Level Wind Speed (V_D)	78.9 mph	80.0	80.9
Ultimate Level Wind Speed (V_C)	90.0 mph	92.8	94.7

* 39% probability of being exceeded in 50 years.

** 10% probability of being exceeded in 50 years.

$$W_D = \text{Damage or Yield Level Wind Load}$$
$$= \overline{P}_T \, (\overline{X} \, V_D)^2 \qquad\qquad\qquad (C\text{-}3)$$

and

$$W_C = \text{Ultimate Level Wind Load}$$
$$= \overline{P}_T \, (\overline{X} \, V_C)^2 \qquad\qquad\qquad (C\text{-}4)$$

then V_D and V_C represent the Yield and Ultimate Level Wind speeds, respectively. By comparing, Eqs. C-2 and C-3, it follows that

$$(W_D/W_{50}) = (V_D/V_{50})^2 \qquad\qquad\qquad (C\text{-}5)$$

and similarly from Eqs. C-2 and C-4,

$$(W_C/W_{50}) = (V_C/V_{50})^2 \qquad\qquad\qquad (C\text{-}6)$$

The wind loads that are established based on the above definitions are applied using a load factor of unity. Also, the Damage Level Design loads should produce member deformations that are at or near minimum yield levels. The Ultimate Level Design wind loads can produce deformations that are larger than yield level deformations but should be less than ultimate level deformations.

Table C-1 shows that the mean and coefficient of variation of the maximum 50 year wind load increase as more randomness is considered. The Damage and Ultimate Level wind loads and corresponding wind speeds are also shown in Table C-1. Case 3 represents the most accurate treatment of the wind problem, and it is clear that when Damage and Ultimate Level wind loads are compared between Cases 3 and 1, the increase in loads is only 5 and 11%, respectively.

Table C-2 presents the results of this illustrative example in an alternate form. We see that for the Damage Level design, the corresponding return period of annual extreme winds is 110 to 145 years. If we compare the wind load ratio (W_D/W_{50}) we see that this ratio varies from 1.17 to 1.23. The ANSI A58 load factor design procedure would use the load factor 1.3 times W_{50} to obtain the design wind. Therefore, ANSI would produce wind loads larger than the Damage Level wind loads for this example because we use a load factor of unity with W_D.

Again, referring to Table C-2, it follows that the ratio (W_C/W_{50}) exceeds the ANSI A58 load factor of 1.3. This is reasonable because deformations induced by W_C can be expected to exceed yield deformations but must be less than ultimate level deformations.

Note that the term "Ultimate Level Wind Load" is used in this paper. This term is not universally used and is often referred to as the "Collapse Level Wind Load".

Table C-2   Damage and Ultimate Level Design

	Case 1	Case 2	Case 3
Damage Level Wind Speed (V_D)	78.9 mph	80.0	80.9
Return Period of Annual Extreme Wind Speed Corresponding to V_D	110 years	128	145
Ultimate Level Wind Speed (V_C)	90.0	92.8	94.7
Return Period of Annual Extreme Wind Speed Corresponding to V_C	556	769	1,000
(V_D/V_{50})	1.08	1.10	1.11
(W_D/W_{50})	1.17	1.20	1.23
(V_C/V_{50})	1.23	1.27	1.30
(W_C/W_{50})	1.52	1.62	1.69

CONSIDERATIONS IN RELATING WIND TUNNEL RESULTS TO DESIGN FOR A SPECIFIC SITE

A. G. Davenport, Director and Professor, Boundary Layer Wind Tunnel Laboratory, Faculty of Engineering Science, The University of Western Ontario, London, Ontario, Canada, N6A 5B9

1. SUMMARY AND INTRODUCTION

In this paper two important considerations will be addressed:

- first the methods whereby wind tunnel test results are combined with meteorological information to define design loading; and
- second, the reliability of the resulting design loads.

The context in which these matters arise is suggested by the make-up of the effective loading due to wind which can be expressed in the manner following

$$
\begin{bmatrix} factored \\ wind \\ load \end{bmatrix} = \begin{bmatrix} reference \\ velocity \\ pressure \end{bmatrix} \begin{bmatrix} exposure \\ height \\ factor \end{bmatrix} \begin{bmatrix} aerodynamic \\ shape \\ factor \end{bmatrix} \begin{bmatrix} dynamic \\ gust\ factor \\ factor \end{bmatrix} \begin{bmatrix} partial \\ load \\ factor \end{bmatrix}
$$

The reference velocity pressure has to be determined from the wind climate for the region; the agglomeration of factors contained in the second, third and fourth terms constitute the principal product of boundary layer wind tunnel testing. The last factor, whose function is to guard against the dangers of overload, is not usually considered explicitly since it is absorbed into conventional safety factors. However, since its value should reflect the uncertainties involved and since there is a current trend towards the use of partial safety factors it is deserving of special consideration. If nothing else it provides some assurance and indication of the improvements in safety brought about by wind tunnel testing.

The two main considerations of this paper explore the interaction of all of these components.

2. WIND TUNNEL RESULTS

In the context of structural engineering, the wind tunnel is in essence an analogue computer for determining the responses of a structure to wind. From the measurements made on aeroelastic and pressure models exposed to turbulent boundary layer flows of varied speed, direction and structure — maps of the response can be obtained. Fig. 1, for example, illustrates schematically a typical measurement of pressure on the exterior of a tall building for different azimuth angles and Fig. 2 shows several polar mappings of the pressure coefficients which result. Fig. 3 shows mappings of the mean, root mean square (rms), maximum and minimum peak bending moments and torque at the base of a tall building. These quantities are each plotted for a range of wind speeds and all compass directions. Finally, Fig. 4a shows the plot of the peak torsional response of a long span bridge.

Plots such as these can be obtained reliably and repeatably using the wind tunnel; however, they only get us halfway in predicting the response of load level appropriate for design. To complete the prediction of the wind tunnel results must be combined with meteorological information on the prevalence of strong winds.

The objective of the exercise is usually to predict the response (i.e. pressure or deflection) having a given return period, that is exceeded only once very 50 years for example. (This choice of return period is probably explained by the continuity it provided with earlier design speeds based on the highest speed on record — generally a period of several decades). We will accept this objective, but return to examine its implications in the latter part of this paper. Alternative procedures for defining the load level corresponding to different return periods are now outlined.

The first approach we should consider makes use of the published information on extreme wind speeds, and assumes that these winds come from a direction aligned with the most severe aerodynamic response. These directions and their associated aerodynamic response values or coefficients are easily identified from the wind tunnel results such as those in Figs. 2 — 4.

This approach has the advantages that it is clearly safe and comparatively simple to apply. Its disadvantages are primarily bound up with the exclusion of any influence the wind direction may have on risk. This can lead to a rather pessimistic estimation of the loading and somewhat uneven margin of safety. There are other difficulties, for example, when combined loads are considered — what value should be used for the second load when the first load is at its maximum value. This question of wind direction has been of concern to the writer dating back to the first studies employing wind tunnel testing, because, unless treated, seemingly unjustified increases in wind loading could be inferred from wind tunnel testing, thus undermining confidence in its usefulness.

One approach to the question of wind direction which was considered at an early stage was the extension of extreme value analysis to the individual comparisons of wind sectors. An example of such an analysis is shown in Fig. 5 for a station in New York City. The approach was rejected, however, still because of the difficulty of developing a satisfactory procedure for combining the extreme wind speeds for each sector with the aerodynamic data. It did not resolve the questions of combined loading and the appropriate treatment of service loading for accelerations, etc.. Above all was the fundamental difficulty that excursions of high wind speed should normally be associated with small directional changes and lower wind

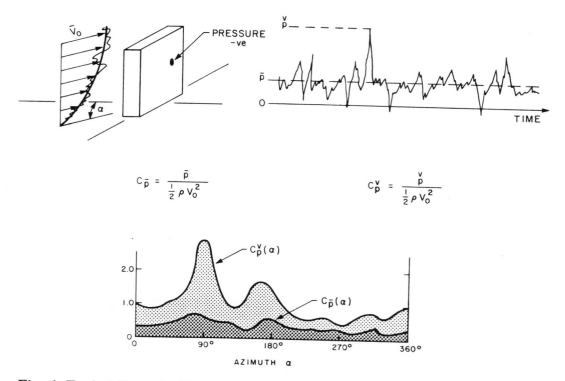

$$C_{\bar{p}} = \frac{\bar{p}}{\frac{1}{2}\rho V_o^2} \qquad\qquad C_p^v = \frac{\overset{v}{p}}{\frac{1}{2}\rho V_o^2}$$

Fig. 1 Typical Record of Pressure and the Azimuthal Variation of the Pressure Coefficients

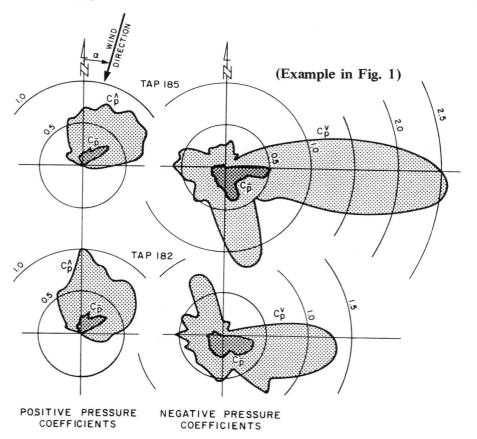

Fig. 2 Azimuthal Variation of Peak and Mean Pressure Coefficients on a Tall Building

46

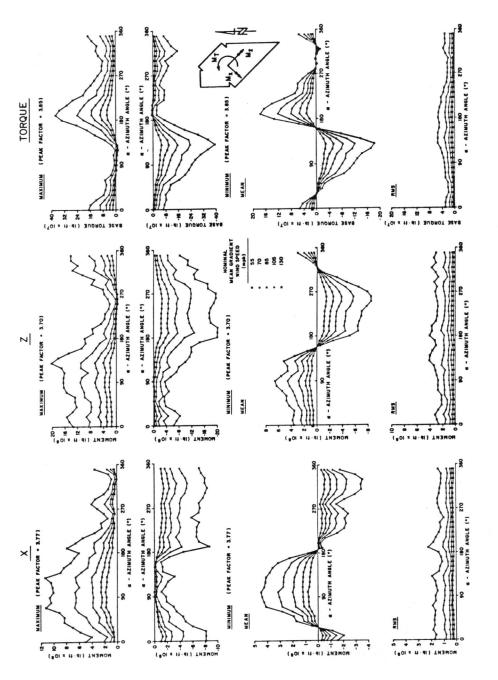

Fig. 3   Mapping of Response of a Tall Building to Wind

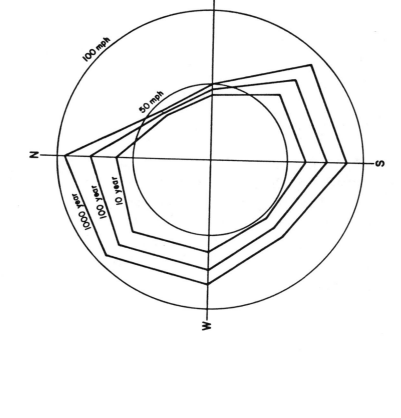

Fig. 5 Distribution of Extreme 5 Minute Average Wind Speed by Octant. (Whitehall Building, New York City; 454 Ft.; 1936–1959)

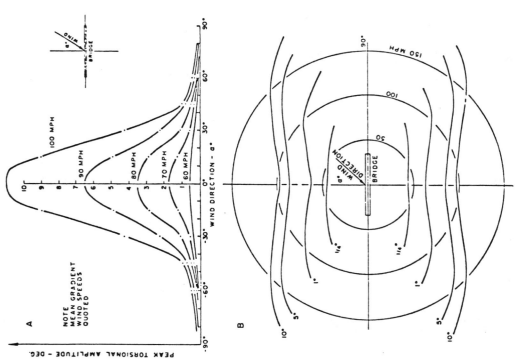

Fig. 4 A. Variation of Peak Torsional Response Amplitude of Suspension Bridge with Wind Direction

B. Development of Response Boundaries

speeds with larger directional changes, neither of which would match up with the fixed angle segment used in the analysis.

An alternative approach to the directional problem is now outlined. Its evolution, to the format now used routinely in the Boundary Layer Wind Tunnel Laboratory has been described in a series of papers (Refs. 1–4).

3. THE PREDICTION OF RESPONSES TO WIND INCLUDING ITS DIRECTION

The first requirement of this approach is to determine the aerodynamic response boundaries from the mappings in Figs. 2, 3 and 4, in which the wind speed at the reference level (gradient or surface) is the dependent variable. These are found from the wind tunnel tests by asking the question, "What wind speed is required from a given azimuth a, to produce a certain response R?". This wind speed is denoted $V_R(a)$. These response boundaries are found by inverting the aerodynamic data mapped for wind speed and direction and they may refer to pressures, deflections, accelerations etc. as required.

If, for example, we refer to the peak pressures of Fig. 2 then

$$V_p(a) = \sqrt{2p/|\rho C_p(a)|} \qquad (1)$$

The response contour corresponding to the suctions at one of these tap locations is shown in Fig. 6 for the value of $p = 1\ kPa$. This shows that for the critical east wind, the wind speed required is comparatively low.

A similar contour diagram has been drawn in Fig. 4b for the suspension bridge response. This shows the penetration of the contours into low wind speed regions near the critical perpendicular directions. Similar response contours can be drawn for aeroelastic response of buildings. These can often be simplified by the use of pwer law dependencies such as

$$R \quad = \quad D(a)\ V^{n(a)}$$

$$V_R(a) = \quad [R/D(a)]^{1/n(a)}$$

where the important point to note is that, unlike pressures for which $n(a) = 2$ in many cases of aeroelastic response, $n(a)$ is often in the range 2.5 to 3.5 and sometimes greater.

It is clear from the above that the basic problem in predicting the likelihood of occurrence of any given response is to predict the likelihood of the wind speed crossing the boundary defined by the response contour. These boundaries, as we have seen, can vary greatly with wind direction and in instances of severe dynamic response can be closely packed.

To attack this problem with any degree of completeness, the description of the wind must be more sophisticated than simply being the extreme speeds in a given time interval. Such wind speeds may not occur from a critical direction and may pass by harmlessly.

The basic requirement for describing the wind climate then becomes the probability distribution of wind speed and direction, preferably at gradient height. This relies on meteorological observations of all winds and is not confined to the extremes. A typical plot of such a distribution is shown in Fig. 7, discounting for the moment the dashed line representing the 1 kPa ($\sim$ 20 psf) response boundary. In this plot, radial distances indicate wind speeds in m/s. Each contour represents those speeds which, on average, are equalled or exceeded or for a specified fraction of time within a 22.5° sector of a chosen azimuth angle. Such a plot is obtained by fitting mathematical expressions to the original meteorological data and extracting the contours. Appropriate weighting may be included to emphasize strong winds.

Generally speaking, these distributions display directional characteristics belonging mostly to the prevailing winds. (Some of the directional effects, however, may be spurious; i.e. due to sampling and extrapolation.) These directional factors clearly have a significance in relation to the directionally sensitive responses discussed above.

Having established the probability distribution of wind speed and direction, it is now possible to relate this to the response. Consider the relationship between the probability distribution and the pressure contour of Fig. 7. Denoting the cumulative wind speed and direction distribution by $F_{V,a}(V,a)$ and the response boundary by $V_R(a)$, then the probability that $R(=1\ kPa)$ is exceeded is

$$ F_R(R) \quad = \quad \int_0^{2\pi} F_{V,a}(V_R,a)\ da $$

i.e., the integral on the contour.

This defines the total fraction of time that R is exceeded. It does not, however, indicate how often this happens, whether for a single long period or many short periods. For this, we must examine the crossing rate of the response contour.

The one dimensional crossing rate has been discussed by Rice. The extension to the two-dimensional situation is shown conceptually in Fig. 8. Fig. 8 depicts a process (in this case the wind vector) which moves randomly in x and y. The possibility that the process crosses

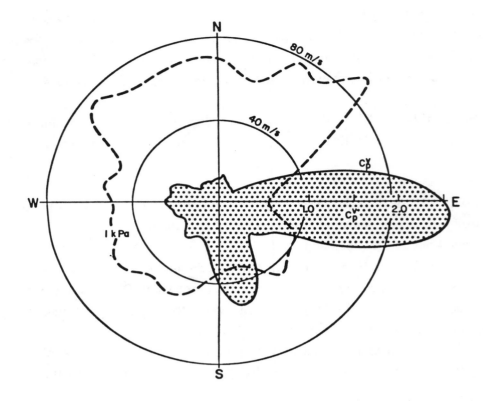

Fig. 6   The Wind Speed Contour Required to Exceed a Selected Pressure

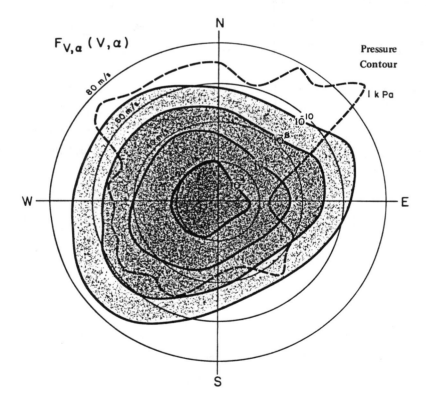

Fig. 7   The Cumulative Probability Distribution $F_{V,a}(V,a)$ of Wind Speed and Direction with a Pressure Contour Superimposed (for Determining Statistics of Extreme Pressures)

a typical element of the boundary in an outward direction has been considered in detail by Davenport[3]. The result is that the average rate of crossing, $N(R)$, of the boundary R (denoted here as $V_R(a)$) can be determined in terms of the joint probability *density* function $f_{V,a}(V_R,a)$ as:

$$N(R) = \sqrt{2\pi}\, \nu\, \sigma \int_0^{2\pi} \sqrt{1 + (\frac{1}{V_R} \frac{d V_R}{da})^2}\; f_{V,a}(V_R,a)\, da$$

$$\simeq \sqrt{2\pi}\, \nu\, \sigma \int_0^{2\pi} f_{V,a}(V_R,a)\, da \qquad (4)$$

In this expression ν is the mean cycling rate of the process. This latter term can be evaluated from the spectrum of the mean wind velocity. For several sites the value of ν turns out to be of the order of 1–2 cycles per day.

If these crossings of the boundary are rare, independent events, their distribution will be Poisson and the cumulative probability distribution of the largest value of R in time T is

$$F_{\hat{R}}(R) = 1 - e^{-N(R)T} \qquad (5)$$

From results such as those portrayed in Figs. 6 and 4b, the distribution of the extreme responses can now be determined from the previous equations, if necessary numerically. A representative distribution of the extreme response is shown in Fig. 9.

The directional effect embodied in this approach can have an appreciable effect on the final result. If the critical direction of response coincides with the prevailing direction of the wind, the extreme prediction will differ significantly from situations in which it does not.

It is also worth noting that through the use of equations (3) and (4) we can find an estimation of the "dwell time" of the response above level R denoted $T(R)$. Specifically

$$T(R) = F(R)/N(R) \qquad (6)$$

This can be useful in estimating the duration of severe effects. For most cases it is found that this time is roughly inversely proportional to R.

Directional effects may become even more pronounced when related to surface conditions. This is illustrated in Fig. 10. The upper diagram refers to the amplification of wind speed over water at the site of a long bridge (relative to that in fairly homogeneous terrain). The lower diagram shows the reduction in wind speed on the coast due to the screening by the buildings in a large city. Both of these imply that surface wind speeds and directions can be profoundly affected in certain situations by the roughness. Against these must play a part in the directional response to wind. Conversely, these diagrams also indicate the potentiality for bias due to surface roughness inherent within surface measurements. They are examples of variation in which has been defined as wind structure. These would be largely taken into account automatically within a wind tunnel test program by the analogue nature of the wind modelling and in a boundary layer wind tunnel, through the selection of the upstream terrain.

4. COMBINED LOADING ACTIONS

Another question arises out of the variable responses for different wind directions, namely the prediction of combined loading actions. Structural analysis of tall buildings usually makes the convenient but unrealistic assumption that the wind always blows perpendicular to the faces of the building or somehow aligned with the building axes. The question arises how to take account of the loading actions arising from combined loads which affect for example the corner columns in a building. The question is important when making full use of aeroelastic studies of building responses in which all wind azimuths are analysed as in Fig. 3. The principles followed can be understood from the following:

Suppose we consider some response R which depends on the primary X and Y responses (i.e. deflections or moments along principal axes) through some relation

$$R(\alpha,\beta) = \alpha X + \beta Y \tag{7}$$

where α and β are influence functions. We could add torsion as well if we wished.

Suppose that the expected peak values of the primary response in time T are $\hat{X}_T$ and $\hat{Y}_T$. Then if either $\alpha = 0$ or $\beta = 0$, we find the values of the peak response in time T are clearly

$$\hat{R}_T(0,\beta) = \beta \hat{Y}_T$$

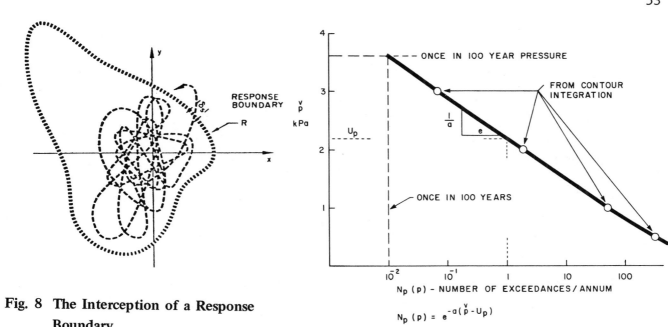

Fig. 8 **The Interception of a Response Boundary**

Fig. 9 **The Distribution of the Number of Exceedances of Pressure Per Annum**

$$N_p(p) = e^{-a(\overset{v}{p} - U_p)}$$

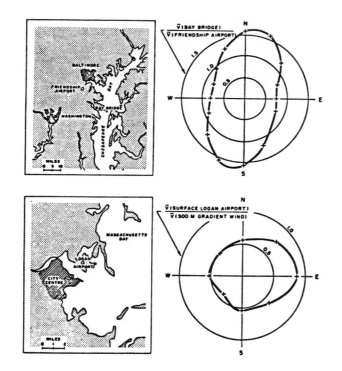

Fig. 10 **Influence of Terrain Roughness on Surface Wind Speeds and Direction**

and

$$\hat{R}_T(a,0) = a \hat{X}_T \tag{8}$$

The problem arises when a and β are roughly the same size. If, with one eye on the designers requirements, we adopt a solution of the type

$$\hat{R}_T(a,\beta) = \theta(a,\beta) [a \hat{X}_T + \beta \hat{Y}_T] \tag{9}$$

the problem reduces to finding the factor $\theta(a,\beta)$ which multiplies the sum of the solutions for $a = 0$ and $\beta = 0$. By considering the extremes of the normalized response function

$$\frac{R_T(a,\beta)}{(a \hat{X}_T + \beta \hat{Y}_T)} = \frac{aX + \beta Y}{a \hat{X}_T + \beta \hat{Y}_T} \tag{10}$$

which depends only on the ratio a/β we can determine $\theta(a,\beta)$ or $\theta(a/\beta)$ directly. This is accomplished by making separate calculations for several values of a/β in the neighbourhood of unity, i.e. $.1 < a/\beta < 10$ and working through the response from the original data. The result is usually about the same; namely that in this range of a/β the value of θ is slowly varying and usually lies between $.70 - .85$. The result is indicated in Fig. 11. Thus the influence of the combined loading can be adequately covered by dealing with the two normal loading cases acting independently and thirdly a loading case consisting of roughly 75% of the joint action of the two. The result of course is not far from the commonly adopted procedure of "wind blowing at the diagonal" but its derivation is quite different (the wind may not be blowing anywhere near the diagonal when the most severe effects are felt) and is much more surefooted.

5. RELIABILITY AND LOAD FACTORS

We have reached the point where we can process wind tunnel test results, amalgamate them with meteorological data on wind speed and direction, and come up with statistical predictions of the anticipated wind loadings. These loadings are customarily defined for specific return periods — once-every-50-years or so — and may be described as the "characteristic design wind load". It will be denoted as W_C.

Although this may appear to be the end product of wind loading prediction there is, in fact, one more nail that should be hammered home. This is to assure that the load or safety factor is adequate. This is particularly important in a relatively new technology such as wind tunnel testing in which there is still some consolidation to be done.

This excercise is not proposed as a requirement for every wind tunnel test, rather it should be reviewed in the overall perspective and compared with conventional practice. The current trends toward load factor design also encourages this.

A useful approach to safety is provided by second moment reliability. This has been the basis for much recent code work and also the present discussion[5,6]:

We will denote the load factor needed to achieve a certain reliability by γ_C and term the product $\gamma_C W_C$ as "the factored design wind load".

There can be, and is, some flexibility and arbitrariness in the definition of γ_C and W_C individually, however, the factored load $\gamma_C W_C$ should provide the required reliability. Second, moment reliability allows us to relate this factored load to the central safety factor, $\bar{\gamma}$, and the expected largest load, $\bar{W}$, through

$$\gamma_C W_C = \bar{\gamma}\, \bar{W} \tag{11}$$

From this we derive the load factor needed with the characteristic load W_C,

$$\gamma_C = \left(\frac{\bar{W}}{W_C}\right) \bar{\gamma} \tag{12}$$

The central safety factor $\bar{\gamma}$ can be estimated approximately but rationally (see Ref. 6) using

$$\bar{\gamma} \simeq exp\,(0.55\,\beta\,V_w) \qquad \text{for wind and dead load acting together, and} \tag{13}$$

$$\simeq exp\,(0.75\,\beta\,V_w) \qquad \text{for wind load only.} \tag{14}$$

In this V_w is the coefficient of variation (C.O.V.) of the largest lifetime load, and β is a measure of safety known as the safety index.

Calibration of present practice and existing codes has indicated that $\beta \approx 3.0$ is representative if the failure mode is ductile and not precipitate, and that $\beta = 2.5$ for servicability states.

The vital components of the reliability analysis now boil down to $\bar{W}$ and V_W, or effectively, the first and second moments of the distribution of the largest lifetime wind load.

To illustrate the problem we will consider the effective design load shown in Fig. 12 which has been deduced from a wind tunnel and meteorological study. Before immediately adopting any one of the curves given in Fig. 12, we should realise that there is some uncertainty arising from the actual damping which will be present in the real structure and from the overall predictive capability of the model. We can express this additional uncertainty through two new factors ϕ_ζ and ϕ_M such that the "true" load W is related to the predicted load W^* by

$$W = W^* \phi_\zeta \phi_M \qquad (15)$$

The required mean value and C.O.V. of the true largest load are then found from

$$\bar{W} = \bar{W}^* \bar{\phi}_\zeta \bar{\phi}_M \qquad (16)$$

and $\qquad (1 + V_W^2) = (1 + V_{W^*}^2)(1 + V_{\phi_\zeta}^2)(1 + V_{\phi_M}^2) \qquad (17)$

Thus to solve the problem we need to know $\bar{W}^*, \bar{\phi}_\zeta, \bar{\phi}_M, V_{W^*}, V_{\phi_\zeta}$ and V_{ϕ_M}.

Experimentally Determined Load — W^*

Fig. 12 shows that the relationship between W^* and return period for 1% damping is

$$W^*(R) = 2.10 + 0.282 \ln (R/100) \qquad (18)$$

If we assume a lifetime of 100 years this defines the mode value of Y, and the dispersion, respectively

$$\text{Mode: } U_{W^*}(100) = 2.1 \text{ kPa}$$

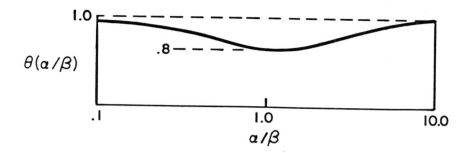

Fig. 11   Interaction Diagram for Combined Loading

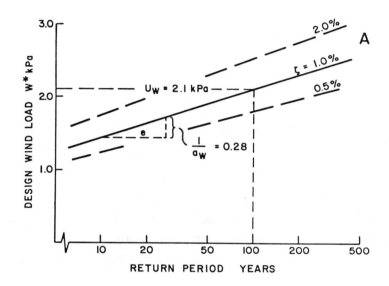

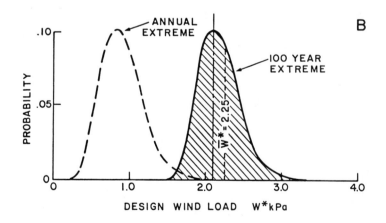

Fig. 12   Predictions of Statistical Distributions of Design
Wind Load

A)   As Function of Return Period

B)   As Probability Density Function of 100 Year

Extreme Load

$$Dispersion: \frac{1}{a_W{}^*} = 0.282 \ kPa$$

Some manipulation then leads directly to

$$\bar{W}^* = U_W{}^* + 0.577 \ [1 \ a_W{}^*]$$

$$= 2.25 \ kPa \ for \ 100 \ year \ lifetime$$

$$\sigma_{W}{}^* = \frac{\pi}{\sqrt{6}} \ \frac{1}{a_W} \quad ; and$$

$$V_W{}^* = 0.161$$

Uncertainty Factor Due to Damping — ϕ_ζ

Available information on damping is not well structured; however a study of damping measurements on about 20 concrete buildings suggests the damping level is on average about 1.3% of critical and the coefficient of variation is 0.32. Steel buildings tend to have a somewhat lower value with similar coefficient of variation. We will assume a mean value of $\bar{\zeta} = 0.01$ with $V_\zeta = 0.32$. We may therefore adopt the 1% curve in Fig. 12 and use $\bar{\zeta} = \zeta_o = 0.01$ as a reference value.

With the definition

$$\phi_\zeta = W(\zeta)/(\zeta_o) \simeq W^*(\zeta)/W^*(\zeta_o) \tag{19}$$

it follows that $\bar{\phi}_\zeta = 1.0$.

The coefficient of variation V_{ϕ_ζ} can be found from the relationship

$$V_{\phi_\zeta} = \left| \frac{\partial \ [W^*(\zeta)/W^*(\zeta_o)]}{\partial \ (\zeta/\zeta_o)} \right| V_\zeta \tag{20}$$

The differential in the above can be found from Fig. 12 and is found to be approximately ¼. Thus $V_{\phi_\zeta} = $ ¼ x 0.32 = 0.08.

Model Uncertainty — ϕ_M

Uncertainty in to the model is due to many different sources; sampling errors in assessing wind speed simplifications in the model, deviations of flow in the model from the true situation, uncertainty in natural frequency estimation, etc.

From an impressionistic assessment (in which full scale verification — e.g. Ref. 7 — plays an important role) we adopt the values

$$\bar{\phi}_M = 1.0$$

$$V_{\phi_M} = 0.10$$

This provides for an excursion of actual predictions from those assumed outside the range ± 20% five percent of the time, which seems pessimistic.

Synthesis

Following the above we have

$$\bar{W}^* = 2.25 \qquad V_W^* = 0.16$$
$$\bar{\phi}_\zeta = 1.0 \qquad V_{\phi_\zeta} = 0.08$$
$$\bar{\phi}_M = 1.0 \qquad V_{\phi_M} = 0.15$$

Thus from equations (16 and (17)

$$\bar{W} = \bar{W}^* \bar{\phi}_\zeta \bar{\phi}_M = 2.25 \text{ x } 1.0 \text{ x } 1.0 = 2.25 \text{ kPa}$$

$$(1 + V_W^2) = (1 + V_W^{*2})(1 + V_{\phi_\zeta}^2)(1 + V_{\phi_M}^2)$$

$$= (1 + 0.16^2)(1 + 0.08^2)(1 + 0.10^2)$$

$$V_W = 0.206$$

If we assume a safety index $\beta = 3$, considered appropriate for normal structural design risks, we find that the central load factor, $\overline{\gamma}$, for wind acting in the presence of dead load is, from equation (13),

$$\overline{\gamma} = exp\ (0.55 \times 3.0 \times 0.206)$$

$$= 1.40$$

The factored load is then

$$\overline{\gamma}\overline{W} = 2.25 \times 1.40 = 3.15\ kPa$$

The load factor in this instance appears quite consistent with load factors specified by some load and resistance factor design codes (e.g. 1.5 by the National Building Code of Canada). If only wind load is acting — as in the design of cladding — the central load factor increases according to equation (14)) to

$$\gamma = exp\ (0.75 \times 3.0 \times 0.206)$$

$$= 1.59$$

This may be somewhat higher than some safety factors specified and it may be prudent to review the implications, particularly in those codes in which increased stresses are permitted under wind. In doing so there are important alleviating factors which should be considered. These include such factors as the appropriateness of a lower safety index, the beneficial influence of ductility and the indirect restraint imposed by deflection criteria.

6. CONCLUDING REMARKS

The insights provided by wind tunnel testing can reach much deeper than simply the provision of numbers to be plugged into conventional design formulae. For example, through the use of aeroelastic models an understanding of the dynamic motion and accelerations of the structure, of the joint action of the wind loading in orthogonal directions and of the torsional forces can be developed.

Through the use of pressure models, the rapidly fluctuating, complex and variable patterns of exterior pressure can be determined. These frequently reveal surprisingly high pressures and suctions near the base of the building — in contradiction to many codes.

Environmental models permit prediction of the convoluted airflow patterns at the base of buildings in pedestrian regions. In many instances, modifications can be made which will alleviate the discomfort caused. The experience of working with the wind tunnel alerts the architectural designer to these additional dimensions which define the quality of the building environment.

These insights into the aerodynamics are not the only ones. In the process of bringing Boundary Layer Wind Tunnel testing into the mainstream of structural engineering, some of the conventional notions of wind loading have been reexamined. This has resulted in new understanding of the ways wind tunnel results connect with the other links of the design chain: In particular this has included the synthesis with the prevailing wind climate and the assurance of safety. This paper has aimed to make specific suggestions in these two areas.

REFERENCES

1. Davenport, A.G., "The Estimation of Load Repetitions on Structures with Application to Wind Induced Fatigue and Overload," Paper presented at the RILEM International Symposium on the 'Effects of Repeated Loading of Materials and Structures,' Mexico City, September 15–17, 1966.

2. Davenport, A.G., "Structural Safety and Reliability Under Wind Action," Presented at the International Conference on Structural Safety and Reliability, Washington, D.C., April 9–11, 1969 (Pergamon Press – 1972).

3. Davenport, A.G., "The Prediction of Risk Under Wind Loading," Presented at the ICOSSAR 1977, Proceedings of the 2nd International Conference on Structural Safety and Reliability, Munich, September 19–21, 1977.

4. Davenport, A.G., "Some Attributes of the Wind Load Requirements of the National Building Code of Canada," Presented at the Third Canadian Workshop on Wind Engineering, Vancouver – April 13–14, 1981 and Toronto – May 7–8, 1981.

5. Allen, D.E., "Limit States Design – A Probabilistic Study," Canadian Journal of Civil Engineering, Vol. 2, March 1975.

6. Ravindra, M.K. and Galambos, T.V., "Load and Resistance Factor Design for Steel," Jnl. Str. Div. Proc. ASCE ST9, 1978.

7. Dalgleish, W.A. and Rainer, J.H., "Measurements of Wind Induced Displacements and Accelerations of a 57 Storey Building in Toronto, Canada," Proceedings of 3rd Colloquium on Industrial Aerodynamics, Aachen, 1978.

PREDICTING EXTREME WINDS AT A SITE FOR EXTRATROPICAL AND TROPICAL STORMS

Emil Simiu, National Bureau of Standards, Washington, D.C. 20234

Structural members are routinely designed to attain a specified stress level under the action of a specified set of loads. For example, in case of a steel beam subjected to gravity and wind loading, the stress level can be specified as the mean yield stress (i.e., the yield stress averaged over a large number of samples), multiplied by a resistance factor smaller than unity. The corresponding set of loads consists of basic gravity and wind loads each multiplied by a load factor larger than unity. The purpose of the resistance and load factors is to account for uncertainties with respect to both the actual yield stress for a given member and the actual loads acting on the member during its lifetime.

In accordance with current U.S. practice, the basic wind load corresponds, in most cases, to a nominal 50-year mean recurrence interval of the wind speed. However, the fact that this load is multiplied by a load factor implies that, on the average, nominal failures of wind-sensitive members will occur under the action of wind loads with mean recurrence intervals, N, considerably longer than 50-years.

Assume that the basic wind load is $W_b = cv_{50}^2$, where c = specified constant and v_{50} = wind speed with a 50-year mean recurrence interval; that the resistance factor is unity; and that gravity loads are negligible. The wind load corresponding to the mean yield stress is then $W_y = L_w W_b$, where L_w = load factor. Since $W_y = cv_N^2$, it follows immediately that $v_N = L_w^{1/2} v_{50}$.

Clearly, the risk of failure depends, in this case, upon the length, N, of the mean recurrence interval of v_N. In turn, N depends upon the probability distribution of the extreme winds at the site under consideration. Thus, a first step in any attempt to estimate failure risks is the probabilistic description of the extreme wind behavior. A first objective of this paper is to discuss ways in which such a description can be achieved.

In reality, the constant, c, has variability due to micrometeorological and aerodynamics (or aeroelastic) uncertainties. Also, owing to uncertainties with respect to material resistance, the resistance factor is less than unity. Therefore, risks of failure do not depend solely upon uncertainties with respect to the wind speeds. This reduces, somewhat, the pressure to obtain "exact" probabilistic descriptions of the wind speeds. The extent to which this is the case can be ascertained from failure risk estimates that involve all the uncertainties in a reliability analysis. This is the motivation for the second objective of the paper, which is to discuss such estimates and their dependence upon the assumed probabilistic description of the extreme wind speeds.

This paper will be confined to a presentation of questions related to extreme wind prediction in the case where wind speeds are considered without taking direction into account. It is shown in Refs. 1 and 2 that current U.S. practice, which ignores the effect of wind directionality, leads to designs that may be grossly inconsistent from a risk of failure point of view, particularly in the case of cladding. It is indeed incongruous to go to great lengths in measuring aerodynamic directionality effects, as is currently done in most wind tunnel investigations, while ignoring climatological wind directionality effects. For practical design procedures suitable for computer use, that account for wind directionality, the reader is referred to Refs. 1 through 4, which deal with both buildings with specified orientation and buildings for which the orientation is unknown.

1. PROBABILISTIC MODELING OF THE EXTREME WIND SPEEDS

Techniques for the probabilistic modeling of extreme speeds depend upon whether the location under consideration can be expected to be subjected to extraordinary winds. These are defined as infrequent winds (e.g., hurricanes) that are meteorologically distinct from and considerably stronger than the usual annual extremes. Climates in which extraordinary winds may not be expected to occur are referred to as well behaved. In a well behaved wind climate, at any given station a random variable is defined which consists of the largest yearly wind speed. If the station is one for which wind records over a number of consecutive years are available, then the cumulative distribution function (CDF) of this random variable may be estimated to characterize the probabilistic behavior of the largest annual winds. The basic design wind speed is then defined as the wind speed corresponding to a specified value P of the CDF, or equivalently, to a specified mean recurrence interval $N = 1/(1-P)$. It is recalled that the largest yearly wind speed data used in the statistical analysis must constitute a micrometeorologically homogeneous set with respect to averaging time, height above ground, and roughness of surrounding terrain.

In a climate that is not well-behaved (e.g., a hurricane-prone region) most of the speeds in a series of the largest annual winds are considerably lower than the extreme wind speeds associated with hurricanes. Because structural design is controlled by hurricane winds, these lower speeds are largely irrelevant from a structural safety point of view. It may be argued that in hurricane-prone regions the series of the largest annual speeds cannot provide useful statistical information on winds of interest to the structural designer, much in the same way as the population of a first grade classroom -- which might include a teacher -- is of little use in a statistical study of the height of adults. It is thus seen that different approaches are necessary in well-behaved and hurricane-prone regions.

1.1 Extreme Wind Speeds in Well-Behaved Wind Climates

1.1.1 Extreme Type I and Rayleigh Distribution as Probabilistic Models of the Largest Yearly Wind Speeds

It is assumed in the American National Standard A58.1-1972 that the largest yearly wind speed data in well-behaved wind climates are best fitted by Extreme Value Type II distributions. However, subsequent research has shown that this assumption is not warranted [5, 6] and the the Extreme Value Type I distribution —which is less severe than the Type II distribution — provides, in general, a better fit to the extreme yearly wind speed data.

As will be shown subsequently, structural reliabilities of failure inherent in current building code provisions are higher for members subjected to wind loads than for members subjected to gravity loads, even if it is assumed that wind speeds are fitted by a Type I rather than Type II distribution [7, 8]. The question thus arises whether a probability model that is milder than the Type I distribution is not warranted for the description of the extreme yearly wind speeds.

In an attempt to answer this question, an analysis was conducted of extremely yearly data sets recorded at over 100 U.S. weather stations and published in Ref. 6. This analysis has shown that at a majority of the stations the data are best fit by Rayleigh distributions or by Weibull distributions with tail length parameter > 2 [9]. (Recall that the Rayleigh distribution is a Weibull distribution with tail length parameter $v = 2$.) From a climatological point of view, this result means that at a majority of stations the extreme wind speeds corresponding to long mean recurrence intervals do indeed have considerably lower values than would be the case if the Extreme Value Type I distribution were valid.

The significance from a structural reliability standpoint of the previously mentioned result will be illustrated by the following example. At Moline, Illinois, the mean and the standard deviation of the extreme yearly fastestmile speeds at 10 m above ground in open terrain recorded in the period 1944 to 1977 are $X = 24.49$ m/s (54.78 mph) and $s = 3.46$ m/s (7.73 mph) (see Ref. 3, p. 73). Assuming that the Moline largest annual fastest-mile wind speeds are best described by the Rayleigh distribution, the estimated wind speeds corresponding to the 50-year and 1,000-year mean recurrence intervals are $V_{50}^R = 32.65$ m/s (73.04 mph) and $V_{1000}^R = 37.50$ m/s (83.89 mph). If an Extreme Value Type I distribution were assumed, the then $V_{50}^I = 3.88$ m/s (75.79 mph) and $V_{1000}^I = 42.31$ m/s (94.67 mph). Note that the difference between V_{50}^I and V_{50}^R is relatively small (of the order of three percent). However, the differences between V_{1000}^I and V_{1000}^R is significant (of the order of 15 percent).

Assume now that a structure is designed to attain the yield stress under the action of a load proportional to $1.6 (V_{50}^I)^2$ (the coefficient 1.6 represents the load factor for wind loads). At Moline, the wind speed which will induce the yield stress is then $(1.6)^{1/2} V_{50}^I = 42.85$ m/s (95.87 mph), to which there corresponds a mean recurrence interval of 1,600-years if the Type I distribution is assumed, and of 70,400-years if it is assumed that the Rayleigh distribution holds. If the mean recurrence interval of the wind speed that induces the yield stress is regarded as an index of the nominal safety level of the member under wind loads, it is seen that the difference between nominal safety levels corresponding to the assumption that a Type I and a Rayleigh distribution holds is of almost two orders of magnitude. Note, however, that these calculations do not take into account uncertainties affecting the failure risk other than those associated with the inherent probabilistic nature of the exteme wind speeds. Such uncertainties, including those associated with the sample size (sampling errors) will be considered in a subsequent section.

For convenience, expressions for the estimation of extreme wind speeds, V_N, corresponding to a mean recurrence interval, N, are given below. These expressions are based on the method of moments.

If a Rayleigh distributon is assumed

$$V_N = \bar{X} + s [2.16 (\ell nN)^{1/2} - 1.91] \qquad (2)$$

If an Extreme Value Type I distribution is assumed

$$V_N = \bar{X} + 0.78 s(\ell nN - 0.5772) \qquad (3)$$

In Eqs. 2 and 3, $\bar{X}$ and s are the sample mean and standard deviation of the extreme wind speed data. It is noted that differences between estimates based on Eqs. 2 and 3 and estimates based, for example, on the least square method have been verified to be generally of the order of 0.5 percent to 2 percent.

1.1.2 Estimation of Extreme Wind Speeds from Short-Term Records

The question has been raised in the literature whether short-term records could provide useful information on extreme wind speeds. If this were the case, short-term records could be used successfully, e.g., for wind climate microzonation purposes. An investigation into this question was, therefore, conducted at the National Bureau of Standards [10]. The following results were obtained. First, the best fitting distributions was determined for sets of maximum weekly data and of maximum monthly data taken at seven locations (Washington, DC; Denver, CO; Great Falls, MT; Bismarck, ND; Chicago, IL; Syracuse, NY; and Detroit, MI). The length of record was three years in all cases. It was found that the sets of weekly data were best fitted by Rayleigh distributions or Weibull distributions with tail length parameter $\gamma \geq 2$ at five locations, and by the Extreme Value Type I distribution at two locations (Syracuse and Detroit). The same results were obtained for the sets of monthly data.

Second, extreme wind speeds with mean recurrence intervals of up to 1,000 years were estimated at the same seven stations using Eq. 2 and data sets consisting of: (a) the largest weekly, and (b) the largest monthly wind speeds over three years of record. In all cases it was found that the estimates based upon the weekly data differed insigificantly, for practical purposes, from those based upon the monthly data. The same result was obtained when Eq. 3 was used for estimating the extreme wind speeds.

Third, a comparison was made at approximately 30 stations between estimates of extreme wind speeds based on: (a) three-year sets of maximum wind speeds, and (b) long-term (i.e., approximately 30-years) sets of maximum yearly wind speeds. At the vast majority of the stations the differences between estimates based on three-years of monthly data on the one hand, and about 30years data on the other hand, were consistent with the sampling errors inherent in the statistical estimates and therefore relatively small for practical purposes. For example, the 1,000-year wind at Detroit based on Eq. 2 was 29.1 m/s (65 mph) if inferred from the maximum monthly speeds recorded in 1968-1970, and 30.4 m/s (68 mps) if inferred from the maximum yearly winds recorded in 1934-1977. The estimated speeds obtained by using Eq. 3 were 32.2 m/s (72 mph) and 34.4 m/s (77 mph), respectively. However, a number of stations were found where the differences between estimates of the 50-year wind based on the short-term and the long-term records were unacceptably large (of the order of 30 percent or so). These large differences were due to nonstationarities of the extreme wind speed time series. For example, in Sacramento, CA, in the period 1949-1955 the maximum yearly wind speeds varied from 56 to 70 mph, whereas between 1960-1973 the range was 31 mph to 42 mph. To avoid errors due to such nonstationarities it is desirable to attempt the development of alternative estimators based on statistically stable measures of the extreme wind climate, e.g., the average monthly speeds, as suggested in Ref. 11. Until such estimators are developed, caution is in order if inferences on the extreme wind climate are attempted on the basis of short-term records.

1.2 Extreme Wind Speeds in Hurricane-Prone Regions

Recently Russel's well-known procedure for estimating hurricane wind speeds was applied for about 50 locations on the Gulf and East Coasts of the United States. The procedure was used in conjunction with data on climatological characteristics of hurricanes given in Ref. 12, and with physical models for the hurricane wind field consistent with Refs. 13 and 14. The results of the calculations are given in Ref. 15 as functions of location along the coast.

Inherent in the estimates of Ref. 15 are physical modeling errors, probabilistic modeling errors, observation errors, and sampling errors. On the basis of an error analysis, it is suggested in Ref. 15 that at locations south of Cape Hatteras, the coefficient of variation of the physical modeling, probabilistic modeling, and observation errors is of the order of 10 percent. However, these errors could be significantly larger at locations north of Cape Hatteras, owing to the possible inapplicability at these locations of some of the physical models used in Ref. 15. Sampling errors, i.e., errors inherent in the fact that the available data may not constitute a representative sample, were estimated using statistical techniques described in Ref. 16. The coefficient of variation of these errors was found to be of the order of 6 percent to 10 percent. It is also shown in Ref. 16 that the precision of the extreme wind speed estimates is increased by about two percent to three percent if tropical cyclone in addition to hurricane data are taken into account in the analysis.

The probability distributions that were found to provide the best fit to the hurricane wind speed data generated by Monte Carlo simulation are of the Weibull type with tail length parameter $\gamma \geq 4$. It is recalled that these distributions have relatively short tails: for example, the increase in the estimated values of the N-year winds was found to be negligible as the mean recurrence interval increased beyond 2,000 years or so.

Finally, it is mentioned that, at locations south of Cape Hatteras, the effect of non-hurricane winds upon the magnitude of the estimated extreme wind speeds is negligible for winds with mean recurrence intervals exceeding 25-years or so.

2. ESTIMATION OF MEMBER RELIABILITIES

In this section, estimates will be presented of member reliabilities, and of the effect upon calculated reliabilities of the assumptions used with respect to the probabilistic model of the extreme wind speeds. It was mentioned in the preceding section that failure risks do not depend on the probabilities of occurrence of extreme wind speeds alone. Additional factors are involved, and these will now be taken into account [17].

The reliability measure used in this paper is the reliability index β. (This index is approximately equal to the distance, in standard deviation units, between the failure surface and a point associated with the mean values of the resistance R and the load Q, the failure surface being defined as the focus of the points (R, Q) such that R=Q.)

The results of calculations reported in Ref. 17 show that the estimated reliability index does depend upon whether the assumed probability distribution of the extreme wind speeds is Rayleigh or Extreme Value Type I. For example, for $s/\bar{X} = 0.14$ and m = 30, if the Rayleigh distribution holds, then the reliability index is equal to 1.74, whereas if the Type I Extreme Value distribution is valid, it is equal to 1.37, i.e., about 20 percent lower. If the calculations are carried out by ignoring variabilities associated with micrometeorological and aerodynamic factors, the discrepancies are found to be about twice as large. These results suggest that the use of a Type I distribution, where in fact a Rayleigh distribution would be appropriate, leads to overestimates of the risk of failure that are significant even when micrometeorological and aerodynamic uncertainties are taken into account.

Similar calculations were carried out to estimate the effect of sampling errors (due to the limited size of the data sample). On the basis of statistical theory (see Refs. 5, 6, 16, and 17) it was found that reliability indices calculated on the basis of 30 years of data are only about 10 percent higher than those based on 10-years only. However, the importance of using as large samples of data as possible (e.g., 30-years of data rather than 10-years) is greater than would be indicated by this result. This is so because larger samples not only reduce sampling errors inherent in the use of stationary time series, but also reduce errors due to possible nonstationarities of the extreme wind speed record.

3. SUMMARY AND CONCLUSIONS

A review of techniques for estimating extreme wind speeds was presented. It was shown that both in well-behaved climates and in hurricane-prone regions extreme wind speeds are in most cases best modeled by probability distributions with considerably shorter tails (for any given sample mean and sample standard deviation) than the Extreme Value Type I distribution.

The implication of this result for the estimation of failure risk was investigated, and it was shown that the reliability of wind sensitive members, as measured by the reliability index, is significantly influenced by the assumed probability distribution of the extreme wind speeds. Calculations were carried out that took into account the effect upon the reliability index of uncertainties pertaining to the material resistance, the aerodynamic properties of the structure, the time dependent effect of the wind pressures, the mean wind profile, and the estimation of extreme wind speeds. It was concluded, on the basis of these calculations, that reliability estimates for wind-sensitive structures, based on the assumption that a Type I Extreme Value distribution best describes extreme wind behavior, appear to be conservative for most geographical locations in the United States.

It was pointed out that it is incongruous to go to great lengths in measuring aerodynamic directionality effects, as is currently done in most wind tunnel investigations, while ignoring climatological wind directionality effects.

For practical design procedures that take these effects into account, the reader was referred to Refs. 1 through 4.

4. REFERENCES

1. Simiu, E. and Filliben, J. J., "Wind Direction Effects on Cladding and Structural Loads," _Engineering Structures_, July 1981, pp. 181-186.

2. Simiu, E. and Batts, M. E., "Wind-Induced Cladding Loads in Hurricane-Prone Regions," National Bureau of Standards, December 1981.

3. Batts, M. E. and Simiu, E., "Hurricane-Induced Wind Loads," Computer Program for Estimating Cladding Loads, National Bureau of Standards, December 1981.

4. Simiu, E., "Aerodynamic Coefficients and Risk - Consistent Design," National Bureau of Standards, January 1982.

5. Simiu, E., Bietry, J., and Filliben, J. J., "Sampling Errors in Estimation of Extreme Winds," _Journal of the Structural Division_, ASCE, March 1978.

6. Simiu, E., Changery, M. E., and Filliben, J. J., Extreme Wind Speeds at 129 Stations in the Contiguous United States, BSS 118, National Bureau of Standards, Washington, DC, 1978.

7. Simiu, E. and Shaver, J. R., "Wind Loading and Reliability-Based Design," Proceedings, Fifth International Conference on Wind Engineering, Fort Collins, CO, June 1979, Pergamon Press, New York, NY, 1980.

8. Ellingwood, B. R., "On Reliability of Structures Against Wind," ASCE Engineering Mechanics Division Speciality Conference, University of Texas, Austin, TX, September 1979.

9. Simiu, E. and Filliben, J. J., "Weibull Distributions and Extreme Wind Speeds," Journal of the Structural Division, ASCE, December 1980.

10. Simiu, E., Shaver, J. R., and Filliben, J. J., "Short-term Records and Extreme Wind Speeds," submitted for possible publication in the Journal of the Structural Division, ASCE, 1981.

11. Thom, H. C. S., "Toward a Universal Climatological Distribution," Proceedings, Second International Seminar on Wind Effects on Structures, Ottawa, 1968.

12. Ho, F. P., Schwerdt, R. W., and Goodyear, H. V., Some Climatological Characteristics of Hurricanes and Tropical Storms, Gulf and East Coasts of the United States, NOAA Technical Report NES15, Washington, DC, May 1975.

13. Malkin, W., Filling and Intensity Changes in Hurricanes Over Land, National Hurricane Research Project, Report No. 34, U.S. Department of Commerce, Washington, DC, November 1959.

14. Revised Standard Project Hurricane Criteria for the Atlantic and Gulf Coasts of the United States, Memorandum HUR7-120, U.S. Department of Commerce, National Oceanic and Atmospheric Administration, June 1972.

15. Batts, M. E., Cordes, M. R., Russell, L. R., Shaver, J. R., and Simiu, E., "Hurricane Wind Speeds in the United States," Building Science Series Report 124, National Bureau of Standards, Washington, DC, 1980.

16. Batts, M. E., Cordes, M. R., and Simiu, E., "Sampling Errors in Estimation of Hurricane Winds," to be published in the Journal of the Structural Division, ASCE, 1979.

17. Simiu, E., Shaver, J. R., and Filliben, J. J., "Wind Speed Distribution and Reliability Estimates," Journal of the Structural Division, ASCE, May 1981.

Session II

Simulation of the Planetary Boundary Layer

Chairman

Nicholas Isyumov

University of Western Ontario
London, Ontario
Canada

SIMULATION CRITERIA BASED ON
METEOROLOGICAL OR THEORETICAL CONSIDERATIONS

Henry W. Tieleman, Professor
Virginia Polytechnic Institute and State University

1.1 INTRODUCTION

Atmospheric boundary layer (ABL) flows near the earth's surface are characterized by conditions which are generally not common to wind tunnel or laboratory boundary layers. Non-uniform boundary conditions, nonstationarities as well as variable atmospheric and thermal conditions, usually encountered in practical situations, make accurate prediction of the atmospheric boundary-layer flow extremely difficult and almost impossible.

The description of turbulent boundary layers has been based primarily on similarity arguments. Using this approach, the statistical flow parameters are described in terms of empirical functions which require a number of parameters which control the flow in different parts of the boundary layer. Specific empirical relations have been derived from experimental data, by organizing the data in accordance with the established turbulent boundary-layer model. However the majority of these experimental data were derived from measurements made in the ABL under near-ideal conditions which include flow over flat terrain with a long upstream fetch of homogeneous surface roughness and under ideal atmospheric conditions. However there are only a few "real world" situations for which these similarity theories can provide realistic predictions of the ABL flow. Few experimental results are available from detailed studies of the response of the turbulent boundary-layer flow to any change in surface conditions.

The asymptotic matching of the similarity laws applicable to boundary layers which are governed by two characteristic length scales, provides for a formal derivation of the logarithmic velocity profile, coupling the flow in the outer layer with the flow in the inner layer. With this development more specific arguments can be made for the conditions for which the application of the logarithmic velocity law is justified.

There are several unanswered questions which can come to one's mind when dealing with the solution of wind-engineering problems. Such problems often require a reasonably accurate description of the actual ABL flow, as the latter is usually developed over a non-homogeneous surface. How well can actual ABL flows and turbulence structure be predicted from empirical relations which were developed on the basis of ideal conditions? Can wind-tunnel boundary layers be used to adequately simulate the ABL flow, and if so what are the requirements for a successful model simulation?

Obviously, no definite answers to these questions will be given in this paper. Instead an attempt will be made to point out some variations in ABL flows under non-uniform surface conditions and under varying atmospheric conditions, based on both actual and wind tunnel experimental results. In addition some basic requirements for an adequate wind-tunnel simulation of the ABL flow will be discussed, with emphasis on the atmospheric surface layer.

2. WIND TUNNEL BOUNDARY LAYERS

2.1 Classical Approach (Surface Layer Similarity)

The most common turbulent boundary layers generated in the laboratory are those with zero pressure gradient, and developed over either a uniform-smooth or a uniform-rough flat surface. Two basic laws have provided some order to the experimental data obtained from these boundary layers. The law describing the flow near the surface is referred to as the "law of the wall" and has been attributed to Ludwieg Prandtl. In dimensionless form this law reads for a smooth wall

$$U/U^*_0 = f(U^*_0 z/\nu), \tag{1}$$

and for a rough wall

$$U/U^*_0 = f(z/k). \tag{2}$$

Here k is a length scale characteristic of the wall roughness. With the introduction of the eddy viscosity and the assumption of its linear variation with height, and with the assumption that the stress is constant and equal to the wall stress, the well known logarithmic velocity profile can be obtained

$$U/U^*_0 = 1/\kappa \; \ln \; (u^*_0 z/\nu) + \text{const.} \tag{3}$$

for a smooth wall, and

$$U/U^*_0 = 1/\kappa \; \ln \; (z/k) + \text{const.} \tag{4}$$

for a rough wall. Similar results can be obtained with the use of either von Karman's or Prandtl's mixing length hypotheses.

At approximately the same time, Theodore von Karman introduced the law governing the flow in the outer layer which is referred to as the "velocity defect law,"

$$U-U_0/U^*_0 = g(z/\delta) = g(\eta). \tag{5}$$

Experimental boundary layer data collected by Clauser [1] show the universality of these two laws (1,5). The results also indicated that the regions of validity of the two laws overlapped. Millikan [2] showed that with the existence of an overlap region, the velocity distribution must be of logarithmic form in this region. It is significant to recognize that by this method one can arrive at this logarithmic velocity distribution without the introduction of an eddy viscosity or a mixing length and without the assumption of a constant stress layer.

2.2 Asymptotic Matching

More recently, Tennekes and Lumley [3] showed the existence of a logarithmic velocity profile with the method of asymptotic matching of the wall layer with the outer layer. For the wall layer the characteristic length scales are ν/U^*_0 for a smooth wall and k for a rough wall, while the friction velocity, U^*_0, is used for velocity scale, representing the turbulence. For large Reynolds-number boundary layers, the boundary layer thickness, δ, is much larger than either one of the wall-layer length scales. Consequently, δ and U^*_0 are the characteristic length and velocity scale for the larger part of the boundary-layer flow (outer layer). If the ratios of the length scales for the outer layer and wall layer, $U^*_0\delta/\nu$ or δ/k, are large enough, it is possible to show the existence of an intermediate layer for which the distance from the wall, z, is simultaneously much larger than ν/U^*_0 and much smaller than δ. In this layer the flow is controlled by the distance from the wall so that dU/dz is proportional to U^*_0/z, which leads immediately to the existence of a logarithmic velocity profile.

Based on the governing equations (momentum and turbulent energy) for the fully developed turbulent flow between two smooth parallel flat plates, Tennekes and Lumley [3] show in detail how with large Reynolds numbers ($U^*_0 h/\nu \to \infty$), the law of the wall (1) and the velocity defect law (5) (functions f and g) are universal. For either law only a single set of boundary conditions are available $f(0) = 0$ and $g(1) = 0$. No required conditions are available for $f(\infty)$ and $g(0)$ other than in these limits f and g should be finite and the extent of each law is limited to its region of applicability. Provided $U^*_0 h/\nu \to \infty$, an overlap region exists where $z \gg \nu/U^*_0$ and $z \ll h$ and the two laws are equally valid.

In this overlap region the nondimensional velocity gradients are not only required to be equal but also independent of any other variables, and therefore must be equal to a universal constant. Integration of both velocity gradients requires that both the law of the wall and the velocity defect law must be logarithmic in this region or

$$U/U^*_0 = 1/\kappa \; \ln \, (U^*_0 z/\nu) + C_1 \qquad (6)$$

$$\text{and} \quad U-U_0/U^*_0 = 1/\kappa \; \ln \, (z/h) + C_2. \qquad (7)$$

For the fully developed flow between parallel walls the governing equations are relatively simple because all streamwise gradients, except the pressure gradient, are zero. For turbulent boundary layers however, the flow in the outer layer changes as a result of downstream development. However for smooth flat-plate turbulent boundary layers with a zero-pressure gradient, solutions to the linearized boundary layer equation satisfying the velocity defect law do exist, provided the Reynolds number is large ($U^*_0\delta/\nu \to \infty$). For this type of boundary layer flow the velocity profiles are self-preserving and the normalized velocity defect, $(U-U_0)/U^*_0$, and the normalized turbulent stress $-\overline{uw}/U^{*2}_0$ are universal functions of z/δ, independent of Reynolds number, downstream distance or any other parameter. Matching of the law of the wall with the velocity defect law in the overlap region where at the same time z >> ν/U_0^* and z << δ, requires the existence of a logarithmic velocity profile [3]. Based on experimental results from several sources, Clauser [1] found the best fit to the data to be

$$U/U_0^* = 2.5 \ln (U_0^* z/\nu) + 4.9 \qquad (8)$$

and

$$U-U_0/U_0^* = 2.5 \ln (z/\delta) - 2.5 \qquad (9)$$

2.3 The Effect of Roughness

The effect of wall roughness on the flow in the boundary layer can only be treated in general terms because of the variation in geometrical shapes and distribution of the roughness elements. The wall roughness does not directly affect the velocity defect law except through the value of the wall shear or friction velocity, U_0^*, provided the average size of the roughness elements, k, is much smaller than the boundary layer thickness, δ. On the other hand, the velocity distribution close to the wall is affected directly by the wall roughness. In a rough-wall surface layer two characteristic length scales can be recognized, namely k and ν/U_0^*, whose ratio is called the roughness Reynolds number, $U_0^* k/\nu$. Consequently the law of the wall may be written as

$$U/U_0^* = f_1(U_0^* z/\nu, \; U_0^* k/\nu), \qquad (10)$$

or

$$U/U_0^* = f_2(z/k, \; U_0^* k/\nu) \qquad (11)$$

These expressions must match the velocity defect law in the overlap region, provided the boundary layer Reynolds number $U^* \delta/\nu$ is large enough. Since the velocity defect law is universal either for smooth or rough surfaces, it is required that in the overlap region the nondimensionalized velocity gradients are equal to the same universal constant as for the smooth-wall case. Integration of these gradients leads to

$$U/U_0^* = 1/\kappa \ln (U_0^* z/\nu) + F_1 (U_0^* k/\nu), \qquad (12)$$

and

$$U/U_0^* = 1/\kappa \ln (z/k) + F_2 (U_0^* k/\nu). \qquad (13)$$

In the limit as $U^* k/\nu \to 0$, the velocity profile (12) has to approach that for a smooth wall[a] (8), so that $F_1(0) = 4.9$. For large values of the roughness Reynolds number, k >> ν/U_0^*, viscous effects become negligible. If simultaneously k is large enough so that the flow about the roughness elements creates eddies whose scale is proportional to the boundary layer thickness, the wall is said to be fully rough. Under these conditions the velocity profile is given by expression (13) with F_2 being a constant.

Between these two extremes (smooth and fully rough) there exists an intermediate range in which only a fraction of the roughness elements disturbs the flow in the region near the wall and where viscosity still dominates (viscous sublayer). Only part of the surface stress is due to the viscous shearing stresses, the remainder is due to the form drag on the roughness elements. Since an increase in roughness size increases the wall shear or U_*, the effect of the roughness manifests itself in the overlap region merely as a shift of the velocity profile in a semi-logarithmic plot.

Subtracting the fully rough-wall profile (13) from the smooth-wall profile (12) an expression results for this velocity shift,

$$\Delta U/U_* = 1/\kappa \; \ell n \; (U_*k/\nu) + A, \tag{14}$$

Experimental results summarized by Clauser [1] indicate that the velocity shift $\Delta U/U_*$ is well correlated with the roughness Reynolds number, U_*k/ν, but that the relation by no means is universal (Fig. 1). In the fully rough range, $U_*k/\nu > 80$, the results follow expression (14) with the same slope but varying intercept, A, depending on the shape and distribution of the roughness elements. The smooth-wall range is approached ($\Delta U/U_* \to 0$) for the uniform-roughness cases and $U_*k/\nu < 5$. In the range, $5 < U_*k/\nu < 80$, the wall is partially rough and the velocity shift is relatively small and obviously does not follow expression (14).

3. ATMOSPHERIC BOUNDARY LAYERS

3.1 Neutral Atmospheric Boundary Layer (Ekman Layer)

Consider the flow in the A.B.L. as being steady and horizontally homogeneous without the effect of any buoyancy forces and over a flat homogeneous surface whose roughness can be represented by a roughness length-scale, z_0. Above the A.B.L., the pressure gradient force must be balanced with the Coriolis force. With the assumption that vertical wind velocities are negligibly small one must have

$$f\vec{k} \times \vec{G} = - 1/\rho \; \vec{\nabla} \; p, \tag{15}$$

where f is the customary Coriolis parameter $f = 2\omega\sin\lambda$, with ω being the earth's angular velocity, and λ the geographic latitude, and $\vec{G} = U_g\vec{i} + V_g\vec{j}$ being the geostrophic wind. In horizontal components,

$$- fV_g = - 1/\rho \; \partial P/\partial x \quad \text{and} \quad fU_g = - 1/\rho \; \partial P/\partial y \tag{16}$$

In the A.B.L. with addition of the turbulence stresses, the balance of force equations are

$$- f(V - V_g) = d(-\overline{uw})/dz.$$

and

$$f(U - U_g) = d(-\overline{vw})dz. \tag{17}$$

The external parameters which control the Ekman-layer flow are f and G as well as the surface roughness length, z_0. Together these parameters form the surface Rossby number, $R_0 = G/fz_0$. Other flow parameters which enter into these considerations are the friction velocity, U^*_0, and the height of the Ekman layer, h. In the absence of buoyancy forces, the turbulence is purely of mechanical origin, and the friction velocity, U^*_0, is the only characteristic velocity scale for the turbulence [4]. Again one can recognize two different length scales, U^*/f which is proportional to the height of the Ekman layer, h, and is characteristic for the flow in the outer layer, and the roughness length z_0 characteristic for the flow in the inner layer.

For increasing values of the Rossby number one can expect the flow in the Ekman layer to become gradually less dependent on the Rossby number, and according to Blackadar and Tennekes [4], the velocity distribution in this layer should take the following universal form (velocity defect law).

$$U-U_g/U^*_0 = f_u(zf/U^*_0)$$

and $\hspace{10cm}$ (18)

$$V-V_g/U^*_0 = f_v(zf/U^*_0).$$

From the nondimensionalized equations of motion (17) with the knowledge that in the atmosphere $fz_0/U^*_0 \ll 1$, one can conclude that in the surface layer the stress must be constant and the Coriolis effect negligible. Then, with the characteristic length scale z_0, and the friction velocity U^*_0, the velocity distribution must follow the law of the wall,

$$U/U^*_0 = f(z/z_0) \quad \text{and} \quad V/U^*_0 = 0 \tag{19}$$

In the limit as $U^*/fz_0 \to \infty$ one expects the flow in the surface layer, if properly scaled with U^*_0 and z_0 to be independent of the Rossby number or any other parameter. The same applies to the flow in the outer layer (18) when scaled with U^*_0 and U^*/f or h. Under these conditions we have these two similarity laws, one for the surface layer and one for the Ekman layer, which must match in the overlap region where both laws are valid simultaneously. This can only work if in this overlap region, with the limits $z/z_0 \to \infty$ and $zf/U^*_0 \to 0$, the two laws are well behaved and are independent of the Rossby number or any other parameters. Under these conditions the nondimensionalized velocity gradients of expressions (18) and (19) should be independent of any other parameters and be equal to the same universal constant, $1/\kappa$ as before. Integration leads then to the well known logarithmic velocity profiles only valid in this region,

$$U/U^*_0 = 1/\kappa \ln(z/z_0) + C_1$$

and $\hspace{10cm}$ (20)

$$U-U_g/U^*_0 = 1/\kappa \ln(zf/U^*_0) + C_2$$

The universal constant κ (the von Karman constant) and the roughness length scale, z_0, cannot be obtained from theoretical considerations. Although at the present time some controversy exists over the numerical value of κ, a value of $\kappa = 0.4$ based on wind tunnel experiments is most commonly accepted. For any natural roughness it is impossible to assign a single numerical value for z_0, taking into account the variation in roughness size, shape, distribution and density. Consequently, the constant of integration C_1 in (20) is not a universal constant, except for the case of the roughness height $k < \nu/U*$ (smooth boundary), but generally depends on the nature of the roughness and the roughness Reynolds number, $U_0^* k/\nu$. In order to circumvent this difficulty, C_1 is often incorporated into z_0 in expression (20), to become

$$U/U_0 = 1/\kappa \, \ln(z/z_0). \tag{21}$$

By doing so, z_0 in (21) does not only characterize the natural surface roughness but also is dependent on the roughness Reynolds number. For all practical situations a roughness length can only be obtained from the velocity profile (21) and should perhaps be referred to as a "profile roughness length" in contrast to the "true roughness length" in (20). Additional complications can be expected with the "profile roughness length" if the terrain roughness is not homogeneous.

3.2 Non-Neutral Atmospheric Boundary Layer

So far the discussion has been limited to a neutral boundary layer with an adiabatic temperature profile. In reality the A.B.L. nearly always has some vertical heat flux either in the form of sensible or latent heat. Dependent on whether the heat flux is upward (unstable) or downward (stable), in addition to the mechanical production, turbulent energy is either produced or suppressed. Consequently the mean flow as well as the turbulence in the diabatic surface layer are different from those in the neutral surface layer.

Stable conditions are usually encountered during night or when warmer air flows over a colder water surface. In the presence of a surface-based inversion possibly with one or more stable layers at higher altitudes, turbulence is generally suppressed in the entire A.B.L.(Fig. 2). Under extreme stable conditions, the air in the different layers becomes uncoupled as a result of reduced mechanical mixing and low-level jets at either the surface or at higher elevations may be observed (Fig. 3). Additional complications may occur when turbulence and gravity waves co-exist in the stable layers(Fig. 4). Under these stable conditions the A.B.L. seldom reaches a state of quasi-steadiness, because conditions vary rather rapidly, and consequently no simple boundary-layer model is available for the description of the flow.

In the case of daytime heating of the earth's surface which results in an upward heat flux, the mechanical turbulence production is augmented with buoyant energy production. Near the ground the mechanical energy production dominates but decreases rapidly with height, while on the other hand the buoyant production is almost constant through the surface layer. Outside the surface layer, the unstable or convective boundary layer undergoes strong vertical mixing due to positive buoyancy forces and the production of large-scale convective turbulence. This well-mixed layer, which extends to the lowest inversion at an altitude of z_i, is characterized by an almost uniform potential temperature and by a constant wind speed and direction [5] (Fig. 5). In addition, downward fluxes of momentum and heat due to entrainment into the convective boundary layer across the inversion complicate matters a great deal. Near the surface the turbulent heat and momentum fluxes exist as a result of temperature and velocity gradients, however in the mixed layer the fluxes are maintained by buoyancy effects and entrainment. In this layer the fluxes become insensitive to existing gradients and near the top of the surface layer the interaction of the gradient-produced turbulence and buoyancy driven turbulence controls the flow.

At the present time no simple universal scaling law is available to describe the mean and turbulent flow in the layer away from the surface, and consequently the method of asymptotic matching of an outer law and an inner law in the overlap region cannot be used.

Instead we have to resort to similarity in the surface layer by considering constant fluxes in this layer, much in the same manner as the original analyses by Prandtl and von Karman for surface layer flow of laboratory boundary layers. It is assumed that the flow in the atmospheric surface layer is uniquely determined by the surface stress, U_0^*, the surface heat flux, Q_0, the buoyancy parameter, g/T, and the height z. Then by dimensional analysis, dependent flow variables, properly nondimensionalized, should be universal functions of the stability parameter z/L, where $L = T(U_0^*)^3/(gQ_0\kappa)$ is the Monin-Obukhov stability length. Accordingly, the dimensionless wind shear should be a universal function of z/L, which can be integrated to give

$$U/U_0^* = 1/\kappa[\ln z/z_0 - \phi(z/L) + C], \qquad (22)$$

with C being a constant of integration. Close to the surface with $z/L \to 0$, the mechanical turbulence dominates and $\phi(0) = 0$. Introducing the roughness length, z_0, as being the value of z near the surface for which both U and ϕ in (22) approach zero, the velocity profile can be written as

$$U/U_0^* = 1/\kappa[\ln(z/z_0) - \phi(z/L)] \qquad (23)$$

Here, z_0 as in (21) must depend on the nature of the surface roughness as well the roughness Reynolds number as is the case for the neutral A.B.L. and laboratory boundary layers.

Experimental results obtained in the surface layer over very uniform and smooth terrain, indicate that vertical turbulence fluctuations follow Monin-Obukhov scaling. The variation of σ_w/U_0^* with z/L is well established, increasing gradually in unstable air and being nearly invariant (approximately 1.25) in neutral and stable air [6]. On the other hand, Monin-Obukhov scaling does not apply to the low-frequency velocity fluctuations of the horizontal turbulence components. Under convective conditions even in the surface layer, the integral scale of the turbulence does not vary with z as is required for Monin-Obukhov scaling, but varies instead with the height of the convective boundary layer, z_i [5] (Fig. 5b). Consequently, under these conditions the low-frequency end of the u and v spectra (Fig. 6) can only be generalized with scales associated with the mixed layer, kz_i and z_i/L, while the high-frequency portion of these spectra follows the Monin-Obukhov scaling [7]. The ratios $\sigma_u/U*$ and $\sigma_v/U*$ follow mixed layer similarity under convective conditions [8] but are approximately constant under stable conditions, provided no low-frequency components associated with gravity waves are included [9]. The neutral values of these ratios vary considerably and depend greatly on how the neutral limit is approached. If approached from the unstable side the values of these ratios are higher than when approached from the stable side, because of the presence of the large-scale turbulence associated with the mixed layer under convective conditions (Fig. 7).

3.3 Atmospheric Boundary Layer Over Complex Terrain

Wind measurements obtained over rolling farm land and in the valleys of the mountainous areas in the Eastern United States, show that fluctuations of the horizontal velocity components vary a great deal with this kind of upstream terrain features (complex terrain) [10,11]. The turbulence ratios $\sigma_u/U*$ and $\sigma_v/U*$ observed over smooth uniform terrain are much lower than those over complex terrain. On the other hand, the ratio $\sigma_w/U*$ seems unaffected by the nature of the upstream terrain (Fig. 8). No single values can be assigned to these ratios of the horizontal velocity components. The complex-terrain observations vary from being 40% to 250% larger than the flat-terrain results, depending much on the orientation of the mountain ridges with respect to the wind direction. Spectral measurements of the horizontal velocity components under slightly stable conditions show a marked increase in low-frequency energy for the complex-terrain spectra in comparison with the flat-terrain spectra [12,13] (Fig. 9). Spectral measurements of the vertical wind component show very little sensitivity to upstream terrain over the entire frequency range. However, the complex-terrain spectra of all three velocity components are similar to the flat-terrain spectra under unstable conditions. Consequently, upstream terrain features affect the horizontal wind fluctuations specially under near neutral conditions approached from the stable side.

3.4 Atmospheric Boundary Layer Over a Change in Roughness

The response of atmospheric boundary layers to changes in surface conditions, including surface roughness and surface temperaure, manifests itself in the development of an internal boundary layer (IBL). Within this layer the flow is adjusting to the new surface conditions, U_o^* and T_o, resulting in a change in turbulence and mean velocity. Obviously it requires some time for the turbulence structure to adapt itself to the new surface conditions, or before the turbulence again becomes in equilibrium with the new surface. The depth of the equilibrium part of the IBL grows very slowly, approximately at a rate of 0.01 times the distance from the change in roughness, while the height of the entire I.B.L. varies approximately as 0.1 times this distance [14].

For flow in the IBL past a change in surface roughness the following successive flow changes can be recognized [15]. First the wind shear changes as a result of the changing momentum loss at the surface. Next the turbulent momentum flux (Reynolds stress) adjusts itself to the change in surface stress. Finally, the turbulent kinetic energy and variances of the different wind components as well as the dissipation are adjusted. In the equilibrium region within the IBL the velocity profile is logarithmic and can be used to obtain values of the surface roughness, z_0, and of the friction velocity U^*. However, the growth rate of this equilibrium layer is extremely low, approximately 0.01. In the remainder of the IBL (transition region) the flow is not in equilibrium with the underlying surface and z_0 and U^* cannot be obtained from a measured velocity profile in this region. Also the turbulence characteristics vary drastically with height and with downstream distance in this region, causing the usual constant-flux layer approximations to be invalid.

Most theoretical and experimental studies in the wind tunnel and in the atmosphere deal with abrupt variations in surface roughness with neutral air flowing at right angles to the line delineating the roughness change. Based on the results of wind tunnel experiments by Antonia and Luxton [16,17] and Mulhearn [18], and the atmospheric experiments by Bradley [19], the growth rate of the IBL is slower in the case of a rough-smooth change than in the reverse situation. Consequently, the ABL recovers slower following a rough-smooth roughness change than following a smooth-rough change. Over natural terrain the roughness varies alternately from rough to smooth and back to rough, requiring continual adjustments of the mean and turbulent flow to new surface conditions. Spectral measurements made at the Riso National Laboratory in Denmark [12] for a smooth-rough surface change from water to land show the following results. High-frequency velocity fluctuations (short wave lengths) respond rather rapidly to changes in surface conditions and are usually in equilibrium with the underlying surface. In the low-frequency energy containing part of the spectra (long wave lengths), spectral values change very slowly after a roughness change and may never be in equilibrium with the underlying surface. A model describing the adjustment of velocity spectra downstream of a roughness change is given by Hojstrup [20].

Some of the experimental Riso results [21,22] show that a variation in surface elevation which is typical for a coastal site, as well as a change in roughness, have a pronounced effect on the surface flow. For on-shore flows encountering not only a change in surface roughness but also an increase in surface elevation in the form of a shallow escarpment, the velocity profiles show two variations. The first variation is a deceleration of the flow near the surface caused by the increased surface roughness, and the second variation is the increase in velocity at higher elevations due to the flow acceleration over the escarpment. An increase of 6% in mean velocity has been observed at Riso [22] at heights of 10 m for an inland distance of 180 m and with an increase in elevation of 2m over a distance of 50 m measured from the water line.

The results from the experimental studies mentioned above clearly indicate that the mean wind and the turbulence in the A.B.L. are not likely to be in equilibrium with the underlying surface for atmospheric flows over most natural terrains. For nonuniform terrain conditions, with or without changes in elevation, the relationships between profiles of the mean quantities and the turbulent fluxes become quite complicated and deviate appreciably from the homogeneous-terrain relations.

4. CONCLUSIONS

4.1. Logarithmic Velocity Profile

Proving the existence of the logarithmic velocity profile by means of the method of asymptotic matching for either a laboratory or neutral atmospheric boundary layer is of profound importance for any scientist who is involved in fluid mechanics and micrometeorology. The consequences of the asymptotic approach as opposed to the classical approach (surface layer similarity) can be summarized as follows.

a. Validity of the logarithmic velocity profile requires the ratio of the length scale for the outer layer (δ or U_*/f) and the length scale for the surface layer (ν/U_*^0, k or z_0) to approach infinity.

b. Also the logarithmic velocity profile is valid only in the range where the universal laws for the surface layer and outer-layer overlap. For this region it is required that the height z is simultaneously much larger than the surface-layer length scale and at the same time much smaller than the outer-layer length scale. At shallow heights above the surface where the first requirement is not satisfied the logarithmic profile can not be valid.

c. Only when the above conditions are satisfied can the nondimensionalized velocity gradients based on the velocity defect law and the law of the wall both be equal and also equal to a universal constant $1/\kappa$. If these conditions are not satisfied the universality of the von Karman constant may be questionable, which would lead to uncertainty in the values of z_0 and U_*^0 obtained from the logarithmic-velocity profile expression.

d. As in the overlap region the flow from the outer layer is matched with the mean flow from the surface layer, it must follow that the turbulence in the two layers is coupled. Consequently large-scale turbulence observed in the Ekman layer can be expected to influence turbulence levels in the surface layer. This observation certainly does not reinforce the necessary classical-theory assumption of a linear relation between eddy viscosity or mixing length versus height. Moreover, these assumptions are not required for the asymptotic derivation of the logarithmic velocity profile.

e. The asymptotic derivation of the logarithmic wind profile does not require any assumption about the stress distribution in the overlap region. Classical theory of course depends on the assumption that the stress be independent of height in this layer. Measurements in laboratory boundary layers show that the logarithmic velocity profile extends to about 10% of the boundary layer height, while the stress at this height is approximately 75% of its surface value.

How the asymptotic wind-profile analysis will be affected by nonhomogeneous surface conditions, nonstationary flow conditions and by the presence of vertical heat flux is difficult to speculate. However, with the presence of still another length scale, namely the Monin-Obukhov length scale, an asymptotic approach for the flow analysis in the non-neutral A.B.L. is extremely difficult and still an unresolved problem at the present time.

4.2 Non-Homogeneous Surface Conditions

The turbulent structure of the A.B.L. and specifically that of the surface layer is very unlikely to be in equilibrium with the underlying surface, except over large regions of horizontally homogeneous surface conditions. In most practical situations the atmospheric flows encounter continuous surface-roughness changes from rough to smooth and vice versa. Changes in roughness are seldom delineated along a straight line perpendicular to the average wind direction and roughness is seldom homogeneous and uniform on either side of this line. Changes in roughness may occur gradually or, abruptly, but usually occur along oblique angles with the mean wind direction. The most common roughness changes include variation in topography, changes in natural and agricultural vegetation and man-made structures. Over rural and suburban terrains these changes may follow each other in rapid succession from rough-to-smooth and vice versa. Based on the experimental evidence, adjustment of the turbulence after a change in roughness takes place rather slowly. Often the distances between successive roughness changes are so short that the equilibrium part of the IBL never has a chance to grow to any appreciable depth. Consequently, for most practical situations the turbulence above a height of several meters is not in equilibrium with the underlying surface, and values of the turbulence ratios, $\sigma_\alpha/U*$ may deviate from their usual magnitudes for equilibrium boundary layers. Moreover, variations in turbulence along a vertical line may be due to several upstream roughness changes. Consequently, nonhomogeneous terrain many kilometers upstream from the observation location may have significant effects on the mean wind profile and turbulence distribution at higher elevations at this location. Under these conditions it is impossible to obtain the appropriate roughness length and friction velocity from tower-measured wind profiles, also because at these heights thermal effects cannot be ignored.

Relatively simple relationships between the mean wind profile and the turbulence are available for the homogeneous case and for a simple surface-roughness change normal to the flow. However for more complicated deviations from these ideal cases, the relationships between the profiles and the fluxes become rather complicated, and turbulent fluxes cannot be deduced from observed mean wind profiles.

4.3 Considerations for the Laboratory Simulation of the ABL

For proper simulation of the atmospheric wind and turbulence in the laboratory, the wind tunnel boundary-layer flow must adhere in theory to many modeling criteria which often cannot be satisfied for reasons of practicality and physical impossibility. Simulation of the stationary, neutral and horizontally homogeneous ABL in the laboratory can only be attempted at relatively small geometric scaling ratios of 1/1000 or less. For simulation of this type of simplified ABL flow, Rossby number similarity in the atmosphere, (18) and (19), has to be replaced by Reynolds number similarity in the laboratory, (13) and (7). The ideal conditions of neutral thermal-stratification with the absence of any vertical heat flux, and of horizontal homogeneity of the surface are only of importance for theoretical studies, but seldom apply to practical situations in the ABL.

Most simulation problems apply to flows near the surface in the lowest 100-150 m, and in order to use larger geometric scaling ratios it is obviously attractive to employ a scheme of part-depth simulation of the ABL. Based on the discussion about the asymptotic matching theory, true simulation of the surface layer flow, which in reality is coupled to the flow in the outer layer, can only be achieved by simulation of the entire ABL. Consequently, limiting the simulation of the ABL flow to that of the surface-layer only, can be achieved on an ad-hoc basis only by relying solely on surface-layer similarity considerations.

In the case the surface layer needs to be modeled for wind-load studies under strong wind (near neutral) conditions on low-rise structures, the basic requirements are the simulation of the mean-flow distribution and the simulation of the wind-velocity fluctuations at both the low- and high-frequency ends of the spectrum. Realistic scale ratios for these kind of problems may vary from 1/25 to 1/100, and with the appropriate upstream roughness in the wind tunnel, scaled values of z_0, based on the logarithmic velocity-profile equation (21), can be established. The geometric scaling of the roughness length may be adequate for the simulation of the mean velocity distribution, U/U_0^*, but certainly cannot quarantee simultaneous simulation of the turbulence in the surface layer.

In addition to the scaling of the mean flow it is necessary to simulate the turbulent flow, because the turbulence greatly affects the surface pressures. Geometric scaling of the integral scale of the turbulence should provide adequate simulation of the low-frequency range of the velocity spectrum. However, the magnitude of the turbulence integral scales in wind tunnels depends greatly on the surface roughness, and model/prototype scaling of these scales with the already established floor roughness may or may not be possible at the prescribed geometric scale ratio. Experimental results have indicated that variation of the turbulence integral scale has little influence on the drag of bluff bodies, provided that the integral scale is larger than the characteristic model length. Consequently exact scaling of the integral scale may be relaxed, provided that the model integral scale is at least as large or preferably larger than the characteristic model size.

Based on Melbourne's "collapsing bubble" hypothesis [23], pressure intensities and peak pressures on bluff bodies near the leading edge are influenced more by the high-frequency content of the velocity spectrum than by the low-frequency content. Consequently, it is of importance that the small-scale motion of the turbulence is scaled properly in relation to the large scale motion. This relation is governed by the turbulence Reynolds number, $U^*\ell/\nu$, where ℓ is the integral length scale of the turbulence. Equality of the turbulence Reynolds number is impossible to achieve, since with any geometric scale ratio an unrealistically high value of the friction velocity, U_0^*, would be required.

In order to satisfy the criterium of properly scaled high-frequency turbulence, it will generally be required to increase the spectral levels in this frequency range. This can be achieved by increasing the size of the upstream floor roughness, which would change the mean velocity profile (z_0) as well as the turbulence integral scale. These are exactly the two flow parameters which we have previously considered to be simulated in an adequate fashion. The problem of not being able to duplicate the exact velocity profile may be avoided by using local pressure coefficients based on the undisturbed mean velocity at the height of the pressure measurement, rather than based on a reference velocity at some reference height [24]. However this approach can only be taken if one is interested in duplication of point pressures rather than wind loads on large areas.

From the theoretical considerations and experimental results discussed in this paper it is obvious that exact simulation of the ABL flow in the laboratory is nearly impossible for many different reasons. Development of a turbulent shear layer in the wind tunnel for which simultaneously the velocity profile, the turbulence integral scale and turbulence Reynolds number are simulated in the atmospheric surface layer, according to established modeling rules, is also impossible. More basic experimental research is required, including full-scale and model scale comparisons, before a better understanding is achieved for the complex flow phenomena which control the mean and fluctuating pressures on bluff bodies.

REFERENCES

1. F. H. Clauser, The Turbulent Boundary Layer, Advances in Applied Mechanics, 4, 1-51, Academic Press, New York, 1956.

2. C. B. Millikan, A Critical Discussion of Turbulent Flow in Channels and Circular Tubes, Proc. 5th Intern. Congr. Appl. Mech., 386-392, Wiley, New York, 1939.

3. H. Tennekes and J. L. Lumley, A First Course in Turbulence, 146-196, MIT Press, 1972.

4. A. K. Blackadar and H. Tennekes, Asymptotic Similarity in Neutral Baratropic Planetary Boundary Layers, J. Atmos. Sci., 25 (1968) 1015-1020.

5. J. C. Kaimal et al., Turbulence Structure in the Convective Boundary Layer, J. Atmos. Sci., 33 (1976) 2152-2169.

6. H. A. Panofsky, Tower Micrometeorology, Workshop on Micrometeorology, Chap. 4, 151-176, American Meteorological Society, 1973.

7. J. C. Kaimal, Horizontal Velocity Spectra in an Unstable Surface Layer, J. Atmos. Sci., 35 (1978) 18-23.

8. H. A. Panofsky et al., The Characteristics of Turbulent Velocity Components in the Surface Layer under Convective Conditions, Boundary Layer Meteorol., 11, (1977) 355-361.

9. N. E. Busch, On the Mechanics of Atmospheric Turbulence, Workshop on Micrometeorology, Chap. 1, 1-65, American Meteorological Society, 1973.

10. H. A. Panofsky, C. A. Egolf and R. Lipschutz, On Characteristics of Wind Direction Fluctuations in the Surface Layer, Boundary Layer Meteorol., 15 (1978) 349-466.

11. H. A. Panofsky, M. Vilardo and R. Lipschutz, Terrain Effects on Wind Fluctuations, Wind Engineering, Vol. 1, 173-177, Pergamon Press, 1980.

12. H. A. Panofsky et al., Spectra over Complex Terrain, Paper presented at the 4th U.S. Nat. Conf. Wind Engen. Res., July 27-29, 1981, University of Washington, Seattle.

13. H. W. Teunissen, Structure of Mean Wind and Turbulence in the Planetary Boundary Layer over Rural Terrain, Boundary Layer Meteorol., 19 (1980) 187-221.

14. E. W. Peterson et al., The Effect of Local Terrain Irregularities on the Mean Wind and Turbulence Characteristics near the Ground, Paper presented at the W.M.O. Symposium on Boundary Layer Physics Applied to Specific Problems of Air Pollution, June 19-23, 1978, Norrkoping, Sweden.

15. E. W. Peterson, N. O. Jensen and J. Hojstrup, Observations of Downwind Development of Wind Speed and Variance Profiles at Bognaes and Comparison with Theory, Q. J. R. Meteorol. Soc., 105 (1979) 521-529.

16. R. A. Antonia and R. E. Luxton, The Response of a Turbulent Boundary Layer to a Step Change in Surface Roughness. Part 1. Smooth-to-Rough, J. Fluid Mech., 48 (1971) 721-761.

17. R. A. Antonia and R. E. Luxton, The Response of a Turbulent Boundary Layer to a Step Change in Surface Roughness. Part 2. Rough-to-Smooth, J. Fluid Mech., 53 (1972) 737-757.

18. P. J. Mulhearn, A Wind-Tunnel Boundary-Layer Study of the Effects of a Surface Roughness Change: Rough to Smooth, Boundary Layer Meteorol., 15 (1978) 3-30.

19. E. F. Bradley, A Micrometeorological Study of Velocity Profiles and Surface Drag in the Region Modified by a Change in Surface Roughness, Q. J. R. Meteorol. Soc., 94 (1968) 361-379.

20. J. Hojstrup, A Simple Model for the Adjustment of Velocity Spectra in Unstable Conditions Downstream of an Abrupt Change in Roughness and Heat Flux, Boundary Layer Meteorol., 21 (1981) 341-356.

21. E. W. Peterson, L. Kristensen and Chang-Chun Su, Some Observations and Analysis of Wind over Non-Uniform Terrain, Q. J. R. Meteorol. Soc., 102 (1976) 857-869.

22. E. W. Peterson et al., Riso 1978: Further Investigations into the Effects of Local Terrain Irregularities on Tower-Measured Wind Profiles, Boundary Layer Meteorol., 19 (1980) 303-313.

88

23. W. H . Melbourne, Turbulence Effects on Maximum Surface Pressures - A Mechanism and Possibility of Reduction, Wind Engineering, Vol. 1, 541-551, Pergamon Press, 1980.

24. R. E. Akins, Wind Pressures on Buildings, Ph.D. Dissertation, Colorado State University, 1976 (available from University Microfilms, Ann Arbor, Mich.).

25. L. Mahrt et al., An Observational Study of the Structure of the Nocturnal Boundary Layer, Boundary Layer Meteorol., 17(1979) 247-264.

26. H. W. Tieleman and S. E. Mullins, The Structure of Moderately Strong Winds at a Mid-Atlantic Coastal Site (Below 75m), Wind Engineering, Vol. 1, 145-159, Pergamon Press, 1980.

27. H. W. Tieleman, On the Wind Structure near the Surface at a Mid-Atlantic Coastal Site, Symposium "Designing with the Wind," Nantes, France, June 15-19, 1981, Proceedings, Part 1, I-5, 1-14. Also the 4th U.S. National Conference on Wind Engineering Research, University Washington, Seattle, July 27-29, 1981, Preprints, 89-100.

28. S. J. Caughey, Turbulence in the Evolving Stable Boundary Layer, J. Atmos. Sci., 36 (1979) 1041-1052.

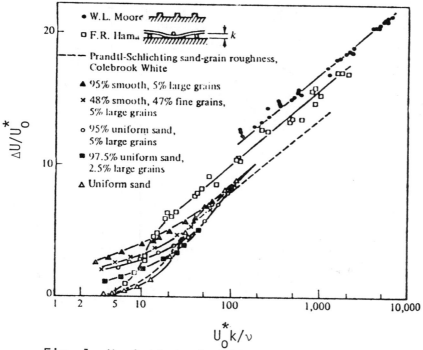

Fig. 1 Variation of shift of velocity profile with roughness Reynolds Number (Ref. 1).

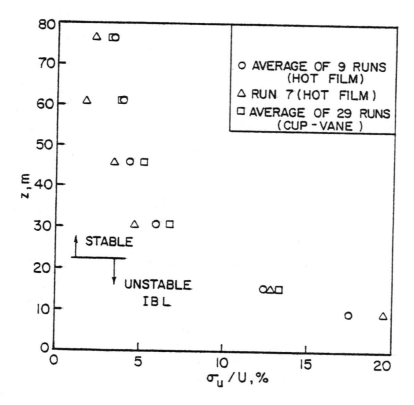

Fig. 2 Turbulence intensity profiles at Wallops Island, Va. (stable on-shore winds, Ref. 26).

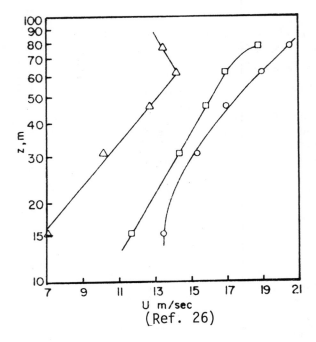

(Ref. 26)

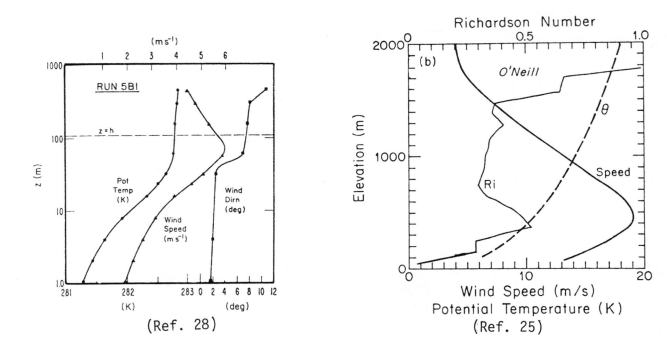

(Ref. 28)

(Ref. 25)

Fig. 3 Profiles of mean quantities under stable conditions.

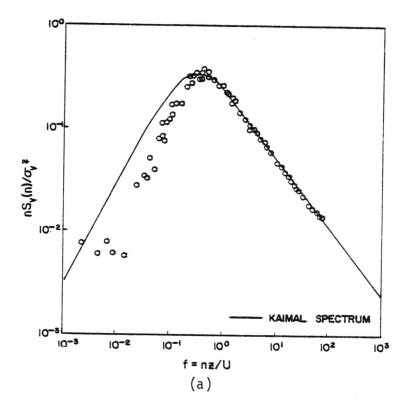

(a)

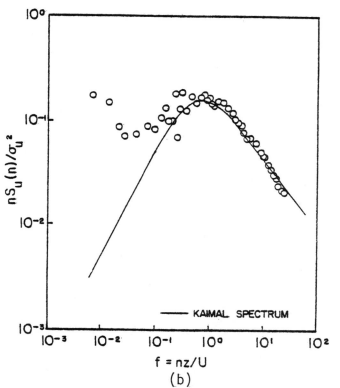

(b)

Fig. 4 Velocity spectra for stable on-shore winds at Wallops Island, Va. (Ref. 27).
    (a) lateral component, without appreciable low-frequency fluctuations
    (b) streamwise component with low-frequency fluctuations (waves)

92

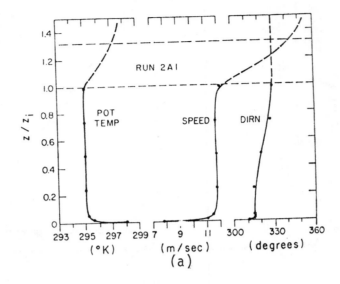

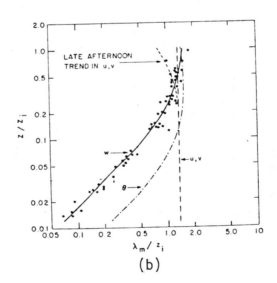

Fig. 5 Profiles of mean quantities (a), and dimensionless peak
wavelengths (b) under convective conditions (Ref. 5).

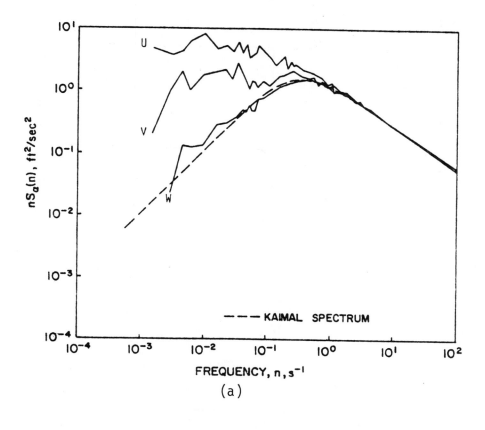

(a)

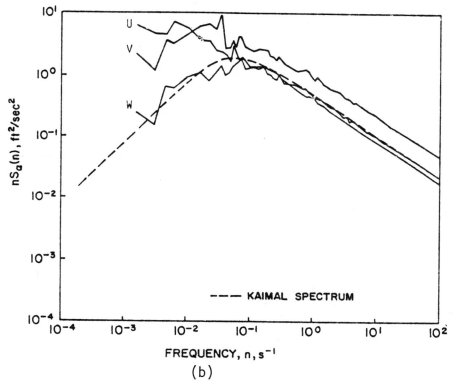

(b)

Fig. 6 Velocity spectra for northwest winds (convective conditions) at Wallops Island. (a) z = 9.1 m (b) z = 76 m (Ref 27).

94

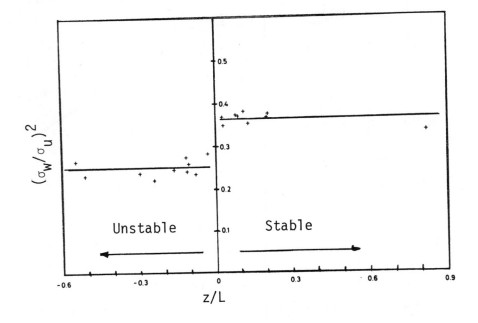

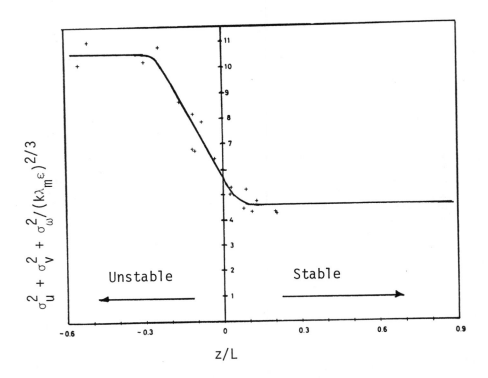

Fig. 7 Variation of σ_w/σ_u and normalized velocity variance with stability (Ref. 9).

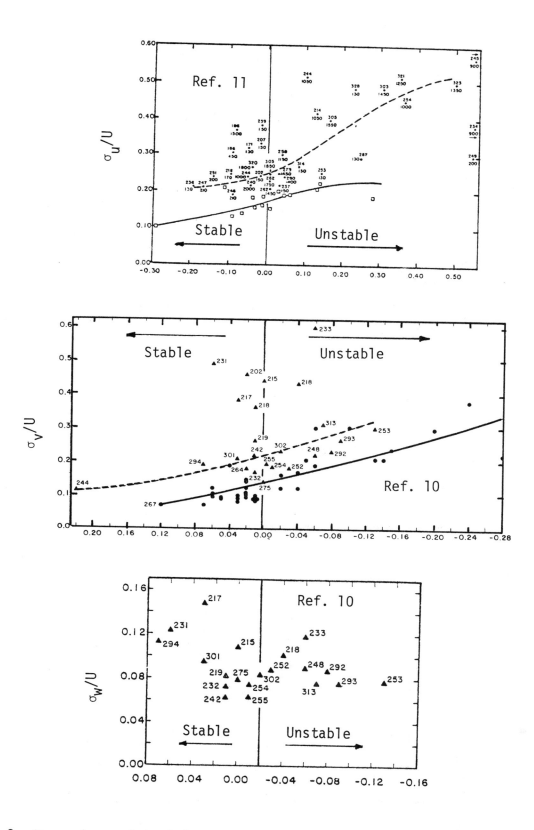

Fig. 8 Comparison of turbulence intensities σ_u/U and σ_v/U for flat and uniform terrain (solid line) and complex terrain (dashed line). The intensity of the vertical component, σ_w/U, is affected very little by terrain variations.

96

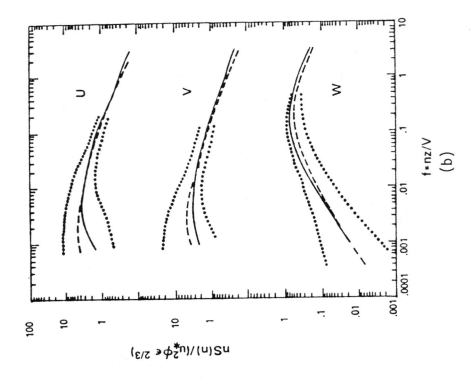

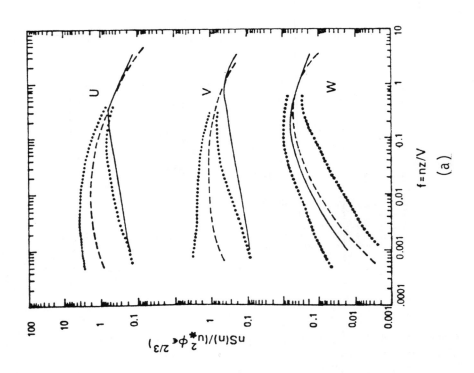

Fig. 9 Normalized u, v and w spectra acquired under (a) stable conditions, and (b) unstable conditions over complex terrain (Ref. 12). Dashed lines represent the average spectral shape, the dots represent the extreme spectra, while the solid line represents flat terrain spectra.

PHYSICAL MODELING OF THE ATMOSPHERIC BOUNDARY LAYER (ABL) IN LONG BOUNDARY-LAYER WIND TUNNELS (BLWT)

J. E. Cermak
Professor-in-Charge,
Fluid Mechanics and Wind Engineering Program
and Director,
Fluid Dynamics and Diffusion Laboratory
College of Civil Engineering
Colorado State University
Fort Collins, Colorado 80523

INTRODUCTION

The earliest scientific efforts specifically directed toward physical modeling of the ABL that utilize BLWT's designed for this purpose are those reported by Rouse (1951), Strom (1952) and Cermak and Koloseus (1953). Test-section dimensions (length, width, height) of these facilities, in the order listed, were as follows: (6.10, 1.83, 1.22 m), (13.72, 2.13, 1.07 m) and (8.84, 1.83, 1.83 m). Studies of momentum and heat transfer were made in the third tunnel, designed and constructed by the writer in 1950, that were reported by Cermak and Spengos (1956). These studies demonstrated that boundary layers growing naturally from the test-section entrance over a smooth boundary did not become fully developed in the test-section length available (8.84 m). Confronted with this finding, a "long" test-section BLWT was designed, Cermak (1958), and constructed, Plate and Cermak (1963). Figure 1 illustrates the completed facility, the meteorological wind tunnel, that serves as the reference BLWT and source of data discussed in this paper.

GENERAL SIMILARITY CRITERIA FOR THE ABL

The atmosphere from ground level up to 300-500 m, where effects of surface drag forces on air flow become negligible, is known as the ABL. In this layer, the wind speed varies from zero at the ground to the geostrophic value U_g when the pressure gradient and the Coriolis forces reach equilibrium at an elevation corresponding to the boundary-layer thickness δ. At this level, the wind direction is parallel to the isobars and varies from the limiting ground-level wind direction by an angle of 10°-30° depending upon the surface Rossby number U_g/fz_o. From ground level to a height of approximately 150 m, the turbulent transport rates (fluxes) of momentum, heat, and mass are essentially constant at values determined by the local surface characteristics. This portion of the ABL, designated as the atmospheric surface layer (ASL), is the region for which physical modeling can be achieved with best accuracy. Most of the micrometeorological data in the ABL has been obtained from measurements made with instruments attached to fixed towers surrounded by open flat terrain with rather uniform surface conditions. Lectures presented in a publication of the American Meteorological Society (1973) provide an excellent source of these data.

Wind at most locations is of the boundary-layer type most of the time. This type of flow serves to define wind characteristics for most wind-engineering applications; however, short-duration phenomena of tornadoes, downbursts, eye walls of hurricanes, downslope flows, and thunderstorms result in local severe winds with distributions that differ from the common boundary-layer structure. These meteorological phenomena offer challenges for physical modeling that are not pursued in this paper.

Fundamental Equations for Description of ABL

An instructive method presented by Ruark (1935) for establishment of appropriate similarity criteria is inspectional analysis of fundamental equations that express basic physical laws that govern the system. Basic physical laws for the ABL and the appropriate equations presented by Cermak et al. (1966) are as follows:

Conservation of Mass: $\quad \dfrac{\partial \rho}{\partial t} + \dfrac{\partial(\rho u_i)}{\partial x_i} = 0$ $\hfill$ (1)

Conservation of Momentum: $\quad \dfrac{\partial U_i}{\partial t} + U_j \dfrac{\partial U_i}{\partial x_j} + 2\varepsilon_{ijk}\,\Omega_j U_k = -\dfrac{1}{\rho_o}\dfrac{\partial P}{\partial x_i}$

$$-\frac{\Delta T}{T_o}\,g\delta_{i3} + \nu_o\,\frac{\partial^2 U_i}{\partial x_k \partial x_k} + \frac{\partial(\overline{-u_j'\,u_i'})}{\partial x_j} \qquad (2)$$

Conservation of Energy: $\quad \dfrac{\partial T}{\partial t} + U_i \dfrac{\partial T}{\partial x_i} = \left(\dfrac{k_o}{\rho_o C_{po}}\right)\dfrac{\partial^2 T}{\partial x_k \partial x_k}$

$$+ \frac{\partial(\overline{-\theta'u_i'})}{\partial x_i} + \frac{\phi}{\rho_o C_{po}} \qquad (3)$$

Equation of State (for a perfect gas): $\quad P = \rho\,RT$ $\hfill$ (4)

Although the foregoing equations are very general, some restrictions are imposed by their form. Because the effect of temperature on density has been linearized in Eq. 2 (Boussinesq approximation) flows are limited to those for which $\Delta T \ll T_o$. This restriction does not impose serious limitations on the use of Eq. 2 for ABL flow. Equation 3 does not allow for an exchange of energy by radiation or phase changes of water. Therefore, a strict interpretation of Eq. 3 would limit consideration to ABL's without cloud content or precipitation. Clearly, the thunderstorm is an extreme meteorological event that cannot be treated by this formulation.

Similarity Criteria

Inspectional analysis of Eqs. 1-3 requires scaling of all variables with appropriate reference quantities. The following dimensionless variables will be used for analysis of Eqs. 1-3:

$$U_i^* = \frac{U_i}{U_o}, \quad u_i^* = \frac{u_i}{U_o}, \quad (u_i')^* = \frac{u_i'}{U_o}, \quad (\theta')^* = \frac{\theta'}{T_o}, \quad x_i^* = \frac{x_i}{L_o}, \quad t^* = \frac{tU_o}{L_o},$$

$$\rho^* = \frac{\rho}{\rho_o}, \quad \Omega_j^* = \frac{\Omega_j}{\Omega_o}, \quad P^* = \frac{P}{(\rho_o U_o^2)}, \quad T^* = \frac{T}{T_o}, \quad \Delta T^* = \frac{\Delta T}{\Delta T_o} \text{ and } g^* = \frac{g}{g_o} \tag{5}$$

Dimensionless forms of the fundamental equations are expressed as follows when transformed by Eq. 5:

$$\frac{\partial \rho^*}{\partial t^*} + \frac{\partial(\rho^* u_i)}{\partial x_i^*} = 0 \tag{6}$$

$$\frac{\partial U_i^*}{\partial t^*} + U_j^* \frac{\partial U_i^*}{\partial x_j^*} + \left(\frac{L_o \Omega_o}{U_o}\right) 2\varepsilon_{ijk} \Omega_j^* U_k^* = -\frac{\partial P^*}{\partial x_i^*}$$

$$-\left(\frac{\Delta T_o}{T_o} \frac{L_o g_o}{U_o^2}\right) \Delta T^* g^* \delta_{i3} + \left(\frac{\nu_o}{U_o L_o}\right) \frac{\partial^2 U_i^*}{\partial x_k^* \partial x_k^*} + \frac{\partial(\overline{-u_i' u_j'})^*}{\partial x_j^*} \tag{7}$$

and

$$\frac{\partial T^*}{\partial t^*} + U_i^* \frac{\partial T^*}{\partial x_i} = \left(\frac{k_o}{\rho_o C_{po} \nu_o}\right)\left(\frac{\nu_o}{L_o U_o}\right) \frac{\partial^2 T^*}{\partial x_k^* \partial x_k^*}$$

$$+ \frac{\partial(\overline{-\theta' u_i'})^*}{\partial x_i^*} + \left(\frac{\nu_o}{U_o L_o}\right)\left(\frac{U_o^2}{C_{po} T_o}\right)\phi^* \tag{8}$$

Inspection of Eqs. 6-8 gives the requirements for kinematic, dynamic and thermic similarity provided that proper boundary conditions are satisfied. Equation 6 is invariant in this transformation that scales all lengths equally; i.e., $x_i^* = x_i/L_o$. Therefore, geometric similarity will result in kinematic similarity. Equation 7 states that dynamic similarity can be achieved if each of the following parameters is equal for two systems: (1) $U_o/(L_o \Omega_o)$, Rossby number (Ro); (2) $(\Delta T_o/T_o)(L_o g_o/U_o^2)$, Richardson number (Ri); and (3) $U_o L_o/\nu_o$, Reynolds number (Re). Thermic similarity will be obtained if, according to Eq. 8, the following dimensionless groups are equal for two systems: (1) $\nu_o \rho_o C_{po}/k_o$, Prandtl number (Pr); (2) $U_o^2/(C_{po} T_o)$, Eckert number (Ec); and (3) Re.

The requirements for geometric, dynamic, and thermic similarities obtained by inspectional analysis must be augmented by additional specifications if requirements for "exact" similarity are to be complete. These external conditions are the following: (1) similarity of surface roughness and temperature distributions at ground level; (2) similarity of flow structure above the ABL; (3) similarity of the horizontal pressure gradient; and (4) sufficient upwind fetch to establish equilibrium of the ABL consistent with external conditions.

100

Even a casual review of the requirements for "exact" similarity of the ABL when the boundary layer in a wind tunnel is one of the flow systems leads to the conclusion that physical modeling of the ABL must be considered on the basis of "approximate" similarity. The primary compromise that must be made is relaxation of the requirements for equal Reynolds numbers and equal Rossby numbers.

The degree to which the similarity requirements can be achieved has been discussed by Cermak (1971, 1975) and will be summarized here. With no rotation of the wind tunnel about a vertical axis, Rossby numbers for the wind tunnel flow will be smaller than for the atmosphere by a factor approximately equal to the length scale. Accordingly, for flow over a plane boundary, the mean velocity will not change direction with height as in the atmosphere. If this deficiency were a serious limitation in applications of the modeled ABL it could be removed by introduction of a cross flow through porous walls of a stationary wind tunnel or by use of a large rotating flow chamber of the type reported by Caldwell and Van Atta (1970). However, because of the increased cost and operational difficulties of such facilities, coupled with the small compromise with exact similitude attainable in a more conventional wind tunnel, effects of the Coriolis acceleration are not simulated in the common BLWT. The ratio of Reynolds numbers for the wind tunnel and the atmosphere will also be approximately equal to the length scale. Fortunately, as shown in Fig. 2, boundary-layer flow characteristics over a surface of length, L_x, become independent of Reynolds number at values readily attainable in a wind tunnel of reasonable length. Through selection of proper combinations of wind-tunnel length, surface roughness and ambient wind speed; boundary-layer flows can be realized that model the micrometeorological features of the high Reynolds number ABL. Typical gross Richardson numbers for the atmospheric surface layer are in the range $-1 < Ri < 1$. In order to achieve Richardson numbers of this magnitude in the BLWT, large values of ΔT_o and small values of U_o must be obtainable to overcome the length-scale ratio. Richardson number equality is essential if the BLWT is to be used for a wide range of diffusion studies; however, for studies of wind forces on structures Ri is set equal to zero to model neutral flow in strong winds. Prandtl number equality is automatically satisfied if air is used as the wind-tunnel fluid. The Eckert number, when Richardson numbers are equal for stratified flow, will be approximately one order-of-magnitude smaller for the wind tunnel than for the ABL. Since the Eckert number can be expressed as a Mach number squared, it will be small compared to unity for both the atmosphere and the BLWT. Therefore, little distortion of the flow field can be attributed to inequality of the Eckert numbers in this context.

The associated external, or boundary, conditions impose additional requirements for physical modeling of the ABL. Similarity of surface roughness distributions can be achieved by addition of roughness elements to the wind-tunnel floor, while surface temperature distributions can be realized by addition of heating and cooling elements. Longitudinal pressure variation can be controlled through slight divergence of the wind-tunnel test-section walls, or installation of a flexible ceiling, or both, that can be adjusted for specific surface roughnesses or models. To accommodate a variety of flow structures above the boundary layer, the facility must develop flow entering the test section that is uniform in mean velocity with low turbulence intensity. Addition of grids at the entrance can add turbulence of a desired scale and intensity over the test area. The requirement that the boundary layer be in equilibrium with external conditions (primarily surface roughness and temperature) does not impose any limitations on simulation capabilities, but suggests the need for a reasonably long test section. This is particularly true if a long length of equilibrium boundary layer is required for either fundamental flow studies or diffusion tests.

The "Long" BLWT

A "long" BLWT is distinguished from other types of wind tunnels by the capability to develop a thick turbulent boundary layer through momentum transfer to the test-section floor without the use of momentum sinks at the test-section entrance. With knowledge that the ABL depth is usually in the range of 300-500 m, choice of minimum boundary-layer thickness for the clean tunnel configuration (smooth boundary and no initial momentum sinks) of 0.5 m results in a practical range for the model length scale of 1:600 to 1:1000. A BLWT designed to achieve this flow condition is extremely versatile in practice since much thicker boundary layers will develop over a rough boundary with or without initial momentum sinks. The required test-section length for neutral flow can be estimated (for the clean configuration and zero pressure gradient) by the following equation from Schlichting (1960):

$$\frac{L_x}{\delta} = 3.47 \left(\frac{\delta U_o}{\nu} \right)^{1/4} \tag{9}$$

Equation 9 was used by Cermak (1958) to establish a test-section length of approximately 28 m when $\delta = 0.5$ m and $U_o = 2$ m/s for the BLWT shown in Fig. 1. The test-section height for this BLWT has a nominal dimension of 2 m to accommodate boundary-layer thicknesses up to about 1.5 m before interacting with the roof boundary layer. Thus, the reference "long" BLWT has a test-section length-to-height ratio of 14. More features of the BLWT shown in Fig. 1, including the heating and cooling systems, are described by Cermak (1981).

The variation of mean velocity profile parameters with distance downstream from the test-section entrance x is shown in Fig. 3. For these measurements flow was "tripped" to initiate the turbulent boundary layer by a gravel-covered surface and saw-tooth fence as shown in Fig. 6. Two significant features of the boundary-layer development are illustrated by data presented in Fig. 3: (1) a distance of approximately 10 m is required for the boundary layer to attain equilibrium (H = 1.3) and (2) the entrance tripping system results in a value of δ at x = 25 m twice the magnitude predicted by Eq. 9.

Comparisons of micrometeorological variables measured in the meteorological BLWT shown in Fig. 1 and in the field have revealed strong similarities. Mean wind speed and temperature distributions in the ASL for both flow systems follow closely the log-linear forms

$$U - U_{ref} = \left(\frac{u_*}{k'}\right)\left[\ln\left(\frac{z}{z_{ref}}\right) + \frac{B_u(z - z_{ref})}{L}\right] \tag{10}$$

and

$$T - T_{ref} = T_*\left[\ln\left(\frac{z}{z_{ref}}\right) + \frac{B_T(z - z_{ref})}{L}\right] \tag{11}$$

in which $L = -u_*^3/[(k'g/T)H_o/(\rho C_p)]$ is the Monin-Obukhov length and $T_* = -H_o/(k'\rho C_p u_*)$ is the friction temperature. Data presented in Fig. 4 are for L ranging from -0.64 to 0.55 m in the wind tunnel (smooth boundary at $x = 25$ m) and -16.90 to 76.20 m in the atmosphere. The reference height, z_{ref}, is 0.07 m and 2.0 m for the laboratory and field data, respectively. Turbulence spectra for the longitudinal component are shown in Fig. 5 for neutral flow over smooth and rough boundaries at $x = 25$ m in the meteorological BLWT and in the atmosphere. Kolmogorov (1941) scaling is used for the comparison. The laboratory and atmospheric spectra are in excellent agreement for a wave-number range that increases with surface roughness. Significantly, this range of wave numbers covers wave numbers at and above the spectral gap associated with turbulence in the ASL. Furthermore, the BLWT spectra are characteristic of high Reynolds number turbulence for which a large set of wave numbers is in the inertial subrange (turbulence energy varies with wave number to the -5/3 power).

BOUNDARY LAYER AUGMENTATION IN THE BLWT

Much effort has been devoted in recent years to development of techniques that will enable thick turbulent boundary layers to be simulated in short wind-tunnel test sections or to thicken boundary layers formed in the "clean configuration" long BLWT. The vortex generators of Counihan (1969, 1973) and the spires of Standen (1972) have been adapted for use in numerous laboratories. Other techniques such as multiple jets of Teunissen (1972), volumetric flow control of Nee et al. (1973) and counter jets of Morkovin et al. (1972) have not found wide acceptance. In an exploratory study Peterka and Cermak (1974) reviewed the various techniques and utilized combinations of spires and roughness arrays to obtain additional measurements for determination of boundary-layer development. Gartshore (1973) investigated the effect of surface roughness on shape of the mean velocity profile and Cook (1978) employed a combination of surface roughness, barrier and vortex generators to simulate the ABL in whole and in part.

Many questions arise regarding the characteristics of augmented boundary layers and how well they simulate characteristics of the ABL. Questions of greatest concern include the following:

1. What distance is required downstream of the augmentation device for disappearance of flow disturbances generated by the device?

2. At what distance downstream of an augmentation device do profiles of turbulence intensity and Reynolds shear stress become similar to those that form over the same rough boundary without augmentation?

3. How do turbulence scales and spectra in an augmented boundary layer differ from those in an unaugmented boundary layer over the same surface roughness?

In an effort to develop answers to such questions, a program of systematic measurements was developed for the meteorological BLWT (Fig. 1). Some findings that are available at this time are given in the following paragraphs.

Augmentation Configurations in the Meteorological BLWT

Nine configurations of three augmentation devices and three surface roughnesses were selected for study. The augmentation devices were a roughened entrance strip and saw-tooth fence shown in Fig. 6, the Counihan (1969) vortex generators shown in Fig. 7 and the Standen (1972) type spires shown in Fig. 8. Surface roughnesses were a smooth boundary, a boundary covered by staggered rows of 1.25 cm cubes spaced 5.08 cm apart and a boundary covered by staggered rows of 2.50 cm cubes spaced 7.62 cm apart. Profiles of mean velocity U, intensity of two components of turbulence (longitudinal and vertical) and the Reynolds shear stress $-\overline{\rho u'w'}$ were measured at locations along the test-section centerline shown in Fig. 9. Spectra of the longitudinal component of turbulence were measured at one height at each measurement location excepting for $x = 18.3$ m where measurements were made at four heights. Data were obtained for two ambient mean wind speeds U_∞ of approximately 9 and 18 m/s. All of the data have not been analyzed in time for presentation at this workshop. However, several sets of data have been prepared to illustrate the effect of surface roughness and test-section length on development of a turbulent boundary-layer downwind of the three augmentation configurations. Table I lists the 18 cases for which vertical profiles of mean velocity, longitudinal and vertical turbulence intensities and turbulent shear stress $(-\overline{\rho u'w'})$ are given in Figs. 10 through 33. Table II identifies the 24 spectra presented in Figs. 34 through 41.

Vertical Profiles of Mean Velocity, Turbulence Intensities and Shear Stress

Each of the Figs. 10 through 33 present vertical profiles for the three augmentation configuration--sawtooth-roughness trip (rectangles), vortex generators (ovals) and spires (triangles)--to facilitate comparison of flow produced by the three devices. The sawtooth-roughness trip remained in place when the vortex generators and spires were added. Data obtained with the sawtooth-roughness trip alone serve as the reference for determination of boundary-layer distortions introduced by the vortex generators and the spires. The mean velocity data (Figs. 10 through 15), longitudinal turbulence intensities (Figs. 16 through 21), vertical turbulence intensities (Figs. 22 through 27), and the turbulent shear stress (Figs. 18 through 32) are presented in pairs for stations $x = 6.10$ m and $x = 18.29$ m to aid in evaluation of test-section length effects. The figure pairs for each of the four types of data are arranged in order of increasing surface roughness--smooth surface ($z_o \cong 0.02$ cm), 1.27 cm roughness ($z_o \cong 0.15$ cm) and 2.54 cm roughness ($z_o \cong 0.35$ cm).

Mean Velocity Profiles

The following trends can be observed by examination of the mean velocity data in Figs. 10 through 15:

1. Profiles for the three augmentation devices are in poor agreement at x = 6.10 m (Figs. 10, 12 and 14) but as x increases to 18.29 m (Figs. 11, 13 and 15) profiles for the vortex generators and spires are in fair agreement with the reference profiles (sawtooth-roughness trip).

2. Agreement with the reference profile generally improves at x = 18.29 m as the surface roughness increases (Figs. 11, 13 and 15).

Longitudinal Turbulence Intensities

Trends similar to those observed for mean velocities can be observed for the longitudinal turbulence intensities.

1. Agreement with the reference profiles is poor at x = 6.10 m (Figs. 16, 18 and 20) but is substantially better at x = 18.29 m (Figs. 17, 18 and 19).

2. Closer agreement with the reference profiles is obtained as surface roughness increases (Figs. 17, 18 and 19)--note differences in horizontal scaling for the figures.

Vertical Turbulence Intensities

General trends of agreement with the reference profiles are parallel to those found for mean velocities and longitudinal turbulence intensities.

1. Profiles at x = 6.10 m (Figs. 22, 24 and 26) are radically different than the reference profiles but tend to better agreement as x increases to 18.29 m (Figs. 23, 25 and 27).

2. Agreement with the reference profiles increase with increasing surface roughness (Figs. 23, 25 and 27).

Turbulent Shear Stress

Comparison of Figs. 28 and 29, Figs. 30 and 31 and Figs. 32 and 33 reveals, as for the previous data, a trend from major disagreement with the reference profiles to fair agreement as both x and surface roughness increase (Figs. 31 and 32).

Spectra of Longitudinal Turbulence

Figures 34 through 41 give spectra for each of the augmentation configurations--sawtooth-roughness trip (circles), vortex generators (triangles) and spires (squares). Figures 34 through 39 appear as pairs for mean freestream wind speeds U_∞ of approximately 9 m/s and 18 m/s (Figs. 34 and 35, Figs. 36 and 37 and Figs. 38 and 39) for x = 18.29 m and a relative height above the surface of $z/\delta = 0.05$. Data for two values of x, 6.10 m and 18.29 m, are presented as a pair of figures (Figs. 40 and 41) for the same freestream wind speed (9 m/s), relative height above the surface (0.25) and surface roughness (smooth).

Comparisons of spectra shown in Figs. 34 through 39 reveal the following general trends:

1. Vortex generators and spires increase energy content at the high frequencies, compared to the reference spectra, for the smooth surface (Figs. 34 and 35) and 1.27 cm roughness (Figs. 36 and 37).

    However, as the roughness increases to 2.54 cm (Figs. 38 and 39), the spectra is dominated by the local roughness at $z/\delta = 0.05$ and the spectra for all augmentation devices are in fair agreement.

2. An increase of freestream wind speed from 9 m/s to 18 m/s causes only minor changes in relationships between spectra for the three augmenation devices.

Examination of Figs. 40 and 41 reveals that the spectra are in better agreement at x = 18.29 than at x = 6.10 m. However, a distance greater than x = 18.29 m is required before spectra for flow over the smooth boundary will be in good agreement with the reference spectra.

Summary Observations

Characteristics of boundary layers generated with the addition of augmentation devices exhibit two distinctive features of significance. These features are identified as follows:

1. Evolution of an augmented boundary layer to an equilibrium state consistent with the surface roughness is a slow process that requires a distance of 18 m or more downwind of the augmentation device.

2. The diffusion rate of flow perturbations generated by augmentation devices toward an equilibrium state is increased as the surface roughness is increased.

CONCLUSIONS

1. Boundary layers developed in the "long" meteorological BLWT without boundary-layer augmentation devices provide a physical model of the ABL that can be utilized for both basic research and for determination of wind effects associated with wind-engineering problems.

2. Boundary-layer thickness can be increased without significant loss of similarity with the ABL by introduction of augmentation

devices at the test-section entrance provided that the test-section is sufficiently long for the flow structure to reach statistical equilibrium.

3. Augmented boundary layers formed over rough surfaces approach statistical equilibrium more rapidly (at shorter distances downwind from the augmentation device) through enhanced diffusion than those formed over relatively smooth surfaces.

4. Extensive basic research is needed to establish diffusion rates for flow perturbations imposed upon turbulent boundary layers and to determine limits of departure from equilibrium boundary layers that are acceptable for physical modeling of various wind effects.

ACKNOWLEDGEMENTS

Boundary-layer data presented in this paper for the various augmentation configurations in the meteorological BLWT were acquired and processed by Dr. K. M. Kothari with the assistance of J. A. Beatty and J. C. Maxton.

NOTATION

B_T, B_U = dimensionless coefficient;

C_p = specific heat at constant pressure;

f = Coriolis parameter, $2\Omega \sin \Psi$;

$f(\kappa)$ = energy spectrum of longitudinal velocity fluctuations;

g = gravitational acceleration;

H_o = surface heat flux;

k = thermal conductivity;

k' = Kármán constant;

K_s = equivalent sand roughness length;

L = Monin-Obukhov length;

L_x = distance downwind from virtual origin of boundary layer;

P = mean pressure;

R = universal gas constant;

T = local mean temperature;

T^* = friction temperature;

ΔT = mean temperature difference (departure from adiabatic lapse rate);

t = time;

$u_i'(u', v', w')$ = ith component of velocity fluctuation;

u_i = local ith component of instantaneous velocity;

u^* = friction velocity, $(\tau_o/\rho)^{1/2}$;

U_i = ith component of local mean velocity;

x = distance downwind from test section entrance;

x_i = ith space coordinate;

z_o = roughness length;

δ = boundary-layer thickness;

δ_{ij} = Kronecker delta;

ε = energy dissipation rate per unit of mass;

θ' = local temperature fluctuation;

κ = wave number;

ν = kinematic viscosity;

ρ = mass density;

τ = surface shear stress;

ϕ = dissipation function;

Ψ = latitude;

Ω_i = ith component of angular velocity;

$\overline{(\ )}$ = time average;

$(\ )_o$ = reference quantity;

$(\ )_g$ = value of quantity at the geostrophic wind height; and

$(\ )^*$ = nondimensional quantity.

108

REFERENCES

1. Barad, M. L., ed., "Project Prairie Grass--A Field Program in Diffusion," Geophysical Research Paper 59, Geophysical Research Directorate, Bedford, Mass., 1952.

2. Caldwell, D. R., and Van Atta, C. W., "Ekman Boundary Layer Instabilities," Journal of Fluid Mechanics, London, England, Vol. 44, Part I, Oct., 1970, pp. 79-95.

3. Cermak, J. E., and Koloseus, H. J., "Lake Hefner Model Studies of Wind Structure and Evaporation," Technical Reports CER54-JEC20 and CER54-JEC22, (A D 1054930), Fluid Dynamics and Diffusion Laboratory, Colorado State University, Fort Collins, Colo., 1953-1954, 275 p.

4. Cermak, J. E., and Spengos, A. C., "Turbulent Diffusion of Momentum and Heat from a Smooth, Plane Boundary with Zero Pressure Gradient," Technical Report CER56-JEC22, Fluid Dynamics and Diffusion Laboratory, Colorodo State University, Fort Collins, Colo., 1956, 77 p.

5. Cermak, J. E., "Wind Tunnel for the Study of Turbulence in the Atmospheric Surface Layer," Technical Report CER58-JEC42, Fluid Dynamics and Diffusion Laboratory, Colorado State University, Fort Collins, Colo., 1958, 31 p.

6. Cermak, J. E., Sandborn, V. A., Plate, E. J., Binder, G. H., Chuang, H., Meroney, R. N., and Ito, S., "Simulation of Atmospheric Motion by Wind-Tunnel Flows," Technical Report CER66-JEC-VAS-EJP-GHB-HC-RNM-SI17, Fluid Dynamics and Diffusion Laboratory, Colorado State University, Fort Collins, Colo., May, 1966, 101 p.

7. Cermak, J. E., "Laboratory Simulation of the Atmospheric Boundary Layer," AIAA Journal, Vol. 9, No. 9, Sept., 1971, pp. 1746-1754.

8. Cermak, J. E., "Applications of Fluid Mechanics to Wind Engineering--A Freeman Scholar Lecture," Transactions, ASME, Journal of Fluid Engineering, Vol. 97, Mar., 1975, pp. 9-38.

9. Cermak, J. E., "Wind Tunnel Design for Modeling of Atmospheric Boundary Layers," Journal of the Engineering Mechanics Division, ASCE, Vol. 107, No. EM3, June, 1981, pp. 623-642.

10. Chuang, H., and Cermak, J. E., "Similarity of Thermally Stratified Shear Flows in the Laboratory and Atmosphere," The Physics of Fluids Supplement, Vol. 10, Part II, No. 9, Proceedings of an International Symposium on Boundary Layers and Turbulence with Geophysical Applications, Kyoto, Japan, Sept., 1966, 1967, pp. S255-S258.

11. Counihan, J., "An Improved Method of Simulating an Atmospheric Boundary Layer in a Wind Tunnel," Atmospheric Environment, Vol. 3, 1969, pp. 197-214.

12. Counihan, J., "Simulation of an Adiabatic Urban Boundary, Layer in a Wind Tunnel," Atmospheric Environment, Vol. 7, 1973, pp. 673-689.

13. Gartshore, I. S., "A Relationship between Roughness Geometry and Velocity Profile Shape for Turbulent Boundary Layers," Report LTR-LA-140, National Aeronautical Establishement, National Research Council, Ottawa, Canada, Oct., 1973, 36 p.

14. Haugen, D. A. ed., Workshop on Micrometeorology, American Meteorological Society, Boston, Mass., 1973, 392 p.

15. Kolmogorov, A. N., "The Local Structure of Turbulence in Incompressible Viscous Fluid for Very Large Reynolds Numbers," Doklady of the Academy of Sciences of the USSR, Vol. 30, No. 4, 1941, pp. 299-303.

16. Morkovin, M. V., Nagib, H. M., and Yung, J. T., "On Modeling of Atmospheric Surface Layers by the Counter-Jet-Technique--Preliminary Results," Technical Report AFSOR-TR-73-0592, Illinois Institute of Technology, Chicago, Ill, 1972.

17. Nee, V. W., Dietrick, C., Betchov, R., and Szewczyk, A. A., "The Simulation of the Atmospheric Surface Layer with Volumetric Flow Control," Proceedings of the Nineteenth Annual Technical Meeting, 1973, pp. 483-487.

18. Peterka, J. A., and Cermak, J. E., "Simulation of Atmospheric Flows in Short Wind Tunnel Test Sections," Technical Report CER73-74JAP-JEC32, Fluid Dynamics and Diffusion Laboratory, Colorado State University, Fort Collins, Colo., 1974, 52 p.

19. Plate, E. J., and Cermak, J. E., "Micro-Meteorological Wind Tunnel Facility," Technical Report CER63EJP-JEC9, Fluid Dynamics and Diffusion Laboratory, Colorado State University, Fort Collins, Colo., 1963, 65 p.

20. Pond, S., Stewart, R. W., and Burling, R. W., "Turbulence Spectra in Wind over Waves," Journal of the Atmospheric Sciences, Vol. 20, 1963, pp. 319-324.

21. Rouse, H., "Model Techniques in Meteorological Research," Compendium of Meteorology, American Meteorological Society, Boston, Mass., 1951, pp. 1249-1254.

22. Ruark, A. E., "Inspectional Analysis: A Method Which Supplements Dimensional Analysis," Journal of the Elisha Mitchell Science Society, Vol. 51, 1935, pp. 127-133.

23. Schlichting, H., Boundary Layer Theory, McGraw-Hill Publishing Co., Inc., New York, N.Y., 1960, 647 p.

24. Standen, N. M., "A Spire Array for Generating Thick Turbulent Shear Layers for Natural Wind Simulation in Wind Tunnels," Technical Report LTR-LA-94, National Aeronautical Establishement, Ottawa, Canada, 1972.

25. Strom, G. H., "Wind Tunnel Techniques Used to Study Influence of Building Configuration on Stack Gas Dispersal," American Industrial Hygiene Association Quarterly, Vol. 13, 1952, p. 76.

110

26. Teunissen, H. W., "Simulation of the Planetary Boundary Layer in a Multiple-Jet Wind Tunnel," Report 182, Institute for Aerospace Studies, University of Toronto, Toronto, Canada, 1972, 162 p.

27. Zoric, D., and Sandborn, V. A., "Similarity of Large Reynolds Number Boundary Layers," Boundary-Layer Meteorology, Vol. 2, No. 3, Mar., 1972, pp. 326-333.

Table I. Velocity Profile Location and Identification

File Name	Surface[1] Roughness	Entrance[2] Condition	x Downwind Distance (m)	δ Boundary Layer Height (cm)	U_∞ Free Stream Velocity (m/sec)
A002	A	1	6.1	40.0	9.48
A004	A	1	18.29	60.0	9.52
A012	A	2	6.1	60.0	9.40
A014	A	2	18.29	100.0	9.28
A022	A	3	6.1	100.0	8.68
A024	A	3	18.29	100.0	8.47
A032	B	1	6.1	40.0	9.34
A034	B	1	18.29	60.0	9.21
A042	B	2	6.1	75.0	8.86
A044	B	2	18.29	100.0	8.83
A052	B	3	6.1	100.0	7.99
A054	B	3	18.29	100.0	7.44
A062	C	1	6.1	40.0	8.74
A064	C	1	18.29	75.0	8.76
A072	C	2	6.1	75.0	8.55
A074	C	2	18.29	100.0	8.97
A082	C	3	6.1	100.0	8.17
A084	C	3	18.29	100.0	8.00

[1] A = Smooth Floor
B = 1.27 cm Roughness
C = 2.54 cm Roughness

[2] 1 = Sawtooth
2 = Counihan Spires
3 = 6 ft Spires

Table II. Spectra Location and Identification

File Name	Surface[1] Roughness	Entrance[2] Condition	x Downwind Distance (m)	z Vertical Distance (cm)	δ Boundary Layer Height (cm)	U Local Velocity (m/sec)
C002	A	1	6.1	10.0	40.0	7.95
C0041	A	1	18.29	3.0	60.0	5.90
C0043	A	1	18.29	15.0	60.0	7.45
C0093	A	1	18.29	3.0	60.0	11.75
C012	A	2	6.1	15.0	60.0	8.13
C0141	A	2	18.29	5.0	100.0	5.69
C0143	A	2	18.29	25.0	100.0	7.19
C0191	A	2	18.29	5.0	100.0	12.72
C022	A	3	6.1	25.0	100.0	7.02
C0241	A	3	18.29	5.0	100.0	5.33
C0243	A	3	18.29	25.0	100.0	6.48
C0291	A	3	18.29	5.0	100.0	10.96
C0341	B	1	18.29	3.0	60.0	4.64
C0391	B	1	18.29	3.0	60.0	9.32
C0441	B	2	18.29	5.0	100.0	4.52
C0491	B	2	18.29	5.0	100.0	9.69
C0541	B	3	18.29	5.0	100.0	4.77
C0591	B	3	18.29	5.0	100.0	9.71
C0641	C	1	18.29	3.75	75.0	3.97
C0691	C	1	18.29	3.75	75.0	8.72
C0741	C	2	18.29	5.0	100.0	4.09
C0791	C	2	18.29	5.0	100.0	8.48
C0841	C	3	18.29	5.0	100.0	3.37
C0891	C	3	18.29	5.0	100.0	8.09

[1] A = Smooth Floor
B = 1.27 cm Roughness
C = 2.54 cm Roughness

[2] 1 = Sawtooth
2 = Counihan Spires
3 = 6 ft Spires

112

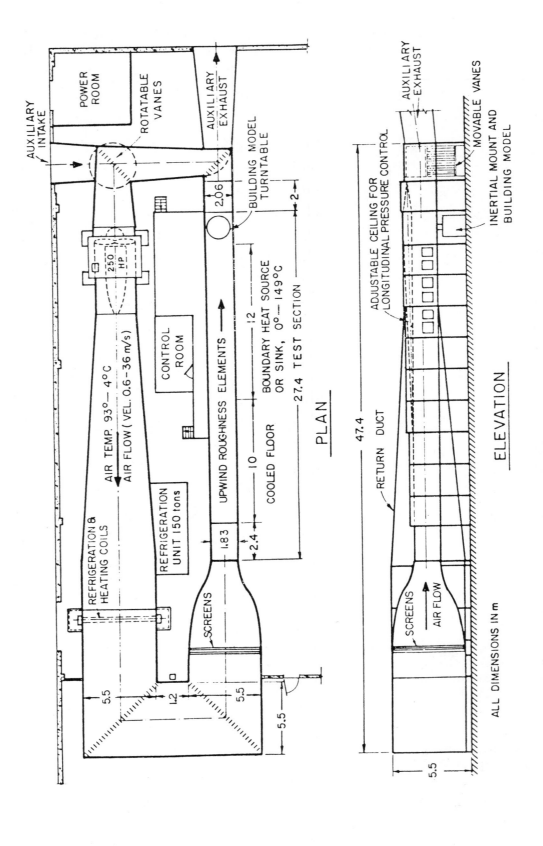

Figure 1. Meteorological Wind Tunnel, Fluid Dynamics and Diffusion Laboratory, Colorado State University

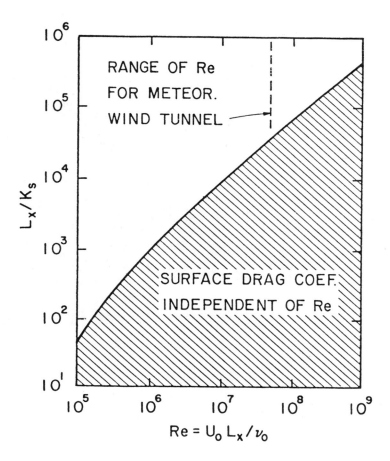

Figure 2. Reynolds Number Independence of Surface Drag Coefficient
as Function of Relative Roughness in Neutral Flow--
Schlichting (1960)

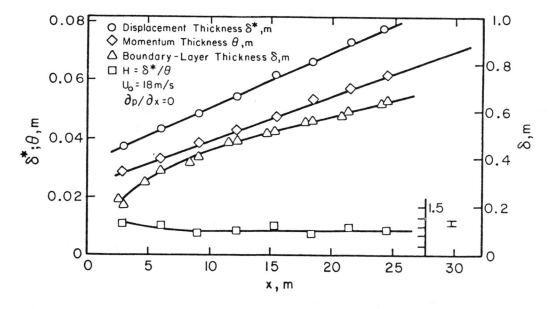

Figure 3. Mean Velocity Profile Parameters for Neutral Flow over
Smooth Boundary in Meteorological Wind Tunnel--
Zoric and Sandborn (1972)

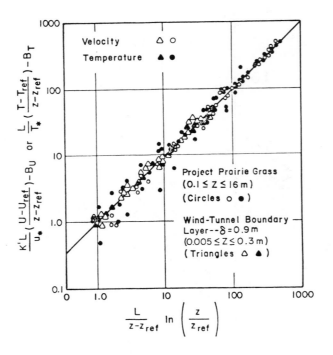

Figure 4. Mean Velocity and Temperature from Meteorological
BLWT (Chuang and Cermak, 1966) and Atmosphere
(Barad, 1952) Compared to Log-Linear Profile

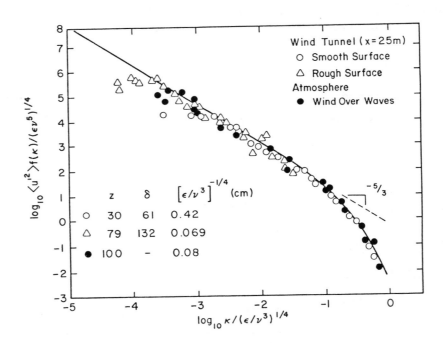

Figure 5. Energy Spectra for Longitudinal Component of Turbulence in
Meteorological BLWT (Cermak et al. 1966) and in Atmosphere
(Pond, et al. 1963)

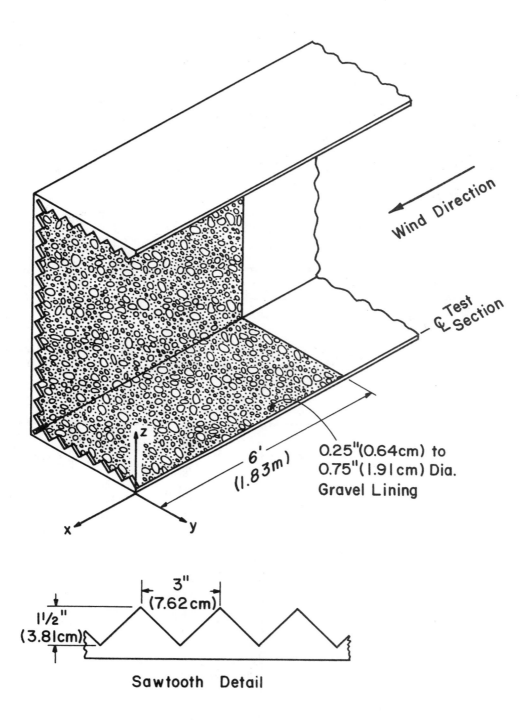

Wind Direction

℄ Test Section

6' (1.83m)

0.25"(0.64cm) to 0.75"(1.91cm) Dia. Gravel Lining

z

x   y

3" (7.62cm)

1½" (3.81cm)

Sawtooth Detail

Figure 6.   Sawtooth-roughness Boundary-layer Trip at Entrance to Meteorological BLWT Test-section

116

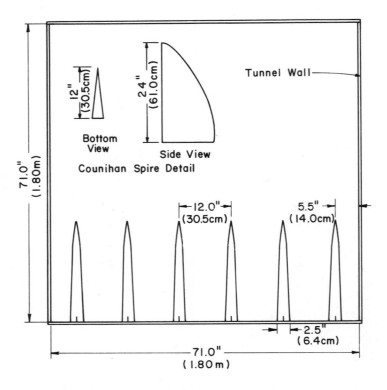

Figure 7. Vortex Generator Array for Boundary-layer
Thickness Augmentation--Counihan (1969)

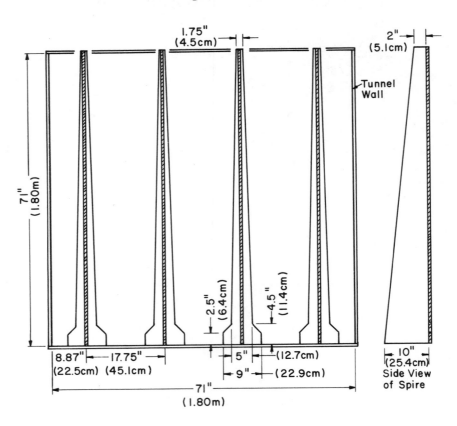

Figure 8   Spire Array for Boundary-layer Thickness
Augmentation--Standen (1972)

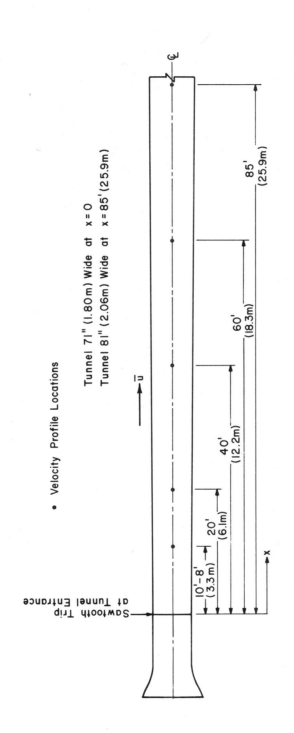

Figure 9. Location of Measurement Stations in the Micrometeorological BLWT

118

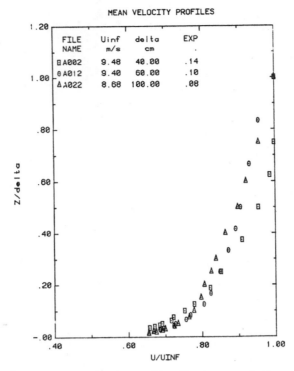

Figure 10. Mean Velocity Distribution--x = 6.10 m;
Smooth Surface

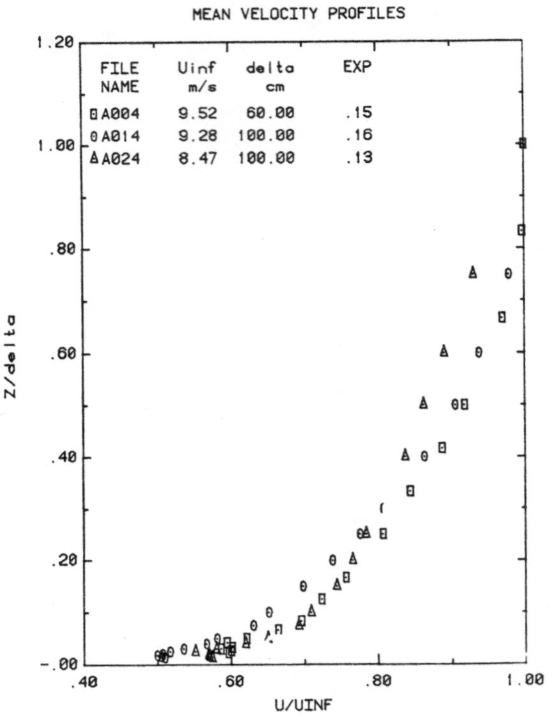

Figure 11. Mean Velocity Distribution--x = 18.29 m;
Smooth Surface

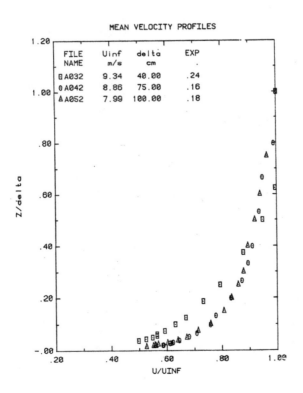

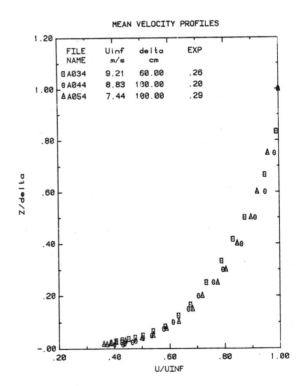

Figure 12. Mean Velocity Distribution--x = 6.10 m;
1.27 cm Roughness

Figure 13. Mean Velocity Distribution--x = 18.29 m;
1.27 cm Roughness

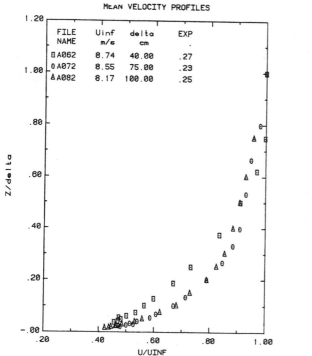

Figure 14. Mean Velocity Distribution--x = 6.10 m; 2.54 cm Roughness

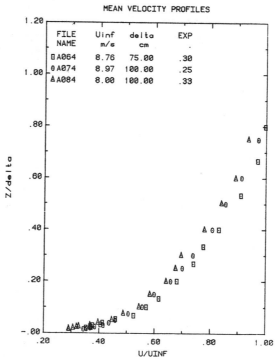

Figure 15. Mean Velocity Distribution--x = 18.29 m; 2.54 cm Roughness

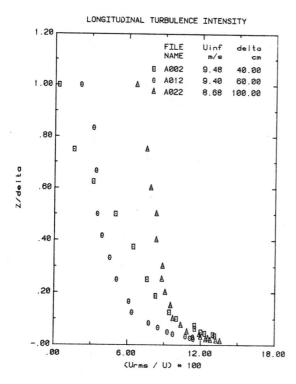

Figure 16. Longitudinal Turbulence Intensity Profile--x = 6.10 m; Smooth Surface

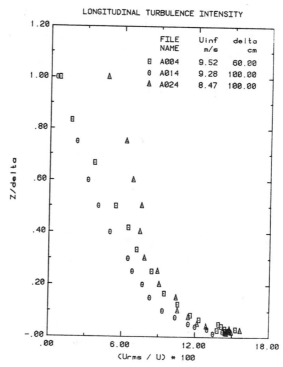

Figure 17. Longitudinal Turbulence Intensity Profile--x = 18.29 m; Smooth Surface

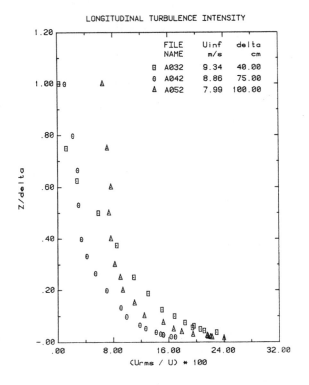

Figure 18. Longitudinal Turbulence Intensity
Profile--x = 6.10 m; 1.27 cm Roughness

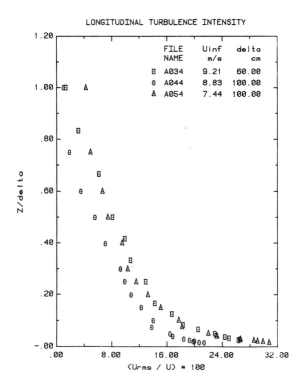

Figure 19. Longitudinal Turbulence Intensity
Profile--x = 18.29 m; 1.27 cm Roughness

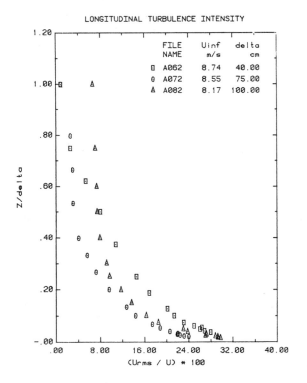

Figure 20. Longitudinal Turbulence Intensity
Profile--x = 6.10 m; 2.54 cm Roughness

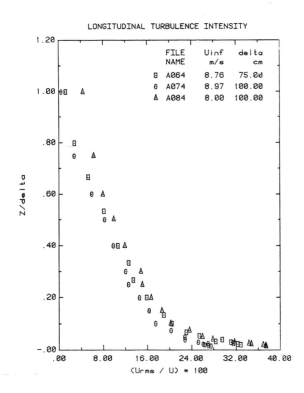

Figure 21. Longitudinal Turbulence Intensity
Profile--x = 18.29 m; 2.54 cm Roughness

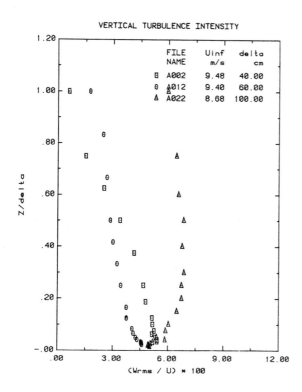

Figure 22. Vertical Turbulence Intensity Profile--x = 6.10 m; Smooth Surface

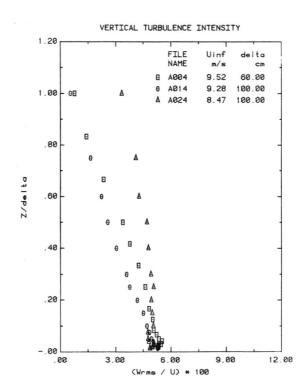

Figure 23. Vertical Turbulence Intensity Profile--x = 18.29 m; Smooth Surface

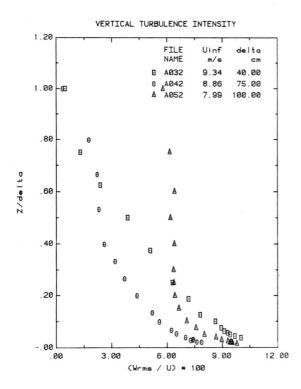

Figure 24. Vertical Turbulence Intensity Profile-- x = 6.10 m; 1.27 cm Roughness

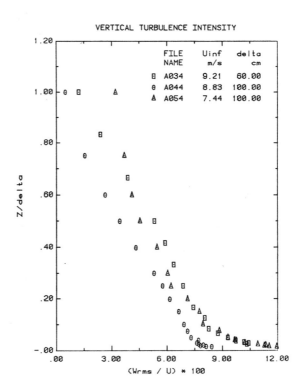

Figure 25. Vertical Turbulence Intensity Profile-- x = 18.29 m; 1.27 cm Roughness

122

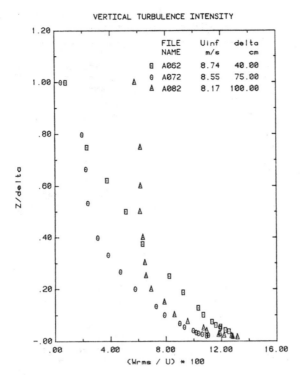

Figure 26. Vertical Turbulence Intensity Profile--
x = 6.10 m; 2.54 cm Roughness

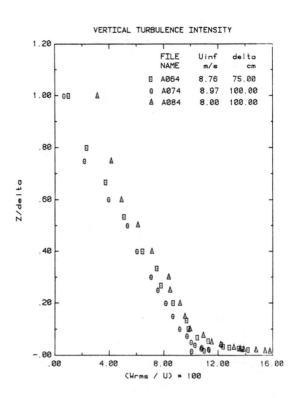

Figure 27. Vertical Turbulence Intensity Profile
x = 18.29 m; 2.54 cm Roughness

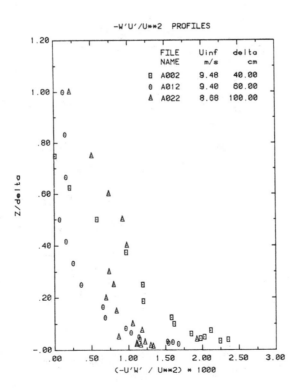

Figure 28. Shear Stress Profile--
x = 6.10 m; Smooth Surface

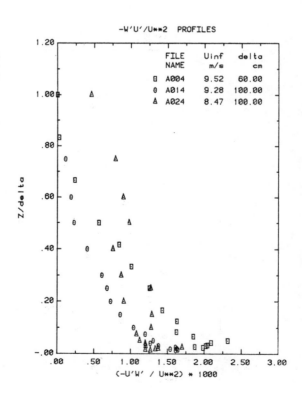

Figure 29. Shear Stress Profile--
x = 18.29 m; Smooth Surface

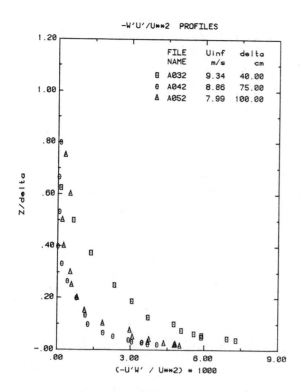

Figure 30. Shear Stress Profile--
x = 6.10 m; 1.27 cm Roughness

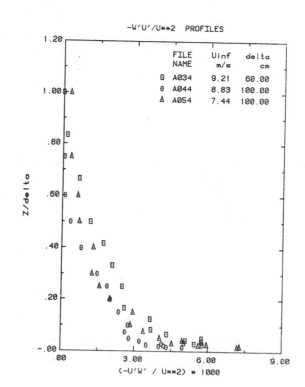

Figure 31. Shear Stress Profile--
x = 18.29 m; 1.27 cm Roughness

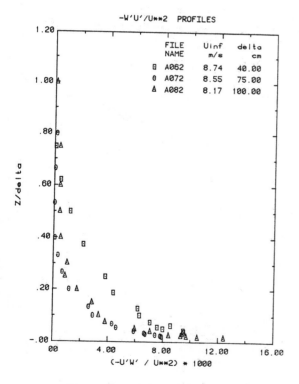

Figure 32. Shear Stress Profile--
x = 6.10 m; 2.54 cm Roughness

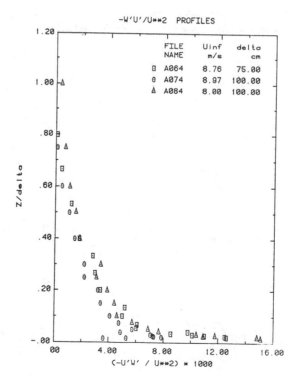

Figure 33. Shear Stress Profile--
x = 18.29 m; 2.54 cm Roughness

124

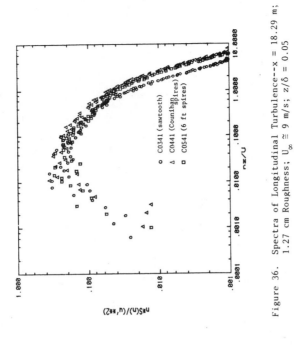

Figure 36. Spectra of Longitudinal Turbulence--x = 18.29 m; 1.27 cm Roughness; $U_\infty \cong 9$ m/s; z/δ = 0.05

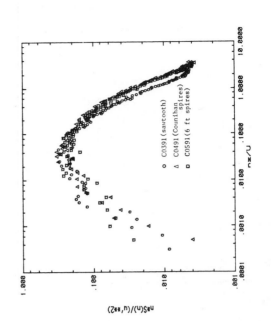

Figure 37. Spectra of Longitudinal Turbulence--x = 18.29 m; 1.27 cm Roughness; $U_\infty \cong 18$ m/s; z/δ = 0.05

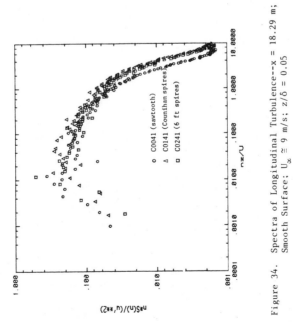

Figure 34. Spectra of Longitudinal Turbulence--x = 18.29 m; Smooth Surface; $U_\infty \cong 9$ m/s; z/δ = 0.05

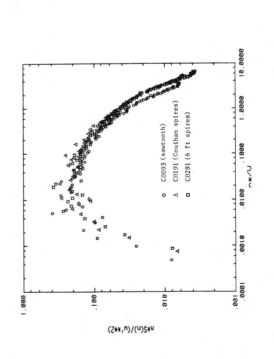

Figure 35. Spectra of Longitudinal Turbulence--x = 18.29 m; Smooth Surface; $U_\infty \cong 18$ m/s; z/δ = 0.05

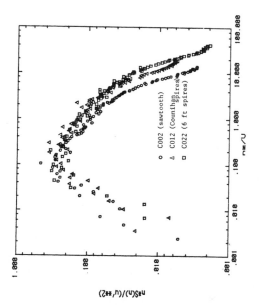

Figure 40. Spectra of Longitudinal Turbulence--x = 6.10 m; Smooth Surface; $U_\infty \cong 9$ m/s; $z/\delta = 0.25$

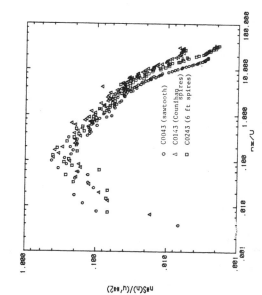

Figure 41. Spectra of Longitudinal Turbulence--x = 18.29 m; Smooth Surface; $U_\infty \cong 9$ m/s; $z/\delta = 0.25$

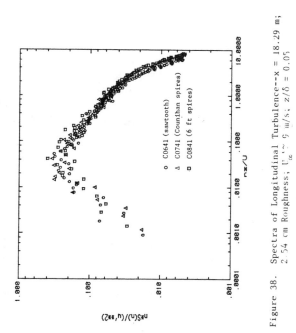

Figure 38. Spectra of Longitudinal Turbulence--x = 18.29 m; 2.54 cm Roughness; $U_\infty \cong 9$ m/s; $z/\delta = 0.05$

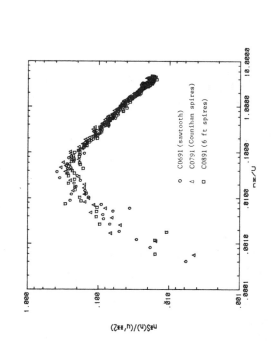

Figure 39. Spectra of Longitudinal Turbulence--x = 18.29 m; 2.54 cm Roughness; $U_\infty \cong 18$ m/s; $z/\delta = 0.05$

SIMULATION TECHNIQUES FOR SHORT TEST-SECTION WIND TUNNELS : ROUGHNESS, BARRIER AND MIXING-DEVICE METHODS.

N J Cook, Building Research Establishment, UK.

1. SCOPE

The preceding paper described the use of naturally-grown rough-wall boundary layers to simulate the atmospheric boundary layer in long test-section wind tunnels, and it is clear that accurate representation of the atmospheric boundary layer can be obtained by these methods. Not every laboratory can afford the cost or space for a boundary layer wind tunnel of sufficient length, neither are such tunnels free from problems of secondary recirculation due to their length. This present paper describes techniques developed in the last decade for accelerating the growth of natural boundary layers by passive methods to obtain simulations of similar accuracy in shorter wind tunnels. The following paper describes some active methods developed for the same purpose.

The philosophy and performance of roughness, barrier and mixing-device techniques was reviewed recently by the author [1]. This present paper will confine its attention to the practicalities of the method, referring to the review and other published papers for the necessary detail.

2. ROUGHNESS, BARRIER AND MIXING-DEVICE METHODS

The hardware of a typical simulation method of this type is shown in figure 1. This is a simulation for urban terrain at 1/500 scale and is one of a standard set of simulations used in the BRE Boundary Layer Wind Tunnel. The three physical components used in these methods are discussed in turn:

2.1 The Roughness

The role of the surface roughness in representing the rough surface of the ground is identical to that in the natural-growth boundary layer simulations of the previous paper. It remains the most important component in this methodology in that it establishes the values of the "law-of-the-wall" parameters z_o, d and u_*. Selection of a suitable roughness is also made in the same manner as previously, ie to reproduce the full-scale values of z_o, d and u_* at appropriate scales.

Parametric analyses of model boundary layers [2,3] have led to rational procedures for selecting the appropriate roughness [4], although this process remains approximate.

In the shorter wind tunnels, the available fetch is too short to grow boundary layers of sufficient depth by natural means, so the growth must be artificially accelerated. The artificial components, the barrier and the mixing device, are invariably inserted at the upwind end of the roughness. The role of the barrier is to provide an initial momentum deficit representing the effect of a longer fetch of roughness upwind. The role of the mixing device is to distribute this momentum deficit through the developing boundary layer. In the ideal case, the flow at the downwind end of the roughness should be characteristic of a boundary layer grown naturally over a much longer fetch of the same roughness, without any additional characteristics from the barrier or mixing device having been superimposed.

2.2   The Mixing Device

Mixing devices may take many forms, but all promote mixing of the initial momentum deficit by the provision of turbulence. Early devices which produced discrete longitudinal vortices were quickly abandoned because the vortices persisted, producing lateral non-uniformities in the flow. A distinction can be made between devices for simulations representing the full depth of the atmospheric boundary layer [5,6,7] and those for the "part-depth" methods which represent only a fraction of the depth (say 1/3 to 1/2) from the ground [8,9]. The intensity of atmospheric turbulence is between 20% to 30% in the lower third of the atmospheric boundary layer, but falls gradually to zero at the top through intermittency between the turbulence of the boundary layer and the smooth external flow. The maximum intensity it is possible to generate by passive means, without also generating non-uniformities or instabilities in the mean flow, is about 15% using a plane turbulence grid of the form shown in figure 1. This is a suitable mixing device for part-depth simulations. Most mixing devices designed specifically for full-depth simulations produce a gradient of intensity in the vertical direction to mimic the gradient of intensity of the boundary layer, resulting in forms like Counihan's elliptic wedge [5,6]. This mimicry is only approximate, however, since the boundary layer turbulence reduces in mean intensity only through a change in the proportion of time it is intermittently strongly turbulent or not turbulent, whereas the mixing devices produce turbulence continuously at a lower intensity. A plane turbulence grid is quite adequate for a full-depth simulation (ie the 1/500 simulation of Table 1), provided the fetch of roughness is sufficient for the excess turbulence in the external flow to decay. There appears to be no rational method for selecting the most appropriate mixing device and the researcher must resort to trial and error. A sole exception to this is the Standen spire [7] which fulfils the dual role of mixing device and barrier in a single component.

2.3 The Barrier

The simplest form of barrier is a plane wall, such as that shown in figure 1, which produces all the initial momentum deficit near ground level. A plane solid barrier produces three flow features not consistent with a naturally grown layer - a separation bubble, a shear layer along the bubble boundary and a pressure gradient down the wind tunnel. The first two features are detrimental but unavoidable; the separation bubble effectively negates the surface roughness between the barrier and the downwind end of the bubble; the shear layer contains the Reynolds stress concentrated at the barrier height and requires mixing into the boundary layer to give the correct distribution. Both these effects are reduced by modifications to the barrier, perforating [8] or castellating [1] the top edge. The latter is the more effective, since a "saw-tooth" form [1] distributes the shear layer and some of the momentum deficit over the depth of the teeth. Standen's spires are the logical end point to this process, in which the initial momentum deficit is correctly distributed through the depth of the boundary layer at source.

The pressure gradient, on the other hand, is advantageous because it is in the "adverse" sense downwind of the barrier and so promotes the growth of the boundary layer. None of the three features is consistent with a naturally grown boundary layer so the fetch of roughness must be sufficient for them to decay and the boundary layer to recover before it is used.

Having chosen a form for the barrier, the main design parameter is its height. In general the effect of raising the barrier is to deepen the boundary layer and increase the scales of the turbulence without changing the roughness-dependent parameters z_o or d. For any given surface roughness raising the barrier, within certain limits, will give a bigger scale simulation of a less rough full-scale surface. This is illustrated in figure 2 by the family of boundary layers grown over an 8m fetch of the roughness shown in figure 1. The upper curve is the mean velocity profile obtained naturally and clearly shows a logarithmic "law-of-the-wall" region with a distinct "velocity defect" or "wake" region above. In each lower curve the barrier is progressively raised so that the boundary layer deepens, first to reach the wind-tunnel roof, and finally to fill the wind tunnel with the logarithmic "law-of-the-wall" region (ie a fully developed duct flow). The family of curves represents the transition from a full-depth simulation to a part-depth simulation, with the scale factor, the equivalent full-scale roughness parameter z_o, and the depth represented, varying as listed in Table 1. Turbulence intensity profiles and spectra for these two simulations are given in reference 1. Determination of the scale factors was made by the method of reference 10.

The process of raising the barrier has two limiting factors which depend on the height of the wind tunnel and the fetch of the roughness. The roof of the wind tunnel limits the growth of turbulent eddies and hence gives lowered values of the integral scales of turbulence in its vicinity, so that only part of the simulation depth is matched to the full scale. (A similar limit is set by the tunnel floor, but here the eddies are distorted in the same manner as the atmospheric turbulence is distorted by the ground surface.) In the 1:300 simulation of Table 1 the lower 30% of the atmospheric boundary layer is represented by the lower 60% of the model boundary layer with the turbulence scales correctly matched. The top 30% of the simulation departs from full scale in having smaller turbulence scales and in being continuously instead of intermittently turbulent.

The second limiting factor has more serious consequences and is the reason that the lower profile in figure 2 and Table 1 did not produce an acceptable simulation. With very high barriers the separation bubble negates too much of the roughness fetch, resulting in flow characteristics from the barrier itself being significant at the downwind end of the roughness. Reference 1 describes how a second z_o derived from the wall dominates the z_o from the roughness itself, so that the flow effectively "loses touch" with the surface. This process is characterised by a sudden change in the zero-plane displacement d to an apparently negative value [1,11] and a Reynolds stress profile which increases with height above the floor [1]. The maximum height of the barrier is primarily dependent on the length of the roughness fetch in terms of roughness heights. The longer the fetch and/or the rougher the surface - the higher the maximum barrier height. The limiting height of the barrier for any given simulation hardware geometry is best determined by raising the barrier by increments to produce a family of boundary layers, as in figure 2, and monitoring the variation of zero-plane displacement.

3 A RUSE FOR MAXIMISING THE SCALE FACTOR

A corollary of the barrier height limit is that good quality large-scale simulations are easier to obtain for rough urban terrain than for smoother rural terrain. It is therefore usually quicker to grow a large-scale rural simulation via a nominal urban simulation instead of directly. The procedure is to use a greatly exaggerated roughness (say X10) over the upwind third of the roughness fetch, together with a high barrier. This will quickly produce a deep boundary layer with large turbulence scales. However, instead of allowing this boundary layer to stabilise over more of the large roughness, the roughness element size is reduced approximately linearly over the middle third of the roughness fetch to the required size, and maintained at that size over the downwind third of the fetch. The surface shear stress responds very quickly to the change on roughness. The strong Reynolds stress profile established by the initial "urban" boundary layer transfers momentum to accelerate the flow near the tunnel floor, reducing its intensity in the process. By the end of the roughness fetch the boundary layer may have adjusted sufficiently to the lower roughness to give an acceptable large-scale rural simulation.

The accuracy of such a simulation is not as good as the smaller-scale simulations described earlier, but the degree of error is under the control of the researcher. The accuracy can be improved at the expense of scale factor by reducing the size of the exaggerated roughness elements and the proportion of the fetch they occupy. The author has achieved an acceptable simulation of rural terrain at 1:100 in a fetch of 12m, using a turbulence grid, a 300mm average height toothed wall, 4m of 180mm high elements, 2m of 90mm high elements, 2m of 50mm high elements and 2m of 25mm high elements (all at 15% area density), followed by 2m of 14mm gravel.

4   CRITERIA FOR ACCEPTABILITY

Until recently, prototype full-scale data were available only through correlation analyses of experimental data. Those of ESDU [12] and Counihan [13] are used principally in the UK. However Deaves and Harris have developed a theoretical model over the last few years which correlates very well with observed full-scale behaviour. This theoretical model was initially for equilibrium boundary layers [14], but has since been extended to cover changes in surface roughness [15]. With this new information, the experimentalist is now in a far better position to judge the quality of his simulations, and may extend the range of his criteria of acceptability.

Mean velocity profile has always been a principal criterion. The original power-law model was only empirical and had no theoretical justification. As theory developed, first the logarithmic law and then the log-linear law began to displace the power law. Deaves and Harris have now shown that a log-quadratic law is correct, but that the simpler log-linear law as used by ESDU [12] is sufficient for the lower third of the boundary layer. It is hoped that the old power-law model can finally be buried at this workshop!

The second principal criterion is a measure of the turbulence scale, either through direct measurement of the integral scale $Lx(u)$ or indirectly through the power spectral density function. Direct measurement of $Lx(u)$, while it appears a simple process, is likely to result in biased values due to finite data length and/or analysis bandwidth. Inferring the value of $Lx(u)$ from the spectral density function is the method preferred by the author, as described in reference 10. It has the advantage that the match to the complete distribution can be assessed. The spectral density functions of the u turbulence component corresponding to the upper (no barrier) and lower (highest barrier) velocity profiles of figure 2 are reproduced in figures 3 and 4, respectively, and show that the required spectral shape is maintained throughout the range of barrier hight. The values of $Lx(u)$ and the roughness parameter z_o are sufficient to estimate the scale factor of a simulation, either by empirical comparison with prototype full-scale data or through a method like that of reference 10.

Turbulence intensity and shear stress profiles through the boundary layer have often been regarded as secondary. Recent work, including that of Hunt [11], has shown that these have more influence on building aerodynamics than previously suspected. Although the overall nature of the flow around a building can be reproduced by just representing the velocity profile, the separation regions on the sides and roof are strongly influenced by the turbulence. The $-uw$ Reynolds stress is particularly important in regard to separation and reattachment on roofs. Turbulence intensity and Reynolds stress profiles for the 1:300 simulation of Table 1 are shown in figure 5, compared with the target values from ESDU [12]. Note that the fit is not as perfect as would be obtained with a naturally grown layer, but is good near the ground surface and must be balanced against the advantage of large simulation scale.

It is suggested that the above criteria are sufficient to judge the acceptability of a simulation for wind loading or wind environment studies. Effluent dispersal studies would require matching the rates of diffusion also.

5 RELEVANCE TO NATURAL GROWTH AND ACTIVE METHODS

A naturally grown layer will always give a better representation of full scale than any accelerated growth method, since the latter methods tamper with the structure of the layer to achieve the accelerated growth. The main question is whether the increase in scale factor and/or the decrease in wind-tunnel length compensates for the loss in accuracy. In the author's opinion the answer is almost invariably that it does! In any given tunnel, a naturally grown layer can be deepened by this method and the extent is under the control of the researcher. If he does not go too far the result will be indistinguishable from a natural layer. Pushed to the practical limits, the result can still be very good and would be otherwise impossible. The early claims by Counihan [5] of development lengths 4 to 5 times the boundary layer height now seem optimistic. Counihan developed his method in a 6 inch high duct. When Robins transferred his method to a 2m high wind tunnel [16], he found that a development length of 7.5 boundary layer heights was required. Part-depth methods can fill a wind tunnel from floor to roof in a fetch 6 to 8 times the wind tunnel height. In comparison, natural boundary layers require a development length 20 to 30 times their height. The advantages of accelerated growth over natural growth are: (1) that the development length for a given scale simulation is reduced by a factor near 4; (2) that the maximum scale factor for a given development length is increased by a factor near 3; without a significant deterioration in quality. These factors can be increased further, as indicated in the previous section, if a reduction in quality is acceptable.

In general, the active methods involving air jets or similar devices perform the same role as the passive methods described here. The major difference, stated by Nee [16] is that turbulence energy can be supplied directly by active methods, whereas passive methods must derive all turbulence energy from the kinetic energy of the mean flow. Both Nee [16] and Nagib et al [17] claim the ability to adjust the velocity profile and the turbulence structure independently of each other using the active methods. The author has previously stated the view [1] that this is evidence that active methods are sometimes permitted to exclude the influence of the roughness, which must result in similar problems to that of an excessively high barrier. When active devices are confined to the same role as their passive counterparts, the advantages of each approach seem evenly balanced.

6 SIGNIFICANCE OF SOME RECENT WORK

Two observations were made recently by Hunt [11] during work in the BRE Boundary Layer Wind Tunnel investigating the use of roughness, barrier and mixing-device methods.

6.1 Observations on Reynolds Number Dependence

Hunt investigated the use of roughness of the form shown in figure 1 and found significant dependence on the roughness Reynolds number $z_o \cdot u_* / \nu$ in the range 40 to 500. As the Reynolds number increases the zero-plane displacement decreases, allowing the flow to 'see' more of the elements, with a consequence that z_o rises. This may explain some of the variability in previous parametric studies [2,3] and the approximate nature of the derived design methods. It would be prudent to check all simulations in current use for similar dependence.

6.2 Observations on Mismatching Model and Simulation Scales

The main aim of Hunt's work was to investigate the effects of mismatching the scale of the building model to the scale of the simulation as an approach to assessing the accuracy required in simulations. Table 2 shows Hunt's results for a simulation at 1:180 scale, comparing measurements on a model that was correctly matched to the simulation to measurements from models that were half and twice the correct size. The effect of a model which is too big for the simulation, the most common mismatch error, is severely to underestimate both the mean and the fluctuating components of surface pressure, particularly in regions of high local suction which are critical for design.

7 ACKNOWLEDGEMENT

The work described has been carried out as part of the research programme of the Building Research Station of the Department of the Environment and this paper is published by permission of the Director.

8 REFERENCES

[1] N.J. Cook, Wind-tunnel simulation of the adiabatic atmospheric boundary layer by roughness, barrier and mixing-device methods, Journal of

Industrial Aerodynamics, 1978, 3, 157-176.

[2] R.A. Wooding, E.F. Bradley & J.K. Marshall, Drag due to regular arrays of roughness elements of varying geometry, Boundary Layer Meteorology, 1973, 5, 285-308.

[3] J. Counihan, Wind tunnel determination of the roughness length as a function of the fetch and the roughness density of three-dimensional roughness elements, Atmospheric Environment, 1971, 5, 637-642.

[4] I.S. Gartshore & K.A. de Croos, Roughness element geometry required for wind tunnel simulations of the atmospheric wind, Fluids Engineering Division ASME, Winter Annual Meeting, December 5 1976.

[5] J. Counihan, An improved method of simulating an atmospheric boundary layer in a wind tunnel, Atmospheric Environment, 1969, 3, 197-214.

[6] J. Counihan, simulation of an adiabatic urban boundary layer in a wind tunnel, Atmospheric Environment, 1973, 7, 673-689.

[7] H.P.A.H Irwin, Design and use of spires for natural wind simulation, National Aeronautical Establishment report LA-233, Ottowa, August 1979.

[8] N.J. Cook, On simulating the lower third of the urban adiabatic boundary layer in a wind tunnel, Atmospheric Environment, 1973, 7, 691-705.

[9] B.E. Lee, The simulation of atmospheric boundary layers in the Sheffield University 1.2 x 1.2m boundary layer wind tunnel, Department of Building Science Report BS38, University of Sheffield, July 1977.

[10] N.J. Cook, Determination of the model scale factor in wind-tunnel simulations of the adiabatic atmospheric boundary layer, Journal of Industrial Aerodynamics, 1978, 2, 311-321.

[11] A. Hunt, Scale effects on wind tunnel measurements of wind effects on prismatic buildings, PhD Thesis, College of Aeronautics, Cranfield Institute of Technology, March 1981.

[12] Engineering Sciences Data Unit, Data Items 72026, 74030, 74031 & 75001, ESDU London, 1972, 1974 & 1975.

[13] J. Counihan, Adiabatic atmospheric boundary layers : a review and analysis of data from the period 1880-1792, Atmospheric Environment, 1975, 9, 871-905.

[14] D.M. Deaves & R.I. Harris, A mathematical model of the structure of strong winds, Construction Industry Research and Information Association, Report 76, London, 1978.

[15] D.M. Deaves, Computations of wind flow over changes in surface roughness, College of Aeronautics, Cranfield Institute of Technology, July 1979.

[16] J.C.R. Hunt & H. Fernholz, Wind tunnel simulation of the atmospheric boundary layer : a report on Euromech 50, Journal of Fluid Mechanics, 1975, 70, 543-559.

[17] H.M. Nagib, M.V. Morkovin, J.T. Yung & J. Tan Atichat, On modelling of atmospheric surface layers by the counter-jet technique, Paper 74-638, AIAA 8th Aerodynamics Testing Conference, Bathesda, July 1974.

Table 1 Scale factor, etc of the simulations shown in Figure 2

Profile	Scale factor	Full-scale z0 (m)	Depth represented
upper	1 : 750	1.03	100%
second	1 : 500	0.63	100%
third	1 : 400	0.49	60%
fourth	1 : 300	0.37	30%
lower	not suitable as a simulation (see text)		

Table 2 Errors due to model scale — simulation scale mismatch

Model size	Front face	Mean pressures Sides and roof	Rear face	rms Pressures
X 1/2	+5%	+25%	+60%	+30%
X 2	−5%	−30%	−40%	−30%

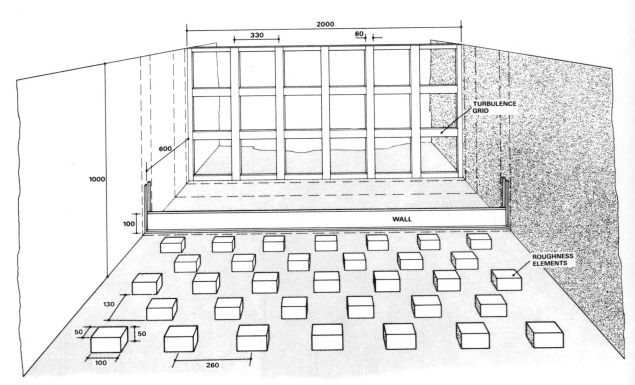

Figure 1 Typical arrangement of simulation hardware

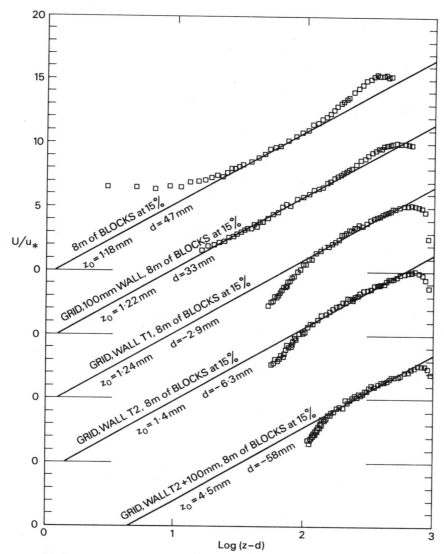

Figure 2 Mean velocity profiles for a family of 'urban' simulations

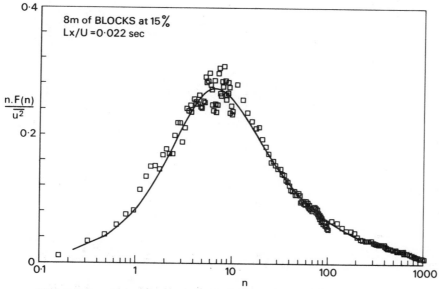

Figure 3 Spectrum of streamwise turbulence at $z = 300$ mm
with no barrier

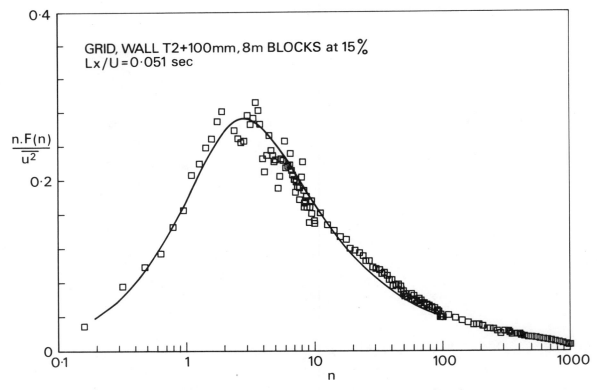

Figure 4 Spectrum of steamwise turbulence at z = 300 mm with highest barrier

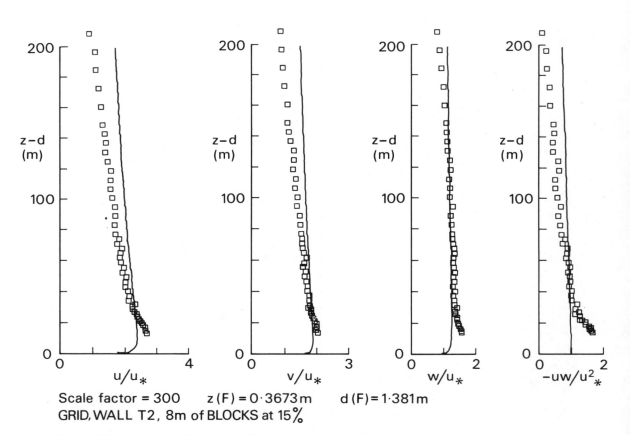

Scale factor = 300   z (F) = 0·3673 m   d (F) = 1·381 m
GRID, WALL T2, 8m of BLOCKS at 15%

Figure 5 Turbulence profiles for 1:300 simulation

CONSEQUENCES OF DISTORTIONS IN THE FLOW INCLUDING MISMATCHING SCALES AND INTENSITIES OF TURBULENCE

D. Surry, Associate Research Director, Boundary Layer Wind Tunnel Laboratory, Faculty of Engineering Science, The University of Western Ontario, London, Ontario, Canada, N6A 5B9

1. PERSPECTIVES

1.1 General

The topic of mismatching of the flow is a challenging one, many aspects of which are beyond the scope of a single paper. Fortunately, in this workshop, the degree of overlap between papers ensures that many of these problems will naturally surface in other presentations.

The primary aim of this paper is to introduce a perspective on the problem: This may assist the reader in judging whether state-of-the-art modelling techniques are being used, and in assessing the reliability of the results. The latter necessarily must also relate to the overall issue of the degree to which models represent full scale, an issue not directly considered here but treated in other presentations. Such a perspective is particularly important for engineers and architects who must translate the model results into practice.

1.2 The Boundary Layer

As earlier papers have indicated, the atmospheric boundary layer, even under neutral conditions when thermal effects can be neglected, has a complex structure which can only be properly defined by a large number of parameters which define the spatial variation of mean speed and the three turbulent velocity components, including the way in which these components are correlated between different points. Even under conditions of homogeneous surface terrain, these relationships have not been defined precisely, but are reasonably well known[1,2].† In non-homogeneous terrains, the variability of the turbulent wind structure is greater and not well-defined. Furthermore, individual severe storms introduce considerable variability into the wind structure, particularly if they are very localized. Nevertheless, it is reasonably well accepted that properly modelling the boundary layer, for most problems of structural interest under strong wind conditions will require modelling the first order properties of:

---

† Although the referencing is reasonably complete, it is not all-inclusive; however, many references are a good starting point for other material, e.g. reference 2, and a short bibliography is provided after the references at the end of the text.

i) the spatial variation of mean velocities;

ii) the spatial distribution of the three components of turbulence intensity (often simplified to just the longitudinal component);

iii) the along-wind and across-wind correlations of the three turbulent components, usually defined by the longitudinal and lateral scales of turbulence (often simplified to just the longitudinal scale).

Some of these are sketched in Figure 1.

Accepting for the moment the above simplified requirements on modelling the flow, then distortions can enter the modelling procedure in two ways: First, because the required flow cannot be defined very well due to inhomogeneities in the terrain surrounding the site to be modelled, and second, because the flow simulation itself may have distortions inherent in its methodology. Both of these items will be discussed further below.

As later detailed, some problems are more sensitive to some boundary layer parameters than to others, and some problems may involve other flow field parameters as well. In particular, the subject of modelling pollutant dispersion is not being considered, nor any like problems which may be sensitive to thermal effects in the boundary layer.

1.3 Scope

It is importnat then to reduce the scope of the discussion and to focus on the most important problems. What are they?

In the context of this workshop, particularly in providing "a resource document for architects, engineers and code-writing bodies" it is important to concentrate on the applied side of the art, rather than the more fundamental. Thus, we should make a distinction between the idealized bluff-body research which is aimed at furthering the understanding of the detailed aerodynamic processes and the applied art which is dealing with real world applications. The distinction is that the former can focus on the detailed aerodynamic sensitivities to the flow modelling in relatively idealized controlled circumstances whereas the applied art is often dealing with much more complex situations within which other factors, such as the presence of nearby structures, may dominate the flow environment. It is the real world situation that this paper primarily addresses. Thus, it is not the subtle special case effects, but rather the common, significant ones that will be of most interest. Nevertheless, the more idealistic experiments do provide insight into potential sensitivities.

1.4 Classification of the Environment

What then is the working environment in the practical application? It is a site to which the wind must come over varied terrains most likely differing with wind direction. In the current state-of-the-art, this is often idealized in the wind tunnel as a wind which develops over a relatively homogeneous terrain (the upstream roughness and/or spires, etc.) and then is modified by the detailed surroundings of the site in question (typically modelled for a radius of 1000 to 2000 feet). In a sense then, there is a 'far field' and a 'near field' simulation as indicated in the sketch of Fig. 1. The far field serves to produce a boundary layer of the 'correct overall characteristics' for that wind direction, while the near field serves to provide the intimate complex interactions which are largely inhomogeneous and intractable to precise definition. It is this near field flow which then interacts with the geometry of the particular structure of interest. The pressures and winds near the structure must then inherently be a function of the far field, the near field and the inherent aerodynamic characteristics of the body itself, which we might term the self field. These categories will be useful in order to assess their relative significance for problems of interest.

1.5 Classification of Structures

Likewise, it is useful to classify problem areas themselves, since different problems have different sensitivities. A useful classification of structures into wind sensitive categories has been made by Harris[3] which is tabulated below with some annotations added. His classification is oriented around the dynamic properties of the structure and hence the significance to the structure of the dynamic properties of the wind. The categories are

		Comments
I	*Class A:* Structures and/or element of structures which are stiff enough for wind effects to be determined by statics, and small enough for the relevant wind information to be specified as a wind speed at a single point	— **corresponds** to simple code approach of providing fixed coefficients for small buildings or parts of larger ones
II	*Class B:* Structures which are stiff enough for wind effects to be determined by statics, but large enough to require the relevant wind information to be specified as multi-point data.	— also amenable to specification of simple coefficients, if appropriately measured
	Class C: As for B, but with the additional complication that the shape of the individual structural load influence lines has to be considered in conjunction with multi-point data.	— as in B but requires measurements aligned with structural requirements which may be specific; e.g. large stiff special structures such as hangars or buildings of unusual geometry

III {

Class D: Structures which are not stiff enough to be treated by static methods but instead require a full dynamic treatment

— amenable to detailed code approaches or wind tunnel tests

— usually have natural periods in excess of one second; e.g. tall buildings, towers

Class E: Structures, generally termed 'aeroelastic', for which the wind, aerodynamics and structural motion are inseparably combined to produce the overall wind effects

— special structures, particularly those subject to aerodynamic instabilities such as bridges, chimneys, some tall buildings, membrane structures

Harris provides charts in his paper, based on the structural damping and non-dimensional frequency of the structure, to assist in this classification.

Several areas of wind engineering do not readily fit into a structural classification, such as the problems of vehicles, wind energy, pollutant dispersal and precipitation. These are areas which this paper also avoids. Another notable exception is the problem of environmental winds which will be discussed further. Otherwise, the above provides a useful framework in which to discuss sensitivity to modelling the flow.

From an aerodynamic viewpoint, the five structural classes tend to reduce to three types with regard to their interaction with the wind as designated I, II and III in the above table. In all cases, the loads are composed of mean and dynamic components; however, the dynamic loading components take on an increasingly complex role for 'higher class' structures, and hence one might intuitively guess that modelling criteria are likely to become increasingly stringent for such structures.

It is interesting to note that the vast majority of special wind tunnel studies are those aimed at classes C, D and E. Design data for classes A and B have often been derived as offshoots of these special studies, as individually they often lack the economic justification for a special study. More recently, suitable load coefficients have been derived by special research studies concentrating on families of such structures.

It is also noteworthy that the natural habitat of almost all structures is with others of their own kind, except perhaps for the very unusual such as long-span bridges, tall chimneys and towers.

We now have a basic framework for discussion of the sensitivity problem, i.e. we have outlined the primary parameters of the flow considered important; these are functions of far, near and self field effects; and there are classes of problems which may be sensitive to different flow parameters.

1.6 Reliability Considerations

Before proceeding to the details of the various permutations, there is one further dimension to the perspective taken in this paper. That is to consider how the reliability of wind tunnel test information fits into the overall reliability of the design wind load. This is a problem considered in detail by Davenport[4] from whose work the following has been extracted.

By considering the formulation of the wind load adopted by the National Building Code of Canada, $W = q\ C_e C_g C_p$ (where W is the design wind load, q is the reference mean velocity pressure, C_e is an exposure factor, C_g is a gust effect factor and C_p is a mean loading coefficient) and the ideas of limit states design, Davenport provides the following estimates of the variabilities associated with the component parts of the process of establishing the load factor or safety factor.

		V_q	V_{C_e}	V_{C_p}	V_{C_g}	V_W	$\overline{W}/W_s$	a_s
Low-Rise Structures	Code Structural Loads	.25	.12	.15		.36*	0.65*	1.18*
	Code Cladding Loads	.25	.12	.25		.45*	0.55*	1.16*
Tall Buildings (Structural Loads)	Code Simple Method	.25	.20	.15	.15	.39	0.7	1.34
	Code Detailed Method	.25	.10	.10	.05	.29	0.85	1.38
	W/T Special Study	.20	.05	.05	.05	.22	1.0	1.43

*these estimates contain an increment due to directional effects

The V's are the coefficients of variation associated with each element of the loading equation, $\overline{W}/W_s$ is the ratio of the expected (or average) design wind loading to that specified, and a_s is the load factor. Although some of the V's are admittedly impressionistic, it is clear that the specified velocity pressure as derived from the wind climate, is the source of a large part of the variability in the resulting wind load in these examples. The operative question is then whether the uncertainties introduced in relaxing certain modelling criteria are significant in changing a_s. This can come about through increasing the variability V_W or by biasing the estimate of the specified wind load, W_s. Note that a_s is porportional to $\overline{W}/W_s$ and to e^{V_W} and that the normally recommended value is 1.5.

It is unlikely that with currently available information it is possible to define alterations in V_W and $\overline{W}/W_s$ associated with relaxation in the matching of the flow parameters; however, it is clearly a useful means for assessing the real importance of such relaxations under realistic environmental circumstances. It is also clear that assessments based only on the basis implied by idealized fundamental studies may tend to be overly pessimistic since they tend to suggest a value essentially for W which may be a far cry from $\overline{W}$ when all envrionmental effects are considered.

It will be one of the recommendations of this paper that the effects of mismatching be put into this context, not through isolated special case experiments but by sensitivity experiments for structures under otherwise realistic situations. For some cases, some experimental indications are already available and simply require careful assimilation and analysis.

2.   COMMON DISTORTIONS

Referring to the properties of the boundary layer listed in Section 1.2 as most important to simulate, distortions arise from not reproducing these parameters properly either through a lack in the simulation technique or due to inadequate knowledge of the most appropriate full scale values. In this case we are discussing primarily the far field simulation. In that list, the parameters are roughly ordered according to their ease of simulation, and essentially correspond to three levels of distortion.

Although up to the early 1960's it was fairly common to test structures in smooth 'aeronautical' flows, it is relatively easy to simulate a desired mean speed profile using a variety of techniques. In fact, the mean profile can be reasonably uncoupled from the turbulence requirements[5]. Next in the hierarchy is to simulate both mean speed and longitudinal turbulence intensity distributions. This is somewhat more difficult, but again can be accomplished through a variety of techniques. Usually, the other components of turbulence also remain reasonably in the right proportions, if the simulation has had sufficient fetch to approach equilibrium. The most difficult aspect of the simulation is to maintain the scales of turbulence correctly as well. This is equivalent to maintaining the correct distribution of the turbulent energy over the frequency domain. Previous papers have discussed various state-of-the-art techniques for simulation, and their relative success at accomplishing these three levels of matching.

If a perfect simulation were attained, the derivation of an appropriate length scale at which models might be tested could be based on any of the length parameters in the flow, such as the roughness length, z_0, a turbulence scale, L, or the boundary layer depth, δ. When the simulation is not perfect, choices have to be made, distortion exists and uncertainty arises as to the appropriate scaling and interpretation of the results.

Difficulties of small scale modelling may also lead to intentional distortions. Instrumentation difficulties for very small scale models may essentially require larger models to allow loads to be measured at reasonable cost. Such deliberate distortion of the modelling length scale can be viewed from two perspectives. It can be seen as introducing errors into the simulated results or it can be viewed as valid testing of a larger building, which then focusses attention on the question in full scale — i.e. for how wide a range of building sizes are the results valid?

Finally, there is the simplest distortion arising from modelling an otherwise valid but inappropriate boundary layer for that site. This is essentially the same as an uncertainty in the required wind model and will be considered as such in this paper.

3. PERMUTATIONS

Within the framework of the previous discussion, it is interesting to first speculate on what one might expect the major sensitivities to be, based on an enlightened but simplistic view of the physical processes. This assumes that one is attempting to model the most appropriate far field conditions, and that the near field is modelled in detail with the structure not unduly isolated.

Class of Structure	Responses of Interest	Most Significant Flow Parameters	Primarily Determined By
1)	Mean Loads Environmental Winds	mean flow distributions + some turbulence	(Far), Near and Self Fields
2) I/II	Unsteady Loads at points and for small areas of buildings	local mean flow + local turbulent Intensities	(Far), Near and Self Fields
3) II/III	Unsteady Loads on large areas or buildings	overall mean flow + local turbulent intensities + relevant turbulent scales	Far, Near and Self Fields

Since these are generalizations, each is arguable, particularly in special cases; however, the trend is that turbulence is most important for unsteady loads and that the scales of turbulence may only be particularly significant when the correlation of loads over large areas of the structure is important. For smaller areas, these correlations are likely to be a self field effect. Furthermore, the influence of the quality of the upstream flow simulation is likely to be less important the more local the responses of interest are and the more the structure is of like size to its neighbours — i.e. the near and self field effects become predominant.

The ordering of the problems into responses of interest, aligned with the simplified Harris classification of structures provides a useful way or ordering the later discussion.

The remainder of this paper presents some experimental evidence to indicate the appropriateness of these perspectives and the nature of the sensitivities.

4. EXPERIMENTS

It is natural, but perhaps unfortunate, that most of the sensitivity experiments described in the literature are for isolated buildings subjected to different far field simulations with no near field included. These are, of course, very useful in determining fundamental sensitivities, but are perhaps overly conservative when used to judge a simulation of a specific structure in a specific setting. Also, they only serve to determine sensitivity to choice of wind models, rather than distortions of the wind model itself.

It is also very difficult to assess the degree of distortion present in a particular simulation. Usually this is again judged in the far field with reference to 'accepted' characteristic profiles, rather than in the near field where both full scale and model characteristics are more difficult to define. Furthermore, it is relatively difficult to uncouple the three levels of flow simulation mentioned. Thus comparative experiments tend to have more than one parameter changing at a time and hence it is difficult to derive precise inferences. This is particularly true of changes within any one level of the hierarchy — say the effect of changing lateral component intensities of turbulence while maintaining longitudinal components.

It is much easier to define what appears to be the most correct procedures (primarily by comparison with full scale experiments) and then to indicate the apparent penalties when relaxed simulation *techniques* are applied, which usually result in a distortion of several of the parameters at once. In unusual structural loading situations, it is most important to review the physics of the particular problem to search for indications of particular sensitivities. This reinforces the idea of organizing the experimental evidence along the lines of the "response of interest".

Little reference is made here to the more fundamental work in which the effects of turbulence intensity and scale on the aerodynamics of isolated bluff bodies has been examined. Such work has centred on simple two- and three-dimensional shapes in homogeneous turbulent flow. The theoretical work of Hunt[6] and the experimental work of Bearman[7,8] and many others (see Bibliography) have been fruitful in leading to an understanding of the basic mechanics at work; however, many of the subtleties are lost in a real three-dimensional site simulation.

It is worth noting, however, that the trend in the fundamental results is to indicate that mean loads on bluff bodies are only weakly dependent on the turbulent characteristics, and that dependency is primarily on turbulence intensities rather than scales. Fig. 2, taken from Bearman's papers[7,8], illustrate some of these trends nicely for mean loads on rectangular cylinders. This composite figure illustrates the effect of turbulence on the base pressures of 2-D cylinders (sensitive to turbulence intensity, but not scale), the strong effect of three-dimensionality, and the differing effects of turbulence on cylinders with varying depths of afterbody. For plates, the primary turbulent effect seems to be to increase entrainment from the wake and thus make base pressures more negative, whereas for larger afterbodies this same mechanism promotes reattachment of the flow and hence less negative base pressures. Unsteady local loads are, of course, dependent on the turbulence intensity but generally appear to be only weakly dependent on scale lengths. Naturally, overall or large area unsteady responses do become dependent on scale because that parameter determines the correlation lengths involved. It was these generalizations and the strong effect of the near and self fields on the site turbulent intensities and scales that led to the previous speculations of Section 3.

4.1 The Far Field Wind

This topic is well represented in other papers; however, it is relevant here as well in that the definition of the target wind characteristics to be simulated form the first source of distortion or uncertainty in the simulation problem. Often the far field is simulated by approximating the terrain by a homogeneous roughness and developing the appropriate boundary layer. In some cases, recourse is made to small scale topographic models which are used to measure site characteristics which are subsequently matched at a larger scale. This technique is particularly useful in complex terrains. Fig. 3 shows such a match for two wind directions in Hong Kong. The larger scale is a reasonable match in mean speed and turbulence intensity for heights of interest (up to about 250m) but the match fails at large heights where mountain wakes are still apparent in the topographic results. Comparative turbulence scales have not yet been analysed, but it is doubtful that the larger scale eddies would be properly matched. This simulation then relies on the adequacy of matching of the primary components of mean speed and turbulence intensity and the best possible turbulence scale over the height of the building in the far field, and on the near field to further polish the simulation. No definitive answer can be given to how good this is, but available model/full scale comparisons in less complex terrain[9] suggest that it is very good. Perhaps the largest uncertainty in this case is how well the topographic model itself represents the strong wind velocity profiles over hilly terrain.

Another example of using topographic matching is shown for less severe terrain in Fig. 4. This is an urban site in Boston. Several wind directions are shown, and wind characteristics from both the topographic and proximity (the larger scale) site models are shown. Here, the match can be seen to be quite good. Furthermore, the complexity of the site profile for different approaching winds is evident. This is typical of the near field effect for most structures. Similar topographic studies carried out for other sites[10,11] confirm the reliability of this technique, including the representation of the turbulence spectra in such terrains.

It is interesting to note that the analog nature of the long fetch boundary layer wind tunnel was found to work very well in developing the most appropriate simplified approach roughness at the larger scales — i.e. introducing appropriate simulation devices at appropriate upstream distances. This is the great advantage of this type of tunnel, particularly when terrains are inhomogeneous and topographic results are not available.

Distortion in simply choosing the incorrect target wind profile can be considered either as a mismatch, possibly deliberately, or an uncertainty in that target profile. Either way, the consequences will be discussed under the following sections on various levels of responses.

4.2 Mean Loads

No review of scaling, particularly for mean loads and environmental winds, would be complete without the classic work of Jensen[12] who, in 1958, published his model law for phenomena in natural wind

> "The correct model test with phenomena in the wind must be carried
> out in a turbulent boundary layer, and the model-law requires that
> this boundary layer be to scale as regards the velocity profile."

Jensen went on to describe his velocity profiles by a logarithmic law and hence required that the model scaling be the same as the scaling of the roughness length parameter, z_0, derived from the log law.

His results comparing full scale mean pressure measurements on a house of height h at $h/z_0 = 170$ with model scale results at values of h/z_0 ranging from 13 to ∞ (smooth flow) are reproduced here as Fig. 5 and dramatically indicate the correctness of this length scaling parameter. In the same paper he presented Fig. 6 which shows the variation with h/z_0 of the coefficient of the averaged pressures on each surface (i.e. the contribution to the mean overall load) as well as the largest individual mean coefficient on the windward roof. From these plots he deduced that to ensure that any of these coefficients be determined within 0.1 (approximately 10%) then h/z_0 must be within a factor of 3.

Fig. 7 indicates approximately what this implies in matching of the parameters of the common power law profiles as defined by the exponent, a, and the gradient height, z_g. In the more severe, rougher conditions, it implies matching the exponent a to within about .05 and z_g to within about 10%.

For tall buildings, a similar fundamental sensitivity of the mean loads to modelling the flow profile has been demonstrated, for example in 1963 by Baines[13] as shown in Fig. 8 and more recently by the comprehensive studies by Akins et al[14] and by Corke and Nagib[15]. Also of note is the recent comprehensive study of intermediate-sized cube buildings in two different boundary layers by Hunt[16]. These researchers have not always directly presented overall load results as functions of h/z_0, but it appears from their data that the overall sensitivity of the mean loads is consistent with Jensen's findings.

These last three experiments contain a great deal of instructive information regarding the sensitivity of the loads to the type of boundary layer being simulated — all of which are considered valid representations of the earth's boundary layer; however, none, including Jensen's shed much direct light on the sensitivity of the mean loads to distortions in the scale and intensity for a particular target boundary layer. Indirectly, they tend to verify the more fundamental studies by showing moderate sensitivity of separations, reattachments and base pressures to turbulence intensities, and lesser sensitivity of these parameters to turbulence length scales. Nonetheless, for the most important normal loading directions, Akins results show little overall sensitivity of mean loading coefficients to the turbulence characteristics. In fact, his results nicely collapse if referenced to a velocity pressure based on the average upstream speed over the height of the building.

It is particularly interesting to note that the results of Baines' early work seem in good agreement with Akins data, despite the fact that Baines used curved screens to generate the velocity profile and hence would not have modelled either the turbulence or the length scales accurately.

All of these experiments, of course, again view the model buildings in isolation — without a variable near field. The Boston site velocity profiles (Fig. 4) indicate the likely variability of the mean load introduced by the near field to be substantial.

For the tall building case, early boundary layer model results led to the establishment of suggested mean loads in the National Building Code of Canada which reflects the observations that the front face pressures vary with height following the incident velocity pressure, and the rear face pressures are roughly uniform and are best referenced to the velocity pressure at half height. Since their establishment about 15 years ago, they have consistently proved well founded for the "worst case" mean drag load for tall prismatic shapes when the building is relatively unshielded by neighbours[17]. It is interesting then to see how these code coefficients appear when integrated to form drag coefficients, and plotted versus h/z_o. They are displayed in Fig. 9, based on Akins best collapsing velocity. As expected, they are relatively invariant except for deliberate distortions introduced by the loading model for low building heights. Moreover, the magnitudes are very similar to those found by Akins for various section geometries, aspect ratios and boundary layers. The range shown corresponds to the values found for each of the 5 cross-sections examined, averaged for the various aspect ratios and boundary layers. For any one section geometry, the effects of aspect ratio and boundary layer depth led to a range of coefficients of typically ±10%. It is doubtful whether it is worth tuning a code for such variations given the differences in geometric detail of full-scale structures and the effects of the near field on the velocity profiles as typified by the Boston site profiles.

While discussing tall buildings, it is also instructive to consider the results of the CAARC standard model comparison by Melbourne[18]. In this case, the exercise was intended to compare experimental results on the same model in different simulation facilities, using nominally the same wind profiles. In fact, the velocity profile exponents were in the range of 0.23 to 0.3, the turbulence intensity at the top of the building varied by about ±10% and there were differences in turbulence length scales of about factors of 2 to 4. There were also more routine differences such as blockage effects and experimental techniques. The results for mean overturning moment, shown in Fig. 10, show a large amount of scatter for some angles, probably associated with turbulence/scale effects on reattachment; however, for the highly loaded normal wind directions, results are typically within ±10%, not all of which, as indicated above, is attributable to the flow differences. As Melbourne concludes, "it might have been more interesting, in addition, to have compared data measured in incorrectly modelled wind flows.".

In summary, then, it appears that as long as an appropriate reference velocity is maintained, mean loading coefficients are relatively insensitive to the modelling of the boundary layer parameters, providing some profile and some turbulence intensities are maintained. If the flow field were badly distorted, then corrections to a more reasonable one might well be made using the ratio of intended to actual $U_A{}^2$ values. However, if the structure is in a near field dominated flow — i.e. if it is surrounded by similar structures, it is likely that this will dominate in forming the incident mean flow field anyway.

4.3 Environmental Winds

For environmental winds, Jensen again showed the way to correct simulation in his early comparisons[12,19] between simulated and full scale measurements of wind shelter near hedges and houses. These were also mean measurements; however, it is common practice in fact to base comfort criteria[20] on the mean wind speeds measured in pedestrian areas. Although turbulence levels do vary at surface level, the mean speed — particularly when averaged over all directions through the wind climate — is probably nearly as sophisticated an indicator of potential pedestrian comfort as can be absorbed in practice. Additional information, such as regional classifications of particularly gusty areas or areas with severe spatial gradients, may be sufficiently assessed qualitatively rather than quantitatively, or at least within a classification system, as suggested by Isyumov[21].

This then strongly suggests that similar criteria for adequately modelling the flow should apply for environmental winds as for mean loads, although there is not the proven potential for simple correction for inadequate mean profile simulation. In this sense the requirements are more stringent. Certainly the physics of the problem suggests much more dependence on the near and self field. Obviously, if a building is relatively isolated or higher than its neighbours then the far field will again become dominant. The experiments by Corke et al[22] demonstrate that different boundary layer simulations certainly alter the flows observed around the building, although similar "flow modules" (Corke's terminology) repeat themselves but in different strengths. Again, neither Jensen's nor Corke's experiments provide insight into effects of distorting the flow parameters within a target simulation, but only the consequences of varying the simulation itself.

In this regard, the experiments by Isyumov[23] in trying to reproduce earlier surface wind speeds measured by Wise[24] are particularly interesting and are displayed in Fig. 11. Wise had simulated a velocity profile but no appreciable turbulence. The results for these relatively isolated structures imply that the effect of the turbulence (presumably on reattachment points to the buildings) has changed the mean flow interaction for these two buildings significantly.

Some confirmation of the dependence of the surface flows on the near field can be seen in Fig. 12 which compares results taken with two dramatically different simulations of surface winds around a building in Hong Kong. In the 1:500 scale model, five different upstream terrain models were utilized over various azimuth ranges as determined by matching with a topographic model as previously described. In the 1:200 scale model, a rough match of the far field flow profile was made for the most exposed and common wind directions and then used for all azimuths, the results being corrected on the basis of maintaining an appropriate roof height speed (at about 180m). Fig. 12 shows examples of both the predicted results, as synthesized with the wind climate, and polar plots of some of the actual aerodynamic characteristics. The results are remarkably good — certainly good enough for the purpose, which was to allow simple modifications to be assessed in relative terms. Since most of the site, particularly for the less well-matched wind directions, is very sheltered by nearby buildings, the results may not be too surprising; however, they tend to confirm the expectation that it is generally the near and self field that is predominant in this kind of problem, and that turbulence length scales are unlikely to be too important.

4.4 Local Unsteady Loads

The problems associated with unsteady local loads are essentially those which are not likely to be directly sensitive to the correlation lengths in the far field. As such, they might encompass both local pressure measurements (i.e. point loads) and the loads on intermediate sizes of areas which are likely to be affected primarily by the self field. That is, they may include spatial averaging effects due to lack of correlation in the unsteady pressures, but these correlation effects are defined primarily by the flows around the body itself rather than by the far field. It is not possible to currently determine how large an area can no longer be seen as primarily a local loading problem. Obviously, it is not a rigid division, and the dependencies on the various components of the flow will change in individual circumstances. In terms of the structural classification of Section 1.5, these are primarily A, with some overlap into B and C.

For both low and high rise structures, the relative success of gust-loading oriented code specifications supports the primary dependence of local loads on the local gust speed. This is accomplished in practice by either specifying a gust speed directly or by providing mean profiles with a further dependence on terrain type (i.e. turbulence level).

Direct experimental evidence is again hazy for the reasons cited before; however, it appears that for high rise structures, modelling the longitudinal components of turbulence becomes as important as modelling the mean profile itself. In this way the local gust speed is about right. It seems unlikely, based also on the more fundamental studies of two dimensional square cylinders, that the scale plays a significant role for most unsteady local loads as long as its within about a factor of 5 (see Melbourne's CAARC data[18] for example).

The variation of high rise building local response with some such parameter as h/z_0 is not established, but recourse to the gust speed variation with h/z_0 would again suggest a relatively insensitive variation with boundary layer parameters as long as a suitable reference gust speed is utilized. This is essentially a result of the fact that as the mean speed decreases in rougher terrains, the turbulent intensity becomes higher, so that the gust speed profile changes much more slowly. Corke and Nagib's data[15] on fluctuating pressures, for example, collapse much better when based on gust speeds.

Hunt's[16] study of cubical buildings of intermediate height generally confirms the sensitivity of the local loads to local conditions, − i.e. profiles and intensities. Again, it would be interesting to see results which had some but not all parameters matched, as by using the velocity profiling techniques of Baines[13] and Sharan[5]. Even though Hunt calls for maintaining appropriate integral scales of turbulence, it is not clear from his results to what degree these should be maintained.

There is some indication from the work of Vicker (P.J.)[25] that scale may play a more important role for low-rise structures. In a variety of simulations for the low-rise experimental house at Aylesbury, there was evidence that maintaining turbulence intensities was not sufficient to correctly reproduce local pressures. In such cases, the intensities can be very large, and the role of the scale is simply as a measure of the distribution of the fluctuations over the frequency spectrum When a large fraction of the structure is directly enveloped by a significant fraction of the turbulent eddies, it certainly suggests scale may be more important. This is also consistent with the ideas of Melbourne[26] who has suggested that extreme local pressures under separation bubbles may be particularly sensitive to the turbulent energy level at higher frequencies since it appears to control the rate of shear layer growth and reattachment. This level of turbulence energy is essentially indicated by both the turbulence intensity and scale. As either is reduced, the spectral density at high frequency will be reduced.

Such mismatching of scales, or high frequency spectral densities, may be part of the explanation for considerable discrepancies currently surfacing for low-rise building results, particularly in comparisons between boundary layer wind tunnel results and partial simulation results using significantly larger models, and particularly near windward corners under separation regions.

The overall sensitivity of local peak pressures and unsteady loads averaged over larger areas has been investigated for low-rise buildings by Stathopoulos, Surry and Davenport[27] by using three scales of buildings in two different flow simulations. Typical data, plotted versus h/z_O, is shown in Fig. 13 taken from reference 28. Significant trends do occur, some of which seem to be perhaps due to other causes — i.e. the data trend is not always smooth and consistent. However, on the whole, the results are not inconsistent with Jensen's original results suggesting that an h/z_O distortion of a factor of 2 should be consistent with other uncertainties in the problem. Note that Fig. 13 provides some information for intermediate size "bay" areas, which seems to follow similar trends. The literature is lacking in the behaviour of the unsteady loading of such partial areas with flow distortions.

In all of these results, we have again been discussing relatively isolated structures. The overall sensitivity of the problems to the near field will be discussed again in Section 4.7.

In summary then, local unsteady loads require modelling the mean profile as in Section 4.2. In addition, turbulence intensities should be maintained accurately, say within 10%. Generally it appears that modelling the scale to within about a factor of 2 to 5 may be adequate, with the larger range appropriate for taller buildings, the smaller for low-rise. There remains, however, some suggestion that the scale, or at least the high frequency spectral densities, may play a more significant role for very local high suction peaks near separation points.

4.5 Overall Unsteady Loads

The problems associated with large scale unsteady loads are essentially those of Harris's classes D and E with again some overlap with Band C.

In the following discussion, we shall consider only sharp-edged bluff bodies. In that case, in addition to the previous discussions on the importance of velocity profiles and intensities of turbulence, there are two additional effects involving the distribution of turbulent energy — i.e. the scales. First, the loads applied to the structure are dependent on how well correlated the gusts are over the spatial extent of the structure. This is usually summarized as an aerodynamic admittance which, for surfaces normal to the wind, has been found to drop off with increases in the ratio of the body size to the turbulence wave length, $\sqrt{A}/(\overline{U}/f) = f\sqrt{A}/\overline{U}$. Since it is the turbulence length scales that determine the distribution of gust energy with inverse wavelength, $f/\overline{U}$, it is obviously important for problems involving loads on large areas to keep the ratio of turbulence length scales to body size correct, even if the structural area under consideration is rigid. It is also intuitive that it is more important for a large area to be scaled properly than a small one.

The second major effect occurs if aeroelastic response is significant. The resonant aerodynamic response, at least in the drag direction, is dependent on the amount of turbulent energy at the natural frequency of the structure. Hence the scale becomes increasingly important.

For drag response, both of these effects can be assessed from detailed building code approaches[24] which describe the peak structural response, W as

$$\hat{W} = \bar{W} \left[1 + g \sqrt{\frac{K}{C_e}(B + \frac{sF}{\beta})} \right] = C_g \bar{W}$$

where $\bar{W}$ is the mean response, g is a peak factor, C_e and K are factors related to terrain roughness, B is a background turbulence factor, s is a size reduction factor, F is a measure of the gust energy at the natural frequency and β is the structural damping Inherent in the formulation is a linear dependence between the unsteady loading and the turbulence intensity — i.e. $\sqrt{K/C_e}$ is essentially proportional to the intensity of turbulence.

For the two effects discussed above, B and F play the most significant roles when a mismatch in turbulence length scale is considered. The effect of increasing the model size in otherwise correctly scaled flow can be assessed from the parameter s. Graphs of B, s and F[29] are shown in Fig. 14. For a rigid structure, the sF/β term is not important. Thus the graph for B can provide a rough guide for vertical surfaces normal to the wind, as to the effect of having only the scale distorted. To do this, we assume the central height of the structure remains constant, so that C_e is the same in both cases. Then, as a fraction of the mean load, this indicates that a 20' x 20' structure experiences about 8% larger unsteady loads then a 40' x 40' structure; however, a 100' x 100' structure experiences about 20% more unsteady load than a 200' x 200' structure. These are both roughly equivalent to a mismatching of the scales in the flow by a factor of 2. Since the unsteady response might be of the order of ½ of the total response, a factor of 2 error in scale might lead to overall errors of half of those quoted above. These changes are in addition to any scale-induced aerodynamic alterations, which are likely to be small.

If the surroundings of the structure are built-up with similar-sized buildings, it is not clear what final role the far field scales play; however, because the scales are indicative of the relationships between gusts some distance apart, the far field is likely to be more pervasive Nevertheless, for rigid structures, it seems tolerable to accept scale deviations of up to a factor of 2. This also seems to be borne out for the low-rise structures' integrated loads in Fig. 13, which are horizontal rather than vertical surfaces.

Note also, that for the above discussion the significant scale is the lateral scale of the longitudinal component, not the commonly measured longitudinal scale. Since some simulation methodologies distort the ratio of longitudinal to lateral scales, this should be kept in mind.

We can now go through a similar argument for a non-rigid structure where the resonant response is dominant — i.e. sF/β — where F plays the primary role. Again, from the graphs of Fig. 14, the indications are that a factor of 2 in scale would alter the dynamic response by as much as about 30% — an overestimate for scales too small. Clearly, this is at the limit of acceptability unless corrections are undertaken or the near field is known to be predominant.

As before, overall loading response may reduce this apparent error by a factor of about 2; however, it should be noted that all of the error would turn up in predicted accelerations.

There is much less information in the literature on the sensitivity of across-wind aeroelastic response to turbulence scaling, but it would be expected to be at least as sensitive as the along-wind response described above, and possibly more so (discounting rounded structures, see Section 4.8). An example[30] is given in Fig. 15 for winds normal to the face of a square building of $H/D = 7$ in various flows and for various dampings as indicated. There is obviously a strong effect of the upstream flow on the RMS lateral response. Since there is strong evidence in many of the papers previously cited that turbulence intensities expecially, and scales, to some extent, affect the size of flow separation regions and where the flow reattaches, it follows that scaling these parameters carefully must be important for aeroelastic response, especially for across-wind excitation.

In summary then, for overall unsteady loads, intensities and scales should be matched as closely as possible, with errors being roughly proportional to intensity. Scales should be maintained within about a factor of 2 for rigid models, and better than a factor of 2 for aeroelastic models.

4.6 Flow/Model Scaling Requirements

Most of the previous discussions regarding sensitivity of loads to differences in flow simulation models can be applied equivalently here if, as Jensen first postulated, h/z_O is the primary parameter on which to scale the boundary layer and if appropriate reference dynamic pressures are adopted. In this case, allowable variations of h/z_O translate directly into allowable distortions of scale. h/z_O would appear to be a valid parameter for the far field in homogeneous terrain cases where the body is well immersed in the boundary layer.

In cases where the far field simulation is non-homogeneous, distortions in the model scale may be more significant since it could be penetrating different layers of the non-equilibrium boundary layer. The effects in this case are not known; however, it is interesting to speculate that this may be the cause of some of Hunt's[16] apparent sensitivity of loads to model size, since the model was essentially in a boundary layer in the process of adjusting from a rough floor upstream to a smooth floor surrounding the model. The particular aspect of Hunt's experiments that are most disturbing is that when he varied cube sizes in two similar boundary layers differing in "size" by a factor of two, he found the larger model/flow combination (1:180) to be considerably more sensitive to building size changes than the smaller (1:360). It seems difficult to attribute this marked sensitivity change to the small component of non-equilibrium flow present, but it is equally difficult to accept the only other apparent explanation of a Reynolds Number effect. Perhaps there are other as yet unknown factors, particular to this experiment.

Again, it is the aeroelastic model that must be viewed with the greatest caution. Here there is not only the alteration in the background response, but also a potential alteration in both the size reduction factor, s, and the turbulent energy level, F, depending on how the final aeroelastic scaling parameters are derived. Furthermore, there is the additional uncertainty of scaling effects on the across-wind response.

In the practical case of particular building investigations, the sensitivity to model scale will also be dependent on the form of the near field. If the near environment is similarly scaled, and is of similar size to the building under test, then the sensitivities are likely to be appreciably less, except possibly for aeroelastic response problems as discussed in Section 4.5.

4.7 Near Field Effects

In this paper, the perspective has emphasized that most of the experimental evidence lacks two important ingredients — the sensitivity to deliberate distortions in the parameters of a particular target profile and the effects of varying the far field when a realistic surroundings or near field is included. The first element will have to await further research; however, some ad-hoc information is available on the latter.

For tall buildings, it is common practice in boundary layer simulations, to use several far field models depending on the azimuth range of interest. Generally, redundant data are taken at azimuth angles separating these different approach roughnesses, to determine sensitivities, worst cases, etc.. These are termed overlap azimuths. A recent study of a 350' high building in New York on the Hudson river provides an example of several different exposure overlaps, with markedly different far field simulations. Since the site is open to the Hudson River, some wind angles have virtually no upstream built-up areas for a considerable distance. Fig. 16 displays velocity and turbulence intensity profiles for the four far field simulations used, just before the flow enters the proximity region, representing a radius of 2000' around the model. Note the different scales for the turbulence intensities. Fig. 17 shows scattergram comparisons between data taken with different far field "exposures". For angles where the site is protected by the surrounding city, the effect of changes in the far field simulation is small. However, for the angles where the site is exposed directly with no built-up environment, considerable sensitivity is displayed in both mean and peak pressure coefficients, all of which are referenced to gradient height (the free stream dynamic pressure).

A number of other high-rise buildings have been investigated similarly, The results are tabulated in Fig. 18 in terms of the slope of the best fit straight line and comments on the overlaps. They indicate similar trends. In all cases, the tendency is towards both lower mean and peak loads for rougher upstream exposures with more effect on the mean than on the peak loads; however, the effect is small if there is a significant near field.

The site in New York is also of interest because an alternate near field was examined, due to the potential development of Battery Park adjacent to the site. In this case most of the relatively open exposure became protected. Fig. 19 shows a comparison for one of the azimuths for which the site protection changed dramatically. Again, the added roughness reduces the mean and peak local loads. Note however, that the character of the loading is changing dramatically towards more dynamic loading. Thus, these results cannot be extrapolated to aeroelastic behaviour.

The overall effect of the change in the near field in the New York site is also included in Fig. 19 in terms of the predicted pressures and suctions, as seen through a synthesis with the wind climate. This may not be the best general approach since the New York wind climate favours certain wind directions but it is a measure of the overall effect of the presence of Battery Park. The predicted loads are all reduced on average by only about 2% for both pressures and suctions. This underlines the requirement that the entire interaction must be considered, rather than just worst angles.

Fig. 20 shows similar sets of predicted values compared for the 180m high building in Hong Kong indicating the effect of present and future city environments. The latter was considerably more densely built-up. Again, little net effect is evident.

In the case of relatively isolated buildings or in special cases, nearby buildings can increase local loads. A particularly dramatic example is included in Fig. 21, showing predicted local suctions on a structure with and without a neighbouring building[31]. Clearly the near/ self field is dominant.

Similar data for the relative importance of near and far field effects on overall loads on aeroelastic models could be made from ad hoc tests; however, it would again be best to repeat the simulation for all wind angles and compare the results in some synthesized fashion akin to the predicted local loads of Figs. 19 and 20.

An example of near field sensitivity of aeroelastic response to a neighbouring structure is shown for interest in Fig. 22. The 600' high structure in question was tested with and without a prospective neighbouring development which was projected as considerably taller than the test structure. It is noteworthy, that the aerodynamic results show potential increases in peak accelerations of about 100% for the worst wind directions, whereas predicted accelerations increased by about 25%. Obviously, the near field can be of critical importance, as found in many examples in the literature of studies of interaction between buildings; however, examining isolated buildings and worst cases tends to overstate the practical case.

For low-rise buildings, the various Aylesbury simulations have dramatically shown the effect of the near field, where the presence or absence of upstream hedges can change local pressure coefficients by typically a factor of 2. This is again for an isolated structure, albeit a real one.

For a low-rise structure surrounded by buildings of similar height, several experiments have particular significance:

The low-rise experiments of Davenport, Surry and Stathopoulos[27] were undertaken using two terrain roughnesses — an open country exposure and a suburban exposure. In the latter, the roughness elements surrounded the structure and were of similar height. It is significant that all of the peak unsteady local and area loads were lower in the built-up terrain, some by as much as 80%. On average, the expected reduction due to surrounding structures of similar size was about 15 to 20%. It is these data that formed the basis of the estimates of expected load in Davenport's uncertainty calculations presented in Section 1.

Likewise, the results of Vickery[32], Holmes[33] and Hussain and Lee[34] on arrays of low-rise structures showed significant shielding effects on the mean loads for most of the locations within arrays of buildings of similar height. A glance at an aerial photograph of a subdivision or industrial area indicates that most low-rise structures have some near field that should be considered.

Notwithstanding these results which are applicable in most cases, it is not difficult to construct proximity effects which can increase the loading significantly by introducing a tall building adjacent to an exposed low-rise building (see (27)).

In summary, near field effects appear beneficial on the whole for local loads but may aggravate overall dynamic loading. Special cases, such as relatively isolated buildings rising significantly above their surroundings, can violate these overall trends.

4.8 Other Effects and Problems

This paper has studiously avoided certain issues which are either less commonplace or best treated elsewhere. For the record, some of these, which are worthy of further discussion somewhere else, include

1) the effects on the trends examined of wind tunnel blockage corrections;

2) sensitivities to the flow simulation of instabilities, particularly those associated with special structures such as bridges and chimneys. The latter, for example, may have critical speeds occurring at wind speeds at which stable thermal effects can drastically reduce atmospheric turbulence levels.

3) the Reynolds number sensitivity of structures with curved surfaces — i.e. without fixed separation points. This is of particular interest to testing of chimneys and cooling towers;

4) the sensitivity of virtually two-dimensional horizontal structures, such as bridges and large suspended roofs and the appropriateness for these of using grid turbulence.

There are probably others as well!

4.9 Reliability Again

As we have tried to illustrate, the assessment of the consequences of flow distortions is very difficult. In many instances, the required information doesn't exist. If this assessment is to go beyond the particular, then the only rational path seems to be the statistical one in which modelling distortions are viewed as another uncertainty in the process illustrated in Section 1.5. In order to do this properly, experiments with distorted simulations must be carried out with sets of realistic near fields. One of these may be an isolated building case, but in the final analysis, the weighting of this case should be appropriate to its practical realization. If huge differences are found, the code forms can be altered to recognize this and hence reduce the uncertainty of the final product.

Some work has already been carried out in this regard. For low-rise buildings, the previously cited work of Davenport[4] draws on the different terrain models investigated in (27) as a basis for the expected loads and their variabilities according to an assessment of the full scale frequency of siting in each terrain. The studies of buildings in groups previously mentioned[32,33,34] provides further data for such an assessment.

For tall buildings, Vickery[35] examined the likely variability of the components of the gust factor, making assumptions about the coefficients of variation of such things as the turbulence scale, the intensity of turbulence and the exposure factor. These are items which could be investigated experimentally in a statistical fashion to determine their sensitivities to both the near and far field.

The statistical analysis of local pressure data[36] arising from many ad hoc tall building tests inherently included near and far field effects and provided a statistical structure within which effective or nominal coefficients could be defined following the ideas of Davenport[37]. A similar concept for nominal coefficients has been presented by Simiu[38]. Similar approaches can be adapted to appropriate experiments to determine the overall impact of flow/model distortions.

The final assessment of the situation must include the variability of the wind. It appears possible to make reasonable estimates of the variabilities of the climate and the mean and intensity profiles on the basis of current knowledge. The matter of scale is less clear. Furthermore, it may not be the right parameter. As indicated throughout this paper, the overall length scale does not seem to be a critical parameter. Instead, it is the energy levels beyond the spectral peak that appear important, both as a potential explanation for observed sensitively of peak local suctions, and for the aeroelastic response problem. It is perhaps the variability of these spectral levels (in non-dimensional form) that need to be defined. It is likely that these are also more closely definable than the scale itself. Simiu[39], for example, quotes uncertainty in scale lengths of between 0.3 to 2.5 times nominal and Tieleman[40] has shown that the turbulence spectra have large variability at significant mean speeds, but that this variability is concentrated below the spectral peak, producing large changes in apparent scale but lesser changes in the energy levels of most importance.

There remains a considerable amount of work to do before the consequences of distortion can put in their proper perspective in all but the simplest cases. Until then, a reasonably conservative approach should be adopted.

5. CONCLUDING REMARKS

It has been the aim of this paper to discuss the consequences of distorting the simulated flow properties, particularly for the problem of wind tunnel testing of common structures. Few hard guidelines can be found. Most codes refer simply to appropriately simulated flow. One worthy of special note is the newly revised ANSI Standard[41] which specifically requires modelling of the mean velocity profile and the longitudinal component of turbulence and requires that the geometric scale of the structural model must be no more than three times the geometric scale of the longitudinal component of turbulence. While this appears reasonable for most problems, it may not be adequate for aeroelastic simulations .

From the discussion in this paper the following general recommendations can be drawn which can be used for further debate:

1) For mean loads and mean surface flows, it is probably sufficient to model the mean far field flow velocities, together with some representative turbulence intensity levels. The relative proportion of the turbulence intensities and the turbulence scale are unlikely to be important.

2) For unsteady loads for classes A and B type structures (local and relatively small areas) it is probably necessary to model accurately both the mean far field flow velocities and the longitudinal turbulence intensities. Turbulence

length scales should be maintained within about a factor of 3. No clear information is available as to the likely effect of the relative importance of cross-component turbulent intensities.

3) For overall unsteady loads for class C and D structures, the conditions of (2) apply with scales being more important. Scales should be maintained within better than a factor of 2, the most important scale being the lateral scale of the longitudinal turbulence component.

(4) For overall unsteady loads on class E structures (excluding the special structures discussed in Section 4.8) as in (2) except even more attention to scales, or at least to the turbulent energy levels near the structure's natural frequencies. (This usually amounts to the same thing.) It is more likely here that cross-component intensities may be important; however, this is not well understood, and may be sufficiently represented naturally by an adequate simulation which fulfills the first conditions.

(5) All of the above may be relaxable, if the near field is dominant, i.e. if the structure is surrounded by like structures. This is particularly true for lower classes of problems. The aeroelastic phenomena remain the most sensitive.

(6) Relaxation of model scales of up to 2 to 3 appear justifiable for mean loads, environmental winds and class A and B structures. Higher order problems should be kept within the requirements for turbulence scale length designated above.

(7) Experiments are required in which typical structural loading cases are examined in realistic surroundings to assess the sensitivity of the various loading classes to different far field simulations.

(8) Similar experiments are required to determine the sensitivity of loads to distortions within target profiles, at least insofar as they might be physically realizable within simple simulation techniques.

(9) The sensitivity of peak local suctions near separation points to high frequency turbulence energy levels needs further investigation, particularly for reference to the low-rise simulation problem.

ACKNOWLEDGEMENTS

The author would like to acknowledge the creative environment sustained by his colleagues at the Boundary Layer Wind Tunnel Laboratory, and particularly the continued encouragement of Dr. A.G. Davenport. Some of the data presented was prepared under considerable pressure by Tracey Upham and Gary Lythe, and put into final form by Paul Lewecki and Brian Allison. The last minute preparation of the manuscript was done very ably by Gwyn Hayman.

REFERENCES

1. A.G. Davenport, The application of statistical concepts to the wind loading of structures, Proc. I.C.E., 19, 449, August 1961.

2. R.I. Harris and D.M. Deaves, Paper 4: The structure of strong winds, Wind Engineering in the Eighties, Proc. of the CIRIA Conf., November 12/13, 1980.

3. R.I. Harris, The classification of structures for the assessment and codification of wind effects, Fifth International Conference on Wind Engineering, Vol. 2, Colorado State University, July 1979.

4. A.G. Davenport, A comparison of wind and earthquake loads and profiles, Paper presented at the Mark Huggins Symposium on Struct. Eng., University of Toronto, September 7/8, 1978.

5. V. Kr. Sharan, Generation of wind tunnel flow with independently varying turbulence level and velocity profiles, Acta Polytechnica Scandinavica, Mech. Eng. Series, No. 69, 1973.

6. J.C.R. Hunt, Turbulent velocities near and fluctuating surface pressures on structures in turbulent winds, Proc. 4th Int. Conf on Wind Effects on Buildings and Structures, Cambridge University Press, 1977.

7. P.W. Bearman, Some effects of free-stream turbulence and the presence of the ground on the flow around bluff bodies, Aerodynamic Drag Mechanisms of Bluff Bodies and Load Vehicles, Proc. Symp. held at G.M. Research Labs, Plenum Press, N.Y.–Lond., 1978.

8. P.W. Beaman, Paper 5: Aerodynamic loads on buildings and structures, (see reference 2)

9. W.A. Dalgliesh, Comparison of model/full-scale wind pressures on a high-rise building, J. Ind. Aerodynamics, 1, 55 (1975).

10. A.G. Davenport, N. Isyumov and T. Jandali, A study of wind effects for the Sears project, Univ. of Western Ontario, Eng. Science Research Report, BLWT–5–71, 1971.

11. A.G. Davenport and N. Isyumov, A wind tunnel study for the United States Steel Building, Univ. of Western Ontario, Eng. Science Research Report, BLWT–5–67, 1967.

12. M. Jensen, The model law phenomena in natural wind, Ingenioren, Vol. 2, No. 4, Nov. 1958.

13. W.D. Baines, Effects of velocity distribution on wind loads and flow patterns on buildings, Proc. Conf. on Wind Effects on Buildings and Structures, held at Teddington, England, 1963, HMSO, 1965.

14. R.E. Akins, J.A. Peterka and J.E. Cermak, Mean forces and moment coefficients for buildings in turbulent boundary layers, J. Ind. Aero. 2 (1977), 195–209.

15. T.C. Corke and H.M. Nagib, Wind loads on a building model in a family of surface layers, J. Ind. Aero. 5 (1979), 159–177.

16. A. Hunt, Wind tunnel measurements of surface pressures on cubic building models at several scales, submitted to J. Ind. Aero.

17. D. Surry and G. Lythe, Mean torsional loads on tall buildings, Proc. Fourth U.S. National Conf. on Wind Engineering Research, U. of Washington, Seattle, July 1981.

18. W.H. Melbourne, Comparison of measurements on the CAARC standard tall building model in simulated model wind flows, J. Ind. Aero. 6 (1980) 73–88.

19. M. Jensen and N. Franck, Model-scale tests in turbulent wind, Part I, Phenomena dependent on wind speed, The Danish Technical Press, Copenhagen, 1963.

20. A.G. Davenport, An approach to human comfort criteria for environmental wind conditions, Colloq. on Building Climatology, Stockholm, 1972.

21. N. Isyumov, Studies of the pedestrian level wind environment at the Boundary Layer Wind Tunnel Laboratory of The University of Western Ontario, J. Ind. Aero. 3(1978), 177–186.

22. T.C. Corke, H.M. Nagib and J. Tan-Achitat, Flow near a building model in a family of surface layers, J. Ind. Aero. 5 (1979) 139–158.

23. N. Isyumov and A.G. Davenport, The ground level wind environment in built-up areas, (see reference 6).

24. A.F.E. Wise, Wind effects due to groups of buildings, Canadian Architect, Nov. 1971.

25. P.J. Vickery, A wind tunnel study on a model of the full scale experimental house at Aylesbury, England, B.E.Sc. Thesis, Faculty of Engineering Science, The University of Western Ontario, 1981.

26. W.H. Melbourne, Turbulence effects on maximum surface pressures — a mechanism and possibility of reduction, (see reference 3).

27. A.G. Davenport, D. Surry and T. Stathopoulos, Wind loads on low-rise buildings: final report of phases I and II — parts 1 and 2, Univ. of Western Ontario, Eng. Sci. Research Report BLWT–SS8–1977, 1977. Also Phase III, BLWT–SS4–1978.

28. T. Stathopoulos, D. Surry and A.G. Davenport, Some general characteristics of turbulent wind effects on low-rise structures, Proc. 3rd Colloq. on Industrial Aerodynamics, Aachen, Germany, June 1978.

29. _____, The Supplement to the National Building Code of Canada, 1980, Assoc. Comm. on the NBC, National Research Council of Canada, Ottawa NRCC No. 17724.

30. P.A. Rosati, The response of a square prism to wind load, Univ. of Western Ontario, Eng. Sci. Res. Report, BLWT–2–68, 1968.

31. D. Surry and W.J. Mallais, Adverse local wind loads induced by an adjacent building, submitted to ASCE.

32. B.J. Vickery, Wind loads on low-rise buildings, presented at D.R.C. Seminar, Darwin, March 1976.

33. J.D. Holmes, and R.J. Best, A wind tunnel study of wind pressures on grouped tropical houses, James Cook Univ. of N. Queensland, Dept. of Civil & Systems Eng., Wind Eng. Rept. 5/76, Sept. 1979.

34. M. Hussain and B.E. Lee, A wind tunnel study of the mean pressure forces acting on large groups of low-rise buildings, J. of Wind Eng. and Ind. Aero., 6 (1980) 207–225.

35. B.J. Vickery, On the reliability of gust loading factors, Proc. Tech. Meeting: Wind Loads on Bridges and Structures, 1969, Bld. Sc. Ser. 30, Nat. Bureau of Standards, NBS BSS 30 (1970).

36. A.G. Davenport, R.B. Kitchen and B.J. Vickery, The reliability of pressure coefficients for the determination of local pressures, Univ. of Western Ontario, BLWT Draft Report, March, 1977.

162

37. A.G. Davenport, The prediction of risk under wind loading, Proc. ICOSSAR Conference, Tech. Univ. Munich, Sept. 1977.

38. E. Simiu, Aerodynamic coefficients and risk consistent design, submitted to the ASCE.

39. E. Simiu, Modern developments in wind engineering : part 2, Eng. Struct., Vol. 3, 1981.

40. H.W. Tieleman and S.E. Mullins, The structure of moderately strong winds at a mid-Atlantic coastal site (below 75m), (see reference 3).

41. _____, Proposed revision to wind load provisions of ANSI A58.1—1972, Revised May 1980, prepared by American National Standards Institute Subcommittee.

PARTIAL BIBLIOGRAPHY

Simiu, E. and Scanlan, R.H., "Wind Effects on Structures", John Wiley and Sons.

Lawson, T.V., "Wind Effects on Buildings", Volumes 1 and 2, Applied Science Publishers.

_____ Wind Engineering in the Eighties, Proceedings of the CIRIA Conference, November 12/13, 1980.

_____ Proceedings of Conferences on Wind Effects on Buildings and Structures : 1963, 1967, 1971, 1975 and 1979.

Cermak, J.E., "Aerodynamics of Buildings," Annual Review of Fluid Mechanics, Vol. 8, 1976.

Davenport, A.G., Mackey, S., Melbourne, W.H., "Wind Loading and Wind Effects," Chapter CL—3, Monograph on the Planning and Design of Tall Buildings, Vol. CL, ASCE. 1980.

Hunt, J.C.R. and Fernholz, H., "Wind Tunnel Simulation of the Atmospheric Boundary Layer : A Report on Euromech 50," J. Fluid Mech, (1975) Vol. 70, part 3, 543—559.

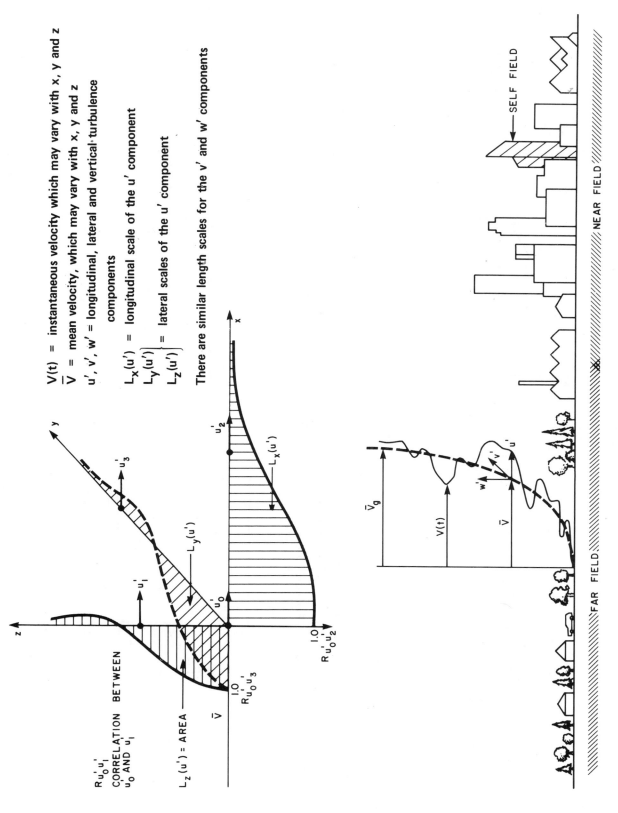

$V(t)$ = instantaneous velocity which may vary with x, y and z

$\bar{V}$ = mean velocity, which may vary with x, y and z

u', v', w' = longitudinal, lateral and vertical-turbulence components

$L_x(u')$ = longitudinal scale of the u' component

$\left. \begin{array}{c} L_y(u') \\ L_z(u') \end{array} \right\}$ = lateral scales of the u' component

There are similar length scales for the v' and w' components

FIG. 1 PARAMETERS OF THE FLOW PROBLEM

164

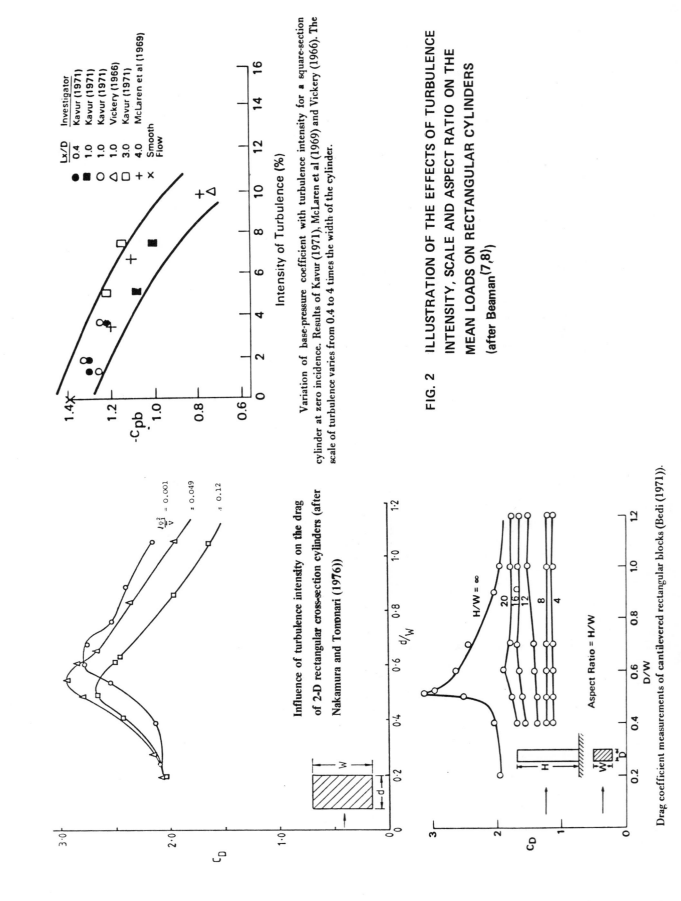

Variation of base-pressure coefficient with turbulence intensity for a square-section cylinder at zero incidence. Results of Kavur (1971), McLaren et al (1969) and Vickery (1966). The scale of turbulence varies from 0.4 to 4 times the width of the cylinder.

	Lx/D	Investigator
●	0.4	Kavur (1971)
■	1.0	Kavur (1971)
○	1.0	Kavur (1971)
△	1.0	Vickery (1966)
□	3.0	Kavur (1971)
+	4.0	McLaren et al (1969)
×	Smooth Flow	

FIG. 2   ILLUSTRATION OF THE EFFECTS OF TURBULENCE INTENSITY, SCALE AND ASPECT RATIO ON THE MEAN LOADS ON RECTANGULAR CYLINDERS (after Beaman[7,8])

$\dfrac{\sqrt{\bar{u}^2}}{V} = 0.001$

$= 0.049$

$= 0.12$

Influence of turbulence intensity on the drag of 2-D rectangular cross-section cylinders (after Nakamura and Tomonari (1976))

Aspect Ratio = H/W

Drag coefficient measurements of cantilevered rectangular blocks (Bedi (1971)).

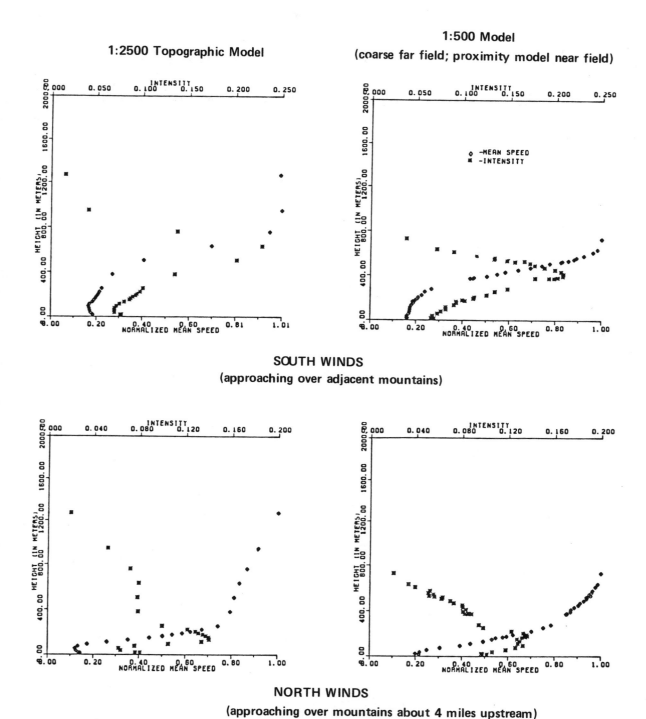

1:2500 Topographic Model

1:500 Model
(coarse far field; proximity model near field)

SOUTH WINDS
(approaching over adjacent mountains)

NORTH WINDS
(approaching over mountains about 4 miles upstream)

FIG. 3   AN ILLUSTRATION OF MATCHING OF VELOCITY PROFILES AT TWO SCALES IN COMPLEX TERRAIN (HONG KONG)

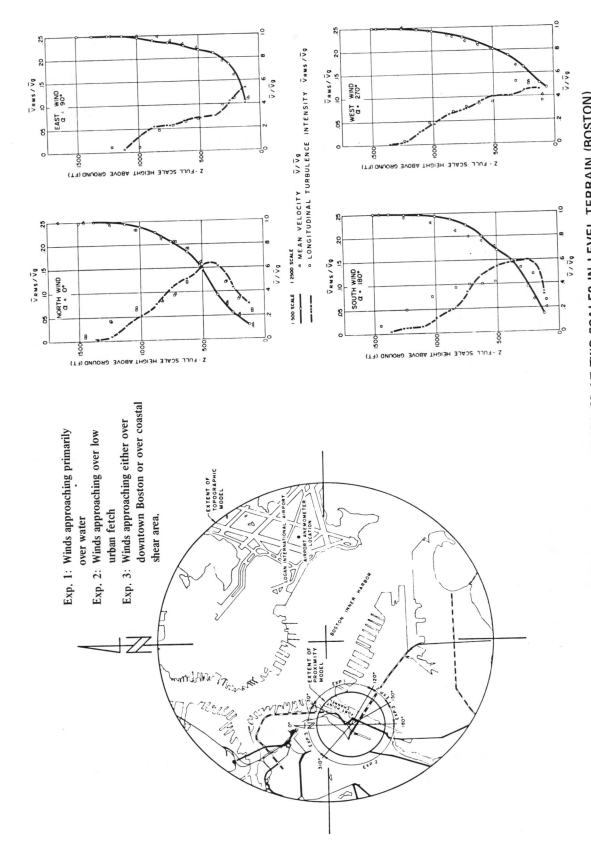

Exp. 1: Winds approaching primarily over water

Exp. 2: Winds approaching over low urban fetch

Exp. 3: Winds approaching either over downtown Boston or over coastal shear area.

FIG. 4 AN ILLUSTRATION OF MATCHING OF VELOCITY PROFILES AT TWO SCALES IN LEVEL TERRAIN (BOSTON)

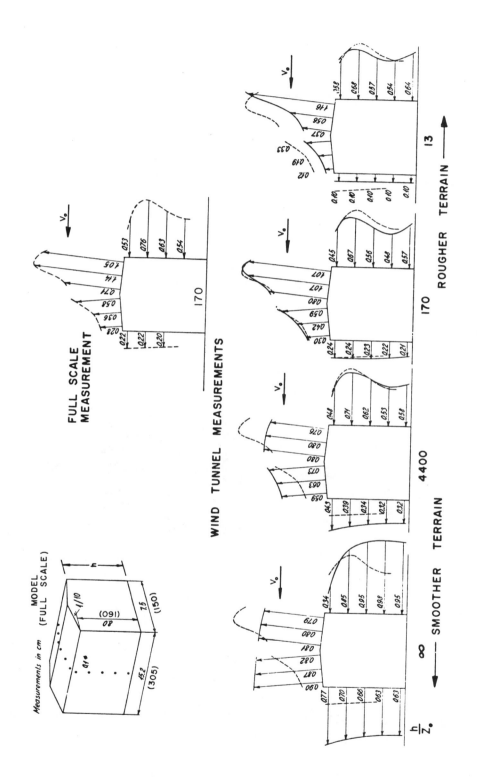

FIG. 5   JENSEN'S[12] COMPARISONS OF FULL-SCALE MEAN PRESSURES WITH MODEL MEASUREMENTS IN DIFFERENT FLOW SIMULATIONS

168

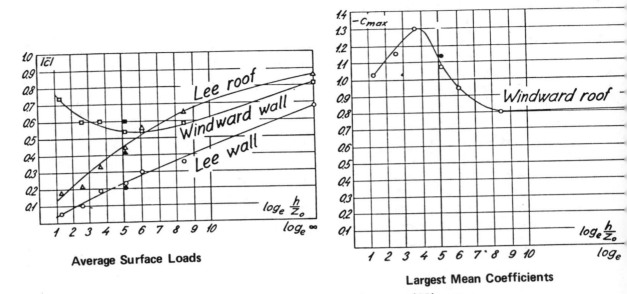

Average Surface Loads

Largest Mean Coefficients

FIG. 6   THE VARIATION OF JENSEN'S DATA[12] WITH h/z_o

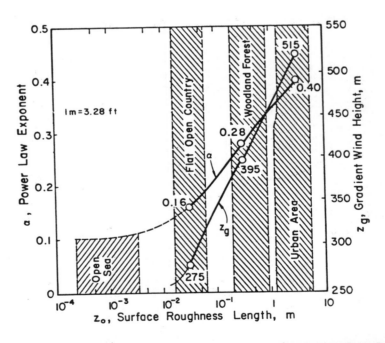

FIG. 7   EMPIRICAL RELATIONSHIPS BETWEEN ATMOSPHERIC BOUNDARY LAYER PARAMETERS (AFTER DAVENPORT[1])

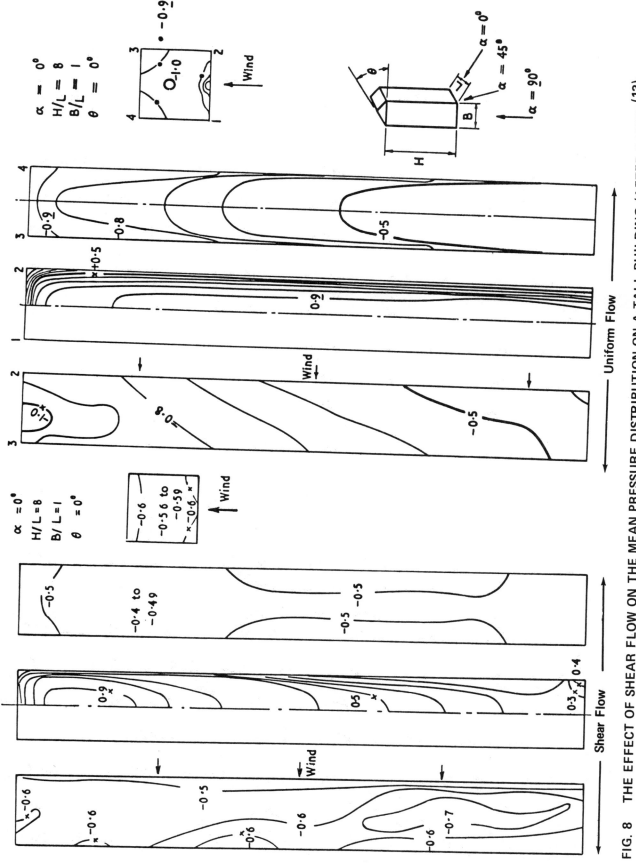

FIG. 8 THE EFFECT OF SHEAR FLOW ON THE MEAN PRESSURE DISTRIBUTION ON A TALL BUILDING (AFTER BAINES[13])

170

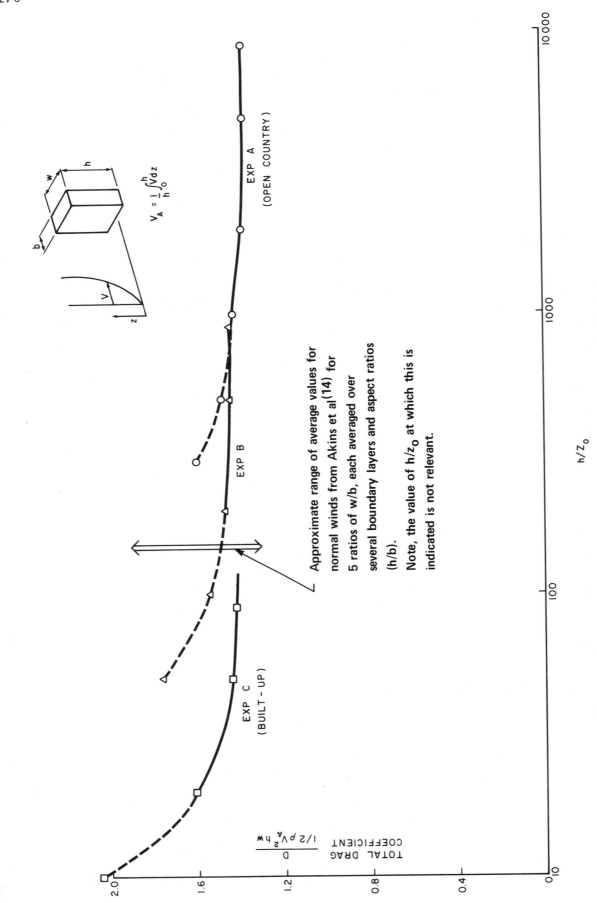

FIG. 9   MEAN DRAG OR TOTAL SHEAR COEFFICIENTS AS DERIVED FROM THE COMMENTARY TO THE CANADIAN NATIONAL
BUILDING CODE VERSUS h/z_0

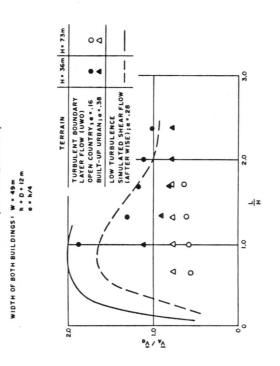

FIG. 11 A COMPARISON OF SURFACE WIND SPEEDS
MEASURED IN A SHEAR FLOW WITH AND
WITHOUT TURBULENCE
(after Isyumov and Davenport[23])

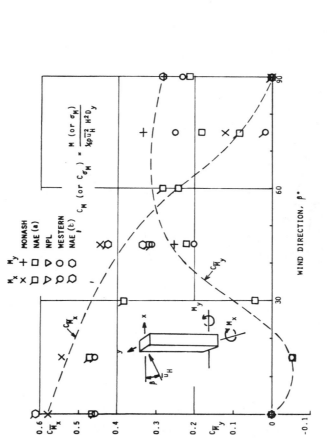

FIG. 10 MEAN MOMENT COEFFICIENTS FOR THE CAARC
STANDARD BUILDING AS MEASURED IN A NUM-
BER OF DIFFERENT WIND TUNNELS
(after Melbourne[18])

172

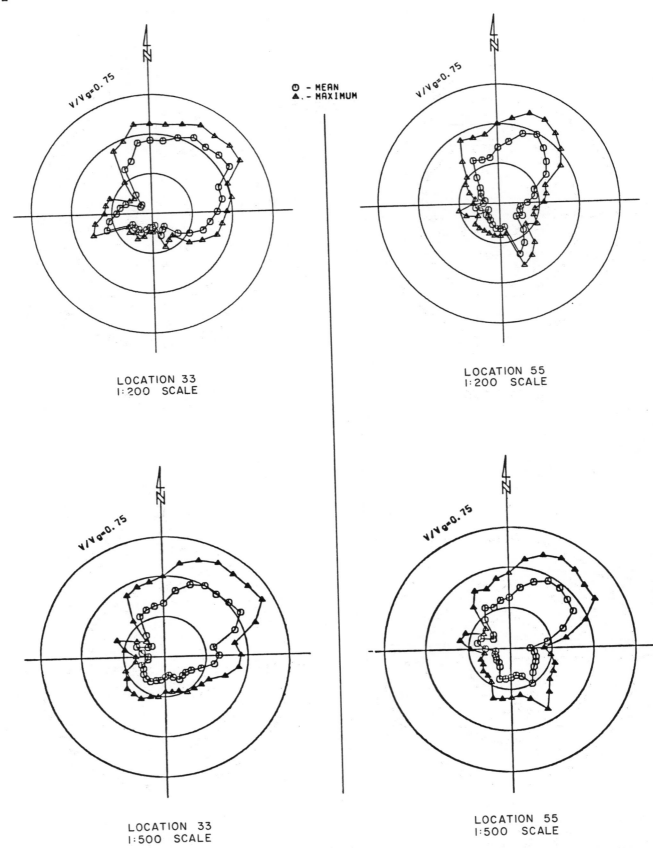

LOCATION 33
1:200 SCALE

LOCATION 55
1:200 SCALE

LOCATION 33
1:500 SCALE

LOCATION 55
1:500 SCALE

FIG. 12a   COMPARISONS OF MEAN AND PEAK SURFACE LEVEL WINDS AS MEASURED IN A DETAILED (1:500) AND AN APPROXIMATE (1:200) SIMULATION

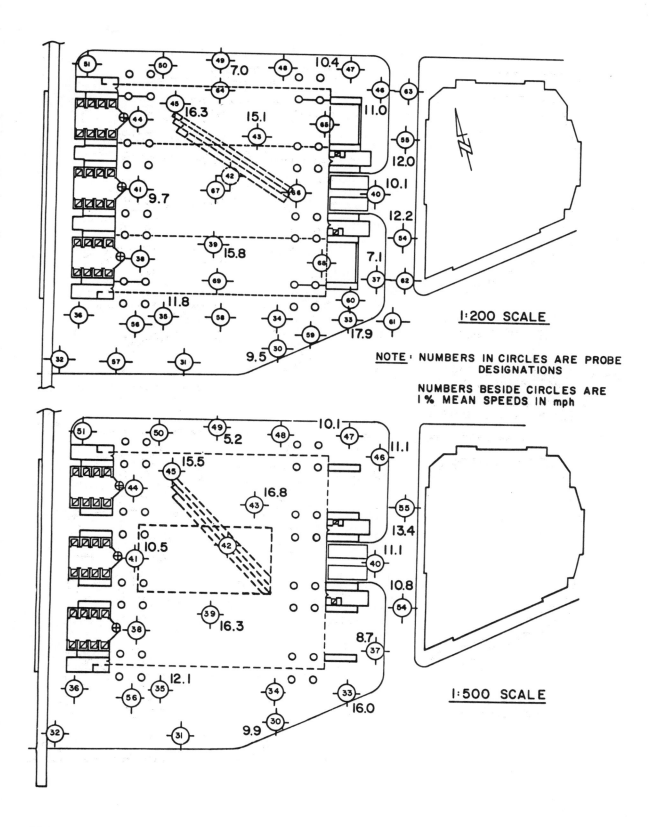

FIG. 12b   COMPARISON OF SURFACE MEAN WIND SPEEDS PREDICTED TO BE EXCEEDED
1% OF THE TIME AS MEASURED IN A DETAILED (1:500) AND AN APPROXIMATE
(1:200) SIMULATION

174

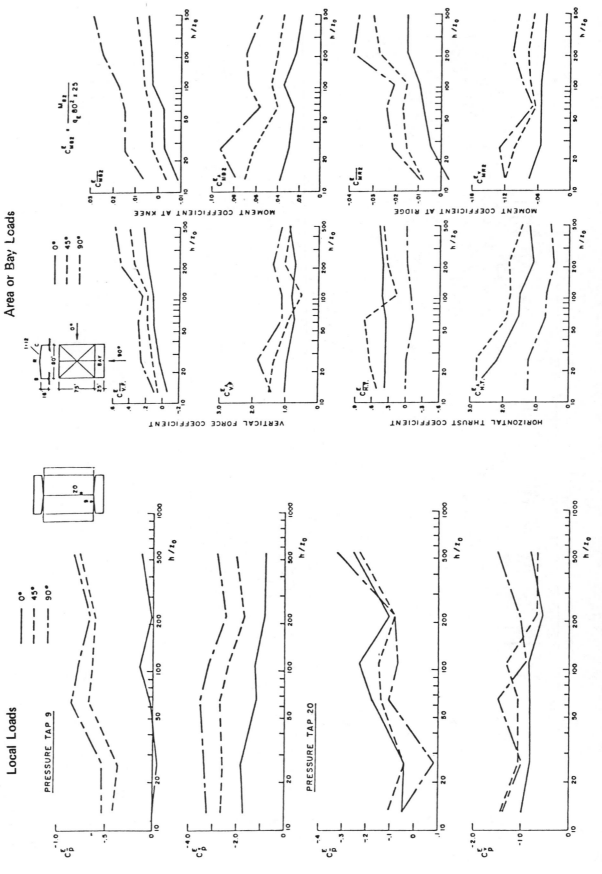

FIG 13   VARIATION OF LOCAL AND AREA LOADS WITH h/z₀, AS MEASURED ON A LOW-RISE BUILDING IN TWO DIFFERENT

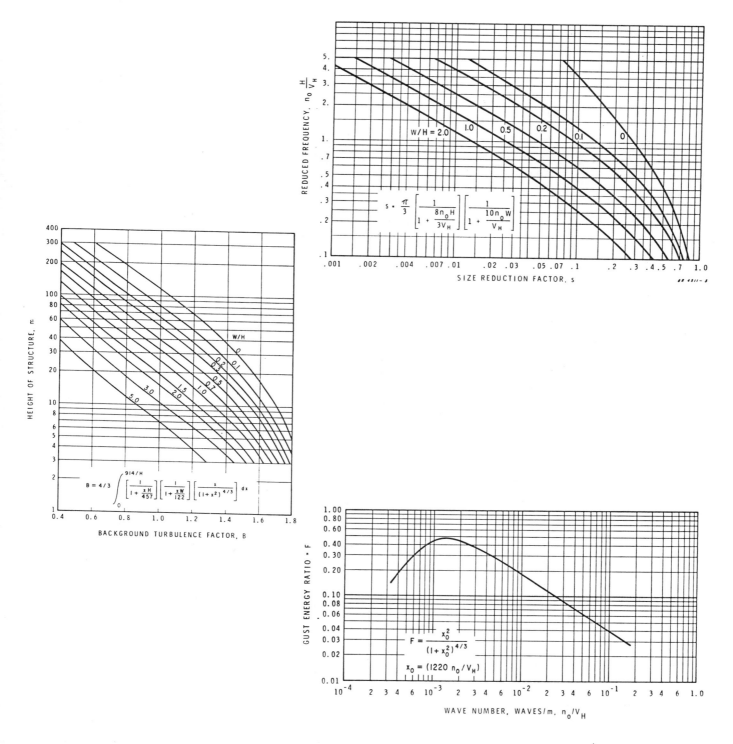

FIG. 14 SOME OF THE FACTORS REQUIRED TO EVALUATE THE ALONG—WIND RESPONSE OF A SLENDER STRUCTURE USING THE DETAILED METHOD OF (29)

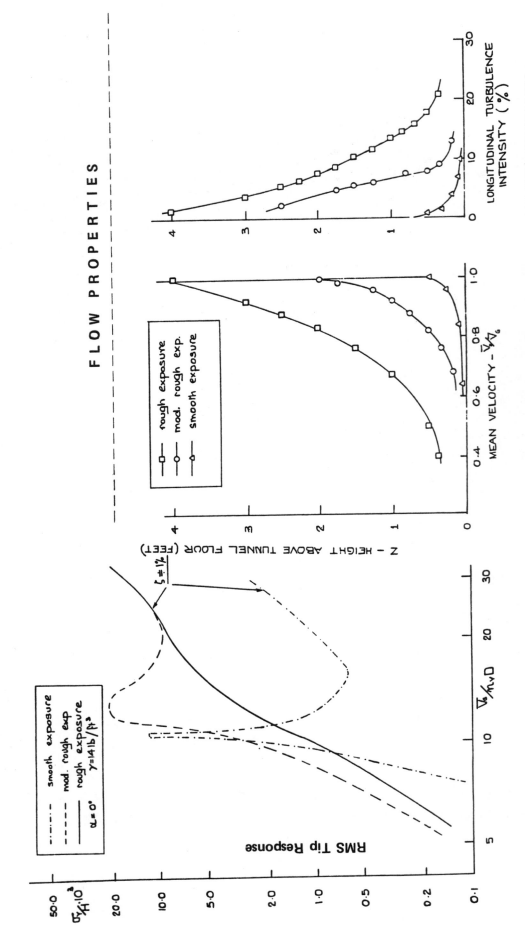

FIG. 15 ACROSS-WIND AEROELASTIC RESPONSE OF A SQUARE BUILDING (H/D= 7) IN TWO TURBULENT BOUNDARY LAYER FLOWS AND IN A SMOOTH UNIFORM FLOW

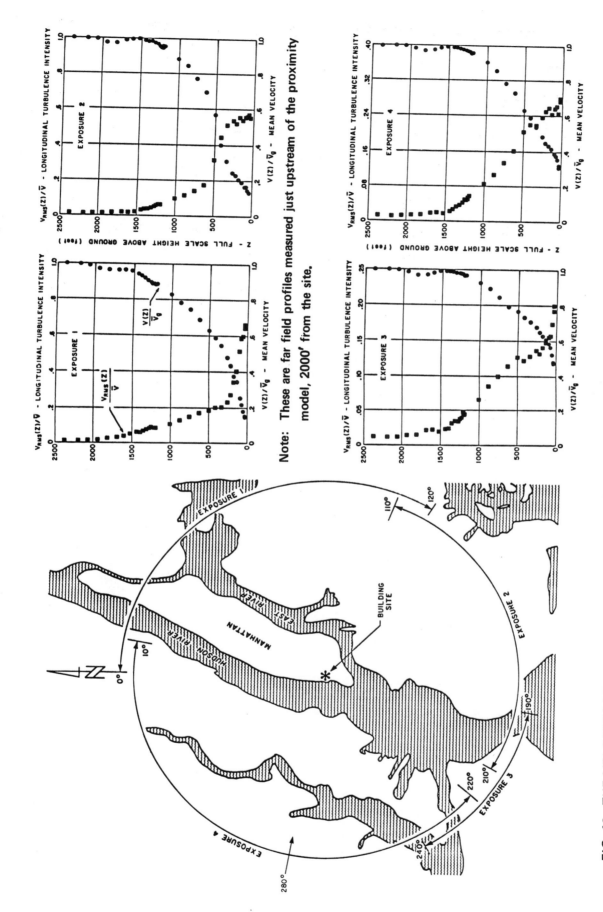

Note: These are far field profiles measured just upstream of the proximity model, 2000' from the site.

FIG. 16 THE VELOCITY AND TURBULENCE INTENSITY PROFILES ASSOCIATED WITH THE FAR FIELD SIMULATIONS FOR A BUILDING SITE IN NEW YORK

178

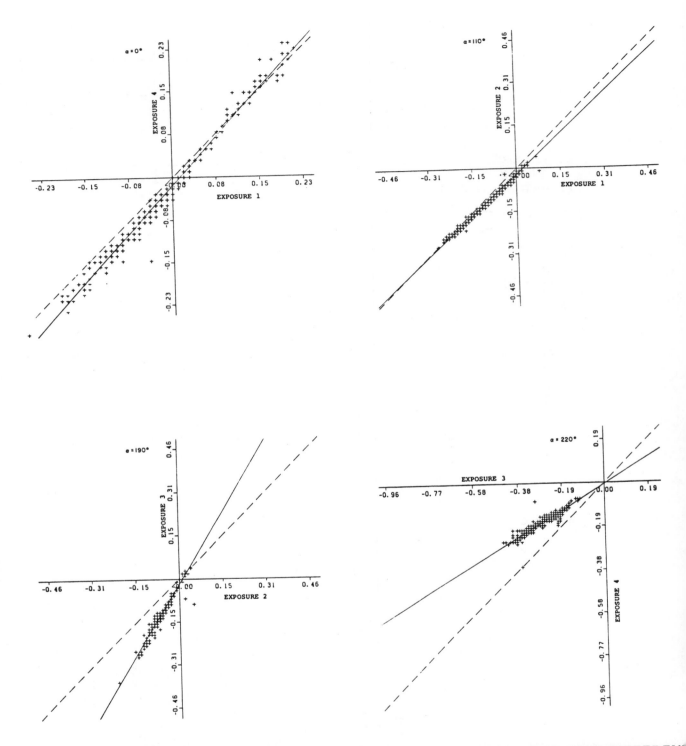

FIG. 17a   COMPARISON OF MEAN PRESSURE COEFFICIENTS MEASURED WITH DIFFEREN
FAR FIELD SIMULATIONS

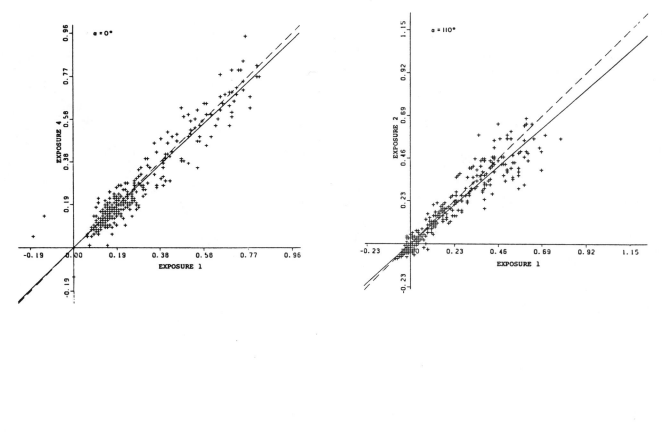

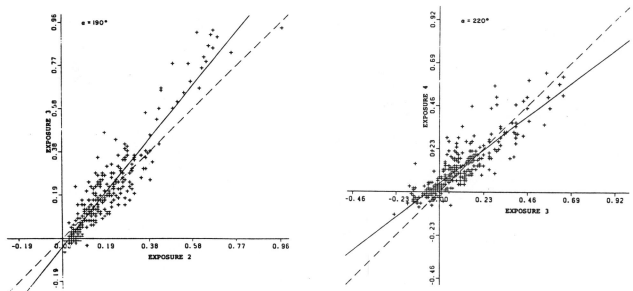

**FIG. 17b COMPARISON OF MAXIMUM PEAK PRESSURE COEFFICIENTS MEASURED WITH
DIFFERENT FAR FIELD SIMULATIONS**

180

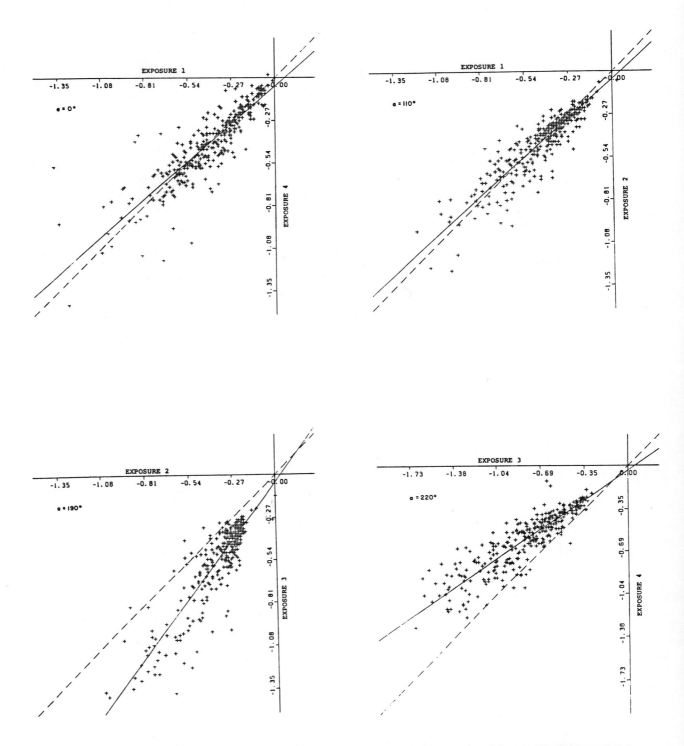

FIG. 17c COMPARISON OF MINIMUM PEAK PRESSURE COEFFICIENTS MEASURED WITH DIFFERENT FAR FIELD SIMULATIONS

Project	Abcissa Exposure Type	Ordinate Exposure Type	Angle	Slope of Max's	Slope of Min's	Slope of Mean's	Comments
420' Tower Jacksonville, Florida	Open Country (5 mi)	Suburban (5 mi)	150°	0.92	0.92	0.84	—proximity of building contains only lowrises resulting in the building being fairly open in all cases
			160°	0.94	0.93	0.85	
			200°	0.93	0.90	0.85	
			210°	0.94	0.92	0.84	—the slightly rougher suburban fetch creates slightly lower loads
620' Tower Boston, Mass.	Open Country (5 mi)	Urban	30°	1.03	0.98	1.01	—blocked by surrounding highrises resulting in similar pressures
	Open Country (5 mi)	Suburban	120°	0.91	0.91	0.90	—both exposures very similar
			150°	0.97	0.95	0.95	
	Suburban	Urban	180°	0.96	0.97	0.94	—urban fetch results in slightly lower loads
			310°	0.96		0.95	—partially blocked by highrises
350' Building New York, N.Y. on Hudson River (without Battery Park)	Urban (Manhattan)	Open Country/ Urban (NJ)	0°	0.97	0.88	1.11	
			10°	0.91	0.93	1.02	
	Urban (Manhattan)	Open Country/ Urban (Brook.)	120°	1.04	1.05	1.09	
			110°	0.90	0.91	0.94	
	Open Country/ Urban (Brooklyn)	Open Country	190°	1.24	1.35	1.65	—dramatic difference in exposures —blocked by highrises as opposed to water
			200°	1.04	1.15	1.65	
			210°	0.94	1.18	1.79	
	Open Country	Open Country/ Urban (NJ)	220°	0.77	0.69	0.61	—roughness of partial urban fetch results in lower loads
			230°	0.77	0.77	0.68	
			240°	0.83	0.84	0.77	
As Above (with Battery Park)	Open Country/ Urban (Brooklyn)	Open Country	190°	1.28	1.18	1.53	—in all cases, BP is not directly upstream
			200°	1.22	1.29	1.78	
			210°	1.00	1.25	1.64	
	Open Country	Open Country/ Urban (NJ)	220°	0.74	0.72	0.64	—urban fetch results in lower loads
			230°	0.77	0.78	0.73	
			240°	0.82	0.81	0.78	
680' Tower Chicago, Illinois	Open Country	Urban/ Open Country	110°	0.92	0.90	0.88	—urban fetch results in lower loads
			120°	0.89	0.91	0.87	
	Urban/ Open Country	Urban	160°	0.83	0.91	0.67	—urban fetch results in lower loads
			170°	0.98	1.11	0.95	
	Urban	Suburban	220°	1.22	0.92	1.65	—suburban fetch results in higher loads
			230°	1.23	0.87	1.57	
	Open Country	Suburban	350°	0.93	0.89	0.91	—suburban fetch results in lower loads
460' Tower New York, N.Y.	Urban (Manhattan) (1 mile)/ Open Country (1 mile)/ Urban (Queens) (3 miles)	Wooded (Central Park)/ Urban (Manhattan)	20°	0.77	0.78	0.26	For most directions, the building is shielded by taller buildings. Loads were small, particularly the means, leading to fits with low correlation coefficients.
			30°	0.91	0.98	0.83	
		Dense Urban (5 mi) (Manhattan)	170°	0.95	1.04	0.86	
			180°	0.99	0.95	0.84	
	Urban (NJ & Manhattan)	Dense Urban (5 mi) (Manhattan)	220°	1.09	1.12	1.18	
			230°	1.06	1.05	1.25	
	Urban (NJ & Manhattan)	Wooded (Central Park)/ Urban (Manhattan)	340°	1.07	0.93	1.29	
			350°	1.05	1.03	1.25	

FIG. 18 COMPARISONS BETWEEN MEAN AND PEAK LOCAL PRESSURES MEASURED WITH DIFFERENT FAR FIELD SIMULATIONS

182

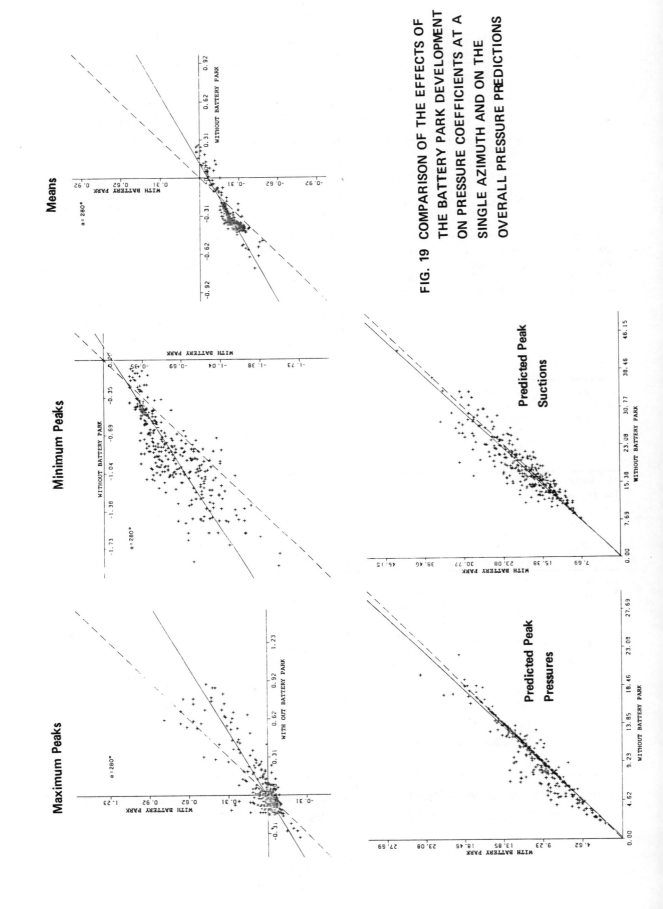

Maximum Peaks

Minimum Peaks

Means

Predicted Peak Pressures

Predicted Peak Suctions

FIG. 19 COMPARISON OF THE EFFECTS OF THE BATTERY PARK DEVELOPMENT ON PRESSURE COEFFICIENTS AT A SINGLE AZIMUTH AND ON THE OVERALL PRESSURE PREDICTIONS

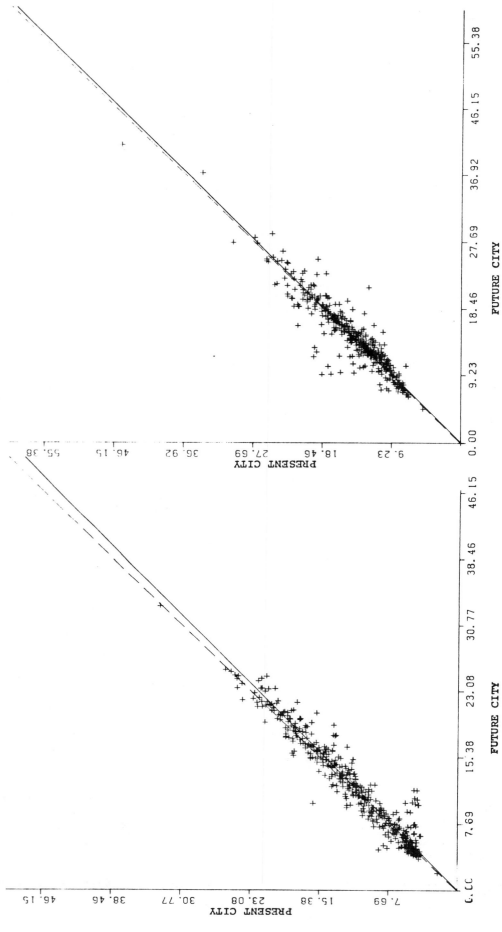

183

FIG. 20   COMPARISON OF THE EFFECTS OF TWO CITY ENVIRONMENTS ON THE OVERALL PRESSURE PREDICTIONS FOR A BUILDING IN HONG KONG

184

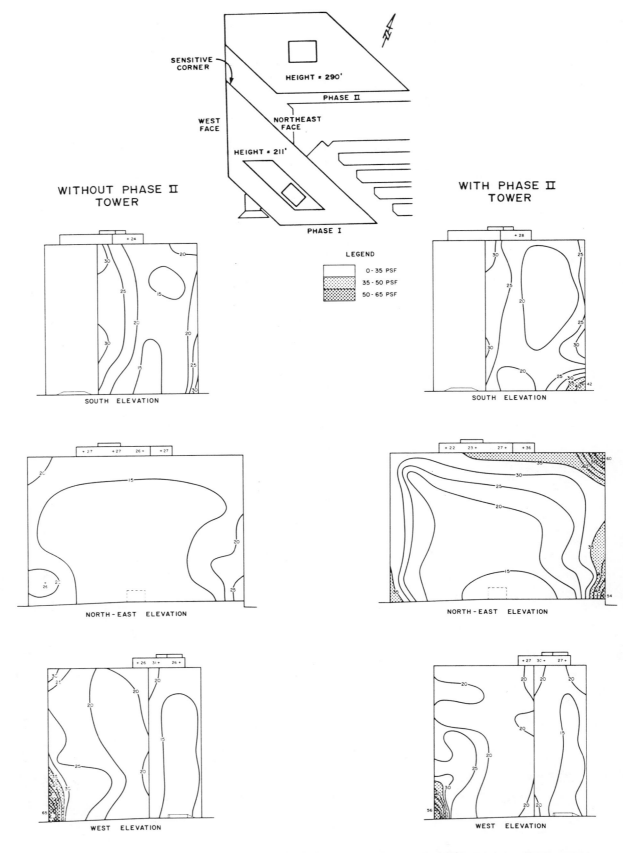

FIG. 21 AN EXAMPLE OF A SEVERE NEAR FIELD EFFECT ON PREDICTED LOCAL SUCTIONS (REFERENCE 31)

185

Predicted Resultant Accelerations at the Worst Corner

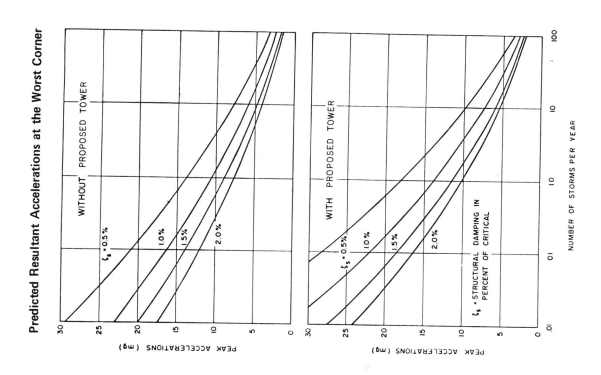

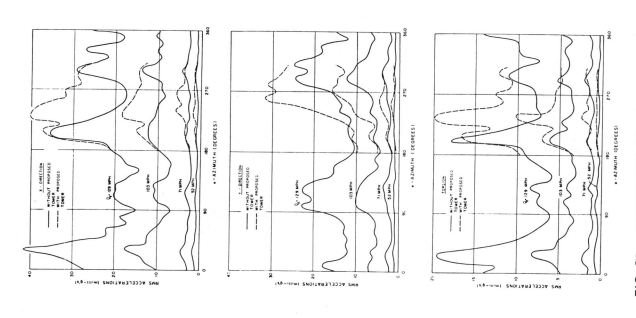

FIG. 22   COMPARISON OF DETAILED AEROELASTIC RESPONSES INDUCED BY A PROPOSED NEARBY TOWER WITH THE PREDICTED OVERALL EFFECTS AS DEDUCED BY SYNTHESIS OF AERODYNAMICS AND WIND CLIMATE

MODELING FLOW OVER COMPLEX TERRAIN AND IMPLICATIONS FOR DETERMINING

THE EXTENT OF ADJACENT TERRAIN TO BE MODELED

Rex Britter
Department of Engineering
University of Cambridge
England

SUMMARY

The study of atmospheric flows in complex terrain is currently very active. Physical modeling of the flow is a common approach for diffusion and wind-loading problems. Recently generic, in contrast to site-specific, studies in the laboratory and field have led to a better understanding of surface roughness and terrain effects.

The paper estimates regions of influence and perturbation magnitudes caused by roughness changes and terrain features. Their neglect in a physical model is justified if either the region of influence does not intersect the area under study or the magnitude has decayed below some acceptable level.

1. INTRODUCTION

A recent report by Snyder (1) has considered, in some detail, guidelines for fluid modeling with a view towards atmospheric diffusion modeling but relevant in a broader context. To ensure dynamical similarity between model and full-scale the fundamental equations of motion require that:-

the Reynolds number $\left(\rho \, \dfrac{UL}{\mu} \right)$

the Rossby number $\left(\dfrac{U}{fL} \right)$

the Richardson number $(g \, \dfrac{\Delta\rho}{\rho} \, L)/U^2$

be equal in model and full scale, in addition to correct geometrical scaling. It is indicative of the "art" of physical modeling that, in general, none, or possibly, only one of the parameters is correctly modeled.

Rossby number modeling is unlikely to be obtainable in laboratory facilities and one is left to argue about what is the largest region modelable or the minimum Rossby number allowable for the neglect of the effects of rotation. It is suggested that a Rossby number of order 0.1 would allow neglect of rotational effects whereas the neglect of a Rossby number of order unity would require justification.

The Richardson number is obviously not relevant in a neutrally stratified atmosphere. With a stable density stratification the densimetric Froude number (inverse square root of the Richardson number) may be usefully defined as

$$U/(g \, \frac{\Delta\rho}{\rho} \, L)^{\frac{1}{2}}$$

or as U/NL, where

$$N^2 = - \frac{g}{\rho} \frac{\partial\rho}{\partial z} ,$$

when the density stratification is continuous. It will be shown later that, for flow over topography, the effects of density stratification are dominant when $U/NL \simeq 0(1)$ where L is horizontal length scale of topography. For many complex terrain problems it is not uncommon for U/NL to be of order unity; however it is uncommon to have a   strong correlation between high wind speeds (externally applied) and strong stable or unstable density stratifications.

Thus for a study of wind-loading the densimetric Froude number is unlikely to be an important parameter. Quantifying this we require $U/NL > 10$ say or, possibly, the hill height to be less than the magnitude of the local Monin-Obukhov length.

We shall consider later situations in which moderate ambient winds and stable density stratification combined with topographic features give rise to locally generated high velocity flows.

The correct modeling of Reynolds number is obviously not possible, however many properties of turbulent flows (and we are concerned with turbulent flows in the full-scale) are independent of Reynolds number provided the Reynolds number is large. The smallest scales of turbulence will not be reproduced in the model; the largest scales reduce with the model scale whereas the smallest scales only reduce as the $\frac{1}{4}$-power of the model scale.

At 10 m above the surface in the neutrally stratified boundary layer with $u_* \simeq 0.5$ m/s, the smallest scale of turbulence is $\simeq \frac{1}{2}$ mm. In a 1:1000 scale model $\eta \simeq 0.1$ mm corresponding to 10 cm in full scale. In essence turbulence scales below 10 cm are lost. It would be surprising if these very small scales were of any great consequence in wind-loading problems.

Not satisfying the Reynolds number in the model essentially implies the neglect of viscosity. However as we are always dealing with "rough wall" turbulent boundary layers the product of the friction velocity u_* and the roughness length z_0, i.e. $(u_* z_0)$, adopts the role of ν. That is, the relevant Reynolds number to model is

$$\frac{UL}{u_* z_0} \quad \text{rather than} \quad \frac{UL}{\nu} .$$

Thus the emphasis changes from trying to model the Reynolds number to modeling u_*/U and z_0/L. In many physical modeling facilities the boundary layer development is accelerated and adjusted until u_*/U, z_0/L together with various other turbulence statistics and length scale ratios are correctly modeled.

The roughness length z_0 is a gross characterisation of the drag force resulting from separation about individual surface structures. Obviously z_0 will depend on both the size and spacing of the surface structures and in typical simulations is about 0.1 of the roughness element height k for sharp edges roughness or about 0.03 k for randomly distributed smooth roughness elements. The former ratio is not dissimilar to full-scale values, e.g. Simiu and Scanlan (2).

To ensure that the flow is fully aerodynamically rough in the model the size of the roughness element must be large compared with any viscous sublayer, that is, $k > 50 \, \nu/u_*$ or $u_* k/\nu > 50$. Mulhearn (3) and Raupach et. al. (4) point out the considerable spatial inhomogeneity in the flow field due to the roughness elements and their spacing; an inhomogeneity in the mean and turbulent fields within (k + 1.5 S) for roughness elements of height k and spacing S.

When large scale reductions are used it may not be easy to satisfy the requirement that z_o/L is constant and

$$\frac{u_* z_o}{\nu} > 5 \, ,$$

say, unless quite large roughness elements are used spread rather sparsely. Such an arrangement may be in conflict with obtaining homogeneous flow within an area of interest close to the model surface.

2. INFLUENCE OF SURFACE ROUGHNESS CHANGES

Although still a matter of argument it appears that the influence of a change in surface roughness along a normal to the wind velocity is usefully described in terms of inner and outer regions characterized by the surface roughness downstream and upstream of the change respectively. For neutral density stratification a perturbation analysis, following Townsend (5), leads to a result for the interface between the inner and outer region ℓ_R as

$$\frac{d\ell_R}{dx} \ln \left(\frac{\ell_R}{z_{o2}} \right) = 2\kappa^2 \, ,$$

which integrates to

$$\frac{\ell_R}{x} \ln \left(\frac{\ell_R}{z_{o2}} \right) = 2\kappa^2 \quad \text{for} \quad \ell_R/z_{o2} \gg 1$$

Thus the influence of a small roughness change "diffuses" out into the shear flow in a manner similar to the plume resulting from a passive source. For $10^2 < x/z_{o2} < 10^4$, ℓ_R/x can be roughly approximated by a constant 0.08 or, more accurately, by

$$\frac{\ell_R}{z_{o2}} = 0.32 \left(\frac{x}{z_{o2}} \right)^{0.8} \, .$$

If there is a large increase in surface roughness then the perturbation analysis is less valid and the flow is better interpreted as a new boundary layer growing in a uniform approach flow. A large decrease in the surface roughness is different as the growth of the inner region will be dominated by turbulence advected in from the outer region i.e. the upstream roughness length will be dominant in determining the growth of the internal layer.

Jackson (6) has reviewed extensive measurements in the field and laboratory concerning the influence of a change of surface roughness over the range $5 \times 10 < x/z_o < 5 \times 10^4$ (where z_o corresponds to the rougher surface). The data are in good agreement (within $\pm 25\%$ for $x/z_o > 200$) of the theoretical expression given the somewhat imprecise measurement of ℓ_R. That this result is valid for large increases in surface roughness (up to $z_{o2}/z_{o1} \simeq 10^2$) is initially surprising but less so when it is realized that the rough wall boundary layer grows characteristically with $\delta \propto x^{0.8}$. A similar correlation appears valid down to $z_{o2}/z_{o1} \simeq 10^{-2}$. Jackson also noted that the region of perturbation is the same for the turbulence quantities as for the mean velocities.

Assuming a self-similar development, the perturbation velocity, as a result of the change of roughness, is

$$\frac{u_{*1}}{\kappa} \frac{\ell n \; z_{o2}/z_{o1}}{\ell n \; \ell_R/z_{o2}} \; \ell n \; z/\ell_R \; .$$

The surface friction velocity varies as

$$\frac{u_*(x)}{u_{*1}} \simeq 1 - \frac{\ell n \; (z_{o2}/z_{o1})}{\ell n \; (\ell_R/z_{o2})} \quad \text{(Townsend, 1965)}$$

or

$$\frac{u_*(x)}{u_{*2}} \simeq 1 + \frac{\ell n \; (z_{o2}/z_{o1})}{\ell n \; (\ell_R/z_{o2})} \; .$$

Thus both the perturbation velocity and the surface shear stress approach their asymptotic values very slowly (as the inverse of $\ell n \; (\ell_R/z_{o2})$).

The turbulent velocities within the inner region scale locally on the surface shear stress and hence on z_{o2}. Hunt (7) indicates several field and laboratory results indicating the correctness of this approach. Flow over an isolated patch of roughness is far less comprehensively studied. Edling and Cermak (8) in a laboratory study confirmed that an axial discontinuity of roughness was able to drive a secondary flow of the second kind of typical magnitude $(u_{*2} - u_{*1})$ and that these motions extend out to about $2\ell_R$. The influence of the change in surface roughness will "diffuse" laterally. The extent of the lateral "diffusion" may be estimated by appeal to the previously mentioned analogy with passive plume behaviour (in this case from an area source).

3. FLOW OVER COMPLEX TERRAIN

3.1 Non-Separating, Neutrally Stratified Flow

The perturbation analysis of Jackson and Hunt (9) for two-dimensional hills of low slopes provides an illuminating framework in addition to the useful quantitative results confirmed by Britter et. al. (10) and others. The perturbed boundary layer consists of an "inner region" where the surface turbulence is in local equilibrium and an "outer" region in which the flow behaves as an inviscid shear flow. The "inner" region is driven by the flow in the "outer" region. The vertical scale, ℓ, of this inner region is given as

$$\frac{\ell}{L} \, \ell n \left(\frac{\ell}{z_0}\right) = 2\kappa^2 \, ,$$

where L is the length scale of the hill (the half-length to the half-height) and is typically about $\frac{1}{4}$ of the overall length of the hill L_0.

The outer region of influence extends out from the hill a distance L_0 in all directions. The flow velocity perturbation varies such as to be a maximum at a distance ℓ above the hill crest and is estimated as

$$\frac{\Delta U_{max}}{U_0(L)} \simeq \sigma \, \frac{h}{L} \, \frac{U_0(L)}{U_0(\ell)}$$

where the subscript o refers to upstream conditions, and where σ is a constant determined by the hill shape and not on the hill slope. The maximum value of σ is typically unity for crested hills but is about half this   for ramps that result in a maintained elevation change.

The perturbation shear stress is zero a distance ℓ above the hill surface but increases to

$$\frac{\Delta\tau}{\rho u_*^2} = \frac{\kappa b}{\sigma} \, \frac{\Delta U_{max}}{u_*}$$

where $b \simeq 0.3$. That is, the surface stress is very sensitive to topography. Note also that by virtue of the assumption of equilibrium in the inner region the turbulence levels will scale on the perturbed wall stresses.

Jensen and Petersen (27) note that close to the ground

$$\frac{\Delta U}{U_0(L)} = \frac{h}{L} \, \frac{U_0(L)}{U_0(\ell)} \, \frac{U_0(z)}{U_0(\ell)}$$

Thus

$$U = U_0(z) \left\{ 1 + \frac{h}{L} \left[\frac{U_0(L)}{U_0(\ell)}\right]^2 \right\}$$

which may be interpreted as an increase in the local friction velocity u_* by

$$u_* \, \frac{h}{L} \left[\frac{U_0(L)}{U_0(\ell)}\right]^2 \, .$$

It is commonly the case that $\ell < h$ and under some conditions (for steep slopes and smooth surfaces) $\ell \ll h$. Under these conditions a simple potential flow solution is appropriate for most of the flow region of interest. For example with $h/L \simeq 0.4$ (typical slope 0.2) and $L/z_0 \simeq 10^5$ the ratio of inner region to hill height is about 0.1. The neglect of the non-linear terms in the perturbation analysis lead to an overestimate of ΔU_{max} and $\Delta\tau_{max}$ for steeper slopes.

Within the inner region the perturbation is best considered relative to the velocity at the same height above the surface well upstream. Thus we consider a fractional speed up ratio ΔS which, for a hill $h/(1 + (x/L)^2)$, is such that $\Delta S(h/L)^{-1}$ varies little from 2 for $0 < z/\ell < 2$ and $10^3 < \ell/z_0 < 10^5$. For ramp shapes $\Delta S \simeq 1.0\ h/L$. In the outer region the velocity is compared with the flow well upstream at the same absolute height. For $h/L < 0.1$ Jackson and Hunt obtained

$$\frac{U(z)}{U_0(h+z)} \simeq 1.0$$

for $10^3 < \ell/z_0 < 10^5$.

It is interesting that the scale ℓ also separates an inner region from an outer region in which the larger turbulence scales decay slowly and their distortion may be calculated using rapid distortion theory (Britter et. al., 1981). Thus, within the inner region, $z < \ell$,

$$\frac{\Delta u'^2}{u'^2} \simeq \frac{\Delta w'^2}{w'^2} \simeq \frac{\Delta \tau}{\rho u_*^2} \simeq 4\,\frac{h}{L}$$

for a bell-shaped hill with low slope, and outside this region

$$-\frac{\Delta u'^2}{u'^2} \simeq \frac{\Delta w'^2}{w'^2} \simeq \frac{\Delta U}{U_0}\ .$$

Panofsky et. al. (12) have recently noted that although the vertical turbulence components within ℓ (or ℓ_R) are in equilibrium the small wave-number longitudinal components may not be.

Qualitatively the arguments and approach presented here may be applied as well to three- as to two-dimensional topography, see Mason and Sykes (13).

In the absence of flow separation the velocity fields return to their upstream values on a scale given by the hill length; for example Khurshudyan et. al. (11) with a smooth, two-dimensional hill with $h/L \simeq \frac{1}{5}$ (their $n = 8$) found no separation and that the flow had relaxed to the upstream conditions by $x \simeq L_0$ ($\simeq 4L$) downstream of the crest.

The influence of the topography in the far field may be adequately estimated by assuming potential flow around suitable simple shapes (e.g. ellipsoids), that is the far field perturbations decay on a scale of L_0, see Milne-Thompson (14). The magnitude of the perturbations depends on h/L.

Snyder and Britter in an unpublished report studied the effect of spanwise aspect ratio on the flow over triangular shaped hills with slopes such that $h/L = 1$. The spanwise aspect ratio varied from ∞ (a ridge) to 2 (a cone). The flow on the axis of symmetry both at crest and upwind base was independent of spanwise aspect ratio (L_s/h) when $L_s/h > 4$. Further data is required but it appears that the influence of the ridge end extends along the ridge less than about $4L$.

3.2 Separating, Neutrally Stratified Flow

Separation of the flow results in severe modification to the mean and turbulent flow fields; modifications in excess of the unperturbed flow. If separation is expected to occur in the full-scale from upwind topography or buildings it is important that this be correctly modeled. As, or more, important is to ensure that flow separation off the terrain of immediate interest is also reproduced in the model. Separation can reduce a bell-shaped hill to, effectively, an escarpment and ΔS may be reduced by half at the crest. It appears that as hills are made steeper separation acts to limit ΔS to about unity. In the absence of significant density stratification the occurrence of separation depends on the hill slope, the state of the boundary layer (and thus the surface roughness) and the existence of any salient edges.

The neglect of the Reynolds number in determining where separation will occur and if so the resulting flow pattern arises from extensive laboratory work on sharp-edged bluff bodies and is apparently legitimate for such flows with $Re > 0(10^4)$. However for topographic features the slopes are far smaller and salient edges less common. Khurshudyan et. al. (11) found no flow separation off a rounded hill with $h/L = \frac{1}{5}$ (average slope $\frac{1}{8}$) but flow separation for a steeper hill with $h/L = \frac{1}{1.7}$ (average slope $\frac{1}{3}$) for $z_0/h \simeq 10^{-3}$. Bouwmeester et. al. (28) found that non-separated flow could not be sustained with $h/L > 0.25$.

These results and others suggest that the flow is unlikely to separate (or that the flow separation region will be of limited extent) when $h/L \lesssim 0.2$ (average slope $\simeq 0.1$) but will lead to a large separated region (characteristic of bluff bodies) when $h/L \gtrsim 0.5$. For larger slopes upstream separation is likely for $h/L \gtrsim 1.0$.

For slopes in the range $0.2 < h/L < 0.5$ (unfortunately a particularly interesting range) it is far less clear as to whether the flow will separate. Using a rounded hill of form $h(1 + (x/L)^2)^{-1}$ with $h/L = \frac{1}{2.5}$, Britter et. al. (10) with $z_0/h \simeq 2 \times 10^{-2}$, and Courtney (15) with $z_0/h \simeq 10^{-3}$ observed intermittent flow separation. Khurshudyan et. al. (11) made a similar observation for a hill with $h/L \simeq \frac{1}{3}$ (average slope $\frac{1}{5}$). Although there was no flow reversal in the mean velocity profiles, (measured with a pulsed-wire anemometer) flow vizualisation suggested strong separation, that is the flow is intermittently separating presumably because temporally-local conditions are favourable for separation. Such flows will be very sensitive to the boundary layer structure in the adverse pressure gradient in the lee of the hill. Both Britter et. al. (10) and Courtney (15) found that flow separation was eliminated by the removal of roughness over the hill. This somewhat surprising observation is a result of the acceleration of the flow near the surface after reduction of the surface roughness. Also Castro and Snyder (16) found that the recirculation region behind a conical hill ($h/L = 1$) in a rough-wall turbulent boundary was very limited in extent for a smooth hill compared with a roughened hill. This extreme sensitivity of the flow to the roughness on the topography appears not to have been noted until recently. It appears essential that the roughness is maintained over the topography when $0.2 < h/L < 0.5$.

A further cautionary note was made by Wilson et. al. (17). The ratio of separation bubble length to hill height for a rather steep ($h/L = 0.7$) triangular ridge decreased from about 6 to zero as the Reynolds number increased. The rough wall boundary layer was not maintained over the topography. Experiments at Cambridge by Allwright and Beazley (unpublished) repeated Wilson et. al.'s experiment but with a smooth-wall boundary layer over a similar shaped hill. As the Reynolds number increased the reattachment length decreased from 8.5 hill heights to about 5 hill heights at the maximum Reynolds number obtainable. Also a small square rod ($h/8 \times h/8$)

placed 6 h upstream of the crest reduced the reattachment length to 3 h. The results with the smooth-wall boundary layer are directly relevant to large scale reduction of flows with small z_o as these are sometimes modeled with a smooth surface.

However the experiments are subject to criticism (e.g. it may not be surprising that a smooth-wall boundary layer is sensitive to the Reynolds number) and other authors have argued for the separation position and extent being independent of Reynolds number. In the opinion of this author the question is still open.

Separation leads to a region of reversed flow though, in the case of three-dimensional flow, not necessarily closed streamlines. We are interested in how rapidly the perturbations to the mean velocity and turbulent velocity fields decay in the wake. There will be no need to model upstream buildings or topography if their influence has decayed below some acceptable level. The recirculating region extends to 2-3 obstacle heights downstream if the obstacle is three-dimensional and 5-10 heights if the obstacle is two-dimensional. It is, therefore, to be expected that the flow will reattach on the lee slopes of typical hill shapes and slopes. Castro and Snyder (16) using a steep (slope 1:2) shape show that the length to reattachment decreases from 8 h to 6 h, 4 h, 2 h, as the spanwise width to length ratio varies from ∞ to 6, 4, 2 respectively.

Snyder (1) used data from Hunt (18) on the turbulent wakes from cubes and cylinders to show that the maximum velocity deficit decays as

$$\frac{\Delta U_{max}}{U_o(h)} \simeq \frac{A}{(x/h)^{3/2}}$$

where A is a function of building shape, orientation, boundary layer thickness and roughness length. A is typically 2.5 (to within a factor of 2).

Thus the maximum deficit in the mean velocity is reduced to about 3 % at $x/h \simeq 20$. A possible exception may occur if the terrain or building induces strong axial vorticity, particularly if the obstacle generates vortices of different strength. This may lead to velocity changes up to 5 % at $x/h \simeq 80$. The velocity deficit reduces to half its maximum value roughly as $y \simeq 0.5 \, x^{\frac{1}{2}} h^{\frac{1}{2}}$, Hosker (19). For essentially two-dimensional topography normal to the wind Counihan et. al. (20) argued that

$$\frac{\Delta U_{max}}{U_o(h)} \simeq B(x/h)^{-1} \, ,$$

with $B \simeq 4$, for bluff shapes. A smaller value of B is appropriate when the separation is intermittent and Courtney (15) finds $B \simeq 1.6$. Courtney also finds that

$$\frac{\Delta(u'^2)_{max}}{u_*^2} \simeq 25(x/h)^{-1} \, ,$$

$$\frac{\Delta(w'^2)_{max}}{u_*^2} \simeq \frac{(\Delta\tau)_{max}}{\rho \, u_*^2} \simeq 10(x/h)^{-1} \, .$$

Larger coefficients are observed for the wakes of bluff bodies, see
Counihan et. al. (20). Thus the turbulence as a result of flow separation
is quite persistent.

4. NON-NEUTRAL DENSITY STRATIFICATION

In the context of wind-loading problems, it is unlikely that very
strong winds will be encountered together with highly convective or very
stable conditions (though see Hunt (21) for an analysis of such flows).
Nevertheless the previous arguments will, in general, be valid for weak
density stratification effects.

An important exception is the generation of strong surface winds by
the interaction of a stable density stratification with topography.
Snyder, Britter and Hunt (22) observed strong lee winds over three-
dimensional topography when $U/NL \sim 1$, similar to results previously obtained
for two-dimensional ridges. In a study of elevated inversions near three-
dimensional topography Lamb and Britter (unpublished) measured strong lee
winds (about twice the upstream wind) when the densimetric Froude number
based on the inversion height was between 0.4 and 0.7 and the inversion
upstream was within $h/2$ of the crest height, h. Snyder (1) stresses
the importance of upstream blocking in model studies if a three-
dimensional terrain feature is reduced to a two-dimensional one by the
tunnel walls. Large wind speeds are also possible if upstream blocking
by a ridge is relieved by a small gap, Baines (23).

The region of influence of a terrain feature under very stable
conditions may be estimated assuming potential flow in horizontal planes
about locally cylindrical geometry. When $U/NL \sim 0(1)$ and lee waves are
formed, their influence will extend many hill lengths downwind.

The modeling of rough wall boundary layers under mildly stable or
unstable conditions has been attempted in a few laboratories with some
success. The correct requirement for an aerodynamically rough turbulent
boundary layer is uncertain when the flow is stratified but larger roughness elements are probably required

5. IMPLICATIONS FOR PHYSICAL MODELING

From the arguments and correlations presented in § 2, 3 and 4 the
perturbations to the flow field and their extent as a result of roughness
changes and terrain changes (in the absence of and including flow
separation) may be estimated. It must then be decided whether neglect of
this perturbation is acceptable. For example the inclusion of a shallow
($h/L \simeq 0.1$) upwind ridge is probably not warranted as the flow will have re-
covered by the downstream base of the ridge. An upstream ridge of the same
height but considerably steeper would cause flow separation. The perturbation velocities caused by this small ridge at the region of interest can
be determined and their relevance decided.

A similar argument will also allow an estimate of the distance
upstream that precise modeling of roughness elements is required in
preference to a gross modeling of the surface roughness. If the maximum
velocity perturbations in the wake of, say, an upwind building have decayed
to a few percent of the base flow by the region of interest then precise
modeling of the building is not warranted.

Physical modeling by many laboratories has led to significant advances in the study of flow over complex terrain. In addition to providing data relevant to specific sites they have indicated what manner of theoretical or numerical analysis may be fruitful and also confirmed, negated or emphasised the restrictions of many theoretical approaches. This symbiotic relationship is extremely valuable both for physical modeling and for analysis. Recent comparison of physical models with field data by Bowen (24), Mereney (25), Holmes et. al. (26) and others gives confidence that the present modeling techniques are, in the main, correct, leading to correlation coefficients of about 0.7 and above between model and field.

ACKNOWLEDGEMENT

The author's interest in, and understanding of, flow in complex terrain are the result of numerous discussions with Dr J C R Hunt.

REFERENCES

(1) Snyder W H (1981). Guideline for fluid modeling of atmospheric diffusion. USEPA Report EPA-600/8-81-009.

(2) Simiu S and Scanlan R H (1978). Wind effects on structures. An introduction to wind engineering. Wiley-Interscience.

(3) Mulhearn P J (1977). Turbulent flow over a very rough surface. 6th Aust. Hyd. & Fluid. Mech. Conference.

(4) Raupach M R et. al. (1980). A wind tunnel study of turbulent flow close to regularly arrayed rough surfaces. Bound. Lay. Met., 18, pp 373-397.

(5) Townsend A A (1965). The response of a turbulent boundary layer to abrupt changes in surface conditions. Jnl. Fluid. Mech., 22, p 799.

(6) Jackson N A (1976). The propagation of modified flow downstream of a change in roughness. Q. J. Roy. Met. Soc., 102, p924.

(7) Hunt J C R (1982). Disturbed boundary layers. In Engineering Meteorology, ed. E Plate, Elsevier.

(8) Edling W H and Cermak J E (1974). Three dimensional turbulent boundary layer flow on a roughness strip of finite width. CSU Report CER 73-74 WH-JEC.

(9) Jackson P S and Hunt JCR 91975). Turbulent wind flow over a low hill. Q. J. Roy. Met. Soc., 101, p 929.

(10) Britter R E et. al. (1981). Air flow over a two-dimensional hill: studies of velocity speed-up, roughness effects and turbulence. Q. J. Roy. Met. Soc., 107, p 91.

(11) Khurshudyan L H et. al. (1981). Flow and dispersion of pollutants over two-dimensional hills. Unpublished USEPA Report.

(12) Panafsky H A et. al. (1982). Spectra of velocity components over complex terrain. Q. J. Roy. Met. Soc., 108, p 215.

(13) Mason P J and Sykes R I (1979). Flow over an isolated hill of moderate slope. Q. J. Roy. Met. Soc., 105, p 393.

(14) Milne-Thompson L M (1960). Theoretical Hydrodynamics, 4th ed. Macmillan.

(15) Courtney L Y (1979). A wind tunnel study of flow and diffusion over a two-dimensional hill. MSc Thesis. North Carolina State Univ.

(16) Castro I and Snyder W H (1982). A wind tunnel study of dispersion from sources downwind of three-dimensional hills. To appear in Atmospheric Env.

(17) Wilson D J et. al. (1980). Reynolds number effects on flow recirculation behind two-dimensional obstacles in a turbulent boundary layer. Proc. 5th Int. Conf. on Wind Engineering. Col. State Univ.

(18) Hunt J C R (1974). Wakes behind buildings. ARC Report 35601, HMSO.

(19) Hosker R P (1982). Flow and diffusion near obstacles. To be published.

(20) Counihan J et. al. (1974). Wakes behind two-dimensional surface obstacles in turbulent boundary layers. J. Fluid Mech., 64, p 529.

(21) Hunt J C R (1981). Turbulent stratified flow over low hills. "Designing with the wind" - Symposium, Nantes.

(22) Snyder W H et. al. (1979). A fluid modeling study of the flow structure and plume impingement on a three-dimensional hill in stably stratified flow. Proc. 5th Int. Conf. on Wind Engineering.

(23) Baines P G (1980). Observations of stratified flow past three-dimensional barriers. Jnl. Geophys. Res.

(24) Bowen A J (1979). Some effects of escarpments on the atmospheric boundary layer. PhD Thesis, Univ. of Canterbury, Christchurch, NZ.

(25) Meroney R N (1980). Physical simulation of dispersion in complex terrain and valley drainage flow situations. 11th Nato CCMS, Amsterdam.

(26) Holmes J D et. al. (1979). The effect of an isolated hill on wind velocities near the ground level - initial measurements. Wind Engineering Report 3/29, James Cook Univ. of North Queensland.

(27) Jensen N and Petersen E (1978). On the escarpment wind profile. Q. J. Roy. Met. Soc., 104, pp 719-

(28) Bouwmeester R J et. al. (1978). Sites for wind power installations: wind characteristics over ridges. Col. State Univ. CER 77-78 RJBB-RNM-VAS 51.

WIND TUNNEL BLOCKAGE EFFECTS AND CORRECTIONS

W.H. Melbourne, Professor of Fluid Mechanics, Department of Mechanical Engineering, Monash University, Australia.

1. INTRODUCTION

The flow about any body in a closed wind tunnel is subject to the constraining effect of the tunnel walls. This constraint effectively prevents streamlines and wakes expanding in the way they would in full scale or unconstrained freestream flows. Knowledge of these constraining, or so-called blockage effects, on forces and pressures measured on models in closed wind tunnels is essential to permit correction of measurements to unconstrained freestream conditions. Blockage corrections mainly depend on the ratio of the model size to wind tunnel working section size, which is typified for bluff bodies by the ratio of the model projected area normal to the flow over wind tunnel cross-section area, S/C.

The derivation of blockage corrections seems, of necessity, to have been one of increasing complexity. Early studies by Glauert (1) in 1933 on wind tunnel interference on streamlined wings and bodies and others summarised by Parkhurst and Holder (2) generally indicated significant but relatively small effects where non-separated flows were concerned. However, it was Maskell who, when studying separated flows on stalled wings in general and delta wings in particular, realised that the blockage effects on bodies with large associated separated wake flows were relatively much greater than for bodies with attached flows and his classic contribution was finally published in the open literature in 1963 (3). For a while Maskell's blockage corrections, expanded and modified by Cowdrey (4) and Gould (5), and which essentially used the simple artifice of a correction to the model test dynamic pressure, were used universally by experimenters in bluff body aerodynamics.

The 1960's saw the advent of boundary layer wind tunnels simulating natural wind flows, and the testing of three-dimensional, bluff body buildings and structures in turbulent shear flows. McKeon and Melbourne (6) in 1971 discussed the limitations of Maskell's methods and gave blockage corrections for drag and base pressures on rectangular building models in a boundary layer flow, and at the same time Modi and El-Sherbiny (7) did much the same in relation to circular cylinders in smooth flow. The problems of ignoring blockage effects became evident at the 1971 International Conference on Wind Effects on Buildings and Structures, in Tokyo, when the conclusion of at least one paper was found in reality to be due to unaccounted blockage effects.

Throughout the 1970's many papers have appeared detailing more aspects of blockage effects in turbulent shear flows relevant to wind engineering studies. The effects of turbulence were noted to complicate the blockage effects and vice versa and even situations where increased blockage resulting in decreased drag have been observed. This paper, whilst by no means a state of the art presentation, will review some of the more recent findings and present blockage correction data relevant to wind engineering studies.

198

2. DRAG OF FLAT PLATES IN LOW TURBULENCE FLOW

The part of Maskell's theory and empirical relations (3) used initially by those in the wind engineering field was the correction formula

$$\Delta q/q \ = \ \varepsilon \ C_D \ S/C \tag{1}$$

where Δq is the effective increase in dynamic pressure due to constraint, q is the test reference, upstream, dynamic pressure, ($= \frac{1}{2} \rho v^2$), ε is a blockage factor dependent on the magnitude of the base pressure co-efficient, C_D is the measured drag coefficient ($=$ drag/qS), S is the model reference area and C is the cross-section area of the closed wind tunnel. The factor ε was shown to range between a value a little greater than 5/2 for axi-symmetric flow to a little less than unity for two-dimensional flow, but the variation from 5/2 was found to be small for aspect ratios in the range 1 to 10 as shown in Figure 1.

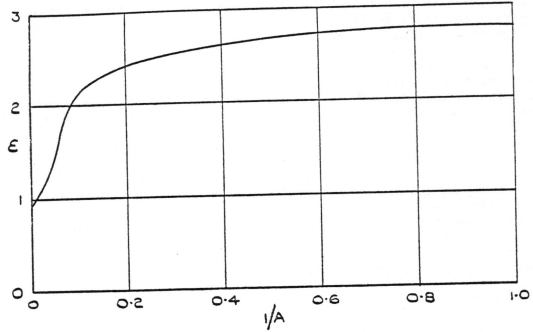

Figure 1. Variation of blockage factor, ε, with aspect ratio for non-lifting (normal to flow) rectangular plates, after Maskell (3).

McKeon and Melbourne (6) confirmed a value of $\varepsilon = 2.52$ for a circular, sharp edged disk normal to low turbulence flow.

3. PRESSURES ON FLAT PLATES IN LOW TURBULENCE FLOW AND INVARIANCE UNDER CONSTRAINT

Maskell's theory from which he derived the relationship $\Delta q/q = \varepsilon \ C_D \ S/C$ is so well known and widely used that it needs little introduction. However the assumptions of an "invariance under constraint" on which this is based does require some elucidation because it is the limitation of this assumption which requires a slightly more strict approach to the blockage corrections to be applied to pressure measurements.

Maskell considered the extent to which wall constraint can be regarded as equivalent to a simple increase in velocity of the undisturbed stream in the following manner.

"Exact equivalence implies that the form of the pressure distribution over the body is invarient under constraint."

This was not demonstrated to be true by Maskell, nor was it necessary to prove it because he was only interested in corrections to total drag and his proof goes on by saying that exact equivalence implies,

"If p is the pressure at any point (y,z) on the surface of the body, and H is the total pressure of the undisturbed stream, then $(p-p_b)/(H-p_b) = f(y,z)$ independent of constraint. And, since $H-p_b = k^2 q$, it follows that C_D/k^2 = constant, independent of constraint."

Where p_b is the base pressure, assumed constant and equal to wake pressure, and k is a base pressure parameter which is defined as the factor by which the stream velocity is increased on the wake boundary. Maskell then demonstrated by experiment that C_D/k^2 was constant and equal to 0.837.

It would seem therefore that the invariance theory is validated. However this is not the case, for although it follows that if $(p-p_b)/(H-p_b) = f(y,z)$, independent of constraint, then C_D/k^2 = constant, the converse does not follow. It will be shown in the next example that the constraint is such that almost the entire effect occurs on the base pressures (via the constraint on the wake), and the change in pressure distribution on the front face due to constraint is relatively small. The use of the concept of invariance under constraint (as demonstrated by C_D/k^2 = constant), although completely artificial, is very useful when only a total drag correction is required. However, the use of such an artifice to correct pressure measurements on a bluff body would be quite erroneous.

Pressure measurements have been made in the 1 metre square wind tunnel at Monash University on a series of flat circular plates set normal to a uniform stream on the wind tunnel centre line. Figure 2 shows the pressure distributions measured on the upstream and downstream faces for blockage ratios, S/C ranging from 1% to 20%, where S is the frontal area of the plate and C is the cross-section area of the wind tunnel working section. It is immediately apparent that most of the wall constraint effect is on the downstream face exposed to the wake pressure and that it is uniform across this face. On the upstream face even for 20% blockage there is no apparent effect on the pressures in the centre of the plate but a distortion of the pressure distribution does occur for the higher blockages towards the edge of the plate.

It is convenient to treat the constraint as it affects the mean pressures on the upstream and downstream faces separately. This does not lead directly to a way for correcting measured pressures at any point, but as can be seen in Figure 2 any correction would be relatively uniform across the downstream face and small on the upstream face.

Values of the mean pressures over the upstream and downstream faces, $\bar{P}_f$ and $\bar{p}_b$ respectively were obtained and added together to give total drag.

In coefficient form $C_{D_t} = C_{\bar{p}_f} - C_{\bar{p}_b}$ (2)

where C_{D_t} is the total measured drag coefficient (Total Drag/qS) and $C_{\bar{p}_f}$

and $C_{\bar{p}_b}$ are the mean pressure coefficients on the upstream and downstream faces respectively ($=(\bar{p}_f - p)/q$ and $(\bar{p}_b - p)/q$ where p is the undisturbed freestream reference static pressure).

The values of $C_{\bar{p}_f}$, $C_{\bar{p}_b}$ and C_{D_t} are plotted against C_{D_t} S/C in Figure 3. These plots show complete agreement with Maskell's momentum approach, in which he determined that the wall constraint effect was directly related to the total drag of the body within the control volume. In particular the incremental correction to the mean pressures and total drag may be expressed as follows, with the suffix c to denote corrected quantity,

Upstream Face
$$\Delta C_{\bar{p}_f} = C_{\bar{p}_f} - C_{\bar{p}_{fc}} = K_f\ C_{D_t}\ \frac{S}{C} \qquad (3)$$

Downstream Face
$$\Delta C_{\bar{p}_b} = C_{\bar{p}_b} - C_{\bar{p}_{bc}} = K_b\ C_{D_t}\ \frac{S}{C} \qquad (4)$$

Total Drag
$$\Delta C_D = C_{D_t} - C_{D_c} = K\ C_{D_t}\ \frac{S}{C} \qquad (5)$$

where for the sharp edged circular plate $K_f = -0.2$, $K_b = -3.2$, $K = 3.0$. The expression for the incremental correction for total drag can be put in one of the forms used by Maskell to show agreement with the blockage factor $\varepsilon = 5/2$ for low aspect ratio bodies. Equation (5) becomes,

$$\frac{C_{D_t}}{C_{D_c}} = 1 + \frac{K}{C_{D_c}}\ C_{D_t}\ \frac{S}{C} \qquad (6)$$

where $\dfrac{K}{C_{D_c}} = \varepsilon$ and which when evaluated from the experimental results for the sharp edged circular plate in Figure 3 gives $\varepsilon = 3.0/1.19 = 2.52$. It should be noted that measurements on disks are very sensitive to disk edge condition and that the circular disks used in these experiments had equal, 45° bevelled edges. Later, unpublished measurements by Laneville and Melbourne showed tha a circular disk with a 45° bevel on the leeward face only, (i.e. sharp edge in plane of upstream face) gave values of $C_{\bar{p}_{bc}}$ $= -0.38$ and $C_{D_c} = 1.245$ compared with -0.46 and 1.19 respectively

for the double sided bevelled edge used in the experiments illustrated in Figures 2 and 3.

4. DRAG AND PRESSURES ON FLAT PLATES IN UNIFORM TURBULENT FLOW

Discussion of blockage corrections for drag and pressure measurements on flat plates in uniform turbulent flow immediately involved one in a very sensitive area of fluid mechanics, where there is not yet general agreement as to the effect of turbulence and or turbulence scale on these measurements. However, the interaction between turbulence effects and blockage corrections are so important that the author proposes to mention some recent, and as yet unpublished, measurements and conclusions of Laneville and Melbourne in this respect.

Firstly, Modi and El-Sherbiny (8) reporting on their lift drag and pressure measurements on 2-D flat plates, concluded as follows:

"Coming to the plate data, it was observed that the static pressure distribution was essentially independent of the Reynolds number in the

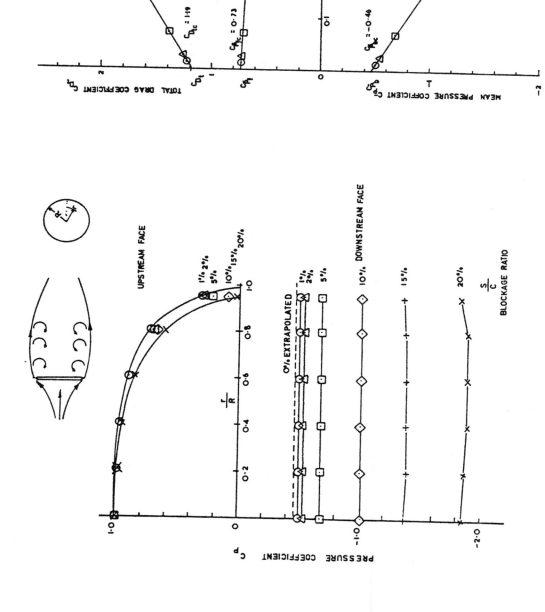

Figure 2 Measured pressure distribution along a radial line of a flat sharp edged (double sided, 45° bevel) circular plate for various blockage ratios. Flow normal to plate face, after McKeon & Melbourne (6)

Figure 3 Mean pressure on upstream and downstream faces and total drag for a sharp edged (double sided, 45° bevel) circular plate as a function of $C_{D_{tC}} \frac{S}{C}$ (i.e. measured drag coef × blockage ratio), after McKeon & Melbourne (6)

range $R = 2 \times 10^4 - 16 \times 10^4$. The effect of turbulence intensity and scale was primarily confined to the downstream face of the plate, where the pressure was found to be uniform as expected. Hence the parameter of interest is the base pressure coefficient and its variation with the plate inclination is shown in Figure 4. It appears that, for a given blockage, the effect of turbulence parameters is to make the base pressure a little more negative at all angles of attack. However, the resultant influence on the drag was observed to be essentially negligible It is interesting to note that the effect of the turbulence scale is confined to the lower blockage at the smaller angles of attack where the projected dimension of the plate normal to the flow is smaller than or of about the same magnitude as the integral scale.

Figure 5 presents typical results on static lift and drag as functions of plate attitude, turbulence parameters and blockage. For all practical purposes the effect of turbulence may be considered to be negligible except for $\alpha \simeq 10^0 - 20^0$ where the lift is significantly higher in the presence of turbulence."

Laneville and Melbourne (9) in recent unpublished work reported in abstract as follows:

"For a flat disk held normal to a turbulent flow, previous research work leads to the conclusion that turbulence had the effect of reducing the mean base pressure (or increasing the drag) as the intensity of turbulence was increased. To correct for blockage effect in turbulent flow, most authors used a semi-empirical method developed for low turbulence flow. The results of an experimental program lead to a completely different conclusion. The mean drag and base pressure coefficients for normal flat disks were observed not to be affected by turbulence. However, blockage effects were observed to be a function of the level of turbulence."

These conclusions are particularly significant; put in other words, the authors are saying that variation in disk drag previously attributed to change in turbulence and turbulence scale, are now shown to be due to blockage effects which, in fact, are turbulence sensitive. Laneville and Melbourne have given blockage correction factors for flat plate circular disks as a function of turbulence intensity.

5. DRAG AND PRESSURES ON RECTANGULAR PLATES/BUILDINGS IN A TURBULENT SHEAR FLOW

The experiments of McKeon and Melbourne (6) previously mentioned in §2 went on to determine blockage corrections for drag and pressures on a series of rectangular plates with little streamwise depth such that reattachment of flow on the side walls did not occur. These tests were conducted in three boundary layer flows which were shown to cause significant changes in upstream face pressure distributions. In particular, the blockage corrections for upstream face pressures were shown to vary with aspect ratio, building height over upstream width, h/b, but that the blockage corrections for the downstream face pressures did not change significantly with aspect ratio. The blockage correction K factors for use in the equations 3, 4 and 5 as given previously,

$$\Delta C_{p_b} = K_b \, C_{D_t} \, \frac{S}{C}$$

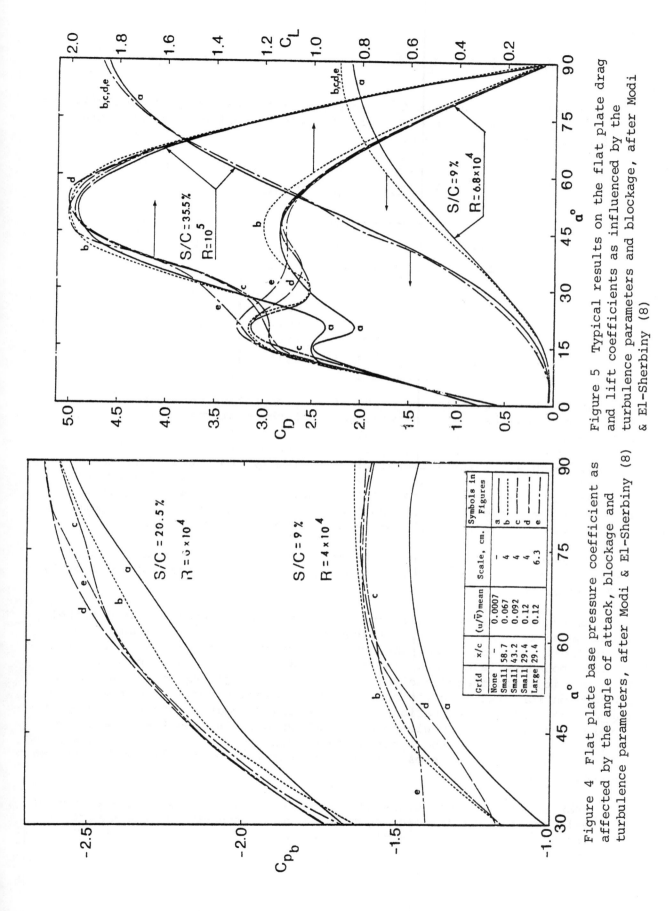

Figure 5 Typical results on the flat plate drag and lift coefficients as influenced by the turbulence parameters and blockage, after Modi & El-Sherbiny (8)

Figure 4 Flat plate base pressure coefficient as affected by the angle of attack, blockage and turbulence parameters, after Modi & El-Sherbiny (8)

Grid	x/c	$(u/\bar{v})$mean	Scale, cm.	Symbols in Figures
None	–	0.0007	–	a
Small	58.7	0.067	4	b
Small	43.2	0.092	4	c
Small	29.4	0.12	4	d
Large	29.4	0.12	6.3	e

$$\Delta C_{\overline{P}_f} = K_f \; C_{D_t} \; \frac{S}{C}$$

$$\Delta C_{D_t} = K \; C_{D_t} \; \frac{S}{C}$$

for base pressures, upstream face pressures and total drag respectively are given in Table 1; an example of the data from which these factors were derived are given in Figure 6.

TABLE 1 Values of the slope of the plots of
C_D, $C_{\overline{P}_f}$ and $C_{\overline{P}_b}$ vs. $C_{D_t} \; S/C$

h/b	K	K_f	K_b
4	2.1	-0.6	-2.7
2	1.9	-0.8	-2.7
1	1.8	-0.9	-2.7
½	1.5	-1.2	-2.7
½	1.6	-1.1	-2.7

As yet the author is not aware of any comprehensive study of blockage effects on pressures on streamwise walls of three dimensional rectangular bluff bodies. It is on the list of jobs to be done, and is particularly relevant to the measurement of peak pressures under reattaching shear layers, which are being increasingly required by designers of large buildings to determine cladding/glazing loads. This study will be very complex; not only are the reattaching shear layers and resulting surface pressures sensitive to freestream turbulence, but it is almost certain that the blockage corrections will be sensitive to turbulence.

Melbourne (6a) showed that the above corrections for isolated plates were also applicable to combinations of rectangular plates mounted side by side, normal to the flow.

Castro and Fackrell (10) have reported on the effects of wind tunnel blockage on two-dimensional, surface-mounted flat plates in a boundary layer flow. Many detailed plots of pressure distributions and drag for various boundary layers and aspect ratios are given in their paper. However, it is in their concluding remarks that a generalised blockage correction for drag is given and some very relevant comments are made, in support of earlier observations by McKeon and Melbourne, about the complex interactions between turbulence and blockage effects, and the effect that wall constraint has on the upstream face pressures when bluff bodies are immersed in a turbulent shear layer. The concluding comments of Castro and Fackrell were as follows:

"For two-dimensional bluff bodies tested in wind tunnels the effect of wall constraint has been shown to be very dependent on the nature of the upstream flow. An empirical correction for the drag coefficient in the form

$$C_{D_o} = C_D (1 - h/D)^m$$

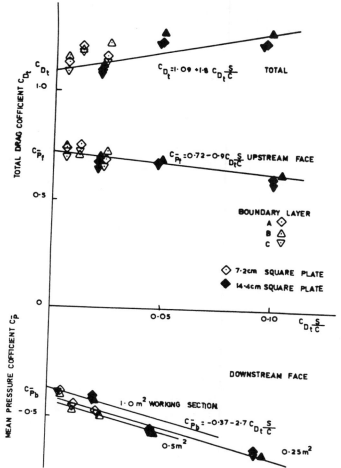

Figure 6   Breakdown of the mean pressure coefficients (based on reference velocity at z=h) for plates attached to one wall, after McKeon & Melbourne (6)

seems to be of fairly universal applicability but the actual value of m is dependent on both body shape (at least for uniform upstream flow) and, if the body is mounted in a boudary layer, on the ratio of boundary layer thickness to body height. In the latter case, for a "fence flow", m increases from about 2.6 at the lowest values of δ/h to about 7.5 for δ/h = 10. It is not clear why the blockage effect on drag should be independent of Reynolds number (the effects on front face and rear face pressures separately are clearly not). Neither is there any obvious physical explanation for the variation of the length of the cavity zone with blockage; this also appears to be determined by δ/h only. For δ/h greater than about 2.3, the reattachment point remains fairly stationary or even moves further downstream as blockage increases, whereas for values less than 2.3 it moves upstream; in both cases the base pressure coefficient decreases (drag coefficient increases) with blockage. It seems that blockage must have a considerable effect on the separation behaviour of the upstream boundary layer, so that when the region of upstream influence is greatest — large δ/h — the effects of blockage are greatest. Detailed understanding of this aspect of the flow requires further measurements. What is clear is that blockage produces additional perturbations, as Owen (1971) intimated that it might, and these can apparently 'scramble' the varing effects of different upstream conditions on the flow around the body itself. It seems that rather than simply keeping

206

blockage ratios fixed in experiments aimed at investigating these latter effects, one should ensure that the blockage effects are negligible, or else accept some uncertainty in the interpretation of the results.

The present work also suggests that correlations between fence drag and either "inner" or "outer" layer parameters describing the upstream boundary layer are not strong. The approach of Raju et al (1976) who state that, for measurements of drag at least only $U_\tau h/\nu$ (or h/y' in the case of a rough wall boundary layer) need be correctly modelled may be too simplistic; this can probably be most easily confirmed by measurements at full scale.

Although we have only investigated fence flows, there is no reason to suppose that these conclusions will not apply to more practical body shapes; further work, particularly for 3D bodies, would, however, be useful."

6. CIRCULAR CYLINDERS

The problems of determining blockage corrections for rectangular sharp edged bodies may seem to have become overly complex, but they pall into insignificance when we look at the problems associated with curved surfaces and the complicating factors of Reynolds number and surface roughness scale effects. Before looking at the trans- and super-critical ranges it helps to set the scene by restricting the problem to sub-critical flows.

Modi and El-Sherbiny (7) in the 1971 paper in Tokyo gave an insight into the problem for a smooth, circular, two-dimensional cylinder in low turbulence flow for sub-critical Rn in the range 10^4 to 10^5. The effect of wall constraint on mean pressures for the above configuration at Rn $= 10^5$ is given in Figure 7.

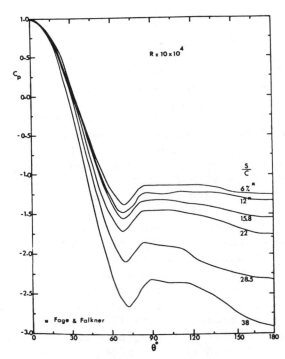

Figure 7 Mean static pressure distribution on a circular cylinder in low turbulence flow as affected by blockage ratio, after Modi & El-Sherbiny (7)

Modi and El-Sherbiny (7) have given the following blockage corrections for a smooth circular cylinder in sub-critical low turbulence flow

drag $\quad\quad\quad\quad C_{DC} = C_D(1 - 1.404S/(C)$

fluctuating list, $\quad C_{L'} = C_{L'}(1 - 1.83S/C)$

Strouhal number, $\quad S_{hC} = S_h(1 - 0.551S/C)$

with average C_{DC}, $C_{L'C}$ and S_{hC} as 1.066, 0.671, 0.189, respectively, and their concluding remarks are as follows:

(i) In general effects of the Reynolds number in the range $10^4 - 1.2 \times 10^5$ is confined to the mean pressure at the higher blockage and the fluctuating lift at the lower end of the bluffness scale.

(ii) The mean pressure in the wake diminishes with increase in the bluffness. The zero pressure point tends to move upstream, however, the minimum pressure point remains virtaully unaffected.

(iii) The effect of blockage is to increase the Strouhal number, first gradually but in a pronounced fashion at the higher value.

(iv) Due to a close similarity between the pressure distribution plots for different bluffness, it is possible to condense the useful information conveniently. In general, the bluffness causes the unsteady pressure to increase.

(v) The wake geometry parameter b/a remains virtually unaffectd by the bluffness.

(vi) The correction procedures by Glauert and Maskell are effective at lower bluffness. The inclusion of higher order terms extends their range of applicability.

In 1975 Modi and El-Sherbiny (8) extended their studies on circular cylinders in sub-critical flow to include turbulence effects. A great deal of experimental information is presented in this paper to assess the effects of blockage, turbulence and Reynolds number. The authors have summarised the overall problem quite neatly, "although, in general, the trends are not well defined, the results clearly establish the importance of turbulence parameters and blockage in the industrial aerodynamics studies involving simulated wind tunnel tests". No generaliseed procedures fo blockage corrections were developed in this paper.

In 1980 Cheung and Melbourne (11) presented the first part of a study of blockage effects on circular cylinders in turblent flow up to $Rn = 10^6$, also covering high sub-critical and critical flow regimes. This work was done primarily to support force and pressure measurements being made on circular cylinders in turbulent flow at super-critical Reynolds numbers in relation to chimney stack loads. The problem of unscrambling the interacting effects of turbulence and Reynolds number were resolved by the application of an interative procedure to first correct the effective

Reynolds number for blockage effects and then the corrections for forces and pressures became consistent. The correction process and examples of the blockage correction factors were given as follows:

"The empirical procedure for obtaining the blockage correction for pressure measurements was by linearly interpolating the data obtained at different blockage ratios. These data for the interpolation had first to be corrected to the same effective Reynolds number, (i.e. the same blockage-corrected Reynolds number).

Since the dynamic head $\frac{1}{2}\rho\bar{u}^2$ of the flow can be reflected by an appropriate pressure coefficient, the ratio of corrected to un-corrected free-stream speed can be estimated by the square root of the ratio of the corresponding pressure coefficients. Then, the Reynolds number can be corrected from this speed ratio. However, this speed ratio can only be calculated when a corrected pressure coefficient is known and the pressure coefficient cannot be corrected until the Reynolds number has been corrected with the speed ratio. Therefore, an iterative procedure as shown in Figure 8 was devised to evaluate this Reynolds number correction factor.

For each nominal Reynolds number, the pressure coefficients obtained from different blockages were adjusted to those of the same effective Reynolds number from cross-plots of pressure coefficients as a function of Reynolds number. These adjusted pressure coefficients are found to vary linearly with blockage ratio. The slope of this linear variation of pressure coefficient with blockage was calculated for each measurement around the cylinder at different speeds and turbu-lence intensity levels. Examples of these correction slopes are plotted in Figure 9. The blockage correction can then be evaluated by multiplying the appropriate correction slope by the blockage ratio (note: the blockage ratio is expressed in percentage of S/C).

Blockage effect on the mean pressure distribution is relatively more significant in the wake region, particularly in subcritical flow. However, this significance is reduced as the turbulence intensity level increases. At higher intensities, both the correction and the sensitivity of the correction to Reynolds number for the minimum pressure coefficient increase.

Similarly, blockage correction slopes for fluctuating pressure coefficients were estimated. They have similar trends and shapes of those for the mean pressure coefficients, except that they are an order of magnitude smaller.

Mean drag coefficients were calculated by integrating the mean pres-sure distribution for each tests. The results show that the drag co-efficients are linearly related to the blockage ratios investigated. This linear relation is represented by a straight line, the slope of which is the correction slope of the drag coefficients. As shown in Figure 10, the correction slopes are positive in low turbulence flow, especially in the low-speed subcritical range. For turbulent flows, negative correction slopes occur. That is, the drag coefficient becomes smaller when the blockage increases. This can be explained from Figure 9. The blockage correction slopes of mean pressure coefficients are more significant in the wake region in smooth flow and so increases the drag coefficient as blockage increases. But, in turbulent flows, comparatively more correction is required near the minimum pressure region which is around 80°. Therefore, the drag coefficient decreases as the blockage ratio is increased.

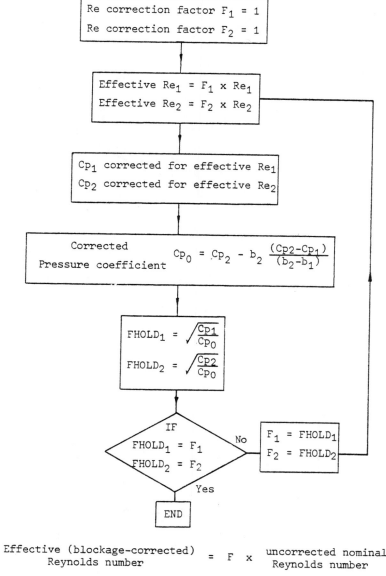

Figure 8 Iterative procedure for evaluating the Reynolds number correction factor, after Cheung & Melbourne (11)

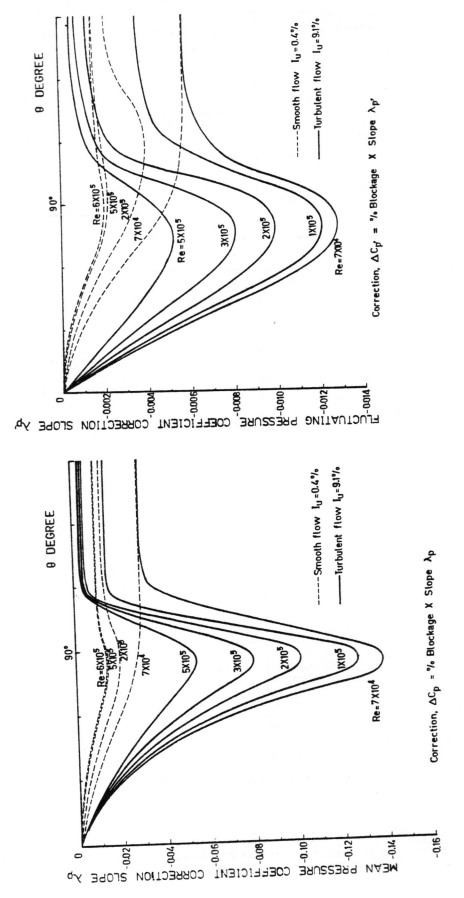

Figure 9   Blockage correction slope as a function of Reynolds number for mean and fluctuating
pressure coefficients, after Cheung & Melbourne (11)

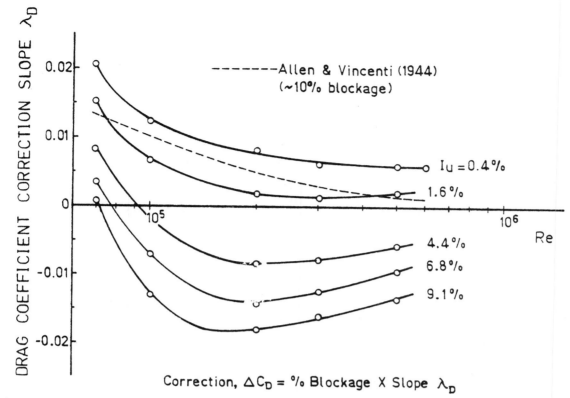

Figure 10 Blockage correction slope as a function of Reynolds number for mean drag coefficients, after Cheung & Melbourne (11)

Cheung and Melbourne (11) demonstrated that blockage corrections for circular cylinders in turbulent flow are very sensitive to turbulence, but that reasonable correction factors can be determined. The two most important conclusions to note are

(a) the severity of the pressure corrections in increasing turbulence, and different sensitivity to sub- and super-critical flow conditions at the lowest pressure regions (e.g. a 2% blockage at 9% turbulence intensity required a correction ΔC_p, to be subtracted, of -0.28 at $Rn = 7 \times 10^4$ and -0.03 at $Re = 6 \times 10^5$) and

(b) that for drag the blockage correction can be negative, that is, to increase drag for increasing blockage ratio for the more turbulent flows.

7. RECTANGULAR BODIES WITH SIGNIFICANT STREAMWISE DEPTH

Although the problem of determining blockage corrections for pressures on streamwise surfaces of rectangular bodies with significant streamwise length has received little attention, Courchesne and Laneville (12) have made a most comprehensive study of the drag and base pressure corrections on such bodies in low turbulence flow and compared their results with Maskell's theory, a free-streamline model developed by Modi and El-Sherbiny (13) and an earlier proposal by Raju and Singh (14).

Based on Maskell's theory and noting that $C_{D_c}/(k_c^2-1)$ is constant for

a given value of H/D, the ratio of streamwise depth, H, over frontal dimension D, they have given drag corrections in several forms as shown in Figure 11.

212

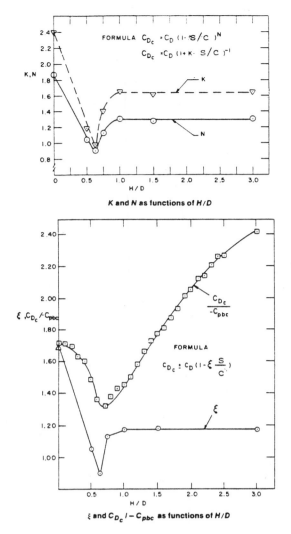

K and *N* as functions of *H/D*

ξ and $C_{D_c}/-C_{pbc}$ as functions of *H/D*

Figure 11 Blockage corrections for two dimensional rectangular bodies in low turbulence flow as a function of the ratio of streamwise depth over frontal dimension, after Courchesne & Laneville (13)

In particular they have discussed the limitations of Maskell's theory and their conclusions (including a comment on the probable effect of turbulence) were as follows:

"From the determination of ξ , K and N as functions of H/D in smooth flow, several observations can be drawn.

(a) An empirical flat plate type correction procedure should not be used for rectangular cylinders. Such a procedure will certainly underestimate the corrected drag coefficient.

(b) For cases in which no Reynolds number influence is present and for values of H/D larger than 1, the independence of ξ, K, and N indicates that th pressure loading remains invariably affected by blockage. Unfortunately, this observation was not verified for values of H/D larger than 3 but the tendency seems well established. On the other hand, this independence of ξ, K, or N has a relevant significance for two dimensional rectangular sections exposed to grid turbulence. Laneville in earlier work has concluded that the behaviour of a given H/D section exposed to grid turbulent flow was similar to the behaviour of a rectangular section of larger H/D exposed to smooth flow. One can thus expect that an empirical formula using the constants determined in smooth flow will achieve adequate drag coefficient correction for values of H/D larger than 1.

(c) For the range of blockage ratios considered, the three empirical formulae gave good corrections. Considering the extension to higher blockage ratios (>13 %), the linear relation will certainly over-estimate the corrected drag coefficient. On the other hand, based on the results of Raju et al., who obtained consistent corrections for blockage ratios up to 24%, the exponential formula should yield a correct value of C_{D_c}. Similarly, based on the results of Modi and El-Sherbiny for flat plates and circular cylinders with blockage ratios up to 35%, the formula with the proper ξ values should satisfactorily correct the drag coefficient for models of larger blockage ratios.

(d) Figure 11 establishes the validity of Maskell's theory for 2D rectangular sections. This theory will generally apply too high a correction especially for sections with H/D > 1.2.

Awbi (15) in 1978 gave results of measurements of blockage corrections on two-dimensional rectangular sections in low turbulence flow and used the linear relationship between the correction increment and C_D S/C to derive values for the slope in this relationship for various H/D ratios. Awbi also measured mean pressures along a streamwise wall shown in Figure 12 which illustrates the complexity of determining blockage corrections when shear layer reattachment is involved.

7. BLOCKAGE REMOVAL BY PROFILING WALLS

In theory it should always be possible for a given model configuration to profile the walls of the wind tunnel to achieve a bounded streamline flow the same as would occur in unbounded flow. The author is aware of this procedure being used to remove blockage effects at several wind tunnel testing centres. However, no references have been found describing how well these techniques work, in particular their validity over a range of model building configurations. Some recent work by Goh (16), as yet unpublished, describes theoretical/empirical methods for determining roof profiles to remove blockage effects for topographical model studies.

8. CONCLUSIONS

For the great majority of model configurations, both two and three dimensional, in flows of various turbulence levels (even boundary layer shear flows), the linear form of blockage correction

$$\mathcal{C}_D = C_D - C_{D_c} = K \, C_D \, \frac{S}{C}$$

or

$$\mathcal{C}_p = C_p - C_{p_c} = K \, C_D \, \frac{S}{C}$$

is applicable. Where C_D is the measured drag coefficient based on reference area S, C is the wind tunnel cross-section area, the subscripted coefficients, C_{D_c} and C_{p_c}, are drag and pressure coefficients corrected to

214

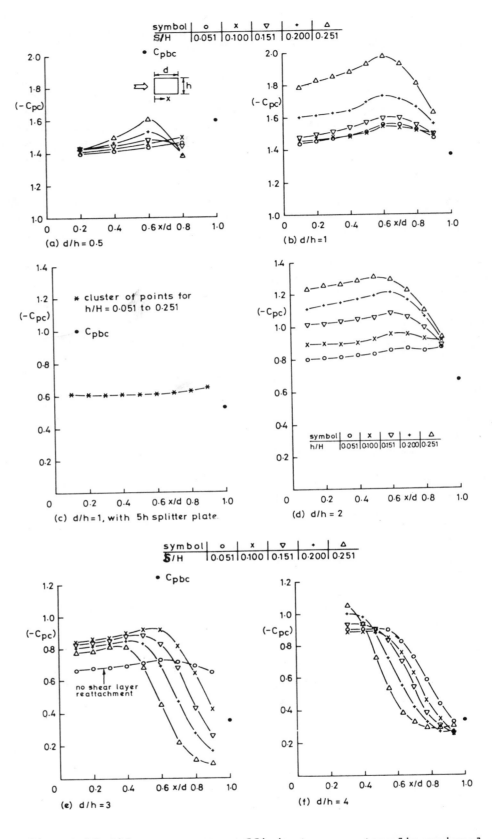

Figure 12 Side pressure coefficients on a two dimensional
rectangular section as a function of blockage ratio, S/C
and depth over frontal dimension ratio, after Awbi (15)

unbounded freestream conditions, and the value of the constant K has to be determined for the particular model/flow/wind tunnel configuration under consideration. The exception to the application of the linear form of correction appears only to occur when shear layer reattachment is involved or when very high blockage ratios are involved. This paper has given values of the constant K for a wide, but by no means exhaustive, range of wind tunnel model configurations.

In conclusion, it can only be stressed that the value of the constant K must be obtained for the relevant model/flow/wind tunnel combination under consideration.

To generalise on the significance of blockage corrections for bluff bodies in a closed section wind tunnel, it can be noted that for blockage ratios of 2% the blockage corrections are likely to be about 5%, in some special circumstances as high as 10%, but rarely less than 3% except on front face pressures, and the magnitude of the blockage correction varies directly with the blockage ratio.

9. REFERENCES

1. H. Glauert, Wind tunnel interference on wings, bodies and airscrews, Research and Memorandum No.1566, HMSO London (1933).

2. R.C. Pankhurst & D.W. Holder, Wind tunnel techniques, Pitman, London (1952).

3. E.C. Maskell, A theory of the blockage effects on bluff bodies and stalled wings in a closed wind tunnel, Research and Memorandum No.3400, HMSO, London (1963) (also RAE Report Aero 2685).

4. C.F. Cowdrey, The application of Maskell's theory of wind tunnel blockage to very large solid models, National Physical Laboratory, Aero Report 1247 (1967).

5. R.W.F. Gould, Wake blockage corrections in a closed wind tunnel for one or two wall-mounted models subjected to separated flow, National Physical Laboratory, Aero. Report 1290 (1969).

6. R.J. McKeon & W.H. Melbourne, Wind tunnel blockage effects and drag on bluff bodies in a rough wall boundary layer, Proc. 3rd. Int. Conf. Wind Effects on Buildings and Structures, Tokyo (1971) 263-272.

6b. W.H. Melbourne, Wind tunnel blockage corrections for mutiple building model tests, Proc. 4th A/Asian Conf, on Hydraulics and Fluid Mechanics, Monash University (1971) 591-598.

7. V.J. Modi & S. El-Sherbiny, Effect of wall confinement on aerodynamics of stationary circular cylinders, Proc. 3rd Int. Conf. Wind Effects on Buildings and Structures, Tokyo (1971) 365-375.

8. V.J. Modi & S. El-Sherbiny, Wall confinement effects on bluff bodies in turbulent flows, Proc. 4th Int. Conf. Wind Effects on Buildings and Structures, England (1975) 121-130.

9. A. Laneville & W.H. Melbourne, Turbulence and blockage effects for normal flat disks. Unpublished paper, submitted for review 1981.

216

10. I.P. Castro & J.E. Fackrell, A note on two-dimensional fence flows, with emphasis on wall constraint, Journal of Industrial Aerodynamics, Vol.3, No.1, (1978) 1-20.

11. C.K. Cheung & W.H. Melbourne, Wind tunnel blockage effects on a circular cylinder in turbulent flows, Proc. 7th A/Asian Conf. Hydraulics and Fluid Mechanics, Brisbane (1980), 127-130.

12. J. Courchesne and A. Laneville, A comparison of correction methods used in the evaluation of drag coefficient measurements for two dimensional rectangular cylinders, ASME Winter Meeting, Paper No./79 WA/FE3 (1979).

13. V.J. MODI AND S. El-Sherbiny, A free-streamline model for bluff bodies in confined flow, ASME Jnl. of Fluids Engineering, Vol.99(1977)585-592.

14. K.G. Ranga Raju and V. Singh, Blockage effects on drag of sharp edged bodies, Jnl of Industrial Aerodynamics, Vol.1 No.3(1976)301-309.

15. H.B. Awbi, Wind tunnel wall constraint on two-dimensional rectangular-section prisms, Jnl. Industrial Aerodynamics, Vol.3 No.4 (1978)285-305.

16. C.B. Goh, Blockage removal in topographical modelling. Unpublished Ph.D. Thesis, University of Auckland (1981).

VALIDATION OF BOUNDARY-LAYER SIMULATION:
SOME COMPARISONS BETWEEN MODEL AND
FULL-SCALE FLOWS

H. W. Teunissen, Atmospheric Environment Service, Environment Canada

1. INTRODUCTION

The previous papers in this session have presented the similarity criteria involved in simulating planetary boundary-layer (PBL) flow in wind tunnels and have described the particular techniques utilized for producing such simulations in various types of tunnel. In most cases, it is inevitable that several simplifying assumptions be made or that certain criteria be relaxed in order to produce a particular simulation, and the consequences of such assumptions have been considered. In the present paper, we attempt to assess the success of some of these techniques and assumptions from the standpoint of their ability to produce wind tunnel flows which accurately reproduce the characteristics of the prototype boundary-layer flows. Thus we are concerned here with validation of the flow simulation per se, rather than the loads or deflections which this flow might produce on a particular building or other structure (which latter is the subject of Session VI of this workshop). Typical questions which are of interest here are:

(i) How good are our flow simulations?

(ii) Are sufficient parameters being compared in evaluating the quality of these simulations?

and (iii) In view of the variability of the real atmosphere, what level of approximation seems acceptable in a wind tunnel simulation?

In this paper, we address such questions by comparing measured results from simulated and prototype flows at several different sites. Full-scale measurements of wind and turbulence structure at various sites have been underway at AES during the past several years expressly for the purpose of comparing actual boundary-layer structure in such flows with the predictions of both idealized engineering models and wind tunnel simulations of them in order to assess the reliability of both types of model. Some typical results from three of these sites are discussed herein and are compared with wind tunnel simulations of the flow at the same sites. Although only a few sample results will be presented here, additional information can be found in the references listed at the end of the paper.

One major simplifying assumption common to all the cases considered here is that of neutral atmospheric stability. This is of course a fairly common assumption when considering problems involving the wind loading of buildings and other structures, and it is generally quite reasonable in view of the high mean wind speeds which are usually involved for such problems. However, as suggested by the results of the first example considered, a priori elimination of potential stability effects may not always be the best approach, although it certainly does seem reasonable in the majority of cases.

Fig. 1   Aerial photograph of Lions' Gate Bridge site.

2.   SITE A: LIONS' GATE BRIDGE, VANCOUVER, B.C.

2.1   Prototype Site and Measurements

The Lions' Gate suspension bridge site is perhaps the simplest of the sites to be considered here in that, for the full-scale wind direction of interest ($\bar{\phi} \sim 260^{\circ}$, see Fig. 1), the upstream fetch is open water effectively to infinity. This is the case for much of the prevailing strong-wind direction at the site, notwithstanding the fact that the bridge is located in otherwise generally mountainous terrain. This terrain tends to funnel winds into the northwest – southeast direction perpendicular to the bridge's length. For the results to be considered here, however, the structure of the approaching wind can reasonably be assumed to have been unaffected by the surrounding terrain.

Five 3-component propeller anemometers were mounted on horizontal booms attached to vertical hangers on the northwest side of the bridge (Fig. 2) at a height of about 78 m above mean water level (about 10 m above the road deck). Details of their installation and subsequent data collection and handling procedures can be found in Ref. 1. Data from these anemometers have been collected intermittently for a number of years, and although much of the information remains to be analyzed, some interesting results have been observed to date. In particular, the results from two specific days are considered here. On the afternoon of April 11, 1978, measurements were obtained during a 4-hour period for which the mean wind speed was about 11 m s^{-1} and the direction 262°. Ambient air temperature was in the range 9 – 11°C and the skies remained essentially clear throughout the afternoon. Conditions were similar during a 2-hour period on June 8, 1977, except that the mean speed was slightly lower (about 8 m s^{-1}) and the temperature slightly higher (about 16°C). The mean wind direction was the same. For both occasions, the general ambient conditions indicated a situation of neutral thermal stability throughout the measurement period.

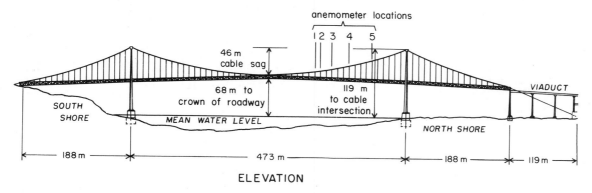

Fig. 2   Schematic layout of bridge (looking northwest).

2.2   Wind Tunnel Simulation

Wind tunnel simulation of the flow at the bridge site was carried out in the 9m x 9m closed-return wind tunnel at the National Research Council of Canada (Refs. 2 - 4). In fact several wind tunnel tests were carried out using different bridge deck configurations and approach-flow characteristics. It was confirmed during these tests that turbulence in the approaching flow plays a major role in the aerodynamic stability of bridges. For example, for one particular deck configuration, a serious flutter instability was observed when the model was subjected to a uniform, nonturbulent approach flow (Ref. 3). This instability was effectively removed and replaced by a much less serious increase in the buffeting response of the bridge when typical boundary-layer turbulence characteristics were simulated in the approach flow. This beneficial effect of turbulence on bridge response has been observed previously (e.g. Ref. 5), although the minimum level of turbulence required to produce such effects is not entirely clear at present.

Two types of boundary-layer flow were simulated in carrying out the above tests, one representing a flow approaching over a pure water fetch and the other assuming some influence of upstream terrain. In both cases the length scale of the simulation was 1:110, based on bridge model length scale. The prototype flows were assumed to be neutrally stable and their mean velocity and turbulence characteristics were estimated using results from some of the available engineering models (e.g. Refs. 6 - 8) for such flows (the wind-tunnel tests were carried out before the full-scale measurements began). The "spire" technique was used to generate the desired boundary-layer flows using six 2.7-m spires placed at the test-section entrance, about 16 m upstream of the model. The main characteristics of the flows produced are summarized in Table 1.

2.3   Results and Implications

The full-scale data were analyzed in blocks of 16.4 minutes' duration (some data were also studied using a block length of 32.8 minutes, with no significant difference in the observations). The results are discussed in some detail in Ref. 1, but for present purposes only two figures need be considered.

Table 1. Comparison of wind and turbulence characteristics predicted, simulated and measured in full scale.

	α	δ,m	Z_o,m	T_u	T_v	T_w	$\tilde{R}_{uw}$	L_u^x,m	L_v^x,m	L_w^x,m	ℓ_u^x,m	ℓ_v^x,m	ℓ_w^x,m
TEUNISSEN (Ref. 6) ("Flat terrain")	0.11	270	0.001	0.089	0.071	0.046	-0.31	96	48	30	-	-	-
ESDU (Ref. 7) ("Off-sea wind in coastal area")	0.12	290	0.001	0.093	0.062	0.043	-0.26	170	72	27	-	-	-
COUNIHAN (Ref. 8) ("Terrain type 1, smooth")	0.12	600	0.001	0.089	0.066	0.045	-0.40	>250	-	30	-	-	-
IRWIN (Ref. 2) (Assuming some land influence)	0.16	-	-	0.117	-	0.075	-0.31	90-180	-	30	-	-	-
IRWIN (Wind Tunnel) Ref.2 (Some Land) Ref.4 (Open Water)	0.15 0.10	220 244	- -	0.120 0.080	- -	0.085 -	- -	79 61	- -	23 -	- -	- -	- -
APRIL 11, 1978, Blocks 1-6	-	-	-	0.052	0.039	0.021	-0.13	114	33	17	209	186	47
APRIL 11, 1978, Blocks 7-13	-	-	-	0.095	0.047	0.026	+0.04	583	40	22	546	290	188
APRIL 11, 1978, (Reference Block 2)	-	-	-	0.031	0.022	0.018	-0.05	54	17	15	58	30	43
JUNE 8, 1977, Blocks 1-6	-	-	-	0.038	0.037	0.032	-	82	66	53	192	216	121

Figure 3 shows the block-by-block variation in turbulence intensities for all three velocity components for the April 11 data set. The two most striking features of these results are their large variability throughout the afternoon and the very low values observed for some of the blocks relative to the expected turbulence levels, particularly for the longitudinal component. It has been suggested in Ref. 1 that the reason for the generally low turbulence levels was the probable existence of a thermally stable layer near the surface of the water throughout the afternoon. This layer was produced within the otherwise generally neutral flow as a result of the water temperature being significantly lower than that of the air, and its existence is not unusual for a site of this type. The large increases in T_u after 1330 h were caused by the superimposition of large, low-frequency gusts on the otherwise small turbulence signals, and it is suggested in Ref. 1 that these gusts may have been caused by thermal plumes generated over tidal flats upstream of the bridge being swept past it by the mean flow. Whatever their origin, the frequency of these gusts was generally lower than those of usual interest for bridge-loading considerations ($n < \sim 0.01$ Hz), and hence they are not particularly important from this standpoint. What is important, however, is the very small magnitude of the turbulence levels observed during parts of the afternoon, and the correspondingly low turbulence levels in the important higher-frequency range of the spectrum throughout the entire afternoon. The latter effect is demonstrated in Fig. 4, where the observed turbulence levels in the inertial subrange are seen to have been well below the expected levels during the entire measurement period. The effect of the low-frequency gusts on the shape of the spectra below $n \sim 0.01$ Hz is also clearly seen in this figure. Since the presence of turbulence in the higher-frequency range is beneficial insofar as minimizing overall bridge response is concerned, the possible absence of normally anticipated (and hence simulated) turbulence levels is of obvious importance when using wind tunnel simulations to predict full-scale response.

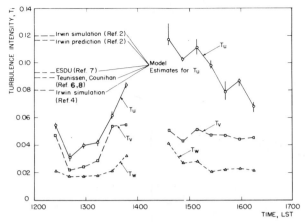

Fig. 3   Variation of turbulence intensities with time (April 11, 1978)

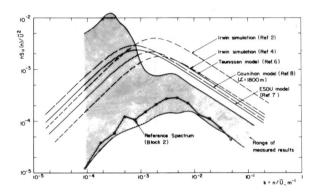

Fig. 4   Comparison of longitudinal-component power spectra predicted,
          simulated and measured in full scale.

Similarly low turbulence levels were observed for at least one other
data set (June 8, 1977 – see Table 1) during which no low-frequency gusting
was observed and for which the power spectra were similar in shape and
magnitude to that of Block 2 in Fig. 4. While it is true that such low
turbulence levels would be expected to increase for mean wind speeds
higher than those encountered here, it is pointed out that Shiotani
(Ref. 9) found some comparably low turbulence intensities for winds approaching his site over open water at mean speeds as high as 18 – 25 m s^{-1},
which are certainly not low insofar as bridge loading is concerned. It is
suggested that these low turbulence levels were likely also the result of
thermal stability in the approach flow.

As for the question of the quality of wind tunnel flow simulations, it
would seem that the main limitation in this case lies not so much with the
capabilities of the wind tunnel itself but rather with the idealized models
used to characterize the prototype flow. The above results demonstrate
that it is entirely possible to encounter prototype flows having large,
stability-induced variations in turbulence characteristics during conditions
which might normally be assumed to be neutrally stable and which would
therefore be easily reproducible in the wind tunnel using idealized models
as guidelines. What is required in such cases is to take these possibilities into account when identifying the characteristics of the prototype
flow which is to be simulated. Although the wind tunnel may not always be

able to reproduce the complete set of desired prototype characteristics simultaneously, this may in many cases not be necessary. As in the present example, satisfactory results may be attainable merely by carrying out individually as many different simulations as are needed to reproduce sequentially the full range of important characteristics of the prototype flow.

Fig. 5   Aerial photograph of Rockcliffe Airport.

3.   SITE B: ROCKCLIFFE AIRPORT, OTTAWA, ONTARIO

3.1 Prototype Site and Measurements

The Rockcliffe Airport site (Fig. 5) is much more complex than the Lions' Gate Bridge site insofar as surface characteristics are concerned. The airport is located in the Ottawa River valley about 5 - 6 km northeast of the downtown core of the city of Ottawa. The surrounding terrain is suburban or light urban in nature and, depending on the wind direction, major changes in surface roughness characteristics can occur immediately upstream of the airfield itself.

Wind and turbulence measurements were obtained at this site during a period of about one year using several different instrumentation systems. The primary source of ground-based turbulence data was a 33-m vertical tower located near the centre of the airfield (labelled 'NAE TWR' in Fig. 5) and instrumented at several levels with anemometers and temperature sensors. An anemometer mounted on a 10-m tower atop an apartment building about 1.5 km south of this location ('APT TWR' in Fig. 5) was used to obtain reference mean winds for the full-scale data sets. This location was chosen because of its height (about 85 m above ground level), in that winds measured here should be relatively independent of most of the detailed surface features of the surrounding terrain. On several occasions, an instrumented helicopter was used to obtain mean wind, turbulence and temperature data throughout the complete PBL.

Most of the results of one complete data set obtained at the Rockcliffe site have been presented and discussed in detail in Ref. 10. For this data set, the mean wind approached the NAE tower from the southwest ($\bar{\phi} \sim 220°$), with the result that the tower was located in the internal boundary layer (IBL) produced by a rough-to-smooth change in surface roughness about 300 m upstream of the tower. Comparisons are presented in Ref. 10 between the data measured on the NAE tower and the predictions of several mathematical models of the flow downstream of a large roughness change. In general, the measured results were found to agree quite favourably with the model predictions, although only limited quantitative comparisons could be made. Comparisons are also presented in Ref. 10 between mean wind and turbulence profiles obtained to 600 m by the helicopter and the predictions of neutral-stability engineering models for a site of this type. Unlike the situation at the Lions' Gate Bridge site, the flow in this case was indeed found to be neutrally stable, even though the reference speed was a fairly modest 11 m s^{-1}, and its gross features (power-law index, boundary-layer thickness, turbulence intensities and most scales) agreed quite well with the neutral-stability model estimates. Close to the surface, however, the measured turbulence structure differed significantly from the idealized estimates. This was of course an expected result of the large change-of-roughness upstream of the tower and emphasized the importance of such local features on wind structure at low levels. Reproduction of such wind structure in wind tunnel simulations may be important for problems such as wind-loading of low-rise structures and pedestrian-level winds, and it is not yet clear whether or not such reproduction can actually be achieved. Thus a wind tunnel simulation of the flow over the Rockcliffe site was carried out to investigate this question. A complete report of this work will be given in Ref. 11 and some sample results are discussed in Section 3.2.2.

An additional noteworthy general feature of the full-scale results from the Rockcliffe site was the large degree of block-to-block variability observed in many of the turbulence characteristics. This variability is perhaps not surprising in view of the complex nature of the surface features at the site. Nevertheless, similarly large variability has been observed in neutral-stability flows at considerably simpler sites (e.g. Ref. 12) where the average characteristics matched the idealized model predictions quite well. The Lions' Gate Bridge results also showed similar variability, of course, albeit for a different reason. Variability of this type is an unavoidable characteristic of most atmospheric flows, and the implications for wind tunnel modelling are essentially the same as those discussed above. That is, it may be necessary, and hopefully also sufficient, to carry out more than one simulation if all the required characteristics cannot be reproduced simultaneously.

3.2 Wind Tunnel Simulations

Two independent wind tunnel simulations were carried out in relation to the full-scale data set discussed above. In one of these (Ref. 13), the objective was primarily to determine the degree of building influence on the reference wind measured atop the apartment tower south of the airfield. In the other (Ref. 11), a full, topographic-model simulation was carried out in an attempt to assess the faithfulness with which the full-scale flow characteristics could be reproduced in the tunnel. A few of the relevant results from each of these studies are given below.

3.2.1 Flow Over an Isolated Apartment Building

Two simple scale models (1:400 and 1:800) of the apartment building and a few surrounding buildings were constructed. Detailed characteristics of the flow over the top of the apartment building were measured in uniform approach flow and in several different types of boundary-layer flow in order to assess the effect of the details of the approach flow on the results obtained. Mean wind speed-up at the reference anemometer location was compared with two fairly crude estimates obtained at the full-scale site and with the result obtained in the complete topographic-model simulation.

In general, it was found that the details of the local flow over the building differed significantly in boundary-layer approach flows as opposed to uniform approach flow, as might be expected. For example, the height of the building-induced shear layer above the building itself decreased in boundary-layer flow and continued to decrease with decreasing ratio of building height to boundary-layer depth. However, since the reference anemometer was located well above the shear layer in all cases, the detailed features of the approaching boundary-layer flow did not have a significant effect on the measured speed-up at this particular location. Had the reference position been somewhat lower, of course, the situation would have been entirely different. Generally speaking, the speed-up value produced by the topographic-model simulation agreed best with the observed full-scale results, although the marginal improvement over the isolated-building results in both uniform and boundary-layer flows would not justify the additional complication of producing a topographic model solely for this purpose. Additional investigations are required to obtain more conclusive results in this area.

3.2.2 Topographic Model Simulation

A 1:1920-scale topographic-model simulation of the prototype flow described in Section 3.1 was carried out in the AES boundary-layer wind tunnel (Ref. 11). The linear scale of the model, which is shown in Fig. 6, was undistorted and considerable care was taken in attempting to represent correctly the detailed features of the local terrain. The model approach flow was adjusted using a trip-barrier and appropriate upstream roughness in order to reproduce the mean velocity and turbulence intensity profiles measured by the helicopter. Good similarity of the approach flows was achieved, following which detailed wind and turbulence measurements were obtained in the wind tunnel at the same locations as those for which full-scale measurements were available.

Fig. 6   Topographic model of Rockcliffe site in AES wind tunnel
         (looking downstream).

In general, it was found that the characteristics of the simulated
flow over the Rockcliffe site agreed fairly well with many of the results
observed in full scale. For example, the relative magnitudes of mean winds
measured at a height of 3 m at several different locations around the prototype site were correctly reproduced in the model flow, even though the
ratios of these speeds to the reference mean wind speed were somewhat lower
in the wind tunnel than they were in full scale. The correct ranking of
surface mean wind speeds according to magnitude on a topographic model has
also been reported by Meroney et al. (Ref. 14). Similarly, most of the
power spectra and integral scales observed in the wind tunnel flow were in
very good agreement with the full-scale results, although some of the
lateral component turbulence intensities were not. Fig. 7 compares the
full-scale and simulated power spectra for the longitudinal velocity component at the NAE tower location, and the agreement is seen to be excellent.

Two additional interesting comparisons from Ref. 11 are presented in
Figs. 8 and 9. In the former, zero-time-delay cross-correlation results
obtained at the NAE tower location for both the u- and v-components in the
wind-tunnel and full-scale flows are compared. The six specific values of
vertical separation for which results are shown were defined by the four
levels on the tower where appropriate anemometry was located. The full-
scale points represent the average of results from 14 consecutive 15-minute
data blocks.   The vertical bars on the u-component data are the $\pm 1\sigma$
scatter bars for the 14 blocks and are indicative of the degree of variability in the individual block values. For the u-component of velocity, the
wind-tunnel and full-scale results are seen to compare quite well. For the
v-component, the wind-tunnel correlation curve lies considerably below that
obtained in the prototype flow. This is presumably a result of the absence
in the wind tunnel of the large-scale, low-frequency variations in the
lateral velocity component which are typically present in the atmospheric

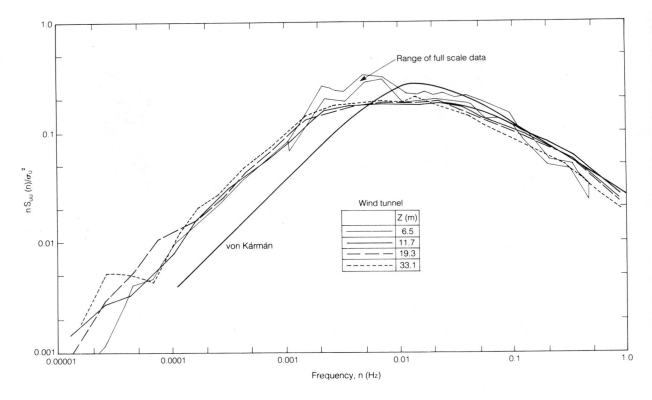

Fig. 7   Comparison of longitudinal-component power spectra at
NAE tower location in full-scale and simulated flows.

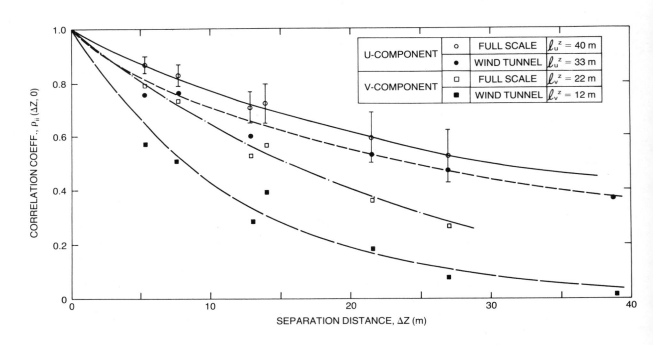

Fig. 8   Comparison of zero time-delay cross-correlations for
vertical separations at NAE tower location.

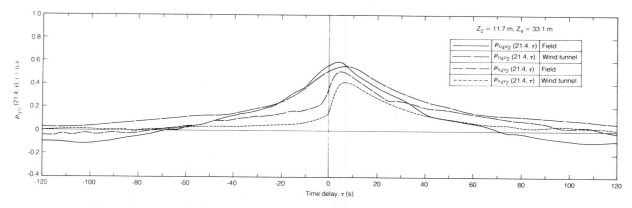

Fig. 9   Comparison of time-delay cross-correlation curves for
large separation on the NAE tower.

flow. The importance of this difference will depend on the problem under
investigation but is probably not too serious for most wind-loading considerations.

The curves of Fig. 9 are time-delayed cross-correlation curves for a
typical value of vertical separation on the NAE tower. Here again, the
u-component results are seen to agree rather well, particularly in the
region of maximum interest (i.e., smallest values of $|\tau|$). The v-component
results are again lower in the wind-tunnel than in full-scale, as expected.
Note however, that the interesting difference in shape between the u- and
v-component curves in full-scale has been reproduced in the wind-tunnel.
Such differences were observed at most separations and were in each case
reproduced quite well in the wind tunnel. Complete details of these and
other comparisons between the simulated and prototype flow at this site
will be available in Ref. 11 and subsequent publications.

4.   SITE C: KETTLES HILL, PINCHER CREEK, ALBERTA

4.1 Prototype Site and Measurements

The Kettles Hill site is intermediate to the Lions' Gate Bridge and
the Rockcliffe Airport sites from the standpoint of complexity of the surface features. The prototype hill (Fig. 10) is located in generally flat
terrain in the foothills of the Rocky Mountains and has relatively little
topographic microstructure superimposed on its general features. It is
generally elliptical in planform (Fig. 11) with major and minor axes about
2000 m and 1000 m in length, respectively. Its height at the summit is
about 100 m, thus producing slopes typically between 0.1 and 0.2, depending
on direction. The ground cover is very uniform both on and around the hill
and consists mainly of short grass, stubble and bare or snow-covered soil
in winter, which is when the full-scale measurements were obtained. There
are a few trees in the area, along with some isolated farm buildings, but
these are not expected to have a significant effect on the results obtained.

This particular site was selected specifically for the purpose of obtaining a full-scale data set describing wind and turbulence structure in
the boundary-layer flow over an isolated, low hill. These data are being
compared with the predictions of wind tunnel and mathematical models of such
a flow for the purpose of assessing the reliability of the models. The field
measurements were obtained during February of 1981 using a set of ten 10-m
towers placed on the hill in the basically straight-line array shown by the

Fig. 10 Telephoto view of Kettles Hill (from $\phi \sim 240°$).

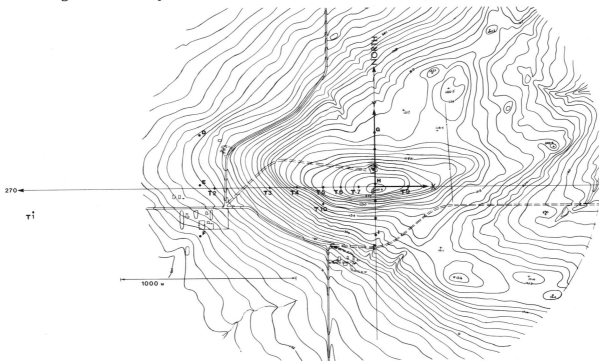

Fig. 11 Contour map of Kettles Hill showing 10-m tower and wind-tunnel reference locations.

numbered locations in Fig. 11. Both mean wind and turbulence data were obtained at three levels on each of these towers. An additional set of ten movable 3-m towers was placed in various configurations on the hill during different phases of the experiment and produced mean speed values at the top of each tower. Several other instrumentation systems were also used at various times during the experiment and are described in detail in Ref. 15.

Most of the full-scale data were obtained for a wind approaching the hill from about 245°, or about 25° south of its major axis (the x-axis in

Fig. 11). Typical mean wind speeds were about 15 - 20 m s^{-1} and available temperature results indicated that the flow was indeed neutrally stable, as desired, during all measurement periods. A total of 15 individual data sets were obtained during the experiment, with each being about 2 hours in length. Complete details of these results will be available in Ref. 15 and subsequent publications.

4.2 Wind Tunnel Simulation

Wind tunnel simulation of the flow at this site is being carried out in the AES boundary-layer wind tunnel at a basic length scale of 1:1000.

Fig. 12. Kettles Hill wind tunnel model (looking downstream from 245°).

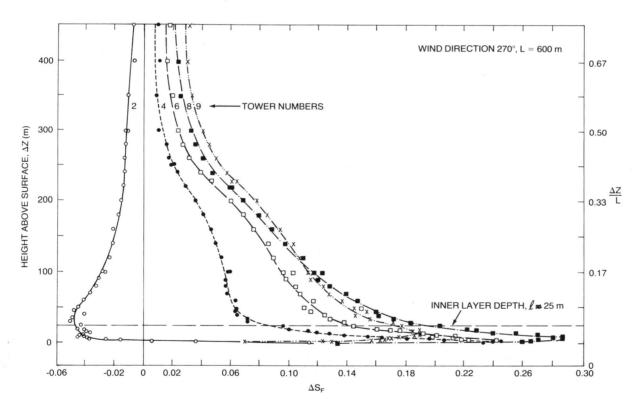

Fig. 13. Fractional speed-up profiles above Kettles Hill wind tunnel model (smooth model).

230

Several different model configurations are being investigated in order to assess the reliability of various modelling techniques insofar as reproducing the full-scale flow characteristics is concerned. For example, different types of surface roughness are being used, and it is intended to investigate the effects of vertical scale distortion on the simulated flow over the hill. A complete description of the experimental techniques being used is given in Ref. 16, along with detailed results obtained from the first model configuration (no scale distortion, rough surface). Fig. 12 gives a view of a typical model (no distortion, smooth surface) installed in the tunnel.

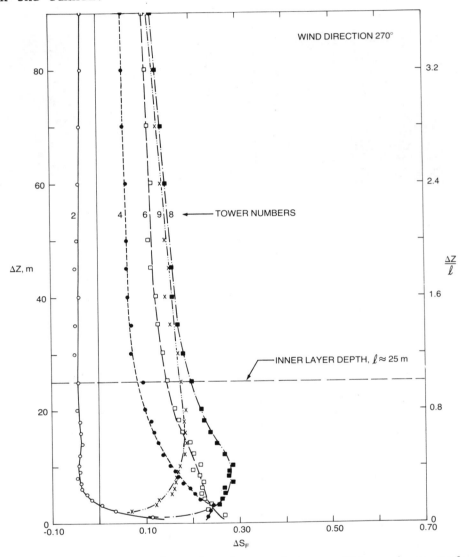

Fig. 14 Near-surface fractional speed-up profiles above model.

Fig. 13 shows some typical vertical profiles of mean flow speed-up at several tower locations for a westerly wind direction over the smooth model. The speed-up is given as a fractional speed-up factor, defined by

$$\Delta S_F = \frac{U(\Delta Z) - U_o(\Delta Z)}{U_o(\Delta Z)} \quad ,$$

where U is the mean speed above the tower location, U_o is the reference speed well upstream of the hill and ΔZ is the height above local ground level. One encouraging feature of these results is the resolution capability of the wind tunnel which they demonstrate. This resolution is estimated to be about $\pm$ 0.005 in ΔS_F in the present case, so that distinct differences could be observed between the results at all tower locations, including those not shown in Fig. 13. Considerable care was required in order to obtain such resolution, since ΔS_F represents a small difference between two large numbers and is therefore very sensitive to small experimental measurement errors (see Ref. 16). The inner layer depth shown in Fig. 13 refers to the 'inner layer' defined by Jackson and Hunt (Ref. 17), above which perturbation shear stresses are assumed to be negligible and the flow behaves essentially as an inviscid shear flow. Fig. 14 is an expanded plot of the speed-ups measured in and above this layer.

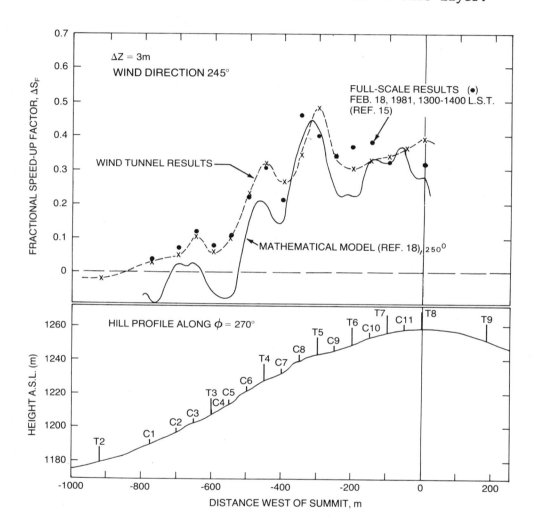

Fig. 15   Comparison of wind tunnel, mathematical and full-scale speed-up results.

Fig. 15 shows a three-way comparison of the mean flow speed-up values at a height of 3 m above the main east-west tower line for a wind approaching the hill from 245°. The mathematical model results shown are those predicted by a three-dimensional version of Jackson and Hunt's analytical model (Ref. 17) which is described in Ref. 18. The full-scale results are those obtained during a one-hour period from the ten 3-m towers and several of the 10-m towers. The wind tunnel results were obtained from profiles of the type shown in Fig. 14. The agreement between the wind-tunnel and full-scale results is seen to be remarkably good. It is particularly noteworthy that the wind tunnel model is able to reproduce most of the local accelerations caused by small, localized 'bumps' on the hill surface. These accelerations were predicted by the mathematical model and are clearly present in the full-scale results as well. Because of the small scale of the wind tunnel model and its very smooth surface (this particular model did not fully meet the requirement of aerodynamic roughness in an attempt to ensure that local roughness effects would not mask the flow characteristics in the inner layer; see Ref. 16), these local terrain perturbations are almost impossible to detect visually, and it is indeed impressive that their effects on the flow are so clearly evident in the speed-up results. Additional comparisons of this type are presently being carried out for both mean speed and turbulence characteristics at various locations on the hill.

5. CONCLUSIONS

Based on the results and comparisons discussed above and the general experience of the author, the following comments and opinions are offered in response to the questions outlined in Section 1:

(i) The quality of neutral-stability wind tunnel flow simulations is in general quite good, at least for the types of flow considered above. While there are obviously certain limitations to the ability of the tunnel to reproduce full-scale characteristics, particularly in the area of the large, low-frequency gusts which are typically found in the atmosphere, there are many problems in which such characteristics may be of little or no importance. In other situations, mere recognition of the limitations of the simulations, in combination with appropriate interpretation and discretionary application of the results obtained, can be sufficient to provide adequate answers to the questions under investigation.

(ii) The question of whether or not sufficient parameters are being considered in evaluating the quality of wind tunnel simulations obviously depends on the particular problem under consideration. Ideally, one might wish to compare as many parameters as possible in each case – for example, mean velocity profiles, turbulence intensities for all velocity components, gust profiles, Reynolds stresses, probability distributions, integral scales, power spectra and/or autocorrelations, two-point cross-correlations or coherences and possibly additional parameters. In Ref. 19, the author considered most of these parameters in comparing the flow simulated in a multiple-jet wind tunnel with the atmospheric prototype, with generally very successful results. The main limitation of such efforts is that the prototype characteristics themselves are usually obtained from the aforementioned idealized engineering models of the flow, which in many cases are derived from very few actual atmospheric data. As has been pointed out in discussing the results above, the characteristics of the prototype flow are extremely variable and are not necessarily adequately represented by the idealized models, even in apparently very simple cases. It is therefore clear that additional, more detailed measurements of prototype flow characteristics are needed in order to maximize the usefulness of more extensive wind tunnel flow measurements, particularly for flows over the more complex and non-ideal terrain types which are important for many typical wind engineering problems. For many of these problems, of course, such additional measurements may not be necessary, and the basic set of parameters which are usually considered (e.g., mean velocity profiles, turbulence intensities and power spectra) should suffice.

(iii) As for the acceptable level of approximation in any wind tunnel simulation, this again must be a function of the particular problem under consideration. The dependence of bridge response on turbulence in the approach flow indicates that gross turbulence levels in the simulated flow must certainly be modelled correctly. It would seem unlikely, however, that differences of the order of 10 - 15% from the values actually observed in the atmosphere would be very important. On the other hand, some response phenomena may indeed be very sensitive to small changes in the flow characteristics and detailed, organized 'sensitivity testing' may be needed to identify the required level of accuracy in such cases. In the example of flow over a model apartment building in Section 3.2.1, the immediate problem under investigation was whether or not the wind speeds measured at the full-scale reference location were influenced by the building itself. Since this location turned out to be well above the building-induced shear layer for all flows (as was hoped would be the case), the results in this case were almost insensitive to even very large changes in the characteristics of the approach flow. For lower potential reference locations, of course, this would clearly not be the case. Thus, as would be expected, the acceptable level of approximation is strongly dependent on the specific problem under investigation and hence must be individually assessed in each case.

ACKNOWLEDGEMENTS

The author wishes to acknowledge the assistance of the many people involved in obtaining the results discussed in this report. In particular, the contributions of Dr. R.G.J. Flay, Mr. I.R. McLean and Mr. K.O. Vanek are gratefully acknowledged.

REFERENCES

1. H. W. Teunissen and C. D. Williams, Full-scale measurements of atmospheric turbulence on a suspension bridge, Report MSRB-78-6, Atmospheric Environment Service, Toronto, December, 1978.

2. H.P.A.H. Irwin, Wind tunnel and analytical investigations of the response of Lions' Gate Bridge to a turbulent wind, Report LTR-LA-210, National Research Council, Ottawa, June 1977.

3. H.P.A.H. Irwin and G.D. Schuyler, Experiments on a full aeroelastic model of Lions' Gate Bridge in smooth and turbulent flow, Report LTR-LA-206, National Research Council, Ottawa, October, 1977.

4. H.P.A.H. Irwin, Further investigations of a full aeroelastic model of Lions' Gate Bridge, Report LTR-LA-221, May 1978.

234

5. A.G. Davenport, N. Isyumov and T. Miyata, The experimental determination of the response of suspension bridges to turbulent wind, Proc. 3rd Int. Conf. on Wind Effects on Buildings and Structures, Saikon Shuppan Co., Japan, 1971.

6. H. W. Teunissen, Characteristics of the mean wind and turbulence in the planetary boundary layer, UTIAS Review No. 32, University of Toronto, October 1970.

7. Engineering Sciences Data Unit, Characteristics of atmospheric turbulence near the ground, Part II, ESDU Item 74031, London, October 1974.

8. J. Counihan, Adiabatic atmospheric boundary layers: A review and analysis of data from the period 1880-1972, Atmos. Envir. 9 (1975) 871-905.

9. M. Shiotani and H. Arai, Lateral structures of gusts in high winds, Proc. 2nd Int. Conf. on Wind Effects on Buildings and Structures, University of Toronto Press, 1968.

10. H. W. Teunissen, Measurements of planetary boundary layer wind and turbulence characteristics over a small suburban airport, J. Ind. Aerodyn. 4 (1979) 1-34.

11. R.G.J. Flay and H.W. Teunissen, Comparison of simulated and full-scale wind structure in the planetary boundary layer over a small suburban airport, Report MSRB-82-1, Atmospheric Environment Service, Toronto, to be published.

12. H.W. Teunissen, Structure of mean winds and turbulence in the planetary boundary layer over rural terrain, Boundary-Layer Meteorol. 19 (1980) 187-221.

13. H.W. Teunissen, Wind tunnel investigation of the air flow over a model apartment building, Report MSRB-80-2, Atmospheric Environment Service, Toronto, April 1980.

14. R.N. Meroney, A.J. Bowen, D. Lindley and J.R. Pearse, Wind characteristics over complex terrain: Laboratory simulation and field measurements at Rakaia Gorge, New Zealand, Report No. TH/FL102/78, University of Canterbury, Christchurch, N.Z., November 1978.

15. P.A. Taylor, R.E. Mickle, J.S. Salmon and H.W. Teunissen, The Kettles Hill experiment: Site description and mean flow results, Report to be published, Atmospheric Environment Service, Toronto.

16. H.W. Teunissen and R.G.J. Flay, Wind-tunnel simulation of planetary boundary-layer flow over an isolated hill: Part I, Rough model, Report MSRB-81-3, Atmospheric Environment Service, Toronto, November 1981.

17. P.S. Jackson and J.C.R. Hunt, Turbulent wind flow over a low hill, Quart. J. Roy. Meteor. Soc. 101 (1975) 929-955.

18. J.R. Salmon, J.L. Walmsley and P.A. Taylor, MS3DJH/2 - Development of a model of neutrally stratified boundary layer flow over real terrain, Internal Report ARQB-81-023-L, Atmospheric Environment Service, Toronto, March 1981.

19. H.W. Teunissen, Simulation of the planetary boundary layer in a multiple-jet wind tunnel, Atmos. Envir. 9 (1975) 145-174.

SOME COMMENTS ON WIND TUNNEL MODELING BY A METEOROLOGIST

Hans A. Panofsky, Pennsylvania State University

1. INTRODUCTION

Atmospheric flows differ from typical other flows in three main ways:

   a)   they occur on a rotating earth

   b)   vertical temperature stratification may become important

   c)   average flows are likely to vary in time and space.

Because of these factors, serious problems may arise in attempting to model atmospheric flows in wind tunnels.

2. THE EFFECT OF THE EARTH'S ROTATION

By convention, in the surface layer, the lowest 10 percent of the atmospheric boundary layer (PBL), the influence of the earth's rotation can be disregarded. This means that, among other things, the wind direction and friction velocity u_* can be taken as constant with height.

With strong winds, the thickness of the surface layer is 100 m – 300 m thick. In this layer, the logarithmic wind profile is a good fit, except that, at the top of the surface layer, heat convection may decrease the wind shear.

Above the surface layer, the wind turns with height. This means that no simple correction to the adiabatic profile is appropriate; further, the "matching theory" discussed by Tieleman in this session requires an additional dimension in the atmosphere, which is not available in the wind tunnel.

3. THE EFFECT OF VERTICAL STRATIFICATION

In many engineering applications, modeling conditions with strong winds are especially important. In that case, it is usually assumed that the influence of heat convection can be disregarded. In such cases, logarithmic profiles prevail for wind speed and scalars; variances of the wind components are essentially constant with height in the surface layer, and spectra have relatively simple properties. If power laws are adapted to the wind profiles, the best fit for the power p is given by

$$p = \frac{1}{\ln \dfrac{z}{z_o}}$$

where z is the geometric mean over the range of heights to which the law is to be fitted, and z_o is the roughness length which, over homogeneous terrain, is constant.

The limits to which these "mechanical" approximations are adequate depends on many factors such as wind speed, roughness, heat flux, and height. The larger the height, the more important convection becomes. Also, the strength of convection increases with vertical flux of heat (to which a moisture term must be added under tropical, humid conditions).

The relative importance of convection and mechanical turbulence is judged by the Richardson number. But since the Richardson number varies with height in an a priori unknown fashion, meteorologists judge the influence of convection by the ratio z/L where L is a parameter which is almost constant in the surface layer. L is defined positive with downward heat flux (usually at night) and negative with upward heat flux.

Typical values of L are -10 m with weak winds in the daytime, -100 m with moderate winds in the daytime, and below -1000 m for very strong winds (wind at 100 m $\sim$ 50 m/sec). $-L$ is approximately proportional to the third power of wind speed.

Thus, it is clear that in extreme winds z/L will generally be much smaller than 0.1 in which case the influence of convection is not important.

But consider the more typical value of $L = -100$ m in moderate winds and moderate heat flux, and a roughness length of 20 cm. If convection is neglected, the ratio of wind speeds at 100 m and 10 m is 1.59; including a correction for convection, the ratio is only 1.39. It is clear that the correction is important under such conditions. In general, we must consider the influence of convection on wind profiles except in case of extreme wind speeds. Cermak's wind tunnel measurements reported earlier in this session confirm the full-scale analysis.

Of course, it is possible to duplicate convective wind profiles artificially, rather than letting them develop naturally, which requires huge vertical temperature gradients. But it is doubtful whether an artifically created wind profile would produce appropriate statistics of the fluctuations, some of which are noted below.

The effect of convection on variances and spectra of horizontal velocity components is dramatic. Spectra with $z/L = 0$ and $z/L = -0.03$ have significantly different appearances (see figure 1). In fact, it was thought for some time that the limits of spectra as $z/L \to 0$ were different when approached from the stable side and from the unstable side. In other words, a very small addition of convection causes an almost explosive growth of turbulence, as first noted by Deardorff.

Vertical velocity spectra scale with z/L, as do wind profiles. But it is now clear that spectra of the horizontal components under convective conditions have two separate portions, which when added yield the observed spectra (Hojstrup, 1981). At the lower frequencies, in the most energetic portion of the spectra, they scale with h/L, where h is the thickness of the boundary layer. The high-frequency portion scales with z/L. Thus, close to the ground, the spectra widen and may show even two peaks. Modeling these characteristics should not present particular difficulties in wind tunnels in which large vertical temperature gradients can be produced. But in the usual wind tunnels, it is very likely that the variances and spectra at low frequencies will be underestimated.

4. HYDROSTATICALLY STABLE AIR

On clear nights, the characteristics of atmospheric flow are complex and not well understood. Typically the temperature increases upward to a level ℓ, and decreases above. Sometimes ℓ is used to define the planetary boundary layer because the atmosphere below ℓ has been cooled as the surface cools. But this cooling is only partially caused by continuous turbulence. It is due, in part, to intermittent turbulence and to infrared radiation. Further, quasihorizontal gravity waves are often superimposed on this flow. Only the very lowest part of the boundary layer (the height depending on wind speed) is continuously turbulent. This region is usually 100 m or so thick. Here, the flow may be the easiest to model; the remainder of the region appears out of reach at this time. This is especially true for the large, slow variation of wind direction typical in stable air, which are important for dispersion. Teunisseun has shown in this session that the effect of hydrostatic stability may be important even with strong winds; in his Lyons Gate experiment, a stable layer just above the water strongly affected the turbulence behavior on the bridge.

5. LACK OF HOMOGENEITY

In most of the previous papers, the roughness length z_o was needed as a parameter to describe the wind profile. It has been discovered recently (see e.g., Korrell, et al., 1982) that z_o may vary with height; e.g., at several towers, the roughness lengths are of the order of 10 cm, or even 1 m. Yet, wind profiles on low masts in the same area may be much smaller, e.g., 1 cm. The reason for this apparent vertical variation is that z_o determined from a low mast reflects the ground cover in the immediate environment. But the z_o computed from tower data is influenced by a wider area, and is enlarged by form drag produced by rolling terrain.

Even though z_o may thus vary with height, it is not influenced by hydrostatic stability, in contrast to the power in a power law.

Other effects of heterogeneous terrain on wind profiles, e.g., changes of terrain roughness and of hills, have been discussed by Britter in this session. These are now quite well understood and seem to produce no special difficulties to wind tunnel model studies. The same conclusion was reached by Teunissen.

6. LACK OF STATIONARITY

Meteorological time series are rarely stationary. In practice, meteorologists separate "slow" and "frequent" variations by quite arbitrary procedures. They may fit lines of regressions to the data and define turbulence as deviations from such lines, or they may define the large-scale flow by moving averages.

Because the procedures are not standardized, the low-frequency portions of horizontal turbulence spectra may vary, depending on the analysis. Hence, also "integral scales" and "variances" will depend on the separation techniques. Therefore, meteorologists avoid integral scales, both lateral and longitudinal. Different attempts to relate integral scales to easily measured parameters (e.g., those by ESDU) differ greatly from each other. This is because integral scales vary more with analysis technique than with meteorological variables. Of course, if only stationary records are used, this problem does not arise. Recent studies of long records at the 300 m tower of the Boulder Atmospheric Observatory (BAO) by Kaimal (unpublished) suggest that hour-long stationary turbulence is not common.

Since the low-frequency portion of atmospheric spectra is poorly defined, it is probably best to avoid total turbulent intensities (variances) discussed in this session by Surry. Instead, the turbulent intensity should be obtained by integrating the spectra over the frequency range of interest.

There are other reasons why the low-frequency portion of the spectra of horizontal wind components is uncertain. As Tieleman pointed out in this session, rolling terrain may increase low-frequency turbulence; and on mountains (see Britter, Hunt and Richards, 1980) the turbulence may be decreased.

Also, we pointed out earlier that very small changes in heating from below can lead to radical low frequency changes, even in strong winds.

Finally, gravity waves affect low frequency variance but this is a problem only in low-wind, stable air.

We can draw two quite general conclusions: one, the low-frequency portion of the spectra of horizontal velocity components are extremely variable and uncertain. Secondly, the frequently used von Karman spectrum is too narrow (see e.g., Teunissen's paper in this session).

7.   THE REYNOLDS NUMBER

The ordinary (molecular) Reynolds number is never simulated in wind tunnels, because molecular friction usually is not important either in the atmosphere or in the tunnel. However, because the frequency extent of the inertial surrange depends on $(Re)^{3/4}$ (Re = Reynolds number) and wind tunnel Re-values are relatively small, wind tunnel turbulence tends to be characterized by considerably smaller inertial ranges than full-scale turbulence. This may be important in problems of resonant response in this frequency range.

As Britter has just suggested, it may be useful to simulate the "turbulent" Reynolds number in which the molecular viscosity ν is replaced by the turbulent viscosity $K \sim u_* z$ where z is the height.

We then have

$$Re_{turb} \cong \frac{UL}{u_* z}$$

Normally, z is scaled in the same way as L, the vertical scale of the model; U/u_* depends only on z/z_o, in strong winds. Hence, if z_o is scaled in the same way as L and z, Re_{turb} is preserved; this is why, e.g., Cook, Britter and others get good results when the ratio of full-scale roughness to model roughness is the same as the ratio of the other length scales.

8. DERIVATION OF LAWS FOR INTERFACES

When terrain changes roughness, the air layer close to the ground is modified immediately, and the air higher up more slowly. Hence, there develops an interface which separates these two air masses.

Miyake (unpublished MS thesis, University of Washington, 1958) has given a simple derivation for a formula describing such an interface. The influence of the roughness change spreads horizontally with the wind speed, U, and vertically with a velocity proportional to the standard deviation of vertical velocities. This, in neutral air is proportional to u_*. Hence

$$\frac{dh}{dx} = A\frac{u_*}{U}$$

where x is downwind distance, h is the height of the interface, and A is a constant of order 1.

Integrating, and neglecting a usually small term, we get

$$\frac{x}{z_o} = B\frac{h}{z_o} \ell n \frac{z}{z_o}$$

where B is another constant.

Note that this is the equation quoted by Britter in this session, derived in a much more complex manner. It also represents the upper boundary of effluent from a line source at right angles to the wind. Almost the same equation (except that x is replaced by L) represents the thickness of the "inner" layer on a hill top.

9. VERTICAL SEPARATION

Meteorologists have found that relatively simple models can be constructed for coherence and phase differences, as a function of frequency, for wind speeds measured at different heights.

Coherence obeys Davenport's similarity (1961) in an approximately exponential manner

$$Coh\ (\Delta z,\ t) = e^{-a\ f\frac{\Delta z}{V}}$$

Here Coh is the "square" coherence, "a" is a constant and Δz is the vertical separation. The quantity "a" increases with increasing z/L and becomes very large for $z/L \sim 0.5$. For more stable air, "a" becomes relatively small, presumably due to vertically coherent gravity waves. For a discussion, see Soucy, et al. (1982).

Unfortunately, Teunissen in his analysis in this session, only shows vertical cross correlations and does not evaluate coherence and phase so no direct comparison with earlier data is possible at this time.

Because of wind shears, wind fluctuations at higher levels precede those at lower levels. Measurements show that this phase delay is approximately proportional to fΔz/V. The factor of proportionality is much larger for the lateral than the longitudinal velocity component. Teunissen's data agree, at least qualitatively.

10. REFERENCES

1. R. E. Britter, C.J.R. Hunt and K.J. Richards, "Airflow over a two-dimensional hill studies of velocity speed-up, roughness effects and turbulence," Quart. J. Roy. M.S., 107 (1980) 91-110.

2. A. G. Davenport, "The spectrum of horizontal gustiness in high winds," Quart. J. Roy. Met. Soc., 87 (1961) 194-211.

3. J. Hojstrup, "A Simple Model for the Adjustment of Velocity Spectra in Unstable Conditions Downstream of an Abrupt Change in Roughness and Heat Flux," Bound. Layer Met., 21 (1981) 341-358.

4. A. M. Korrell, H. A. Panofsky and R. J. Rossi, "Wind profiles at the Boulder Tower," Bound. Lay. Met., 22 (1982) 295-312.

5. R. Soucy, R. Woodward and H. A. Panofsky, "Vertical Cross Spectra of Horizontal Velocity Components at the Boulder Atmospheric Observatory," accepted by Bound. Lay. Met. (1982).

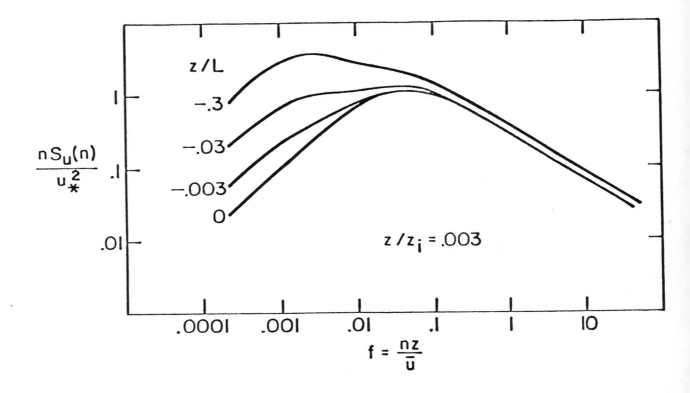

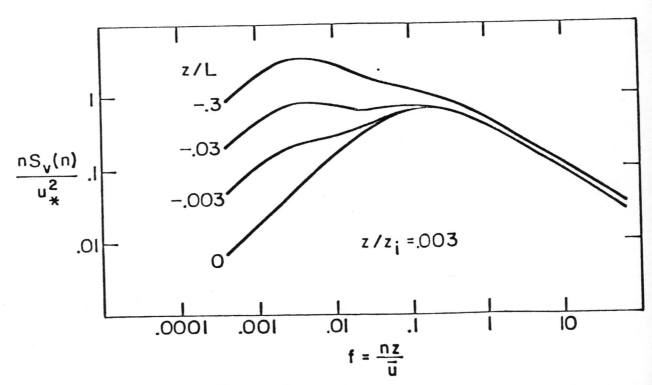

Figure 1. Influence of Convection on Spectra of Horizontal and
Vertical Velocity Components.

Session III

Pressure and Force Measurements

Chairman

Jon A. Peterka

Colorado State University
Ft. Collins, Colorado
USA

TECHNIQUES AND MODELLING CRITERIA FOR THE MEASUREMENT OF EXTERNAL AND INTERNAL PRESSURES

J D Holmes, CSIRO Division of Building Research (Australia)

1.   INTRODUCTION

The introduction of the boundary layer wind tunnel with the reproduction of the turbulent flow in the atmosphere, and the development of statistical techniques for wind loading, in the nineteen sixties and seventies, was accompanied by parallel developments in instrumentation and techniques for the measurement of fluctuating pressures and forces on wind tunnel models of structures. There is now a well-established fund of technology for such measurements from which the newcomer to wind engineering can drawn. Information on some of these techniques is not to be found in the open literature, however, and it is timely to survey the current state of the art; this is being attempted in the present paper.

In this paper, an attempt has been made to overview the techniques required to measure fluctuating pressures on rigid models of structures, i.e. models for which no attempt is being made to reproduce resonant dynamic structural behaviour in wind. It is inevitable that there will be considerable overlap between this paper and other papers of this Volume.

The topics covered are: flow simulation (Section 2), model design (Section 3), instrumentation (Section 4), and reference pressures and corrections (Section 5).

2.   FLOW SIMULATION

The difficulties in modelling the atmospheric boundary layer satisfactorily, increase as the available fetch length/test section height ratio decreases. Unfortunately low values of the latter ratio (5 or less), are a reality for operators of ex-aeronautical wind tunnels, and for those with limited building space available. However, operators of even the longest tunnels today, apparently cannot rely on natural boundary layer growth over floor roughness alone, when modelling buildings for wind loads at scales of 1/500 or greater. The use of additional drag-producing devices such as grids, spires and fences is necessary. The techniques seem to have reached the greatest sophistication (and complexity!) in the United Kingdom where mainly short test section tunnels are available. The difficulties in using short wind tunnel test sections to model the boundary layer for low-rise building wind load studies were also shown by Tieleman, Reinhold and Marshall[1].

At James Cook University and at other laboratories in Australia, fetch lengths of 6-10 times the test section height are available. The most common technique in Australia, is to mount a single plain fence spanning the floor at the start of the test section. This fence performs two functions: firstly it acts as a boundary layer 'trip' providing a momentum deficit, and giving a much thicker boundary layer at the end of the test section than occurs with surface roughness alone. Secondly for the higher fences, the turbulence scales in the decaying wake behind the fence are an important contribution to the low frequency end of the turbulence spectrum. These large scales are important in the testing of low-rise buildings, when geometric scaling ratios of 1/300 or greater are necessary to model the buildings in sufficient detail. Vertical members, such as spires, appear to be less suitable for simulations at large scale for low-rise building tests, due to the small scales of turbulence they produce. However, a grid upwind of the barrier or fence, as described in this Volume by Cook, may assist in promoting earlier reattachment of the separated shear layer downwind of the fence, and allow satisfactory simulations to be achieved in a shorter fetch length.

In using geometric scaling ratios of 1/300 or larger, clearly only the lower part of the atmospheric boundary layer can be modelled in wind tunnels of 1-2 m height. This should cause no problems for wind loading studies of low-rise buildings, as the reference height is normally taken either as the building height or a height equivalent to 10 m in full scale.

Fluctuating and peak point pressures on low- and high-rise buildings are found to be closely related to the upwind turbulence intensity (e.g. Marshall[2]). It is therefore important to model this parameter correctly as well as the mean velocity profile. Figure 1 shows mean velocity and turbulence intensity profiles obtained at James Cook University with a 200 mm high fence upwind of about 12 m of carpet roughness. The cross-section dimensions of the test section are: height 2 m, width 2.5 m. Reference curves based on the logarithmic law, with a roughness length of 20 mm at a scaling ratio of 1/200, are found to match the measured data well in the lower part of the simulated boundary layer. Figure 2 shows the longitudinal turbulence spectrum at a height of 50 mm in the wind tunnel (equivalent to 10 m in full scale at the scaling ratio of 1/200). The Harris-von Karman spectrum with the length scale appropriate to the roughness length of 20 mm suggested by the Engineering sciences Data Unit (E.S.D.U.[3]) is also shown. Good agreement with this reference curve is apparent.

At larger geometric scales of 1/100 and 1/50, less good agreement is achieved with the E.S.D.U. curves for the longitudinal spectrum, although the fence height is increased for these cases, and the turbulence intensity can be matched well. At James Cook University, ratios of achieved to desired turbulence scales down to 0.5 or so, are tolerated for low-rise building measurements where local cladding pressures only are of concern.

If the determination of area averaged loads and structural effects is required, the smaller ratio of 1/200 is preferred, where possible. A similar ratio of 1/250 was used for the majority of tests in the recent study of low-rise buildings at the University of Western Ontario (Davenport, Surry and Stathopoulos[4,5]).

Exact agreement with standardised turbulence spectra such as those recommended by E.S.D.U., should be judged in context with the substantial scatter in the available full scale data on which these curves are based. Also these data were invariably obtained from large scale depression systems occurring at temperate latitudes. In Australia at least, most of the damage due to wind occurs during storms of much shorter duration – tropical cyclones, thunderstorms, tornadoes and frontal squalls. There is a need for more full scale data for storms of these types, but it would be surprising if the turbulence scales for these events are found to be larger than those occurring in storms caused by the larger pressure systems.

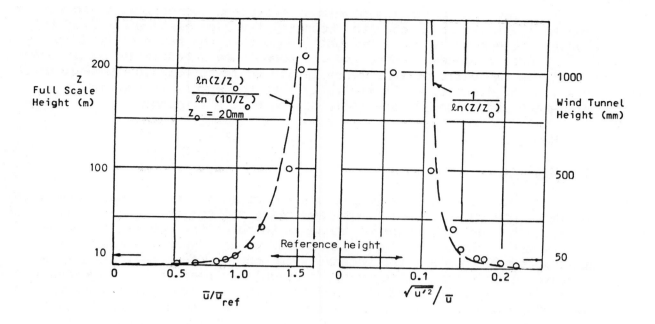

Figure 1. Mean velocity and turbulence intensity profiles for 1/200 rural boundary layer simulation

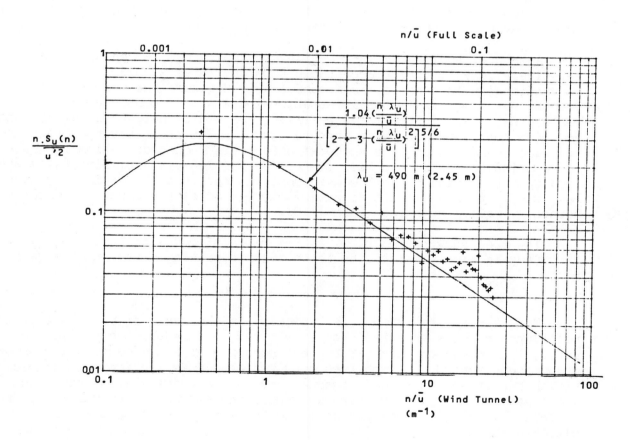

Figure 2. Longitudinal turbulence spectrum for 1/200 simulation z = 10 m (50 mm)

The practice of testing models at different geometric scaling ratios from that which best matches the boundary layer parameters can lead to significant errors as shown by Hunt[6]. Note that the ratio of building dimension to roughness length and the turbulence intensities are in error, as well as the ratio of turbulence length scale to building dimension, in this type of distortion. It is likely that the errors in the former parameters had a significant influence on Hunt's findings.

3. MODEL DESIGN

3.1 Scaling

The selection of a geometric scaling ratio for wind tunnel models to determine pressures on buildings requires a compromise between a number of factors. The most important of these is to match the geometric scaling of the atmospheric boundary layer simulation as discussed in Section 2. The other major factors are the necessity to model surface features on the building with sufficient accuracy, particularly near those areas where flow separations may occur, and the need to match the frequency response at the available pressure measurement system to the desired full scale frequency response. The question of wind tunnel blockage effects should also be considered at the largest scales. This topic is discussed in detail by Melbourne in this Volume; however, blockage corrections are normally not made if area ratios less than 5% are maintained. A further constraint on the scaling ratio for particular buildings in urban and/or hilly environments, may be the need to fit surrounding large structures or topography within the available test section dimensions.

A natural scaling ratio for high-rise buildings, at which it is relatively easy to obtain acceptable boundary layer simulations in conventional-size tunnels, is around 1/500. However, the scaling ratios used by participants in the international C.A.A.R.C. model tests (Melbourne[7]) ranged from 1/240 to 1/690. For medium height buildings, and low-rise buildings for which integration of point pressures to obtain area-averaged loads and other structural effects is required, ratios in the range 1/100 to 1/250 have been preferred. For low-rise building tests where local point pressures are the main concern, ratios as large as 1/50 have been used. It is difficult to satisfactorily model the important details on houses at ratios much less than 1/100.

The frequency ratio, n_r, i.e. the ratio of model to full scale frequency is simply related to the geometric scaling ratio and the velocity ratio

$$\text{i.e.} \quad n_r = u_r / l_r$$

The velocity ratio, u_r, is typically in the range of 0.25 to 1. This relationship shows that the frequency ratio, n_r, reduces as the length ratio, l_r, increases, i.e. it becomes possible with a wind tunnel pressure measurement system of fixed frequency response (see Section 4.2) to measure higher equivalent full scale frequencies as the model scaling ratio is increased. This is of considerable importance as short term suction peaks in separated flow regions are believed to be a major cause of local cladding failures on buildings.

The question of Reynolds Number similarity has been ignored in the scaling discussion in the previous paragraphs; it is of course not possible to approach full scale Reynolds Numbers in conventional wind tunnels with even the largest geometric scales. However, intuition and the available experimental evidence suggests this is very much a secondary effect for sharp edged buildings where points of flow separation are fixed. For circular cylindrical buildings and cooling towers and curved roof buidings where the separation points are not well-defined, a commonly used approach is to roughen the surface of the wind tunnel models to attempt to force the flow into the 'fully-rough' regime of higher Reynolds numbers. This topic seems to be a fruitful area of research for the few compressed air wind tunnels in existence, in which higher Reynolds Numbers can be achieved; however, it is essential that a full atmospheric boundary layer simulation be achieved in such tests.

3.2 Model Construction

Although wooden or metal (usually aluminium alloy) models are occasionally used, it is most common to manufacture rigid models of buildings for pressure measurement, from methyl methacrylate sheet ('Perspex', 'Plexiglass', 'Lucite'). This material can be easily and accurately drilled and machined, and is transparent, facilitating observation of pressure tubing inside the model. However, it is prone to chipping or cracking if not handled with care. Because it is a thermosoftening type of plastic, this material can also easily be formed into curved shapes if required, by heating to about 200°C. Often models are cemented together, but it is preferable to join the model panels together by means of flush mounted screws to enable easy access to pressure taps. Care should be taken to avoid air leakage between the joints however.

Conventionally, pressure tappings have been made from pieces of metal tubing with 1.5 mm external and 1.0 mm internal diameters. This allows for convenient connections via pressure tubing to common sizes of connections on 'Scanivalves' and pressure transducers.

Gumley[8] has reviewed the design of pressure tappings and concluded that an internal diameter of 1 mm is a maximum to avoid errors due to curvature in boundary streamlines. Quoting work of Rayle and Armentrout and Kicks, Gumley[8] advocated countersinking the pressure tapping to a depth of one-eighth of the orifice diameter. If flush tappings are used, all burrs on the edge of the tapping should be removed.

3.3 Location of Pressure Tappings

A disadvantage of the single point pressure tapping technique for the determination of wind loads, is that large numbers of pressure taps are required to adequately cover the building surface. Also, even for cladding loads, the single pressure tapping overestimates pressure peaks, since its area is generally less than the tributary area of a cladding panel. There is now an increasing tendency to use manifolding techniques by means of which fluctuating pressures from arrays of pressure taps are 'pneumatically averaged' to obtain area-averaged loads. This topic is discussed in detail by Stathopoulos in this Volume.

However, the single point pressure tapping technique gives a better definition of the gradients of mean and fluctuating pressures over the building surfaces, and enables the upper limit to the peak pressures to be determined. There will continue to be a place for this basic technique for fundamental studies such as full scale/model comparisons.

Pressure tap locations should be more concentrated in regions of high pressure gradients, particularly in regions under separating flows near the corners of walls, and at the leading edges and ridges of roofs. Symmetry can often be used to advantage either to limit the number of pressure taps, or the number of wind angles tested. Figure 3 shows a distribution of pressure taps used for recent tests carried out on gable-roofed houses with various roof pitches at James Cook University. Note that pressure tappings above and below the eaves were provided on these models. In this case, the wind directions tested were confined to a single quadrant.

3.4   Internal Pressure Models

Internal pressures can be measured in the same way as external pressures, although, because of the need to provide a clear internal space, it is desirable to provide double walls on the model to allow for the pressure tubing to be brought in without interfering with the internal air flow.

Internal pressure fluctuations have been found to be highly correlated within the internal volume, particularly when there are only one or two wall or roof openings (Davenport, Surry and Stathopoulos[5]). This is compatible with the Helmholtz resonator model for internal pressure fluctuations (Holmes[9], Liu and Saathoff[10]). This can be used to advantage to limit the number of pressure taps required within an internal volume. Pressure taps can also conveniently be located on the internal floors.

Since density changes are involved in the air flow in and out of buildings with openings of small size relative to the total surface area, Mach Number scaling emerges as a requirement when modelling fluctuating internal pressures (Holmes[9]). Mach Numbers are equal in model and full scale only if the wind tunnel testing is carried out at full scale wind speeds, which is not normally possible. A computer simulation study by Holmes[9] showed that the error when testing at one third of full scale velocity is not likely to be large for small buildings. However, for larger buildings with lower Helmholtz resonance frequencies, the effect could be of some significance. However, Holmes also showed that correct scaling could be maintained if the parameter $u^2 V_0$, where V_0 is the internal volume, is maintained constant in full and model scale. This can be achieved by providing additional volume open to the inside of the model beneath the test section.

4.   INSTRUMENTATION

4.1   Pressure Transducers

The performance of miniature electronic pressure transducers has been improved greatly over the last two decades. It is now possible to obtain instruments with a flat frequency response up to several thousand Hertz with a very stable sensitivity. The zero offset voltages may still be found to be temperature dependent but can be controlled in a wind tunnel environment. The desirable range for wind engineering applications is approximately ± 1000 Pa, which is at the lowest end of the range of most manufacturers. Less sensitive transducers with a larger range can be used with larger amplifications, at the expense of greater drift.

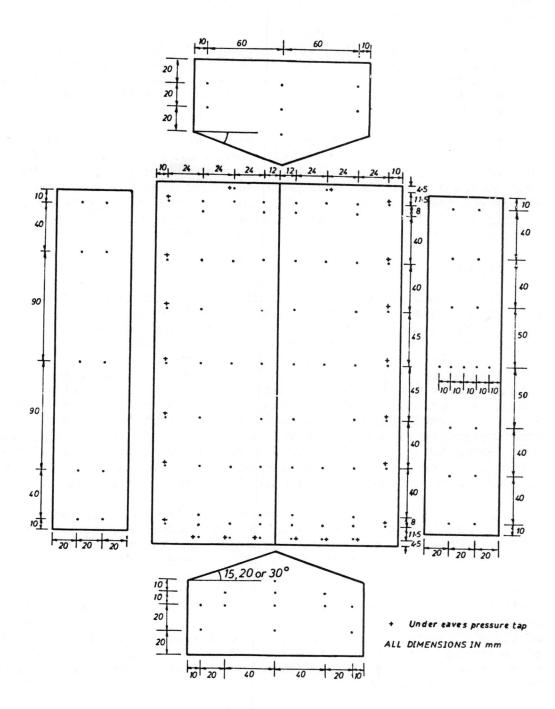

Figure 3. Distribution of pressure tappings for a
low-rise building study

The capacitative-type Setra 237 transducer is probably the most widely used instrument in wind engineering, at present. It has the advantages that it can be used without any additional amplification, and that it is made to fit inside the Type D 'Scanivalve' pressure scanning switch. It has a disadvantage in that it has a low frequency response for pressure fluctuations applied to the back of the diaphragm so that it cannot be used for accurate pressure difference measurements when both pressures are fluctuating, e.g. for the pressure across the overhanging eaves of a low-rise building model.

Other types of transducer in common use are: Statham, Validyne, Bell and Howell (C.E.C.), Kulite, Druck, Scanco and D.I.S.A. A recent development has been the emergence of cheap integrated circuit-based transducers such as the range produced by National Semiconductor. However, at present they require high amplification. Cheap microphones made for acoustics measurements are also available, and can be used if they can be adapted electronically to read down to zero Hertz.

4.2 Pressure Tubing Systems

For reasons of space and economics, it is not usually feasible to mount a separate pressure transducer adjacent to each pressure tapping on a building model. Most commonly the pressure transducer is mounted within a pressure scanning switch such as a 'Scanivalve' enabling up to 48 pressure tappings to be connected as required to the single transducer. Normally the 'Scanivalve' must be mounted beneath the wind tunnel floor so that tubes of lengths in the range 250-500 mm must be used for the connections to the model. Transparent, flexible, vinyl or polyethylene tubing is usually used. Tubes of internal diameters of 1-1.5 mm with lengths in this range, exhibit strong amplification of pressure fluctuations in the range of frequencies above about 50 Hz, with fundamental peaks in the range 100-200 Hz, due to 'organ pipe' resonances. These frequencies are in the range of interest for wind pressures on buildings, and modification or corrections are required.

A variety of methods have been used to attenuate the resonant peaks to produce a flat frequency response. The most common approach is to use some form of mechanical constriction to produce additional energy losses and hence damping in the system. Early methods used bent metal tubes or pieces of cotton wool inserted partway along the tubing (Dreher and Cermak[11]). More recently the most common approach has been to insert one or more sections of small diameter tubing in the line usually near the halfway point. These may consist of pieces of steel capillary tubing glued within sections of larger metal tubing to which the vinyl tubing is connected, or brass tubing compressed on to a small diameter 'keeper' wire by a specially made press (as used at the University of Western Ontario and the Building Research Establishment, U.K.).

The theoretical work of Bergh and Tijdeman[12] has been adapted to the design of pressure tubing with small diameter restrictors by Gumley[8] (Gumley also studied the response of parallel input manifold systems). However, it is advisable to check the response of systems experimentally, using either sinusoidal or random excitation. The system used at James Cook University uses wind tunnel generated fluctuating internal pressures in a cavity, with a single windward circular opening. As discussed in Section 3.4, well correlated pressure fluctuations are generated within the cavity and the output of experimental pressure tubing systems can be compared with the reference output from a transducer connected by a short tube, or flush mounted to the cavity. However, with this technique, it is necessary to ensure that the cavity volume is large in comparison with that of the tubing, and also that the Helmoltz resonance frequency is well above the frequency range of interests. At other laboratories, special calibration rigs have been constructed (e.g. Gorove and Cermak[13], Kao[14], Irwin, Cooper and Girard[15]).

Alternative approaches to the modification of the response of pressure tubing systems are the use of analogue or digital filters (Irwin, Cooper and Girard[15], Vimal du Monteil[16]); in the latter case, use is made of the efficient Inverse Fast Fourier Transform Technique. A hard-wired 'Pressure Measurement Equaliser' has recently been marketed commercially by M.H.T.R. Ltd.

4.3 Data Processing

Most wind tunnel laboratories in the last decade have moved in the direction of digital data processing, usually with a dedicated on-line minicomputer system. Standard analogue/digital conversion systems normally have sampling rates which exceed the frequency requirements in wind engineering applications, except perhaps when multiplexing a large number of input channels. Such systems allow for standard statistical analysis of fluctuating pressure records for mean, standard deviation and peak values, for spectral analysis, and for the extreme value analysis techniques described by Cook and Peterka in this Workshop.

The availability of high speed disc storage is a desirable feature if extensive off-line processing of pressure signals is envisaged. A growing trend is to use separate analogue/digital conversion and memory, controlled by a microprocessor at the experiment. Later processing is carried out on a remote and more powerful machine with plotting facilities, etc.

4.4 Inclined Manometer

Despite the development of the sophisticated techniques described in the foregoing sections, there is still an occasional use for the older technique of measuring mean pressures only using an inclined multi-tube manometer. Long (about 1 m) connecting tubes and the high inertia of the manometer fluid effectively damps out the fluctuations in pressure enabling mean values to be read fairly easily. This, of course, was the technique used in the pioneering boundary layer wind tunnel studies of Jensen and Franck[17], and have recently been used by the author, when mean pressure coefficients were required for a large number of configurations in a wind loading code application.

5. REFERENCE PRESSURES AND CORRECTIONS

5.1 Static and Dynamic Pressure Reference

Fluctuating pressures on buildings are most usefully dealt with in a non-dimensional form as follows:

$$C_p(t) = 2(p(t) - p_0)/\rho \bar{u}^2$$

where p_0 is the reference static pressure
$\bar{u}$ is the reference mean velocity.

254

Note that the pressure <u>coefficient</u> has been treated as a time varying quantity to be subjected to statistical analysis. Figure 4 shows part of a typical wind tunnel pressure record showing mean, r.m.s., maximum and minimum values of the pressure coefficient.

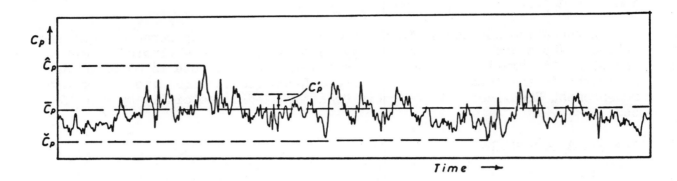

Figure 4. Typical fluctuating pressure coefficient record

The static reference pressure, p_0, ideally should be a mean static or ambient pressure in the vicinity of the building. Logically this should be measured at the same position as for the reference mean velocity, u. If this position is in a region of high turbulence, as is often the case with low-rise building tests, then corrections for turbulence must be applied. These corrections depend upon the scale and intensity of turbulence, and are difficult to determine (e.g. Wood[18]). In most wind tunnels, the static pressure reference is obtained from the static pressure holes of a pitot-static tube, or wall pressure tappings, located in a low turbulence region near the roof of the wind tunnel. However, if there are leaks in the tunnel wall or roof in the vicinity of this point, there may be small differences in the static pressure here compared with the value near the building model, and some correction may be required. The Aylesbury model experiments (discussed in another paper in this Volume) indicated differences in static pressure reference when results from different wind tunnels were compared so that this is a question of some importance.

Two approaches are used to obtain the reference mean velocity, $\bar{u}$, or dynamic pressure, $\rho \bar{u}^2/2$. One is to use the dynamic pressure determined by a pitot-static probe mounted above the boundary layer. For tall buildings, this may also be a logical reference position to use for computing pressure coefficients. In other cases, a correction is made using the mean velocity profile to relate the dynamic pressure or mean velocity at the pitot-static tube to that at the required reference position - usually at the building height or at the 10 m equivalent height. This method requires an accurate determination of the mean velocity profile. An alternative procedure is to mount a hot-wire or hot-film anemometer probe at the desired reference position. The anemometer output is sampled at the same time as the fluctuating pressures, and the required mean velocity is determined.

5.2 Superfluous Acoustic Pressures

The occurrence of superfluous pressure fluctuations of acoustic origin in wind tunnels has long been recognised as a problem, and curative measures proposed (e.g. Batchelor[19], Wills[20]). However, they remain a potential source of error in measurements of fluctuating pressures. Clearly the generally incompressible pressure fluctuations generated by the turbulent and separated flows around building models should be separated from compressible pressure fluctuations associated with standing or travelling waves generated as a result of the wind tunnel duct work or other components. Longitudinal standing waves ('organ pipe' resonances) and transverse modes can be generated, but more common sources are unsteady flow in diffusers, the tips of fan blades, or the vibration of turning vanes. Acoustic wave pressure fluctuations often manifest themselves as isolated sharp spikes on frequency spectra or as high correlations obtained at certain frequencies for fluctuating pressures across the face of a building model.

Corrections to the measured mean square fluctuations or spectra can be attempted if the ambient sound level can be determined in the absence of turbulence or bluff bodies. However, it is much better to determine and remove the source of the problem if possible.

6. CONCLUSION AND ACKNOWLEDGEMENTS

The contents of this paper represent the author's interpretation of current technology for the measurement of fluctuating pressures in wind tunnel studies of wind loading on buildings. In many cases, overlapping with other papers in this Volume will occur, and in some cases, the interpretations may disagree with those of other authors.

The author wishes to thank the large number of wind tunnel modellers he has visited, had discussion with, and learnt from, over the last several years. No doubt if he has offended by misinterpretation, misrepresentation or ommission, the hospitality he has received will be less cordial on future occasions.

7. REFERENCES

1. H.W. Tieleman, T.A. Reinhold and R. D. Marshall, On the wind tunnel simulation of the atmospheric surface layer for the study of wind loads on low-rise buildings, J. Ind. Aerodyn., 3 (1978) 21-38.

2. R.D. Marshall, A study of wind pressures on a single-family dwelling in model and full scale, J. Ind. Aerodyn., 1 (1975) 177-199.

3. Engineering Services Data Unit, Characteristics of atmospheric turbulence near the ground, Part II : single point data for strong winds (neutral atmosphere), E.S.D.U. Item No. 74031, 1974.

4. A.G. Davenport, D. Surry and T. Stathopoulos, Wind loads on low rise buildings : Final report of Phases I and II, University of Western Ontario, Report BLWT-SS8-1977, August 1977.

5. A.G. Davenport, D. Surry and T. Stathopoulos, Wind loads on low rise buildings : Final report of Phase III, University of Western Ontario, Report BLWT-SS4-1978, July 1978.

6. A. Hunt, Scale effects on wind tunnel measurements of surface pressure on model buildings, Proc. Colloque 'Construire avec le Vent', Nantes, 1981, pp. VIII-8-1 - VIII-8-15.

256

What is this?

7. W.H. Melbourne, Comparison of measurements on the C.A.A.R.C. standard tall building model in simulated model wind flows, J. Wind Engg. and Ind. Aerodyn., 6 (1980) 73-88.

8. S.J. Gumley, Tubing systems for the measurement of fluctuating pressures in wind engineering, University of Oxford, Report OUEL 1370/81, 1981.

9. J.D. Holmes, Mean and fluctuating internal pressures induced by wind, Proc. 5th International Conference on Wind Engineering, Ft.Collins, Co., 1979, Pergamon Press, 1980, pp.435-450.

10. H. Liu and P.J. Saathoff, Building internal pressure : sudden change, J. Eng. Mech. Div., A.S.C.E., 107 (1981) 309-321.

11. K.J. Dreher and J.E. Cermak, Wind loads on a house roof, Colorado State University, Report CER72-73 KJD-JEC 22, March 1973.

12. H. Bergh and H. Tijdeman, Theoretical and experimental results for the dynamic response of pressure measuring systems, National Aero- and Astronautical Research Institute, Holland, Report No. NLR-TRF.238, Jan. 1965.

13. A. Gorove and J.E. Cermak, Dynamic response of pressure transmission lines to pulse input, Colorado State University, Report CER-66-67AP-JEC51, June 1967.

14. K.H. Kao, Measurement of pressure/velocity correlation on a rectangular prism in turbulent flow, University of Western Ontario, Report BLWT-2-1970, January 1970.

15. H.P.A.H. Irwin, K.R. Cooper and R. Girard, Correction of distortion effects caused by tubing systems in measurements of fluctuating pressure, J. Ind. Aerodyn., 5 (1979) 93-107.

16. J. Vimal du Monteil, Optimisation et automatisation de la technique des mesures de pression instationnaire, Proc. Colloque 'Construire avec le Vent', Nantes, 1981, pp.IX-7-1 - IX-7-17.

17. M. Jensen and N. Franck, Model-scale tests in turbulent wind, Part II Phenomena dependent on the velocity pressure, Danish Technical Press, Copenhagen, 1965.

18. C.J. Wood, On the use of static tubes in architectural aerodynamics, J. Ind. Aerodyn., 3 (1978) 374-378.

19. G.K. Batchelor, Sound in wind tunnels, Australian Council for Aeronautics, Report ACA-18, June 1945.

20. J.A.B. Wills, Spurious pressure fluctuations in wind tunnels, J. Acoust. Soc. Amer., 43 (1968) 1049-1054.

TECHNIQUES AND MODELING CRITERIA FOR MEASURING AREA AVERAGED PRESSURES

Theodore Stathopoulos
Centre for Building Studies
Concordia University
Montreal, Québec
Canada H3G 1M8

INTRODUCTION

The evaluation of the appropriate wind loads for the design of structures depends primarily on wind tunnel measurements either directly through model tests or indirectly through coefficients suggested by codes and standards. Methodologies for wind tunnel testing of models of buildings and structures have been developed relatively recently - primarily within the past twenty years. The importance of modeling the atmospheric boundary layer characteristics (mean speed gradient and turbulence properties) and the significance of the determination of the dynamic component of wind loading are now generally accepted.

Wind tunnel measurements are typically, however, often limited either to overall structural loads, such as base moments, total drag forces etc., or to very local loads such as surface point pressures. The latter are useful for the design of small elements of cladding, such as fasteners, but it would be very conservative to use them for the design of areas of intermediate size such as glass or cladding panels, or the tributary area of structural elements such as purlins or girts. The conservatism arises from the fact that the individual peak pressures, as recorded at each surface point, will overestimate the peak area load because the individual maxima do not occur simultaneously at all points. In other words, the average of the peak pressures acting at various points of a particular area is bound to be larger (equal in the rare case of fully correlated pressures) than the actually required peak of area averaged pressures.

Various techniques and methodologies have been developed and used for the measurement of area averaged pressures and integrated loads acting upon structures exposed to wind. Some of these methodologies have also been applied in full scale tests for the evaluation of wind loading of structures. It is the purpose of this paper to review these experimenal techniques and to discuss their advantages and limitations in practical applications. Typical experimental results will be presented to indicate the wind load reductions found when considering the area averaging effect of peak pressures.

AREA AVERAGED PRESSURE MEASUREMENT TECHNIQUES

In the past, unsteady spatially-averaged loads acting on the area of design interest, such as a roof, or a section thereof, have been measured by using special purpose area transducers, an example of which is the flush diaphragm pressure transducer. The particular area of interest was thus modeled as a load cell within the larger structure. The difficulty encountered with this technique was to maintain an aerodynamic seal while also ensuring negligible load transfer from the element to the rest of the

structure. This method was also expensive and lacked flexibility for a
number of measurements on tributary areas of various structural elements.
An example of application of the technique for the evaluation of loads on
segments of a large hangar roof wind tunnel model is given in ref. 1.

An alternative approach for measuring area averaged pressures, consists of electronic summation of the output of numerous pressure transducers recording simple point pressures in the area of interest. This
approach is very flexible and gives very accurate results but the difficulty here is that the cost of sufficient instrumentation generally prevents its use in practice. Nevertheless, the technique has been applied
to a number of studies. In particular, Marshall (2) has used this methodology for the full-scale measurement of area pressures acting on the roof
and the walls of a mobile home. Also, Kim and Mehta (3) have applied this
technique in an experiment to measure the total lift force acting on a
flat roof by simultaneous summation of the output signals of four load
cells attached on the roof.

A compromise between the above two techniques has led to the development of a flexible and economical approach, the so-called "pneumatic
averaging" technique which has been investigated at the University of
Western Ontario and has been described in detail in ref.4. The technique
consists of pneumatically averaging the pressures from a number of tappings connected in a carefully controlled fashion through a multi-input
manifold to a pressure transducer. The resultant output pressure has been
shown to be a good approximation to the instantaneous arithmetic
mean of the component input pressures. The idea behind this
technique is not new. Following Baker's ideas and observations that the
mean wind pressure on a large area must be less than that on a small area
because "*threads of the currents moving at the highest velocity will strike
an obstruction successively rather than simultaneously*" (5), Stanton (6)
invented the wind pressure recorder shown in Fig. 1. This recorder, which
can be considered as a mechanical mean pressure transducer, consists of a
series of chambers, each containing a flexible corrugated diaphragm, the
displacement of which, due to the difference of air pressure on its sides,
is followed by a guide-rod carrying a light pulley. A thin platinum wire
is fixed at one end and passes over the system of the pulleys shown, the
free end carrying a pen in contact with a drum actuated by clock-work.
The motion of the pen integrates the separate motion of the diaphragms,
so that a record of the average pressure differences in the chambers is
obtained.

The pneumatic averaging device is also similar in principle to the
concept of interconnected Pitot-static tubes for air-flow measurement (7)
or to the common practice of interconnecting reference static pressure
holes around circumferences of wind tunnels. In the pneumatic averaging
technique, however, the device (manifold) has been designed and tested
with the specific intent of obtaining the unsteady area loads as well as
their mean values. Such unsteady information has not usually been developed
in the past studies with the exception of the work of Banister and Holley
(8) who have applied a similar approach in full-scale, combining it with
the use of monometer dynamics to produce an analogue simulation of the
structural behaviour.

Typical forms of pneumatic averaging manifolds used for measurements
of area loads on wind tunnel models of buildings and structures are shown

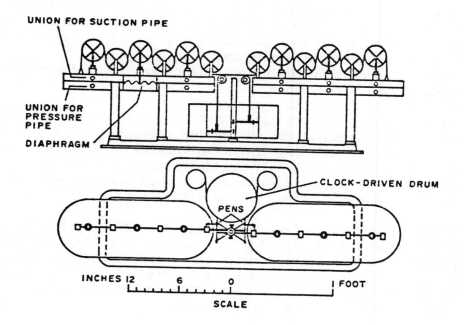

UNION FOR SUCTION PIPE

UNION FOR PRESSURE PIPE

DIAPHRAGM

CLOCK-DRIVEN DRUM

PENS

INCHES 12   6   0   I FOOT

SCALE

Fig. 1   Wind Pressure Recorder

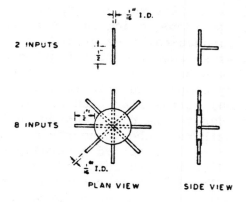

$\frac{1}{16}"$ I.D.

2 INPUTS

$\frac{1}{2}"$

8 INPUTS

$\frac{1}{4}"$ I.D.

PLAN VIEW   SIDE VIEW

Fig. 2   Pneumatic Averagers

in Fig. 2. The apparatus is extremely simple but the mathematics of the response of the multi-tube system is complex and has not been studied until very recently (9). As has been discussed in ref. 4. pneumatic averaging techniques would be expected to be valid under certain limiting conditions, at least for time averaged pressures. These conditions are:
-Establishment of laminar flow in all tubes within the system (i.e. pressure drop directly proportional to tubing length).
-Length of tubes is such that end effect losses are small compared to frictional losses within the tubes.
-Tubing lengths are of identical geometry.
-Flow through the system is negligibly small compared to the external flow. For all practical purposes and for the tubing geometry used in the Boundary Layer Wind Tunnel Laboratory of the University of Western Ontario (flush surface pressure tap of 0.035" inside diameter, a two foot length of 1/16" ID tubing with an embedded restrictor to add damping) it can be easily shown that these conditions are satisfied (4).

The frequency response of the pneumatic averaging system for sinusoidal inputs is presented in Fig. 3 for a variety of configurations. As can be easily noticed, the distortion of the output signal is small and appears only in the higher frequency inputs. Similar results have been found in the case of random signal inputs (4). Manifolds of up to 16 inputs have been used for area load measurements.

Pneumatic averaging has been applied successfully in a number of studies in Canada (10 - 12), Great Britain (13) and Australia (14). Of particular interest are some initial experimental studies comparing pneumatic averaging with other techniques. For instance, pneumatic averaging has been applied in a project (10) to determine area loads for various regions of a square roof. Parallel measurements using large diaphragm transducers set flush with the flat roof areas have verified the validity of this technique. Conventional point pressure measurements on a 9-point grid have also been made. The comparison of these various experimental data stresses the significance of the tributary area for the assessment of the extreme wind loads acting on flat roofs. Typical results are shown in Fig. 4 for peak area loads (suctions) acting on inner roof regions. The comparison indicates that the pneumatically-averaged results compare well with those of the direct area measurements of the flush-diaphragm transducers; although, they are always high. This is due in part to some overestimation of the high frequency components (finite grid effect which will be discussed in the next section) as has also been confirmed through spectral measurements. It is also due to the different weighting functions, which are inherent in the two methods - the pneumatic averaging technique being uniform and the transducer being weighted according to the deflection shape of the diaphragm. Comparison between pneumatic averaging results and those obtained by averaging local peak pressures measured independently at individual points located inside the area considered shows the significant overestimation which occurs when the lack of correlation of action (or synchronization) of the localized peaks is neglected. Results appear more dramatic in areas closer to flow separation where the fluctuating character of pressure is more apparent.

LIMITATIONS OF AREA AVERAGED PRESSURE MEASUREMENT TECHNIQUES

Modeling criteria and limitations applied for the measurement of local

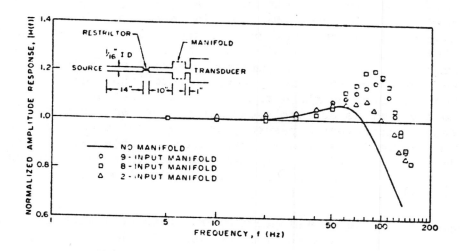

Fig. 3   Pneumatic Averaging Effect on the Response of the
Pressure Measuring System

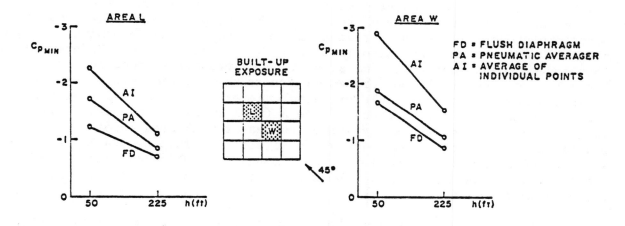

Fig. 4   Peak Suction Coefficients from Different Measurement
Techniques

(point) pressures hold also, in general, for the measurement of area loads. It is pertinent however, to discuss some additional limitations which appear in measuring area averaged pressures. One of these limitations, which exists in the application of the pneumatic averaging technique is the finite grid or spatial sampling effect which arises from the implicit assumption that the pressure measured at a point applies over the entire tributary area of this point (1/N of the area in a N-point symmetric grid). In fact this assumption breaks down for the highest frequency components of the load. Thus, although the reduction in overall load on an area due to the difference in correlation or phase from one point to the next is correctly estimated, the contribution of the highest frequency component pressures, which are not expected to be correlated even over the area associated with a single grid point, is overestimated. This effect has been examined in detail (10) and has been found to be generally small and conservative; nevertheless, corrections, if desirable, are possible.

The simplified theory developed in ref. 10 to deal with the finite grid effect provides a guide to grid spacing. It has been suggested that in order to minimize this effect, the average distance between individual tappings of the grid should be less than about 1/3 the wavelength of the highest frequency of interest. In practice, the finite grid effect becomes significant only after the effective mean square pressure on the area has already reduced to less than 20% of the individual point values. Furthermore, such high frequency excitation is expected to be considerably weaker for reasonable choices of grids.

Wind-tunnel modeling scaling criteria for measurements of area averaged pressures have been discussed in the literature very recently (15). For local point pressure measurements, Jensen's famous experiments (16, 17) have shown that a factor of two or three in the determination of model height leads, under the worst possible conditions, to a maximum error of 10% in the evaluation of mean pressures. For instantaneous peak pressures, reference 11 has shown that a factor of two in the determination of scale leads to an error also less than 10% on the average. Reference 15 however, points out that, for area averaged pressures, modeling criteria for scaling should be more severe due to the drastic susceptibility of area loads to integral scale or turbulence effects. For example, peak area loads measured for an area of size A (full scale) on a model scale larger by a factor of two would indeed represent peak area loads corresponding to an area of actual size 4A (full scale). This generally results to unconservative results since from spatial coherence considerations, area averaged pressures are expected to decrease as the size of the area increases.

Despite the validity of the above argument, it is possible that some small flexibility in the determination of scale could be tolerated in practice. Figure 5 shows worst instantaneous area averaged peak pressure coefficients measured for the evaluation of roof purlin loads acting on low-rise buildings. Pneumatic averaging has been applied for the measurement of pressures acting on tributary areas of 100–200 ft^2 on three different building models representing an actual gabled roof (1:12) building (80 ft wide, 100 ft long, 16 or 24 or 32 ft high) in three different scales, namely 1:500, 1:250 and 1:100. The experimental measurements

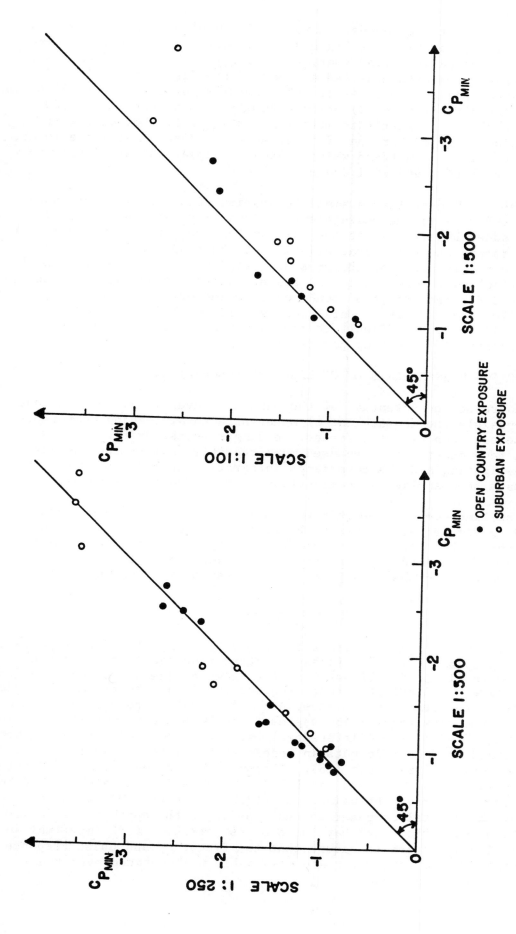

Fig. 5 The Effect of Scaling on Area Averaged Pressure Coefficients for Roof Purlins

have taken place at the Boundary Layer Wind Tunnel of the University of Western Ontario, in which the 1:500 scale appears to be the most representative of the natural wind characteristics. Each of the diagrams of Fig. 5 compares the data measured at each of the distorted scale models with the ideal scale model. As it can be easily noticed, the 1:250 scale (scale factor distortion of 2) shows a fair agreement with the 1:500 scale (in fact, it overestimates the loads slightly), whereas the 1:100 scale (scale factor distortion of 5) generally underestimates the loads, as expected. This is a trend which typically appears in all cases examined for both roof and wall areas.

The above suggest that a small relaxation of scale will not induce significant errors in the measurement of area averaged pressures. It should be added however, that the size of the area, the location of the area with respect to the characteristics of the flow (e.g. separation points) and the location of the pressure tappings inside the area may be factors affecting the relaxation of scale. There are no data available, for instance for the effect of scale on averaged pressures acting on areas smaller than 100 ft^2. Until such experimental data become available it would be very difficult to make a concrete statement for the general relaxation of scaling parameters.

TECHNIQUES FOR MEASUREMENTS OF GENERALIZED (GLOBAL) WIND LOADS

To increase the potential of the pneumatic averaging technique and to make possible the measurement of other structural loading components and parameters such as total forces, bending moments, drifts etc. two different techniques have been used more recently, namely the on-line computer-weighting and the covariance integration method. Both these methods have been applied in combination with pneumatic averaging.

The on-line computer-weighting technique (11) is diagrammaticaly presented in Fig. 6 for a low-rise building. The model design is such that the entire bay load is approximated by eight "purlin loads" on the roof and two "wall loads". The required combination of these loads depends on the end purpose. Any combination can be written in the form

$$B_i = \sum_{j=1}^{10} \eta_{ij} P_j$$

where P_j are purlin and wall loads for a bay (each monitored by a pneumatic averager) and η_{ij} are influence factors required to provide the generalized bay load of interest B_i. For example, one set of η_{ij}, $j=1$ to ten would give total uplift force; another total bay shear; another bending moment in the frame at the knee, etc. A computer program has been developed for the data acquisition system to perform this summation "on-line". This technique (with minor differences) has also been used for measurements of generalized load coefficients in models appropriate to simulate both external and internal pressures (18). A simplified pressure tap array for purlin and wall loads was used and the differences of the external and internal pressures derived by using both sides of the transducers were sampled by the digital computer and multiplied by the appropriate influence factors to provide the total bay loads B_i as before. It is important in this type of application to ensure that the frequency and phase response of the two sides of the transducer are reasonably well-matched over the frequency range of interest.

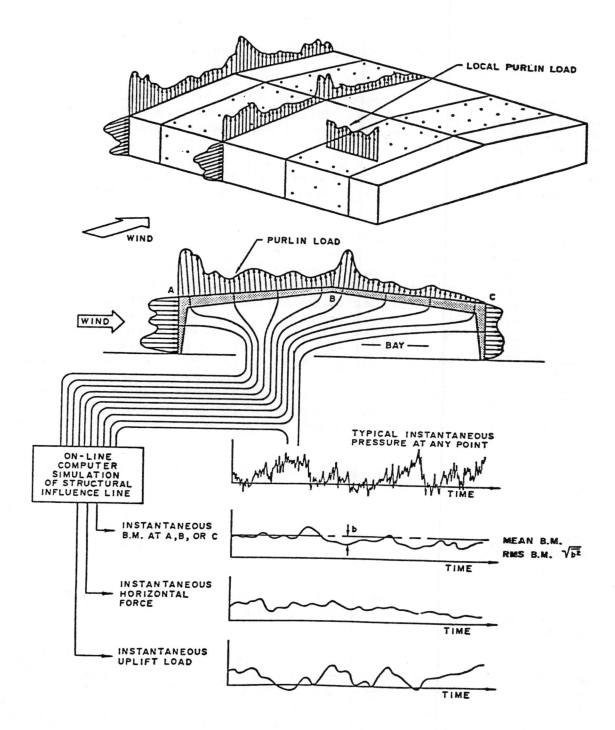

Fig. 6 Measurement of Unsteady Wind Loads on Low-Rise Buildings

The on-line computer-weighting technique provides the fluctuating values of structural effects of interest directly and, implicitly, full account of the effect of partial correlation of unsteady pressures is taken. However, there are a number of disadvantages in the procedure and these have been identified in ref. 14 as follows:

i) The large number of pressure transducers required to be operating simultaneously;

ii) The simultaneous sampling of a large number of input channels and multiplication by the appropriate influence factors at a suitable rate – which may be beyond the capabilities of many computer systems associated with wind tunnels; and

iii) The fact that further wind tunnel tests will be required if it is necessary to monitor further structural effects, unless the unweighted load records are stored.

The covariance integration technique (14) has been developed to obtain peak values of generalized structural effects with much less instrumentation. The wind tunnel test data required can be obtained relatively easily with only two pressure transducers and two data channels. Pneumatic averaging is again used and the fluctuating pressure coefficients are recorded for each "panel" (area averaged pressures). The cross-correlation coefficient of any two such "panel" loads is also measured and the rms value of the fluctuating structural effect is calculated analytically by using the appropriate influence factors. Then the peak value of the structural effect of interest can be found assuming Gaussian probability distribution for its fluctuations. This methodology has been applied in some projects dealing with low-rise buildings (19,20) but no comparisons have been carried out with the results obtained by the on-line computer-weighting technique.

Disadvantages of the covariance-integration technique can be identified as follows:

i) The assumption of Gaussian probability distribution for the structural effects of interest may not be valid in all cases. Unconservative peak values may thus be derived for global loads influenced by small areas of building surface.

ii) Separate aerodynamic data, including cross-correlation coefficients are required for each wind direction which makes the procedure for the evaluation of structural effects lengthy.

Despite the fact that both techniques have been developed and first applied for the evaluation of generalized wind loads on low-rise buildings, the on-line computer-weighting methodology has been applied to other projects dealing with wind loads on unusual roof systems such as cable-stayed roofs (21,22). For the application of these techniques however, a quasi-static structural behaviour must be assumed.

The effect of relaxation of scale has also been examined for generalized wind loads. Figure 7 shows typical data for bay uplift force and bay horizontal thrust coefficients obtained by using the on-line computer-weighting technique. A 25 ft long bay of an 80 ft wide low-rise building

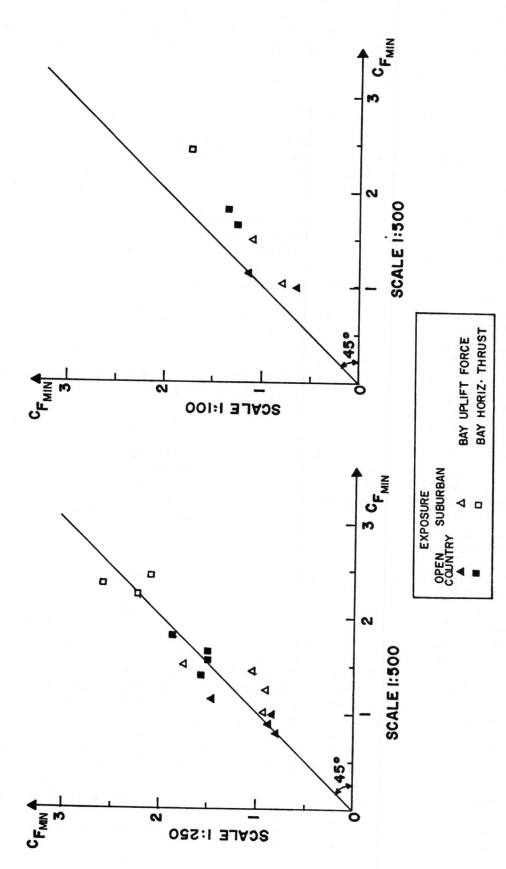

Fig. 7 The Effect of Scaling on Generalized Force Coefficients
for Low-Rise Building Frames

similar to that presented in Fig. 6 has been considered. Data are
organized in the same way with those of Fig. 5 and the comparison indicates that a small relaxation of scale (by a factor of 2) will
not induce serious errors in the evaluation of the generalized wind forces.
A significant underestimation of peak force coefficients occurs, however,
in case of a large violation of scale (by a factor of 5).

EXAMPLES OF AREA LOAD REDUCTIONS

Some typical examples of the load reductions measured by using area
averaged pressures instead of peak values of local (point) pressures are
presented in this section. These examples have been taken from a project
of the evaluation of wind loads acting on low-rise buildings (11,12).

It has been found that roof corner loads depend significantly on the
size of the corner area considered. This feature is shown in Fig. 8
for three heights of a model building exposed to open country terrain.
Peak suction coefficients of the large (0.25W x 0.25W) roof corner are
highly reduced in comparison to the values measured for the small
(0.10W x 0.10W) roof corner. This is due not only to the lower pressures
acting on points farther from the separation point but also to the weaker
correlation of pressures acting over larger distances. Data are shown
for two azimuths (225° and 270°) corresponding to the worst peak loads
recorded. The load reduction due to area averaging effect has been found
to be larger for these critical azimuths than for other wind directions.

Figure 9 shows the variation of peak suction coefficients measured
for a region of constant size area (0.10W x 0.10W) when this area is taken
at different distances from the verges. Results are presented again for
the two critical wind directions and for two terrain roughness exposures.
Peak suction coefficients show the maximum reduction when the region on
which they apply is taken a small distance from the edge. No further significant reduction occurs for distances from the edge which are larger than
5% of the width of the building. The reduction is also larger for the
rougher exposure and the arithmetic average of the worst local peaks than
for the instantaneous spatially averaged coefficients.

Finally, the general trend of reduction of loading with increasing
area is shown in Fig. 10. In this diagram, the worst uplift coefficients
have been determined for each low-rise building configuration (i.e. each
combination of length, height and roof slope) in each exposure for the
various tributary areas associated with point pressures, purlin loads and
bay loads. Then, for each configuration, these worst cases have been
normalized by the worst bay load for the configuration. The band shown
in Fig. 10 contains the normalized curves for all the basic configurations
examined. Some of the alleviation of these peak suctions with increasing
area arises from the lack of simultaneous action of the component peak
local suctions; however, much of the reduction is undoubtedly because the
highest suctions are very localized, particularly near edges. It should
be noted that the ratios shown in Fig. 10 cannot be directly applied to
determine the loading of any particular roof area, since this depends on
the location of the area, but merely illustrate the trend in the data.

Further examples of load reduction due to area averaging can be found
in references 2, 10, 11, 12, 14, 21 and 22.

269

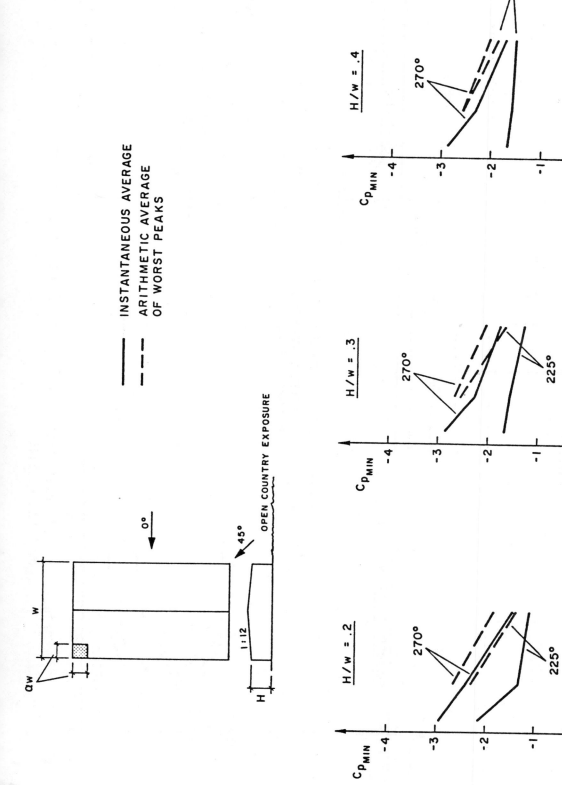

Fig. 8 Area Averaged Peak Suction Coefficients for Roof Corners

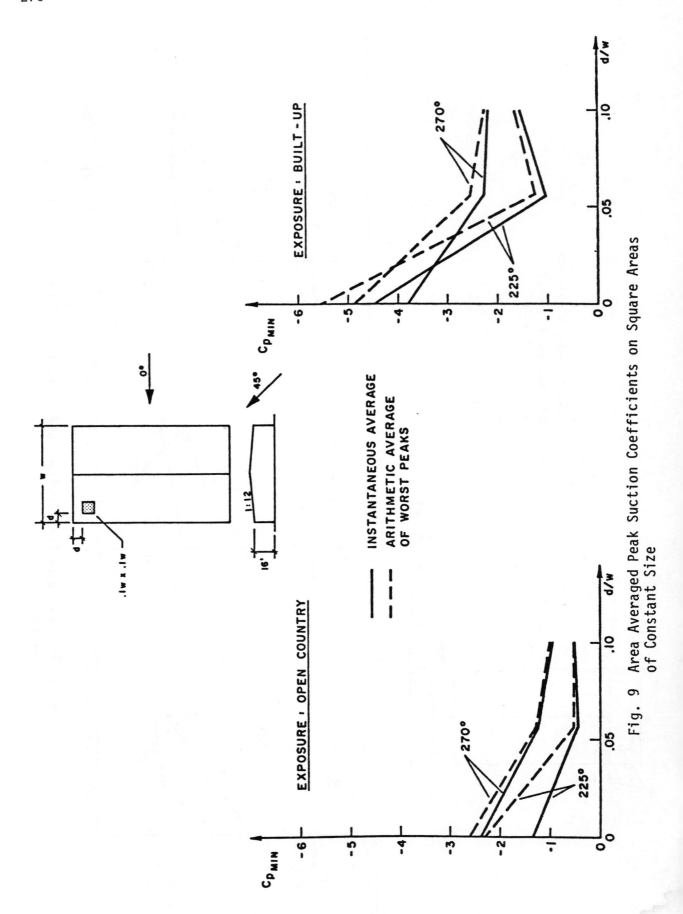

Fig. 9   Area Averaged Peak Suction Coefficients on Square Areas of Constant Size

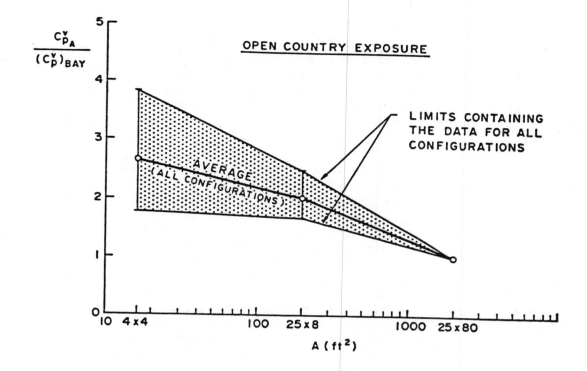

OPEN COUNTRY EXPOSURE

$C^v_{P_A} =$ WORST INSTANTANEOUS PEAK SUCTION COEFFICIENT
MEASURED FOR ANY AREA "A" FOR A PARTICULAR
CONFIGURATION

$(C^v_p)_{BAY} = C^v_p$ FOR A = 25' x 80'

A CONFIGURATION HERE IS ESSENTIALLY ANY COMBINATION
OF LENGTH, HEIGHT AND ROOF SLOPE

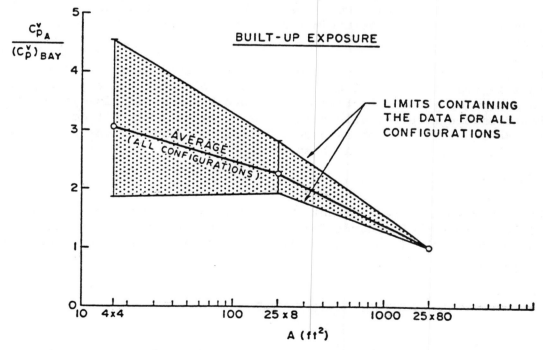

BUILT-UP EXPOSURE

Fig. 10 Effective Uplift Loading Reduction with Increasing
Area Size

CONCLUDING REMARKS AND FUTURE TRENDS

The various techniques developed and used for the measurement of area averaged pressures have been reviewed and discussed. The pneumatic averaging technique by itself or in combination with either the on-line computer-weighting or the covariance integration method provides a powerful tool to determine effective peak wind loads on structural areas of design interest.

The pneumatic averaging technique can be further developed to produce weighted average values if variable weighting applies to individual pressure taps. Although this could be attained by using different tubing lengths, since the weight given to any tube is inversely proportional to its length (4), it has been preferred to distribute the tappings non-uniformly over the surface. In principle, for example, the overall model sway load of a tall rectangular cross-sectioned tower could be measured directly by differencing manifolded pressure taps on the front and rear faces. Each face would have a density of pressure taps that varied over the faces proportionally to the modal displacement. Hence the result would be an estimator for the total dynamic modal load appropriate to the first sway mode. Naturally, this scheme would require a large number of taps to be manifolded; however, this requirement can be reduced by incorporating some of the weighting into the on-line integration program. Similar approaches are being investigated to allow measurement of appropriate torsional loads on entire buildings, and modal loads on tributary components of structures.

An advantage of this approach is that estimates of large-scale structural loads can be determined from the same rigid model built to investigate local point pressures. The method does not of course, provide any direct estimate of the potential dynamic magnification due to structural resonance. For modest structures however, resonant application may not be significant, or may simply be estimated by conventional gust factor approaches.

The most important feature, however, of the development of these techniques is the highlight of the significant role that the tributary area plays in determining the effective design load in many wind loading cases. This should enable more reliable data bases to be determined from which realistic design loads can be assessed, both for individual cases and for general application to codes and standards.

ACKNOWLEDGMENTS

The writer would like to thank the Centre for Building Technology of the National Bureau of Standards for the kind invitation to write and present this paper to the International Workshop on wind tunnel modeling criteria and techniques.

REFERENCES

1. B.J. Vickery, Wind Loads on Low-Rise Buildings, presented at D.R.C. Seminar, Darwin, Australia, March 1976 (unpublished, 11 pp. 19 figs).

2. R. D. Marshall, The Measurement of Wind Loads on a Full-Scale Mobile Home , National Bureau of Standards, Washington, D.C., NBSIR 77-1289 (Sept. 1977)

3. S.I. Kim and K.C. Mehta, Wind Loads on Flat Roof Area through Full Scale Experiment, Institute for Disaster Research, Texas Tech. University, Lubbock, Texas 79409 (Sept. 1977).

4. D. Surry and T. Stathopoulos, An Experimental Approach to the Economical Measurement of Spatially-Averaged Wind Loads, J. of Industrial Aerodynamics, 2, 1978, pp. 385-397.

5. B. Baker, The Forth Bridge, Engineering, August 29, 1884, p. 213.

6. T. E. Stanton, Report on the Measurement of the Pressure of the Wind on Structures, Proceedings, The Institution of Civil Engineers, Paper No. 4513, 1924.

7. J.M. Gasiorek, The Use of Inter-Connected Pitot-static Tubes for Air Flow Measurement, Building Services Engin. J., 41 (Aug. 1973) 106-111.

8. J.R. Banister and W.L. Holley, Manometer Measurements of Wind Pressure Loads, Invit. Symposium on Full-Scale Measurements of Wind Effects on Tall Buildings and Other Structures, BLWT Laboratory, The University of Western Ontario, June 23-29, 1974.

9. S.J. Gumley, Tubing Systems for the Measurement of Fluctuating Pressures in Wind Engineering, O.U.E.L. Report No. 1370/81, University of Oxford, 1981.

10. T. Stathopoulos, Wind Pressure Loads on Flat Roofs, BLWT-Report-3-75, The University of Western Ontario, London, Ontario, Canada, Dec. 1975.

11. A.G. Davenport, D. Surry and T. Stathopoulos, Wind Loads on Low-Rise Buildings: Final Report of Phases I and II - Parts 1 and 2, BLWT-SS8-1977, The University of Western Ontario, London, Ontario, Canada, Nov. 1977.

12. A. G. Davenport, D. Surry and T. Stathopoulos, Wind Loads on Low-Rise Buildings: Final Report of Phase III, Parts 1 and 2, BLWT-SS4-1978, The University of Western Ontario, London, Ontario, Canada, July 1978.

13. N.J. Cook and F.J. Heppel, Report of Model Scale Tests in the BRE Boundary Layer Wind Tunnel on the UWO Low-Rise Building Model, Building Research Establishment Note N146/80, Nov. 1980.

14. J.D. Holmes and R.J. Best, An Approach to the Determination of Wind Load Effects on Low-Rise Buildings, J. of Wind Engineering and Industrial Aerodynamics, 7, 1981, 273-287.

REFERENCES Cont'd

15. E. Simiu, Modern Developments in Wind Engineering: Part 2, Engineering Structures, 1981, Vol. 3, Oct. pp 242-248.

16. M. Jensen, The Model-Law for Phenomena in Natural Wind, Ingenioren International Edition Vol. 2., No. 4. 1958.

17. M. Jensen and N. Franck, Model-Scale Tests in Turbulent Wind, Part II, The Danish Technical Press, Copenhagen, 1965.

18. T. Stathopoulos, D. Surry and A.G. Davenport, Internal Pressure Characteristics of Low-Rise Buildings Due to Wind Action, Proceed. of 5th Intern. Conf. on Wind Engineering, Colorado State University, July 1979.

19. J.D. Holmes and G.J. Rains, Wind Loads on Flat and Curved Roof Buildings - Application of the Covariance Integration Approach, Proceed. of Symposium Designing with the Wind, Nantes, France, June 1981.

20. R.J. Roy and J.D. Holmes, Total Force and Moment Measurement on Wind Tunnel Models of Low-Rise Buildings, Proceed. of Symposium Designing with the Wind, Nantes, France, June 1981.

21. D. Surry, R.B. Kitchen and A.G. Davenport, Wind Loading of Two Hangars Proposed for Jeddah Airport, Saudi Arabia - A Rigid Model Study, BLWT Report to be published, The University of Western Ontario, London, Canada.

22. D. Surry and D. Bailey, A Study of Wind Loads on the Recreational Pavilion for the Loyola Marymount University, Los Angeles, California, BLWT Report to be published, The University of Western Ontario, London, Canada.

MEASUREMENT OF TOTAL LOADS USING SURFACE PRESSURES

Ahsan Kareem
Department of Civil Engineering
University of Houston

INTRODUCTION

It is important to recognize the unsteady nature of wind-loading in the design of buildings in an effort to ensure structural safety and service performance. Although the knowledge of wind effects on buildings has significantly improved over the past decade, an understanding of the mechanism that relates the fluctuating atmospheric flow field to various wind-induced effects on structures has not been developed sufficiently for the functional relationship to be formulated. Davenport (1) introduced the concept of the gust loading factor through which the dynamic part of a building's alongwind (in the direction of wind) response can be calculated. A lack of a viable transfer function for the fluctuations in the approach flow and the nonhomogenous pressure fluctuations on the faces of a building with separated flow has prohibited any acceptable formulation to date of acrosswind (in a direction perpendicular to the wind) and torsional response based on a gust factor approach.

Physical modeling of wind-structure interaction in a wind tunnel, therefore, continues to serve as the most accurate and practical means of relating aerodynamic loading on a structure to properties of local wind climate. Aeroelastic models of buildings can be used in a boundary layer wind tunnel to estimate the dynamic response under the action of wind. These models can represent two lateral and a torsional modes of vibration (2). More sophisticated models can be constructed to represent lumped floor masses as a multi-degree-of-freedom system (3). However, these studies are expensive and the exact dynamic characteristics of a building are not generally available in the early design stages.

A quantitative description of wind loading would permit the numerical estimation of a building response at the preliminary design stages. These estimates allow early assessment of the structural requirements to resist oscillatory response to ensure occupancy comfort, and also to assess the need for detailed aeroelastic wind tunnel tests. Quantitative description of wind loading derived from experimental measurement of aerodynamic forces on a scale model can be introduced in lieu of the solution of equations of fluid motion around a building model (4,5,6,7,8,9,10,11,12,13). This paper presents a new methodology for the understanding and quantification of wind loads on buildings in urban and suburban flow conditions.

MEASUREMENT OF UNSTEADY WIND FORCES

The aerodynamic loading on a structure results from unsteadiness in the approach flow (atmospheric turbulence), wake excitation and forces associated with structural motion. Unsteadiness in the approach flow causes gust loading which is primarily responsible for the alongwind structural loading. The wake excitation and unsteadiness in the approach flow induces acrosswind loads. Torsional loads on a building result from a possible aerodynamic and/or inertial coupling in various degrees of freedom. Motion-induced aerodynamic loads generally result from a combination of negative aerodynamic damping and an increase in the correlation of fluctuating pressures on the building surface. Other motion-induced effects, e.g., aerodynamic stiffness and mass, normally become significant at wind speads which are far beyond the normal design range of typical buildings. Discussion and development of wind forces in this paper is limited to the quasi-static loading.

The quasi-static wind loading can be determined by the following experimental methods:

1. direct measurement of force spectra
2. indirect development of force spectra from response measurements, and
3. pressure measurements.

A discussion of some basic concepts related to the first two approaches will be presented here. A more detailed discussion of these topics is available in the proceedings of this workshop. This will be followed by a detailed outline of the pressure measurement technique.

Measurement of a quasi-static force spectrum which is independent of the dynamic characteristics of the model poses a problem of sensitivity and frequency response. In order to obtain a reliable measurement of fluctuating forces and moments on a stationary model mounted on a force balance, the resonant frequency of the entire system must be higher than any expected forcing frequency and at the same time the transducer should have a high sensitivity.

Whitbread (11) developed such a system using carbon fibre reinforced plastic and semi-conductor strain gauges. Davenport & Tschanz (12) have recently developed a five component high sensitivity force balance for measuring quasi-static forces on building models. Ellis (4) used a novel approach in which the outer shell of the model representing the architectural shape was connected to a flexible cantilever core and fitted with accelerometers and strain gauges. Aerodynamic forces were derived from the acceleration and strain measurements. This technique provides total integrated load, segmental loads along the model height, and inter-segment relationship. Spectra of wind loads can be determined from aeroelastic models by measuring the response spectra and appropriate mechanical admittance function (8). This is an inverse approach and the uniqueness of the solution is not necessarily guaranteed and the accuracy depends on how accurately the dynamic properties of the model were determined. All of the techniques described above with the exception of Ref. 4 can be used to obtain only the resultant fluctuating loads on the building model as a whole and it is not possible to determine the manner in which the wind loads are distributed over the building surface.

SURFACE PRESSURE MEASUREMENTS

Direct measurement of wind pressures have been made since the early wind tunnel test programs. This involves a model with a number of pressure taps on its exterior surface. Mean forces and moments on the model can be determined by simply integrating mean pressure values at numerous locations over the entire model. A simple integration of RMS values of pressure over the model will result in an upper bound estimate of the fluctuating forces and moments, since perfect correlation is implied in the scheme. However, spatio-temporal nature of pressure fluctuations on the surface of a building exposed to an atmospheric boundary layer precludes the assumption of perfect correlation. Therefore, it is necessary to measure sufficient number of two-point correlations of pressure fluctuations. Subsequently, a statistical integration scheme can be used to obtain overall fluctuating wind loads on a building. Such a procedure has been used in basic fluid mechanics research for evaluating lift and drag coefficients on two-dimensional circular cylinders (14). The application of this tecnhique to a building model immersed in a simulated atmospheric layer would generally require a large number of pressure transducers, for simultaneous monitoring of pressure fluctuations, to account for the inhomogeneity of the pressure field on the model surface. However, the limitation on the availability of a large number of pressure transducers, and the subsequent data-acquisition and reduction problems rendered this procedure cumbersome and impractical in the past. This methodology can be streamlined by making measurement of pressure at various levels throughout the building model height using a limited number of transducers (5,15). Subsequently, by monitoring the relationship between pressure at various levels, an expression for the integral wind loading function can be developed through statistical integration using a digital computer. This approach not only determines the multi-level and integral fluctuating loads on buildings but also provides information on the local spatio-temporal distribution of pressure for the cladding design. A detailed outline of this procedure is presented in this paper.

Szechenyi (16) measured fluctuating lift force on circular cylinders by means of pressure transducers, placed on the circumference of the cylinder in the same cross-sectional plane, through a real-time analog summation of the lift components of the pressure at each transducer. A weighting coefficient, representing a function of the position of the transducer and the arc length over which it was assumed to measure pressure, was applied individually to the output signal of each transducer. Finally the signals were summed to determine the fluctuating lift force.

More recently Reinhold (7,17) used direct pressure measurements for evaluating total wind loads on building models. The problem of accumulation of a large mass of data, representing simultaneous time histories of pressures fluctuations at each pressure tap location, was alleviated by performing an analog summation of the transducer output signals to obtain individual signals representing the resultant fluctuating loads. This technique does not require the evaluation of correlations, thus reduces considerably data manipulation and statistical computations. However, description of the spatio-temporal distribution of local pressure fluctuations is not available in this scheme.

In the following sections a detailed outline is presented for the statistical integration (digital) and analog procedures for determining total loads using surface pressure.

STATISTICAL INTEGRATION METHOD

Development of Forcing Function

The mean square value of the wind loads on a building due to fluctuating pressure on face 2 (Fig. 1), using an averaging procedure (5), is given by the following expression:

$$\overline{F^2} = \int_{-B/2}^{B/2} \int_{-B/2}^{B/2} \int_0^H \int_0^H \rho(y_p,z_p,y_q,z_q;0)\sigma_p\sigma_q dz_p dz_q dy_p dy_q, \quad (1)$$

in which $\rho(y_p,z_p,y_q,z_q;0)$ is the normalized cross-correlation function of pressure fluctuations between points p and q at zero time lag; σ_p and σ_q represent rms values of pressure fluctuations.

The mean square value of torsional moment due to pressure fluctuations on Face 2 can be derived from:

$$\overline{M^2} = \int_{-B/2}^{B/2} \int_{-B/q}^{B/2} \int_0^H \int_0^H \rho(y_p,z_p,y_q,z_q;0) \, \sigma_p\sigma_q y_p y_q dz_p dz_q dy_p dy_q, \quad (2)$$

where y_p,y_q are the respective moment arms.

In the frequency domain the power spectral density of wind load fluctuations can be derived from Eqs. 1 and 2 by the Wiener-Khintchine transform;

$$S_F\left(\frac{nb}{v}\right) = \int_{-B/2}^{B/2} \int_{-B/2}^{B/2} \int_0^H \int_0^H Co(y_p,z_p,y_q,z_q;\frac{nb}{v})$$

$$\sqrt{S(y_p,z_p;\frac{nb}{v})} \ \sqrt{S(y_q,z_q;\frac{nb}{v})} \ dz_p dz_q dy_p dy_q, \quad (3)$$

where $Co(y_p, z_p, y_q, z_q; \frac{nb}{v})$ is the normalized co-spectrum of pressure fluctuations between two poitns p and q; and $S(y_p, z_p; \frac{nb}{v})$ is the power spectral density of pressure fluctuations at p. Similarly the torsional moment in frequency domain is given by:

$$S_M(\frac{nb}{v}) = \int_{-B/2}^{B/2} \int_{-B/2}^{B/2} \int_{0}^{H} \int_{0}^{H} Co(y_p, z_p, y_q, z_q; \frac{nb}{v})$$

$$\overline{\sqrt{S(y_p, z_p; \frac{nb}{v})}} \ \overline{\sqrt{S(y_q, z_q; \frac{nb}{v})}} \ y_p y_q dz_p dz_q dy_p dy_q \ . \qquad (4)$$

Expressions for the wind loads caused by the pressure fluctuations on a single face can be extended to synthesize the contribution of all the building faces (5).

In order to determine the loading function it is necessary to define the cross-correlations and the co-spectra on various faces. For the stagnation face the cross-correlation function is best estimated by an exponentially decaying function (1,5). Similarly for frequency domain analysis, the co-spectra can be curve-fitted by an exponentially decreasing function of separation distance between the locations under consideration and frequency. However, in the case of side faces inhomogeneity in the pressure field arising out of complexities of turbulent shear flow interaction with a bluff body, separation and reattachment process prohibits any simple analytical description of cross-correlation or co-spectrum for the entire face. Therefore, measured values of co-spectra between locations at one level and various other levels are used in lieu of the analystical description in order to formulate the overall loading function. Recently, analytical expressions for the single-point spectra and co-spectra have been developed, which provide a good estimate of the spectral characteristics of fluctuating pressure field on a side face of a building (13). An example of acrosswind loading is presented here, for the wind approaching normal to one of the faces, to illustrate the procedure.

Acrosswind Forcing Function

The acrosswind forcing function is obtained by synthesizing multiple random inputs resulting from the fluctuating pressure on faces 1 and 3 (Fig. 1):

$$S_F\left(\frac{nb}{v}\right) = 2 \int_{-D/2}^{D/2} \int_{-D/2}^{D/2} \int_{0}^{H} \int_{0}^{H} Co\left(x_p, z_p, x_q, z_q; \frac{nb}{v}\right)$$

$$\sqrt{S\left(x_p, z_p; \frac{nb}{v}\right)} \ \sqrt{S\left(x_q, z_q; \frac{nb}{v}\right)} \ dz_p \, dz_q \, dx_p \, dx_q$$

$$-2 \int_{-D/2}^{D/2} \int_{-D/2}^{D/2} \int_{0}^{H} \int_{0}^{H} Co\left(x_p, z_p, x_r, z_r; \frac{nb}{v}\right)$$

$$\sqrt{S\left(x_p, z_p; \frac{nb}{v}\right)} \ \sqrt{S\left(x_r, z_r; \frac{nb}{v}\right)} \ dz_p \, dz_r \, dx_p \, dx_r \ , \tag{5}$$

where p, q are two arbitrary points on face 1, and r is a point on face 3. The first term of Eq. 5 represents the contribution from fluctuating pressure on faces 1 and 3, whereas, the second term represents interface contribution. Equation 5 can be written in a discrete matrix form so that it is compatible with experimental measurements:

$$\left[G_{F_{ii}}\left(\frac{nb}{v_h}\right)\right] = \sum_{j=1}^{2} \left[\{A_i\}_j^T \left[\sqrt{S_i}\left(\frac{nb}{v_h}\right)\right]_j \left[Co\left(\frac{nb}{v_h}\right)\right]_j^i \left[\sqrt{S_i}\left(\frac{nb}{v_h}\right)\right]_j \{A_i\}_j -$$

$$\sum_{j=1}^{2} \{A_i\}_j^T \left[\sqrt{S_i}\left(\frac{nb}{v_h}\right)\right]_j \left[Co\left(\frac{nb}{v_h}\right)\right]_{jk}^i \left[\sqrt{S_i}\left(\frac{nb}{v_h}\right)\right]_k \{A_i\}_k\right] \ , \tag{6}$$

$$k = 2,1$$
$$i = 1,2,3,4,\ldots.$$

where $\{A_i\}$ is the vector of effective areas around a pressure tap, $\left[Co\left(\frac{nb}{v}\right)\right]_j^i$ and $\left[Co\left(\frac{nb}{v}\right)\right]_{jk}^i$ are the co-spectral matrices of intraface and interface pressure fluctuations at the i-th level, and $[\sqrt{S_i}\ (\frac{nb}{v})]$ is the diagonal matrix. Equation 6 represents diagonal terms of the cross-power spectral density matrix of acrosswind loads at various levels on the building. In order to formulate an inter-level relationship which represent the off-diagonal terms in the cross-power spectral density matrix the following relationship is used:

$$[G_{F_{il}}\ (\frac{nb}{v_h})] = \sum_{j=1}^{2}\ [\{A_i\}_j^T\ [\sqrt{S_i}\ (\frac{nb}{v_h})]_j^T\ [Co\ (\frac{nb}{v_h})]_j^{il}\ [\sqrt{S_1}(\frac{nb}{v_h})]_j\ \{A_1\}_j\ -$$

$$\sum_{j=1}^{2}\ \{A_i\}_j^T\ [\sqrt{S_i}\ (\frac{nb}{v_h})]_j^T\ [Co\ (\frac{nb}{v_h})]_{jk}^{il}\ [\sqrt{S_1}\ (\frac{nb}{v_h})]_k\ \{A_1\}_k],\qquad (7)$$

$$i \neq 1 \qquad\qquad\qquad\qquad\qquad\qquad\qquad k = 2,1$$
$$\qquad\qquad\qquad\qquad\qquad\qquad\qquad\qquad i,1 = 1,2,3,\ldots.$$

where $(Co\ (\frac{nb}{v})]_j^{il}$ and $[Co\ (\frac{nb}{v})]_{jk}^{il}$ are the co-spectral matrices of intraface and interface pressure fluctuations between the i-th and the 1-th levels. The quadspectrum of pressure fluctuations has not been included in the above formulations, since being an odd function it is eliminated by double integration (5).

In the time domain the mean square acrosswind force can be obtained as:

$$\overline{F^2} = 2\ \int_{-D/2}^{D/2}\ \int_{-D/2}^{D/2}\ \int_{0}^{H}\ \int_{0}^{H}\ \rho(x_p,z_p,x_q,z_q;0)\ \sigma_p\sigma_q dx_p dz_q dx_p dz_q$$

$$-2\ \int_{-D/2}^{D/2}\ \int_{-D/2}^{D/2}\ \int_{0}^{H}\ \int_{0}^{H}\ \rho(x_p,z_p,x_r,z_r;0)\ \sigma_p\sigma_r dz_p dz_r dx_p dx_r \qquad (8)$$

Equation 8 can be expressed in a discrete matrix form:

$$\overline{F^2} = \{A\}^T[\sigma][\rho][\sigma]\{A\} \qquad (9)$$

where $[\rho]$ is a cross-correlation matrix, $[\sigma]$ is a diagonal matrix of rms values of pressure fluctuations and $\{A\}$ is a vector representing the respective area around each pressure tap. In order to obtain modal mean square force, Eq. 9 should be replaced by

$$\overline{F^2} = \{\Psi\}^T [A][\sigma][\rho][\sigma][A]\{\Psi\} \;, \tag{10}$$

in which $[A]$ is a diagonal matrix of respective areas and $\{\Psi\}$ is the modal vector.

EXPERIMENTAL PROCEDURE

Placement of Transducers

Selection of number and the location of pressure transducers along the chordwise and spanwise direction on a building model is important for the synthesis of wind loads from direct pressure measurements. A large number of transducers at each level of all the faces is the most desirable configuration; however, limited number of available transducers at most of the facilities precludes this option. If a very small number of transducers is used, then the output of each transducer would represent a relatively large fraction of the model surface determined by dividing the width by the number of transducers in the associated configuration. This would imply that the pressure fluctuations monitored at a tap represent the pressure fluctuations over the entire fractional area around the tap. This may underestimate or overestimate the associated force or moment depending on the relative orientation of a face with respect to the approach flow. It is also important to monitor areas near the edges of a model, especially on the faces parallel or near parallel to the wind direction, where separation and reattachment processes may significantly influence the fluctuating pressure field.

Reinhold (7) conducted a series of tests to select the optimum number and location of transducers to obtain reliable measurement of force and moment fluctuations through analog integration. He concluded that the maximum estimated error obtained by using the four transducers on a side as opposed to seven was less than 5 percent for the normal force and less than 15 percent for the torsional moment associated with any wind direction. The errors commonly encountered were considerably lower than these maximum values. Therefore, at least four transducers on each side, of a typically scaled building model, should be used to sufficiently represent the instantaneous characteristics of fluctuating pressure field.

Frequency Response of Transducers

The frequency response of pressure transducers should be flat within the frequency range of interest for typical buildings. It is generally desirable that the frequency response of pressure transducers, used for a single-point pressure measurement, be flat at least to 200 hz. This helps to include the contribution of high frequency pressure fluctuations. These fluctuations are partially correlated over small surface areas, thus their contribution to overall integral loading is not very significant. Therefore, high frequency response requirements may be relaxed for transducers used to measure total loading. However, it is recommended to use high frequency response transducers, since the results of total loading measurements also provide information on the nature of local pressure flcutuations which may be useful for the cladding design.

Model

A 5-inch (12.7 cm) square and 20-inch (50.8 cm) tall prism made of "Lucite" was modeled for this study. A typical pressure tap was 0.06-inch (1.5 mm) in diameter and normal to the building face. Flexible "Tygon" tubing of short length was used to connect a pressure tap to electronic transducer. The model was instrumented on two faces and the pressure tap locations are shown in Fig. 3. All measurements were conducted on two side faces with flow normal to the model face in two approach flows, boundary layer 1 (BL1), and boundary layer 3 (BL3) with the building model mounted on a turntable with a large inertial mass at downwind end of the test section (5). The BL1 flow represents suburban and BL3 matches the conditions for urban flow environment. In Fig. 2 the variation of mean speed and turbulence intensity is presented for BL1 and BL3.

A total of fourteen configurations of pressure fluctuations were measured. Eight pressure taps were monitored simultaneously in each configuration. The building model was divided into six levels at various heights, each level contains eight taps, four in either side face, located according to the layout given in Fig. 3. These configurations provide information of the dynamic pressure fluctuations in the chordwise direction. In order to obtain a relationship between pressure fluctuations in the spanwise direction, pressure taps were monitored along four vertical elements on the side faces.

The output of pressure transducers was fed through a signal conditioning unit for amplification. An on-line digital data acquisition system with analog-to-digital converter, a minicomputer, and a digital tape was used to monitor simultaneously eight channels of pressure data. The data were measured and recorded on digital tapes for duration of 300 seconds at 500 samples per second. Methods of digital time series analysis were used to reduce the spatio-temporal pressure fluctuations (3,5).

The normalized co-spectra and power spectral density of pressure fluctuations at five levels were used to generate a normalized cross-spectral density matrix of the acrosswind forcing function (Eqs. 6 & 7). A multi-step computer program (5) was developed to evaluate Eqs. 6 and 7. In Figs. 4 and 5 each element of the upper triangel of the cross-power spectral density matrices are presented for BL1 and BL3 flows.

These matrices are symmetrical and each element is a function of reduced frequency.

Generalized modal force spectra for acrosswind loading were obtained from Eqs. 6 and 7 using the fundamental mode of vibration as a weighting function. This is given by:

$$S_{F1}(\frac{nb}{v}) = \{\Psi^1\}^T \; [G_F \; (\frac{nb}{v})] \; \{\Psi^1\}$$

where $\{\Psi^1\}$ is the fundamental mode of vibration. Similarly, the generalized forcing spectra in the higher modes can be derived using the respective mode shapes. The generalized force spectra in the fundamental mode of vibration for BL1 and BL3 are presented in Fig. 6 along with the estimates made by Reinhold (17) and Saunders and Melbourne (8). The results are in general agreement. However, there is a slight deviation in the spectral description, which could be attributed to the variations in the model aspect ratios and in the characteristics of the approach flows.

Fluctuating Force Coefficients

The areas under the diagonal elements of Figs. 4 and 5 represent the non-dimensional RMS acrosswind force coefficients at various levels. Similarly, the areas under the curves in Fig. 6 represent the modal acrosswind force coefficients in the urban and suburban conditions. The non-dimensional force coefficients are reported in Table 1.

In the time domain analysis the cross-correlation matrices were used to compute the fluctuating acrosswind force in accordance with Eq. 9. The non-dimensional force coefficients are reported in Table 1, and are in good agreement with the frequency domain results. Vickery (9) determined load fluctuations in a turbulent flow using a dynamic balance. For a square cross-section model he reported a value of 0.41 for the acrosswind force coefficient. Evidently, Vickery's estimate is comparatively higher. This could be attributed to differences in flow conditions especially the uniform flow in Vickery's experiment which promotes higher spatial correlation of pressure fluctuations due to lack of mean velocity gradient and consequently, higher acrosswind force coefficient.

ANALOG METHOD

The output of pressure transducers can be added and substracted, using inverting summing amplifiers (7,17) to obtain a time series of fluctuating force. A typical schematic of the network used to sum the signals is shown in Fig. 7. Reinhold (17) used four summing amplifier networks to produce the resultant fluctuating loads at each level. One network is used to produce each of the normal loads and two networks are needed to obtain the moment. By referring to Fig. 8, the details of the procedure can be easily explained. The alongwind force, F_x, is synthesized by connecting transducers 1, 2, 3, and 4 to input A, while transducers 9, 10, 11, and 12 are connected to imput B. The acrosswind force, F_y, is synthesized in a similar manner using a second network and connecting transducers 13, 14, 15, and 16 to imput A, and transducers 5, 6, 7 and 8 to imput B. The torsional moment is produced in two segments. The first segment, M1, is obtained by connecting transducers 1, 5, 9, and 13 to imput A, and transducers 4, 8, 12 and 16 to input B of the third network. The second segment, M2, is produced by connecting transducers 2, 6, 10, and 14 to input A, and transducers 3, 7, 11, and 15 to input B of the fourth network. Signals from segments M1 and M2 are multiplied with appropriate lever arms. These products were then added using an analog computer to obtain the torsional moment.

In this manner fluctuating force and moment coefficients can be computed for each level on the model from the output of the respective analog circuitry. Interlevel correlations can be incorporated, for computing the overall force coefficient, by simultaneously monitoring at a minimum of two levels. One of these levels serves as a reference level. The reference level should be included for all subsequent measurements for inter-level correlations. These measurements can be carried out expeditiously by monitoring simultaneously more than two levels, but a limited number of available pressure transducers may preclude this option. Reinhold (17) monitored pressure simultaneously at three levels. The records of fluctuating forces at each level can be analyzed to determine the power spectra of the loads and the cross-spectra between loads at different levels. In Fig. 9, modal force spectra measured by Reinhold (7), on a square cross-section (1:1:8.33) prism exposed to a typical urban boundary layer, in the acrosswind, alongwind and torsional directions, are presented for various angles of approach flow.

DYNAMIC RESPONSE

The dynamic response of a building, e.g., displacement, acceleration and jerk, can be predicted from the forcing function using methods of random vibration theory (5,15,19). The mean square value of response σ_y^2 at the i-th node of a multi-degree-of-freedom system is given by

$$\{\sigma_{y_i}^2\} = \sum_{j=1}^{n} [\phi][\int_0^\infty G_{y_{ij}}(f)df][\phi]^T \ , \tag{11}$$

$$[G_y(f)] = [H(i2\pi f)]^*[\phi]^T[G_F(f)][\phi][H(i2\pi f)] \tag{12}$$

in which $[H(i2\pi f)]$ = matrix of frequency response function, $[G_F(f)]$ = cross-power spectral density matrix, and $[\phi]$ = normalized modal matrix. For the response in the fundamental mode the above expression can be simplified,

$$\sigma_{y_1}^2 = \int_0^\infty S_{y1}(f)df \ ,$$

$$S_{y1}(f) = \frac{S_{F1}(f)}{(2\pi f)^4\{(1-(\frac{f}{f_n})^2)^2 + 4\xi_n^2(\frac{f}{f_n})^2\}} \ , \text{ and}$$

$$S_{F1}(f) = \{\phi^1\}^T[G_F(f)]\{\phi^1\} \ . \tag{13}$$

The higher derivitives of displacement response required for occupancy comfort can be evaluated as

$$\sigma_{y_{(r)}}^2 = \int_0^\infty (2\pi f)^{2r} S_{y1}(f)df \ , \tag{14}$$

where r represents higher derivatives.

The above integral can be evaluated using the residue theorem if the excitation is idealized to be a white noise. The mean square response for various response components is given by

$$\sigma^2_{y_{(r)}} = \frac{\pi f_n S_{F1}(f_n)(2\pi f_n)^{2r}}{4(2\pi f_n)^4 \xi_n} , \tag{15}$$

in which $S_{F1}(f_n)$ = the spectral density of excitation evaluated at the fundamental frequency. The above expression can be extended for a multi-degree-of-freedom system

$$\sigma^2_{y_{(r)}} = \sum_{n=1}^{N} \frac{\pi f_n S_{Fn}(f_n)(2\pi f_n)^{2r}}{4(2\pi f_n)^4 \xi_n} \tag{16}$$

in which subscript n represents the n-th mode. More relevant for design purposes is the peak value of structural response over a specified time period T. The expected value of the largest peak response is given by

$$y^{(r)}_{max} = k^{(r)}\sigma^{(r)}_y \tag{17}$$

in which $k^{(r)}$ = the peak factor and it varies between 3 to 4. A more detailed treatment can be found in references 15 and 20.

The top floor displacement of a typical tall building was computed, using Eqs. 13 and 14 based on the spectra given in Fig. 6, and compared with the response of an aeroelastically-modeled building in BL1 and BL3 flow conditions. The experimental values and predicted values are plotted in Fig. 10. The predicted and experimental values show good agreement at lower reduced velocities which is generally the expected range of wind speeds for typical tall buildings. Furthermore, the predicted response of the example building shows good agreement with values computed in accordance with the Australian (21) building code (Fig. 10). At high reduced velocities the predicted values underestimate the response. This trend can be attributed to the motion-induced loading which is not included in the forcing function model discussed in this paper. Modified estimates of response made by introducing measured values of amplitude-dependent aerodynamic damping (5,15) in the mechanical admittance function enhances the viability of predicted response estimates at high values of reduced velocity (Fig. 10).

SUMMARY AND CONCLUSIONS

A method for evaluating the fluctuating wind forces and their distribution on buildings has been developed and applied to the understanding and quantification of acrosswind loads on a squre cross-section building. Wind loads are determined by integrating simultaneously monitored pressure fluctuations on a building model

288

surface in a boundary layer wind tunnel. An example is presented to illustrate that a suitable combination of a limited number of pressure transducers and the use of a digital or analog computer can streamline the manipulation of data for the development of instantaneous local and integral wind loads. This methodology also provides information on the local spatio-temporal variation of pressure fluctuations for the design of cladding systems. Acrosswind force spectra are formulated for the urban and suburban flow conditions. The effects of added turbulence, in the case of urban flow condition, on the spectral peak of acrosswind forcing funciton, apart from a slight reduction in the Strouhal number, are primarily broadening and lowering of the peak. Therefore, an increase in the incident turbulence has a dampening effect on the wake fluctuations which is reflected in the force spectra.

The forcing function presented in this paper permits the acrosswind response to be estimated at the preliminary design stages to allow early assessment of the structural requirements. The predicted values of acrosswind response show good agreement with the aeroelastic measurements within generally expected design range of wind speeds for typical tall buildings. At higher reduced velocities the inability of a quasi-static forcing function to account for the motion-induced loading results in underestimation of the predicted values of response. This can be compensated by incorporating amplitude-dependent aerodynamic damping values in the mechanical admittance function.

ACKNOWLEDGEMENT

The writer would like to gratefully acknowledge many helpful conferences with Drs. T. Reinhold and J. E. Cermak.

Appendix 1 -- References

1. A. G. Davenport, Gust loading factors, Proc. Am. Soc. Civ. Eng., J. Struc. Div., Vol. 93, No. ST3, June, 1967.

2. J. E. Cermak, J. A. Peterka, A. Kareem, Wind-engineering study of block-259, (First International Plaza, Houston, High-Rise), Dept. of Civil Engineering, Colorado State University, Rept. No. CER78-79JEC-JAP-AK18, January, 1979.

3. A. Kareem and J. E. Cermak, Wind-tunnel simulation of wind-structure interactions, ISA Transaction, Vol. 18, No. 4, 1979.

4. N. Ellis, A new technique for evaluating the fluctuating lift and drag force distributions on building structures, Proceedings of the Fourth International Conference on "Wind Effects on Buildings and Structures," Heathrow, England, 1975.

5. A. Kareem, Wind-excited motion of buildings, Thesis presented to Colorado State University, at Fort Collins, Colo. in 1978 in partial fulfillment of the requirements for the Doctor of Philosophy.

6. A. Kareem, Wind-induced torsional loads on structures, J. of Eng. Structs., Vol. 3, No. 2, April 1981.

7. T. Reinhold, Measurement of simultaneous fluctuating loads at multiple levels on a model of a tall building in a simulated urban boundary layer, thesis (Ph.D.), Virginia Polytechnic Institute and State University, Blacksburg, Virginia, Nov. 1977.

8. J. W. Saunders and W. H. Melbourne, Tall rectangular building response to cross-wind excitation, Proceedings of the Fourth International Conference on "Wind Effects on Buildings and Structures," Heathrow, England, 1975.

9. B. J. Vickery, Load fluctuations in turbulent flow, Proc. Am. Soc. Civ. Eng., J. Eng. Mech. Div., Vol. 108, No. EM1, Vol. 94, 1968.

10. B. J. Vickery and A. W. Clark, Lift or cross-wind response of tapered stacks, Proc. Am. Soc. Civ. Eng., J. Struct. Div., Vol. 98, No. ST1, 1972.

11. R. E. Whitbread, The measurement of non-steady wind forces on small-scale building models, Proceedings of the Fourth International Conference on "Wind Effects on Buildings and Structures," Heathrow, England, 1975.

12. A. G. Davenport and T. Tschanz, The response of tall buildings to wind effects of wind direction and the direct measurement of dynamic force, Proceedings 4th U. S. National Conference on Wind Engineering Research, July 27-29, 1981, Seattle.

13. A. Kareem and Chi-Ming Cheng, Closed form description of fluctuating pressure field on the side faces of a square cross-section building model and the development of acrosswind loading, University of Houston Department of Civil Engineering, Structural Aerodynamics Laboratory Report, AK-CC-CEUH81-1.

14. D. Surry, Some effects of intense turbulence on the aerodynamics of a cricular cylinder at subcritical Reynolds number, J. of Fluid Mech., Vol. 52, 1972.

15. Kareem, A., Acrosswind response of buildings, Proc. Am. Soc. Civ. Eng., J. Struct. Div., Vol. 108, No. ST4, 1982.

16. E. Szechenyi, Supercritical Reynolds number simulation for two-dimensional flow over circular cylinders, J. of Fluid Mech., Vol. 70, 1975.

17. T. Reinhold, A technique for measuring fluctuating wind loads on a tall building model irrespective of model motion, presented at the October 27-31, 1980, ASCE, Annual Convention and Exposition, held at Hollywood , Florida, USA.

18. A. Kareem, Fluctuating wind loads on buildings, Proc. Am. Soc. Civ. Eng., J. Engr. Mech. Div., to appear 1982.

19. A. Kareem, Wind-excited response of buildings in higher modes, Proc. Am. Soc. Civ. Eng., J. Struct. Div., Vol. 107, No. ST4, 1981.

20. A. G. Davenport, Note on the distribution of the largest value of a random function with application to gust loading, J. Inst. Civ. Eng., 24, 1964.

Appendix II -- Notations

The following are symbols used in this paper:

$\{A\}$ = area vector

B, b = building breadth

$Co(x_p, z_p, x_q, z_q; \frac{nb}{v})$ = co-spectra of pressure between points p and q

D = building width

$\overline{F^2}$ = mean square value of force

F_x = fluctuating force in x direction

F_y = fluctuating force in y direction

$G_F(\frac{nb}{v})$ = power spectral density of across-wind forcing function

H = building height

H_i = building segment height

n = frequency

$S(x_p, z_p; \frac{nb}{v})$ = power spectra density of pressure fluctuations at point p

$S_F(\frac{nb}{v})$ = power spectral density of forcing

$S_{F1}(\frac{nb}{v})$ = generalized force in fundamental mode

U_h, V_h = mean wind speed at building height

$\{\Psi^1\}$ = fundamental eigenvector

$\{\phi^1\}$ = fundamental normalized eigenvector

$\rho(x_p, z_p, x_q, z_q; 0)$ = cross-correlation of pressure between point p and q at zero time lag.

ρ_a = air density

Table 1. Acrosswind Fluctuating Force Coefficients

$$\sqrt{\overline{F^2}}/(\tfrac{1}{2}\rho_a V_h^2 BH_i)$$

Level	Boundary layer 1		Boundary layer 3	
	Time Domain	Frequency Domain	Time Domain	Frequency Domain
1	0.2214	0.2183	0.2336	0.2283
2	0.3525	0.3522	0.3168	0.3120
3	0.4201	0.4185	0.3302	0.3335
4	0.4622	0.4572	0.3331	0.3952
5	0.4748	0.3336	0.2661	0.2679
Overall	0.3422	0.3423	0.2702	0.2703

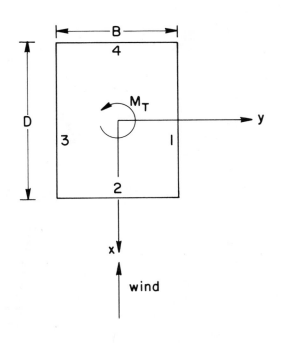

Fig. 1. Coordinate System and Building Plan.

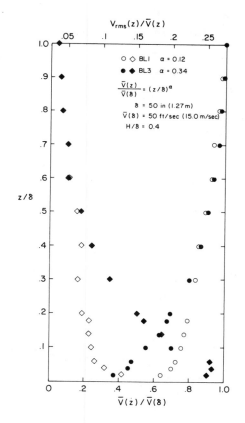

Fig. 2. Mean Speed and Turbulence Intensity Profiles for Both Terrains Considered.

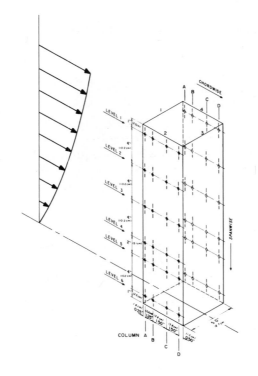

Fig. 3. Location of Pressure
Taps on the Model Faces.

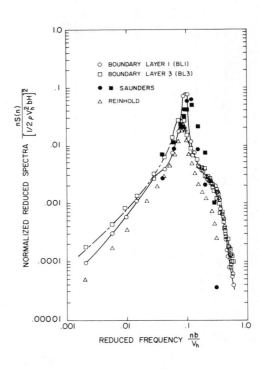

Fig. 4. Cross Power Spectral Density
Matrix of Acrosswind Forcing Function
in BL1 Flow.

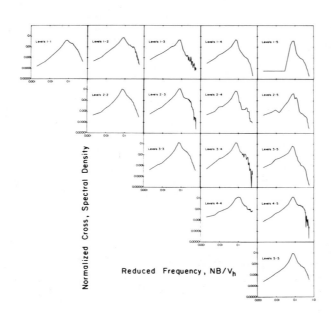
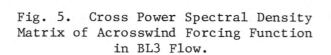

Fig. 5. Cross Power Spectral Density
Matrix of Acrosswind Forcing Function
in BL3 Flow.

Fig. 6. Generalized Spectra of
Acrosswind Forcing Function in BL1
and BL3.

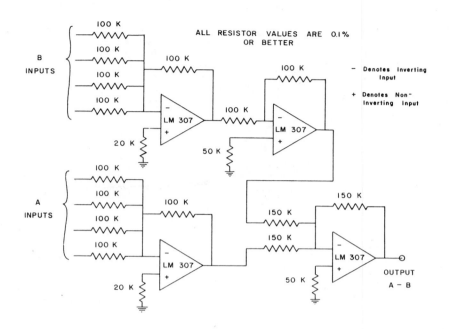

Fig. 7 Schematic of Summing Amplifier Network (Ref. 7)

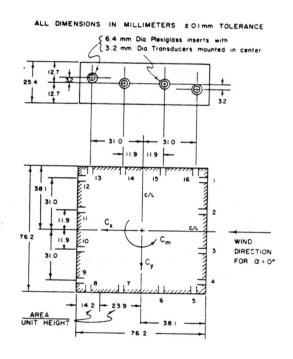

Fig. 8 Placement of Transducers (Ref. 7)

294

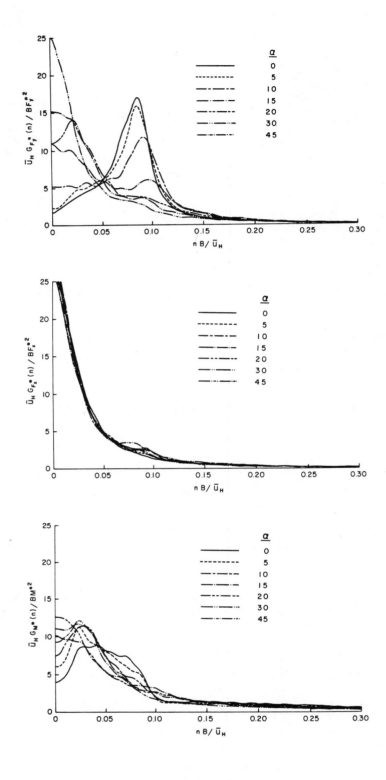

Fig. 9 Modal Force Spectra for F'_y Force Component, F'_x Force Component, and Twisting Moments (Ref. 7)

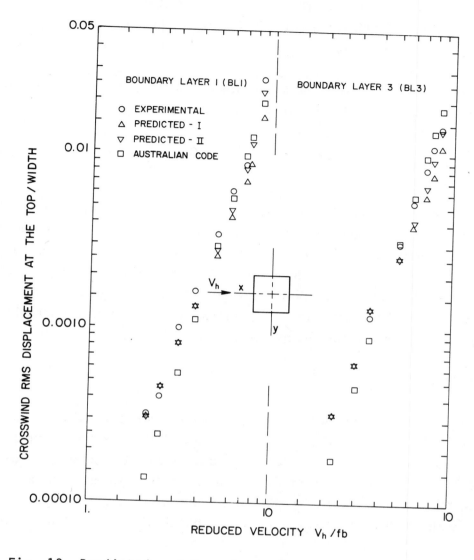

Fig. 10 Predicted and Experimental Acrosswind Response Estimates

MEASUREMENT OF TOTAL DYNAMIC LOADS USING ELASTIC MODELS WITH HIGH NATURAL FREQUENCIES

Tony Tschanz, Boundary Layer Wind Tunnel Laboratory, The University of Western Ontario, Faculty of Engineering Science, London, Ontario, Canada   N6A 5B9

SUMMARY

This paper describes an approach to the prediction of the dynamic response of structures to wind loads using rigid foam models mounted on a sensitive, high frequency balance. The requirements and design of the force balance, models, and instrumentation are described in detail, and applications are shown to determine the fundamental translational response of two tall buildings. Discussed are advantages, as well as shortcomings of this method, and comparisons with related approaches by other authors. It is shown that high frequency model tests can be an efficient way to predict the response of structures to wind.

1.   INTRODUCTION

Steady progress has been made in the description of the response of taller buildings to wind (Davenport, 1966; Davenport, Mackay and Melbourne, 1980). The main components of the response come from each of the fundamental modes, excited by the modal forces. The mechanical transfer function, relating the load function to the response, is straightforward. The aerodynamic transfer function, relating the gust structure to the forces, however, has to be determined experimentally in most cases. A further complication exists if body motional effects interact with the load function (aerodynamic damping).

Multi-degree-of-freedom aeroelastic models are believed to provide the most truthful results, but they are expensive and time consuming, and extrapolations for different building properties than those modelled are difficult. Rigid models, pivoting at the base, are less expensive, but they do not provide information on torsional response and are modelled for particular structural properties as well.

If aerodynamic damping effects are negligible, which is usual for most buildings at practical wind speeds and practical structural damping values, and if a high frequency model can be made, then it is possible to measure the load function directly. The modal loads are a function only of the wind structure and aerodynamic shape of the building. This paper describes the development of a sensitive high frequency balance at the Boundary Layer Wind Tunnel Laboratory of the University of Western Ontario (referred to as BLWT balance below). The development and construction of the five-component balance was an expensive, but one-time investment. Valuable attributes, in addition to the high stiffness and sensitivity, are the flat top of the balance i level with the wind tunnel floor, and measurement of the forces are done with respect to this level. Simple foam models, cut only to the correct size, without involving other scaling factors, now provide an economical and rapid description of the modal forces.

The response predictions for actual structures is analogous to the gust factor approach (Davenport, 1966, for example). Typical force spectra measured with the BLWT balance have been presented previously (Davenport and Tschanz, 1981). More emphasis is given in this paper to reviewing the technique and describing the balance in some detail.

2. CONCEPT OF THE FORCE METHOD

The response of a tall building to wind loads essentially consists of a mean response plus resonant responses in each of its fundamental modes. Computation of the modal responses is complicated because of the need to include factors such as the mode shape, correlation and size of the gusts, and aerodynamic effects. Prediction of the mean response is fairly straight-forward.

The generalized forces — its mean value $\overline{F}$, its root-mean-square σ_F^2, and its spectrum $S_F(f)$ are defined from the pressures by

$$\overline{F} = \int_A \overline{p(z,t)} \, \mu(z) \, dA \tag{1a}$$

$$\sigma_F^2 = \int_A \int_A \overline{p(z,t) \, p(z',t)} \, \mu(z) \, \mu(z') \, dA \, dA' - \overline{F}^2 \tag{1b}$$

$$S_F(f) = \int_A \int_A S_p(z,z',f) \, \mu(z) \, \mu(z') \, dA \, dA' \tag{1c}$$

In these $p(z,t)$ and $\mu(z)$ are the instantaneous pressures and mode shapes at point z, respectively. $S_p(z,z',f)$ is the cross spectrum of pressures at points z and z' at frequency f, and dA and dA' are elementary areas at positions z and z' of the projected area A.

Tall buildings have a fundamental mode shape which is more or less a straight line. Vickery (1970) derived expressions for estimating the errors in calculating the response by assuming a straight line. Even for large deviations of the actual mode shape the errors are roughly 1% to 3%, making the effect of moderate deviations from a straight line mode shape insignificant.

This enables the generalized force in the fundamental overturning modes to be measured directly, using a straightforward base moment balance. Torsional forces and aerodynamic effects need special consideration and will be described at greater length below.

The difficulties in the experimental determination of the dynamic forces lie primarily in the demanding frequency response of the measuring system. The requirements of a force balance are now described.

3. REQUIREMENTS FOR A FORCE BALANCE

The ideal balance capable of measuring the complete time history would require the following attributes:

a)   infinite sensitivity;

b)   infinite rigidity;

c)   stability and free of temperature and environmental drift; and

d)   uncoupled measurement of the orthogonal force components.

The sensitivity has to be high, since the processing of the output from the balance introduces errors which are in general smaller for larger outputs from the balance. The quasi-static force which has to be resolved is approximately one hundred times smaller than the resonant response of a conventional aeroelastic model. The wide range of forces from large buildings with high wind speeds, to small buildings with low wind speeds, also require high output to resolve small forces accurately.

The rigidity has to be high in order to reduce kinematic effects. This is contradictory to the sensitivity requirement, since, in general, high sensitivity is achieved using flexible transducers. The dynamic response of the balance model combination with high natural frequencies can be shown from Figure 1.

The response is greatly distorted for input frequencies near the fundamental frequency of the measuring system. It is, hence, desirable to have high rigidity which gives a high fundamental frequency. The quality of a balance, or for that matter any strain gauged force transducer for rapid measurements, could primarily be assessed by a high product of stiffness times strain output, with units of strain/length for forces and strain/radian for moments.

Good stability is required to accurately measure the loads in the lower force ranges, since drift is independent of the load level. Great attention to mechanical and electronic design can provide better stability, but drift errors can best be masked by high output levels from the balance. For durations of typical wind tunnel tests, the stability can be maintained within roughly one micro strain. For short times, i.e. purely dynamic measurements the stability requirements are less severe. If the same balance has to measure the mean as well as the dynamic loads, the stability and related sensitivity is a critical requirement.

Coupling of orthogonal force components can be minimized by careful mechanical design and especially by good workmanship. First order effects can be corrected by multiplication with a matrix containing all coupling terms, however computer time is increased, and flexibility from the independent monitoring of components is reduced.

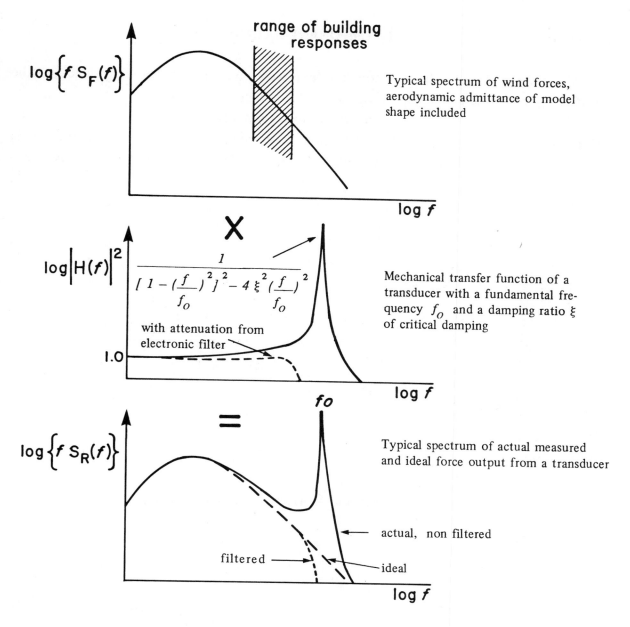

Fig. 1   Dynamic Response of Balance—Model Combination

The ideal balance is not realizable, therefore compromises have to be made, expecially between sensitivity and rigidity. The following secondary requirements should also be taken into account; output from six components available simultaneously, preferably referred to a common origin at the centre of the model at the base elevation, ease of mounting different models on balance and economy of building models, low hysteresis and creep, ruggedness, repairability in case of breakdown, low mass, high damping, basic design applicable to different load ranges, and cost of the balance.

4.   DESCRIPTION OF THE FIVE-COMPONENT BALANCE

A number of different balance configurations were investigated. The design chosen is believed to represent the best compromise for the load ranges to be used. In order to achieve maximum rigidity, the measurement of vertical forces was forgone. The remaining five components are measured with respect to the turntable centre and at the level of the wind tunnel floor, as shown in Figure 2. A photograph of the balance with a typical model is shown in Figure 3.

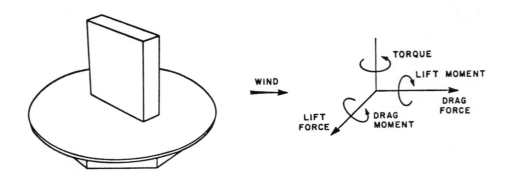

Fig. 2   Force Balance and Measured Force Components

Fig. 3   Photo of Balance With Typical Building Model

The balance system is basically a statically determinate frame with force links and miniature load cells equipped with semi-conductor strain gauges forming a full Wheatstone bridge. A cross section showing schematically the measurement of the base bending moment and shear force is shown in Figure 4.

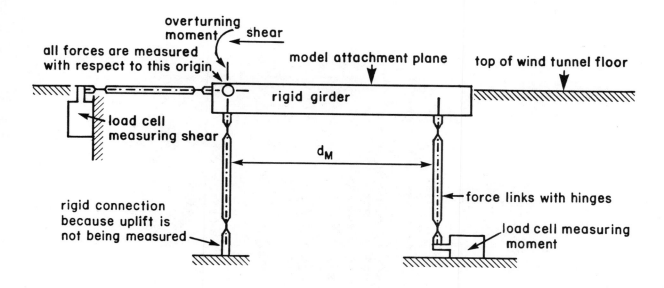

Fig. 4   Schematic Cross Section for Measuring Forces in One Direction

The output from the shear load cell is equal to the shear force, and the output from the moment load cell is equal to the bending moment divided by the force link spacing d_M.

The cross section in the perpendicular direction is similar, except that two load cells are required to measure the shear force. This is shown schematically in the plan view of the balance, Figure 5.

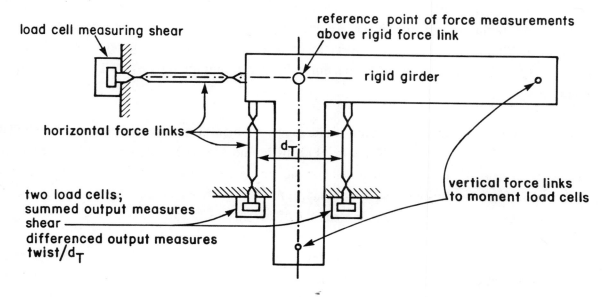

Fig. 5   Schematic Plan View of Balance

The detailing and machining requires extreme care. Space requirements do not permit further discussion, however, a complete description will be available in Tschanz (1982). A photo of the balance with the top plate removed is shown in Figure 6. The girder, the three horizontal force links and the three load cells are visible. The load cells are mounted on elaborate overload protection devices which link out as soon as a load cell gets near its critical load in any direction. Mechanical stops are not possible for such rigid transducers.

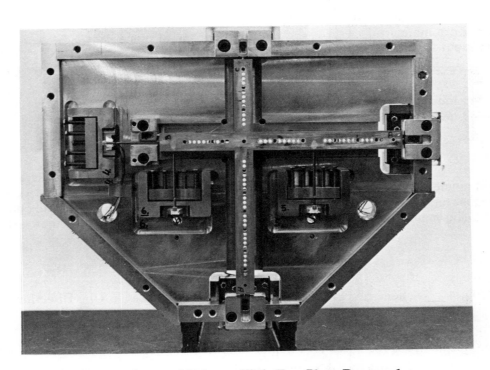

Fig. 6   Photo of Balance With Top Plate Removed

The most critical components of the balance are the load cells. For forces encountered by building models, only the bending beam type is applicable. Optimized for maximum strain versus stiffness demands: short length and great depth (i.e. a low modulus of elasticity). In addition, desirable material properties are good linearity to high strain after the cement is cured at high temperature; a low coefficient of thermal expansion and good thermal conductivity. A typical miniature load cell, machined from Aluminium alloy 7075-T6 is shown in Figure 7.

Prototypes using semi-conductor strain gauges or highest quality foil gauges gave similar performance. The measured stiffness is 2900 lbs/in and the strain output is 600 micro strain per lb. applied load. Using the load cell for measuring shear gives a stiffness-strain product of 1.7 strain/inch, and using it as a moment transducer with a spacing of 4 inches for the force links results in a stiffness-strain product of 6.7 strains/radian.

For a rigid balance such as the one described in this section, the frequency of the balance-model combination depends almost entirely on the material and construction of the building model. Only light-weight models can be considered, for these the frequency is virtually identical to a cantilever on a rigid support. For a homogenious model this frequency is proportional to the velocity of sound in the material ($c = [E/\rho]^{-\frac{1}{2}}$, where E is the modulus of elasticity and ρ is the density of the material). Models made from light styrofoam, reinforced with a fibre tape as shown in Figure 2 are presently being used, however, new materials are being sought. The resulting lowest frequency of a 6 x 6 x 1½ inch model is 280 Hz as measured on the balance. A number of different shapes and sizes have been tested to date.

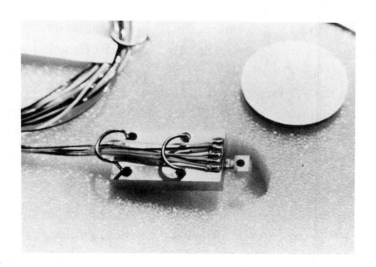

Fig. 7   Photo of Miniature Load Cell

Calculated frequencies for Rohacell 31 are higher, but no material was available for experimentation. More elaborate fabrication using carbon fibre shells, such as used by Whitbread (1975) and Evans and Lee (1981), can be expected to give significantly higher fundamental frequencies at the expense of much greater cost and time required to fabricate the models.

5.   DISCUSSION OF OTHER METHODS OF MODAL FORCE MEASUREMENTS

5.1   Applications Relying on a Force Balance

The technique of obtaining the modal wind-load spectra and computing the corresponding dynamic spectra for those combinations of modal frequency and damping values appropriate to a particular building or structure has been known for a long time and is gaining popularity.

One of the first attempts at direct force measurements of quasi-static wind load spectra has been described by Cermak et al (1970). Their approach used a sting type balance, which is predominantly used in aeronautical wind tunnels for testing air planes. The 5 inch long beam protruded into the wind tunnel, and was covered by a shell of the building model of the correct dimensions (20.75 x 7 x 3.9 in. for height, width and depth respectively). The natural frequency of 200 Hz is impressive, but the stiffness-strain product is only 0.2 strain/ radian assuming a reasonable guess of the missing details in their paper. Compared to the value of 6.7 strain/radian for the five component BLWT balance described above, it is obvious that the sting type balance does not provide the best possible stiffness-sensitivity combination. Other shortcomings are that it measures only one component, and the relative difficulty of mounting different building models. An advantage is the cheap fabrication of the sting balance.

Whitbread (1975) measured the shear force at the base and the base overturning moment about the weak axis for a 12 inch tall model of the CAARC standard building model (Wardlaw and Moss, 1970). Using a carbon fibre model with integrated silicon sensors at the base, he achieved a natural frequency apparently above 250 Hz and a strain output of .332 μin/ in/lb-in for the moment and .71 μ in/in/lb for the shear force. Not enough details are given in the paper to compare the performance of this balance with the five component BLWT balance. The balance-model combination is designed for one single use only, and measures two components. Whitbread noted some difficulties with temperature and calibration. The frequency response is excellent with the carbon fibre model, but is gained with a large compromise of the sensitivity.

English and Durgin (1981) constructed a sting type balance measuring five components with a model from a balsa wood shell. The lowest natural frequency for the 8 x 12 x 3 inch model was only 65 Hz, but considerable emphasis was given to increase the damping. From the transfer function shown in Figure 1, it can be seen that for sufficiently high damping the high frequency requirements are somewhat less severe, but it is questionable if sufficiently high damping values can be practically achieved. Their paper does not show any spectra, the time history figure appears to show some periodicity and the rms force coefficients show some inconsistencies. Details are not available to compare the stiffness-strain product with the five component balance described in this paper, but it is obvious that a sting balance with two links per direction is too flexible to give a high frequency response, and the models involve considerable fabrication time. Torsional moments may be difficult to measure because of insufficient sensitivity and coupling effects.

A different approach, using low frequency models and inferring the force from response measurements, was used by Saunders and Melbourne (1975). Knowing the frequency and the damping of the transfer function shown in Figure 1, a correction can be made at each measured frequency to compute the spectrum of the force. The main difficulty is that the damping has to be known exactly, including the aerodynamic damping contribution, because of the sensitivity to the damping value near the natural frequency of the balance-model combination.

Evans and Lee (1981) use a one component balance, measuring the base bending moment on models. The frequency and damping of the model is scaled to the values required by the scaling laws. Their frequencies and damping can apparently be varied independently, but this makes their method more comparable to the conventional pivoting aeroelastic models. The mass has to be scaled as well, or taken into account when converting to full scale values.

Noteworthy are their carbon-fibre reinforced plastic models but with the low frequencies required for pivoting models it might be economical to use a more efficient balance and simple balsa wood models.

5.2 Application Without a Force Balance

Modal forces can be directly determined using rigid models, instrumented with pressure transducers, without employing a balance. This approach was followed by Reinhold and Sparks (1979) to compute the modal forces for the fundamental translational modes and the fundamental torsional mode of a building, by summing the products of pressure times area times mode shape throughout the height of the building according to equations (1). The stringent frequency response requirements are eliminated, as the response for pressure measurements is flat to beyond 100 Hz for routine test set-ups at the BLWT. This is the only method besides conventional aeroelastic tests, theoretically permitting arbitrary mode shapes, including realistic ones for the torsional modes.

In practical applications, however, a number of compromises are required. The ideal case would be to use the same pressure model which is routinely being used in wind engineering studies to determine the cladding pressures, with one pressure transducer, signal conditioning, sampling and digitizing port in the computer for each pressure tap, and to perform simultaneously all operations according to equations (1). While the signal conditioning, the digitizing and the computers are getting cheaper and more powerful, the total cost, including the transducers and labour involved, makes this approach not feasible. Practical approximations include proportional spacing of pressure taps according to mode shape and pneumatic averaging or, like Reinhold and Sparks (1979), selecting as many taps as pressure transducers are available, and summing and multiplying the contributions to the modal forces with analogue circuits as required.

5.3 Discussion

All methods discussed, including the five component balance of the BLWT, make certain approximations, and none of the methods can take body motion dependent forces into account. The method using pressure transducers enables realistic torsional mode shapes to be used, but it is estimated that the limited number of transducers which can be used in practical applications, do not improve the accuracy.

Judging by the simplicity of mounting the economical foam models on the balance, and by the direct availability of all five components, the BLWT balance-model combination is most attractive.

6. INSTRUMENTATION OF THE BLWT BALANCE

The amount of instrumentation required to measure the modal forces with the BLWT balance is relatively small. Inexpensive integrated chips are used for bridge supply and signal amplification, giving drifts of 0.1 $\mu V/^{o}c$ or 1.0 μV/month,referred to input. These strain gauge amplifiers have provisions for a low pass filter, enabling a simple first order correction to be built in to acheive, for typical models, a unity gain of the transfer function shown in Fig. 1 up to a higher frequency.

Summing and differencing amplifiers are easily added to the amplifier cards to form the shear force and torque moment output indicated in Figure 5.

Appropriate notch and low pass filter modules were used to remove fan induced acoustic pressures at approximately 60 Hz harmonics, which, with a balance of this sensitivity, contributed significantly to the measured signal. This signal conditioning is suitable to the majority of models tested at the BLWT.

Additional filtering may be necessary, i.e. to prevent aliasing in spectrum measurements, as is standard practice for digital data acquisition.

The instrumentation required for one channel is shown in Figure 8. The electronics to the left contain the filters, those to the right contain the bridge supply and signal amplification. In the foreground is a load cell and in the background are some typical foam models.

Fig. 8   Instrumentation for the Force Balance

7. LINEAR ELASTIC RESPONSE CALCULATIONS WITH THE BLWT BALANCE

The linear response of an elastic system can be expressed in the general form:

$$R = \bar{R} + g\,\sigma_R \qquad (2)$$

where $\bar{R}$ is the mean response, σ_R the rms response and g a dimensionless time varying factor. A particular response of interest is the peak response $\hat{R}$, in which case the corresponding value of g is roughly 3 to 4.

The quasi-static response can be broken down into two components: a quasi-static component σ_B; and a resonant component σ_{R_e} which can be written

$$\sigma_B = \frac{\sigma_F}{k} \qquad (3)$$

$$\sigma_{R_e} = \sqrt{\frac{\pi}{4}\,\frac{1}{\zeta_s + \zeta_a}\,\frac{f_o\,S_F\,(f_o)}{k^2}} \qquad (4)$$

in these σ_F is the rms generalized force, k the generalized stiffness, ζ_s and ζ_a the structural and aerodynamic damping and $S_F(f_o)$ the power spectrum of the generalized force at the natural frequency f_o. Using mean square addition of the components gives:

$$\sigma_R = \frac{\sigma_F}{k}\sqrt{1 + \frac{\pi}{4}\,\frac{1}{\zeta_s + \zeta_a}\,\frac{f_o\,S_F(f_o)}{\sigma_F^2}} \qquad (5)$$

This representation has been given in various references (Davenport, 1966, for example) and is the basis of the gust factor approaches. The difficult problem in evaluating equation (5) has always been the determination of the rms generalized force, σ_F, its spectrum $f_o\,S_F(f_o)/\sigma_F^2$ and the aerodynamic damping ζ_a.

The aerodynamic damping is body motion dependent, and can only be measured with conventional aeroelastic models. There are, however, indications (Davenport and Tschanz, 1981) that the value is small and positive for typical buildings in the drag direction, and positive too in the lift direction up to the velocity where vortex shedding has a major contribution to the response. This velocity is usually higher than the design wind speeds, and is indicated in the force spectra by a narrow band peak. The result is that equation (5) gives a slightly conservative value, except for lift forces at high velocities.

The generalized forces and its spectra are easily measured using the BLWT balance, and are shown in Figs. 9 and 10 for the drag and lift direction of two tall buildings. A similar building as shown in Fig. 9 has been tested by Rosati (1968) using an aeroelastic model. A

308

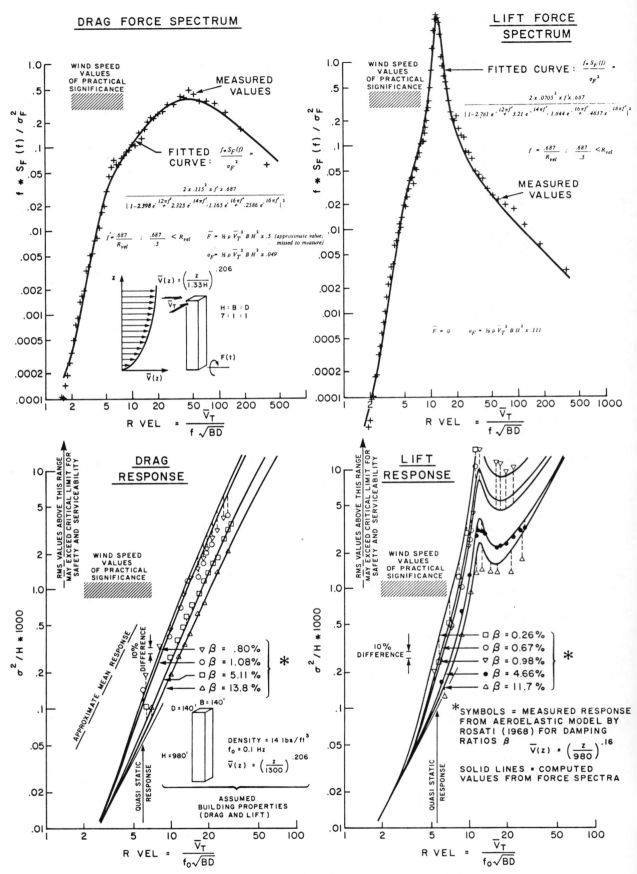

Fig. 9 Force Spectra and Computed Response Compared to Aeroelastic Tests of a Tall Square Building for Open Country Exposure

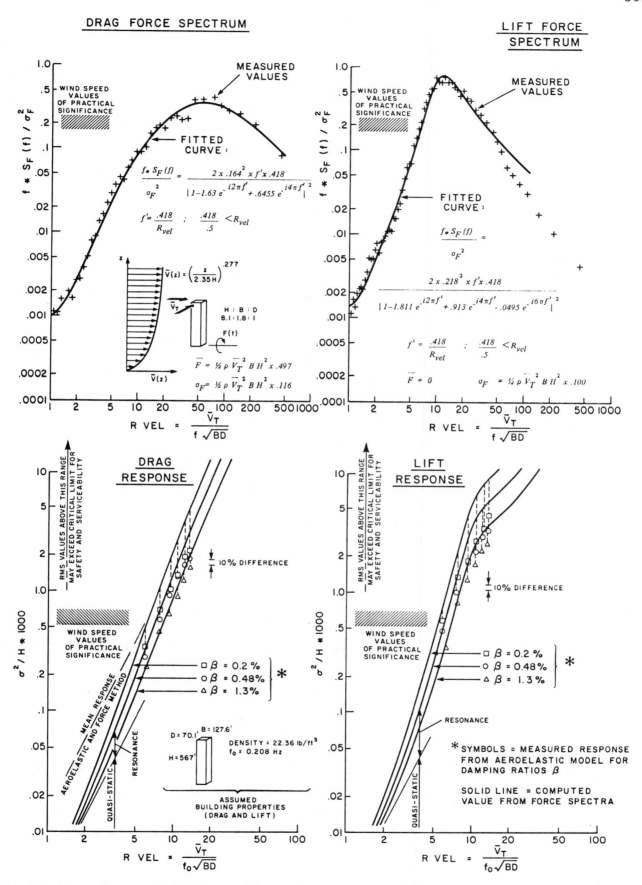

Fig. 10 Force Spectra and Computed Response Compared to Aeroelastic Tests on a Rectangular Building for Suburban Exposure

comparison of the response predictions using equation (5) with Rosati's test is shown in the lower graphs of Fig. 9. Some deviations should be expected because important variables, such as the boundary layer flow and turbulent conditions, can not be matched exactly in tests conducted at different times. Most of the aeroelastic test results are for higher wind speeds than are encountered by the building and are of little practical significance. The comparisons at lower velocities are, hence, more important to assess the validity of the force method. Comparisons at the higher velocities give an indication of the degree of overestimate in case of positive aerodynamic damping, or underestimate in case the aeroelastic damping becomes negative.

A comparison of response predictions using the force method and from an aeroelastic test under identical conditions, carried out by the author, is shown in Fig. 10. Again it is shown that the force method overestimates the response in all cases, even at high velocities in the lift direction, indicating that the aerodynamic damping may be positive for all tested wind speeds. It should be noted that this aeroelastic test was also conducted at high wind speeds. At lower wind speeds the signal would be contaminated by noise and would not be accurate for comparative purposes. Very low damping values are also difficult to keep at a controlled level and the sensitivity to the damping is immense. Comparison at practical velocities and practical damping ratios, or by allowing for some aerodynamic damping for very low structural damping values, show that the force method approach can serve as an economic and valid method to predict the response of structures to wind loads.

Coupling between mode shapes is easily incorporated and has been applied in tests of actual buildings at the BLWT.

8. TORSIONAL RESPONSE

The motions due to the twisting moment can be particularly disturbing to occupants. Of the buildings causing problems, full scale or wind tunnel measurements often show that the torsion is the culprit. It is, hence, important to assess the torsional response, especially if there are eccentricities in mass or stiffness. Simple pivoting models only give information on the fundamental translational response, and expensive multi-gree of freedom models are required for truthful assessment of the torsional response.

The BLWT force balance-model combination gives an apprxoimation to the torsional response, having in effect a constant magnitude mode shape in equation (1), instead a mode-shape varying with height, and with a maximum value at the top of the structure. The response depends on the degree of co-ordination of the eddies in the wind, multiplied by the mode shape. The correlation, decreasing with increasing separation distance, is likely overestimated by a uniform mode shape, resulting in a non-conservative modal force spectrum at higher frequencies.

Several methods of applying a correction factor could be used and these will be discussed in a forthcoming paper.

9. CONCLUSIONS

Approaches are discussed to measure the forcing function from buildings exposed to the turbulent boundary layer. A new sensitive, high frequency balance is described in some detail. enabling the direct measurement of modal forces, using simple foam models. Response predictions from modal force spectra are straightforward and are shown for two tall buildings. These are compared to results obtained from conventional aeroelastic tests.

Discussed are also related approaches by other authors, evaluating advantages and shortcomings compared to the method proposed in this paper. The force method described can be considered an economical and practical approach to predict the response of conventional structures subjected to realistic wind load conditions. The method assumes that aerodynamic damping effects can be safely neglected or otherwise allowed for.

ACKNOWLEDGEMENT

The balance requiring a high level of workmanship was built at the University of Western Ontario's Co-Ordinated Machine Shop. The efforts of Messrs. Peter Teunissen and Robert Kager are acknowledged with thanks. This study was supported by a grant in aid of research from the Natural Science and Engineering Research Council of Canada awarded to A. G. Davenport under whose general supervision the work was carried out.

REFERENCES

Cermak, J. E., Sadeh, W. Z. and His, G., "Fluctuating Moments on Tall Buildings Produced by Wind Loading", Proc. Technical Meeting Concerning Wind Loads on Buildings and Structures, Nat. Bureau of Standards, Building Sci. Series 30, Nov. 1970, pp. 45—59.

Davenport, A. G., "The Treatment of Wind Loading on Tall Buildings", Symposium on Tall Buildings at the University of Southampton, 13—15, April 1966, Pergamon Press, pp. 3—44.

Davenport, A. G., Mackey, S. and Melbourne, W. H., Council on Tall Buildings, Committee 9, 1980, "Wind Loading and Wind Effects", Chapter CL—3, Vol. CL of Monograph on Planning and Design of Tall Buildings, ASCE, New York.

Davenport, A. G. and Tschanz, T., "The Response of Tall Buildings to Wind: Effects of Wind Direction and the Direct Measurement of Force", Proceedings, Fourth US National Conference on Wind Engineering Research, July 27–29, 1981, Seattle, Wa.

English, E. and Durgin, F. H., "A Wind Tunnel Study of Shielding Effects on Rectangular Structures", Proceedings, Fourth US-National Conference on Wind Engineering Research, July 27–29, 1981, Seattle, Wa.

Evans, R. A. and Lee , B. E., "The Assessment of Dynamic Wind Loads on a Tall Building: A Comparison of Model and Full Scale Results", Proceedings, Fourth US National Conference on Wind Engineering Research, July 27–29, 1981, Seattle, Wa.

Reinhold, T. A. and Sparks, P. R., "The Influence of Wind Direction on the Response of a Square-Section Tall Building", Proceedings of the Fifth Int. Conference on Wind Engineering, Colorado, U.S.A., July 1979, Pergamon Press, Oxford, pp. 685–698.

Rosati, P. A., "The Response of a Square Prism to Wind Load", Research Report BLWT–II–1968, M.E. Sc. Thesis Supervised by A. G. Davenport.

Saunders, J. W. and Melbourne, W. H., "Tall Rectangular Building Response to Cross-Wind Excitation", Proceedings of the Fourth Int. Conference on Wind Effects on Buildings and Structures, Heathrow 1975, Cambridge University Press, London, pp. 269–379.

Tschanz, T., Ph.D. Thesis, The University of Western Ontario, London, Ontario, Canada. To be published in 1982.

Vickery, B. J., "On the Reliability of Gust Loading Factors", in Proceedings of the Technical Meeting Concerning Wind Loads on Buildings and Structures, Building Science Series 30, National Bureau of Standards, Washington, D.C., 1970.

Wardlaw, R. L. and Moss, G. F., "A Standard Building Model for the Comparison of Simulated Natural Winds in Wind Tunnels", Commonwealth Advisory Aeronautical Research Council, cc 662, Tech. 25, 1970.

Whitbread, R. E., "The Measurement of Non-Steady Wind Forces on Small-Scale Building Models", Fourth International Conference on Wind Engineering, Heathrow 1975, Cambridge Univeristy Press, pp. 567-574.

PREDICTING PEAK PRESSURES VS. DIRECT MEASUREMENT

Jon A. Peterka
Fluid Mechanics and Wind Engineering Program
Colorado State University, Fort Collins, CO

1.   INTRODUCTION

Development of procedures for determination of peak local pressures on buildings for design purposes has included wind tunnel tests for large or unusual structures in which the peak pressures are measured directly. The probability distribution of the largest peak pressure achieved during a design storm at a given location on a building has a long tail extending toward higher loading [1]. Recent analysis by Cook and Mayne [2] has provided a simple calculational method in which the probability distribution of largest pressure can be used to provide a consistent level of design pressure. In this method

$$P_{50} = 0.5 \rho U_{50}^2 C_p^* \tag{1}$$

$$C_p^* = U_o + 1.4/a \tag{2}$$

where P_{50} is the 50-year recurrence pressure, ρ is the air density, U_{50} is the 50-year recurrence velocity, C_p^* is the effective peak pressure coefficient, U_o and $1/a$ are the mode and dispersion of the Type I extreme value distribution $P(S)$ for the largest peak pressure during the 50-year wind event.

It is desirable to obtain the probability distribution $P(S)$ of the largest peak as a matter of routine during wind-tunnel testing. In the past, it has been common to use a single sample of peak pressure to estimate C_p^*. A more complete description of the probability distribution can be obtained by measuring 16 to 20 samples of largest peak and fitting them to a Type I distribution to obtain U_o and $1/a$. This method requires several times as much wind-tunnel time as is required for measurement of one sample even if shortened records are used. To overcome this time requirement, a method was proposed [3] to estimate the properties of $P(S)$ using all independent peaks in a single one-hour record. While the method worked well for some cases, the prediction of U_o and $1/a$ was not as consistent as might be desired. A modification of this procedure appears to provide a more consistent prediction of probability distribution parameters.

2.   PROCEDURE FOR ESTIMATING MODE AND DISPERSION

Consider a time record of length T, approximately one hour in length full-scale, of pressure coefficient C_p obtained at a point on a building. Each entry in the record can be nondimensionalized by

$$s = \frac{C_p - C_{p_{mean}}}{C_{p_{rms}}} \tag{3}$$

If p(s) is the probability density of all negative going (minima) peaks in the record (for simplicity, only the maxima will be considered in the analysis; analysis of the minima is the same if negative s values are used), then P(s), the probability density of the largest of the peaks, will be

$$P(s) = N[1-q(s)]^{N-1} p(s) \qquad (4)$$

if the peaks of p(s) are independent. In this equation, N is the number of independent peaks in time T and

$$q(s) = q \ (S \geq s) = \int_{s}^{\infty} p(s) \ ds \ , \qquad (5)$$

the cumulative distribution of all peaks.

From theoretical considerations [4,5] and from empirical experience [6], q(s) can be approximated by an extreme value distribution of Type I:

$$q(s) = 1 - e^{-e^{-y}} \qquad (6)$$

$$y = a(s-U) \qquad (7)$$

where U and 1/a are the mode and dispersion of the distribution. The expression, y, is called the "reduced variate."

Thus $\qquad p(s) = - \dfrac{dq}{ds} = a(e^{-e^{-y}}) \qquad (8)$

and $\qquad P(s) = aN(e^{-y})(e^{-e^{-y}})^{N} \qquad (9)$

P(s) has the same shape as p(s) but is shifted along the s axis to larger values. The mode U_{o} of P(s), from the maximum of (9), is

$$U_{o} = U + \frac{1}{a} \ln N \qquad (10)$$

The form of P(s) should be that of an extreme value distribution of Type I. Using the probability density form of (8)

$$P(s) = a_{o} (e^{-y_{o}})(e^{-e^{-y_{o}}}) \qquad (11)$$

where $y_{o} = a_{o}(s-U_{o})$, U_{o}, and $1/a_{o}$ are the reduced variate, mode and dispersion of P(s). Equating (9) and (11) at a particular point, say s = U_{o}, and incorporating (10) for the value of U_{o},

$$a_{o} = a \qquad (12)$$

The preceding analysis shows that the probability distribution of the largest peak in time T can be represented by an extreme value distribution of Type I whose mode and dispersion, given by Equations (10) and (12), are functions of the mode and dispersion of the Type I extreme value distribution and the number, N, of independent peaks in T.

N can be represented by

$$N = \nu T \tag{13}$$

where ν is the number of independent peaks per unit time. Rice [7] showed that the number of extrema per unit time could be found from ratios of moments of the spectral distribution of the parent time series. Because this technique requires a large amount of computer time if many cases are to be studied, an alternate scheme can be used to estimate the value of ν. The autocorrelation, R_k, of the sequence of peaks is formed, where k is the delay in peaks. Peaks become essentially uncorrelated, and hence independent, when $R_k < 0.2 - 0.3$. k_c is the difference in k between independent peaks and T_M is the length of the record to obtain M peaks. Then

$$\nu \cong \frac{M}{k_c T_M} \tag{14}$$

Two factors act to make (14) an acceptable estimate for ν: 1) the probability distribution P(s) is not highly sensitive to the value of ν, and 2) peaks become uncorrelated rapidly so that only a few values of R_k need by computed with a reasonably small sample, M, of peaks.

A modification of this procedure appears to provide a superior estimate U_o and 1/a and require less computation. The M largest peaks are recorded from the time series of length T. A Type I distribution is fit to these peaks to obtain a value for U and 1/a. The probability distribution P(s) then has

$$a_o = a$$

$$U_o = U + 1/a \ln M$$

3. EXPERIMENTAL VERIFICATION

Eleven tap locations with negative means were selected at random (except that two taps in regions influenced by vortex flows were required) from several model studies under investigation in the Fluid Dynamics and Diffusion Laboratory at Colorado State University. The values of $C_{p_{mean}}$ and $C_{p_{rms}}$ are listed in Table 1. Taps 501 and 509 represent locations where vortex flows were present. One hundred records, each approximately one full-scale hour in length, were recorded for each tap. The largest peak pressure coefficient was selected from each record for analysis of U_o and 1/a by direct fit to a Type I distribution. The values thus obtained are listed in Table 1 and serve as the "exact" answer for comparison. Each of the 100 records was used individually for prediction of U_o and 1/a from a single record by both methods discussed above. Values of U and 1/a averaged over the 100 records are listed in Table 1. The corresponding value of predicted U_o are also shown in Table 1. A value of M of 100 was used in the analysis.

It is evident from Table 1 that prediction using 100 largest peaks in the record provides a better approximation to the actual values of U_o and $1/a$ than does the prediction using all N independent peaks. Figure 1 shows a typical tap with a Type I fit to all N independent peaks where prediction of the actual U_o and $1/a$ from a single record is not particularly good. For a case where prediction of U_o and $1/a$ from N peaks in a single record was satisfactory, a fit to a Type I distribution is shown in Figure 2.

Values of C_p^* given by Equation (2) were calculated for the 100-record case and for the single record case and are shown in Table 1. The mean and standard deviation of the C_p^* obtained from each of the 100 records are shown. These values are shown in Figure 3 to illustrate how closely the single-record approximation of C_p^* is to the actual value.

4. CONCLUSIONS

The data presented in this paper show that the probability distribution of peak negative pressure on a building can be estimated with acceptable accuracy with a single time record of approximately one hour full-scale duration. Further optimization of the technique should provide increased precision and/or reduced computational cost.

5. REFERENCES

1.   Peterka, J. A. and J. E. Cermak, "Wind Pressures on Buildings--Probability Densities," Jl. Structural Div., ASCE, Vol. 101, pp. 1255-1267, 1975.

2.   Cook, N. J. and J. R. Mayne, "A Refined Working Approach to the Assessment of Wind Loads for Equivalent Static Design," Jl. Wind Engrng. and Indust. Aerodyn., Vol. 6, pp. 125-137, 1980.

3.   Peterka, J. A., "Probability Distributions of Local Peak Pressures," Fourth U.S. National Conference on Wind Engineering Research, Seattle, WA, July 1981.

4.   Gumbel, E. J., "Statistical Theory of Extreme Values and Some Practical Applications," U.S. Dept. of Commerce, Nat. Bur. Stand. Applied Mathematics Series 33, 1954.

5.   Mayne, J. R. and N. J. Cook, "On Design Procedures for Wind Loading," Building Research Establishment Current Paper, CP25/78, 1978.

6.   Lou, J. J., "Extreme Value Analysis of Peak Wind Pressures on Buildings," M.S. Thesis, Fluid Dynamics and Diffusion Laboratory, Colorado State University, Fort Collins, CO, 1981.

7.   Rice, S. O., "Mathematical Analysis of Random Noise," Bell System Technical Journal, Vol. 23, pp. 282-332, 1944 and Vol. 24, pp. 46-157, 1945.

TABLE 1. TYPE I EXTREME VALUE DISTRIBUTION FITTED PARAMETERS

Tap Number		101	102	403	404	405	501	509	518	528	567	704
$C_{p_{mean}}$		-0.05	-1.05	-0.49	-0.48	-0.10	-1.08	-1.53	-0.45	-0.29	-0.27	-0.39
$C_{p_{rms}}$		0.06	0.17	0.38	0.09	0.05	0.43	0.43	0.08	0.12	0.08	0.07
Largest peak from each of 100 records	U_o	2.61	3.80	2.70	4.92	3.49	3.89	4.04	5.06	6.60	5.28	4.24
	$1/a$	0.23	0.41	0.20	0.78	0.39	0.37	0.32	0.72	1.02	0.68	0.46
100 largest peaks from one record-average of 100 records	U_o	2.71	3.67	2.67	4.10	3.14	3.65	3.96	4.70	6.30	4.14	3.81
	$1/a$	0.36	0.36	0.28	0.50	0.38	0.51	0.53	0.56	0.83	0.72	0.61
N independent peaks from one record-average of 100 records	U_o	4.47	3.35	3.19	3.73	3.33	3.66	3.80	3.73	4.61	3.79	3.63
	$1/a$	0.71	0.66	0.69	0.78	0.69	0.77	0.81	0.79	0.95	0.81	0.78
Largest peak from each of 100 records	C_p^*	-0.21	-1.78	-1.61	-1.03	-0.29	-2.96	-3.43	-0.95	-1.23	-0.76	-0.72
100 largest peaks from one record-average of 100 records	C_p^*	-0.23	-1.75	-1.64	-0.92	-0.27	-2.95	-3.52	-0.90	-1.17	-0.67	-0.70
	$C_{p_{rms}}^*$	0.08	0.08	0.06	0.13	0.08	0.09	0.10	0.13	0.13	0.18	0.10

318

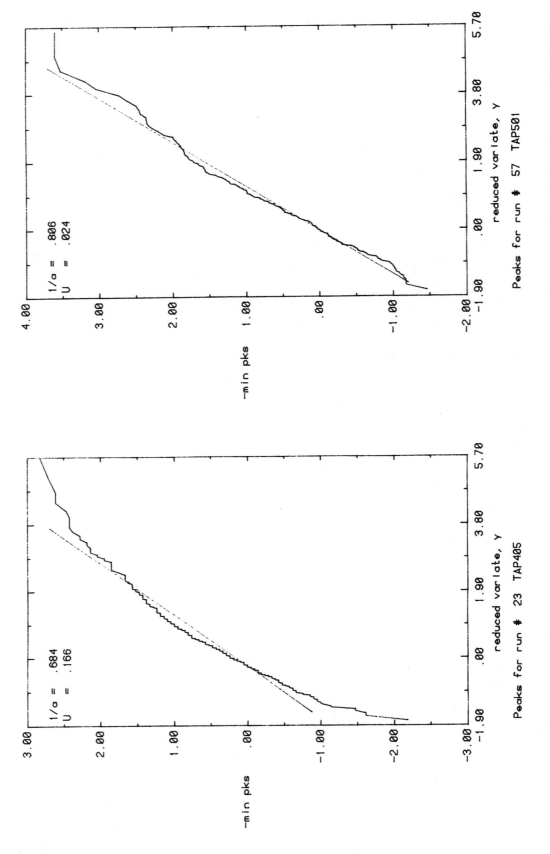

Figure 2.   Type I Extreme Value Fit to 265 Peaks

Figure 1.   Type I Extreme Value Fit to 333 Peaks

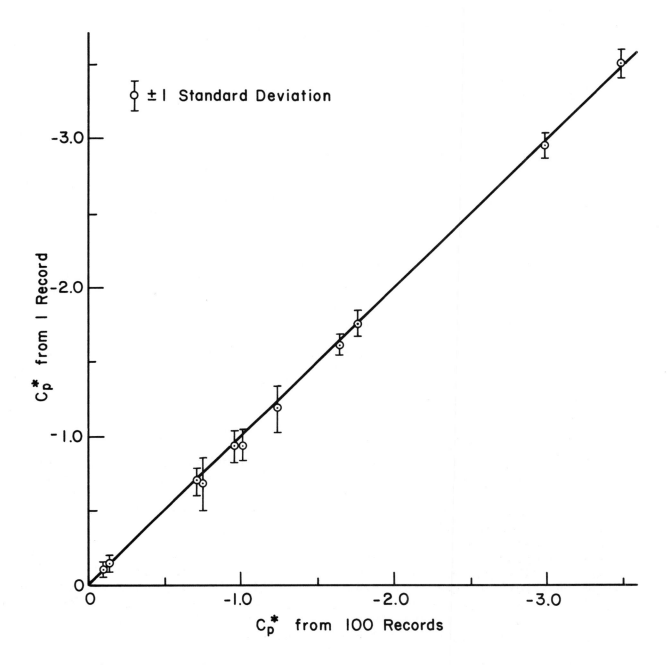

Figure 3.   Comparison of C_p^* obtained from 1 record to that
obtained from 100 records.

CALIBRATION OF TECHNIQUES FOR THE PREDICTION OF PEAK PRESSURES

N J Cook, Building Research Establishment, UK.

1   INTRODUCTION

Techniques for the assessment of wind loads for the design of static structures have developed considerably over the past two decades. Purely static assessment of loads was used in the UK Code of Practice until the 1970 revision, when a quasi-static approach was adopted. The design wind speed became a short-duration gust with a given probability of exceedence, but static values were retained for the pressure coefficients. The assumption was made that the flow around structures responded to short duration gusts in the same manner as to the mean flow, ie 'quasi-statically'. It quickly became clear that the quasi-static mechanism assumed by this approach worked well only in the attached flow region on the windward face of buildings. The short-duration pressures in the separated flow regions on roofs, sides and leeward walls were shown to be due to turbulence generated by the buildings themselves.

Selecting the level relevant for design from within the observed range of pressure variation was a major problem. Attempts were made to relate peak values to the more easily measured mean and rms values through defining a peak factor, g, as :-

$$g = (peak - mean)/rms \qquad \ldots\ldots(1)$$

and searching for some consistent behaviour. Various design values of peak factor were proposed, ranging from g = 1.5 to 4. Later, the advent of digital data processors enabled, not only the peak, but also the second-highest, third-highest etc. to be measured. In effect the 'tails' of the probability distributions were put under close scrutiny. Wind tunnel tests showed that these tails were different in the attached and separated flow regions [1]. Attached flow regions showed tails which were nearly Gaussian, whereas separated flow regions showed exponential tails. These observations were confirmed in full scale [2]. A popular alternative to the peak factor approach was to select the pressure corresponding to a certain probability of occurrence [3]. But the original problem of the appropriate gust factor or probability quantile remained.

The evolution of a fully-probabilistic approach, the method of Cook and Mayne (C-M method), which solves this problem has been described in a series of papers [4-9]. The C-M method works through the extreme value distributions of the wind speed and of the loading (pressure, force or moment) coefficient jointly, allowing the risk of exceedance of the design load to be quantified, and thus removing the elements of arbitrary selection inherent in the previous methods. The way is open to back-calibrate the earlier approaches, so testing the assumptions made in their application and quantifying their differences.

The subject of this paper is the calibration of the quasi-static and the peak-factor approaches against the C-M method, using data from the Building Research Establishment's programme of data collection for the revision of the UK Code of Practice.

2 MODEL DATA USED FOR THE CALIBRATION

Data from six models, listed in Table 1, were used for the calibration. Each model was placed in a simulation of the atmospheric boundary layer generated by the roughness, barrier and mixing-device technique [10] and matched to the scale of the model by the process proposed by Cook [11]. The nominal scaling parameters for each model are also given in Table 1. The reference mean wind speed was measured at the ridge height of each building to be compatible with the UK Code [12].

The data were collected in the form of point-pressures averaged in time over periods equivalent to 1, 4 and 16 seconds in full scale. These averaging periods are equivalent to the loading Classes A, B & C in the UK Code of Practice [12], which takes temporal averaging of point-pressures as being equivalent to spatial averaging. The maximum and minimum values for each time-averaging period were subjected to the extreme value analysis described in references 4 and 6 to determine the mode U_{Cp} and dispersion $1/a_{Cp}$ of the resulting best fit Fisher-Tippett Type I distribution, as required by the C-M method. The rms value for each averaging period and the mean value were also determined for implementing the peak-factor and quasi-static methods.

An overview of the C-M method and its use was given at the CIRIA seminar "Wind engineering in the Eighties" [9]. The full C-M method is dependent on both the site and the structure, through the mode and dispersion of the wind speed and of the loading coefficient, repectively. The current UK Code of Practice simplifies the model for wind speed behaviour by assuming a constant value of $a_V \cdot U_V = 10$. By adopting this assumption, the C-M method simplifies to Equation 18 of reference 8, namely :-

$$P_{50} = 1/2 \; \rho \; V_{50}^2 \; (U_{Cp} + 1.4/a_{Cp}) \qquad \ldots (2)$$

where P_{50} is the 50-year return-period design pressure, V_{50} is the 50-year return-period hourly-mean wind speed, and U_{Cp} & $1/a_{Cp}$ are the Fisher-Tippett Type I parameters of the pressure coefficient for a one-hour observation period. The simplification leads to slightly reduced accuracy over the full C-M method, so the value of the constant in equation 2 was selected to give less than a 1% risk of underestimation, which results in an average overall safety factor of 1.04.

The data will be presented in terms of pressure coefficients based on the 50-year return-period hourly-mean wind speed at the height of the building, V_{50}. The mean pressure coefficient will be denoted here by $\overline{Cp}$ and the rms pressure coefficient by Cp'. The quasi-static method uses the mean pressure coefficient $\overline{Cp}$ directly, but combined with a short-duration gust velocity. The peak-factor method uses a peak pressure coefficient, denoted here by $Cp\hat{}$, combined with an hourly-mean wind speed. The peak pressure coefficient is estimated from equation 1 by :-

$$Cp\hat{} = \overline{Cp} + g.Cp' \qquad \ldots\ldots(3) \text{ for maxima, and}$$

$$Cp\hat{} = \overline{Cp} - g.Cp' \qquad \ldots\ldots(4) \text{ for minima.}$$

The choice of whether the maximum or minimum peak is relevant in any design application is usually made by examining the sign of the mean Cp [3, p46]. The simplified C-M method of equation 2 results in a pressure coefficient, denoted here by $Cp*$, given by :-

$$Cp* = U_{Cp} + 1.4/a_{Cp} \qquad \ldots\ldots(5)$$

where the choice of whether the maxima or minima are relevant is again set by the sign of the mean $\overline{Cp}$. The loading durations, Class A = 1 sec, Class B = 4 sec and Class C = 16 sec, will be denoted in brackets {}, eg. $Cp*\{B\}$.

3 CALIBRATION OF THE QUASI-STATIC APPROACH

The quasi-static approach was calibrated by means of a regression analysis of the values of mean pressure coefficient, $\overline{Cp}$, with the values from the C-M method, $Cp*$. The fit obtained for model 1 and a 1-second loading duration (Class A) is shown in figure 1. If the quasi-static assumption held exactly true, the slope of the regression $Cp*/\overline{Cp}$ would equal the ratio of the gust dynamic pressure, $q\hat{}$, to the mean dynamic pressure, $\overline{q}$, at the reference height and for each loading duration. This comparison is made in Table 2, where the dynamic pressure data have been derived from the ESDU data set [13]. The scatter of the measured pressure coefficients is also given in terms of the standard deviation about the regression line, ie. the standard error. Compared with the C-M method, the quasi-static method underestimates 16-second pressures (Class C) by 5%, 4-second pressures (Class B) by 13%, and 1-second pressures (Class A) by 21% on average. This trend indicates the increasing contribution of fluctuations additional to those of the atmospheric turbulence as the loading duration decreases. However, the simplified C-M method has an average safety factor of 1.04, thus the actual contribution of fluctuation additional to the quasi-static response - but still proportional to the mean pressure coefficient - is approximately 1% for 16 seconds, 9% for 4 seconds and 16% for 1 second durations. These systematic departures from quasi-static response are accountable by adopting the effective 'gain' factors $Cp*/\overline{Cp}$ of Table 2 in place of the purely quasi-static parameters $q\hat{}/\overline{q}$, thus making a significant improvement. However the large random departures from quasi-static response expressed by the standard error will still remain. The scatter of these errors is skewed in the direction of increasing value for both maxima and minima. The largest deviations correspond mainly to the areas bordering the peripheries of the models, where deviations from the regression line exceeding a pressure coefficient of 2 occur for Class A. The scatter reduces with increasing loading duration, showing that the departures from quasi-static response are primarily due to the higher-frequency building-generated eddies.

4 CALIBRATION OF THE PEAK-FACTOR APPROACH

A rational choice of the peak factor value must be made before the method can be used in design. One approach is to choose a value equivalent to a certain probability, having first assumed a shape (often Gaussian) for the probability distribution. Values of peak factor between 1.5 and 4 are commonly used in design. However, the range of peak factor found in full-scale experiments is very large, $g = 4.8$ to 7.1 on a slab block [14], $g = 2.2$ to 4.3 on a tower block [15], and $g = 4$ to 10 on a low-rise building [2]. The problem of selecting a suitable peak factor becomes much more amenable when the peak values have been assessed probabilistically by the C-M method. The value of peak factor that gives the same result for $Cp\hat{}$ as for $Cp*$ can be found by combining equations 3 or 4 with equation 5, thus:-

$$g = (U_{Cp} + 1.4/a_{Cp}) - \overline{Cp}) / Cp' \qquad \ldots (6)$$

Values of the 'calibrated' peak factor so obtained for Class A pressures on model 1 in a skewed wind are plotted in figure 2. Inspection shows a coherent pattern to the variation, the value remaining between 4 and 5 over much of the model, but increasing dramatically along the reattachment lines behind the pair of conical roof vortices and again, less dramatically, at the reattachment line on the long side.

Values for the calibrated peak factor averaged for all locations and wind directions, $\overline{g}$, and the standard deviation about this value, g', are given for each model in Table 3, together with overall values for all six models. The calibrated peak factor for the Class C pressures is close to the value of 4 commonly used for design. Values increase as the averaging time decreases, even though the rms pressure coefficients also increase, indicating an increase in the peakiness of the data. The variation of $\overline{g}$ with structural form, ie. between models, is small compared with the variation between Classes.

The calibrated values of $\overline{g}$ were adopted for a regression analysis. The fit obtained for model 1, Class A pressures is shown in figure 3, and is a considerable improvement over the fit of figure 1. The scatter about the regression line is reduced by a factor of 2. The most outlying values now correspond, not to the peripheral regions, but to the regions of high peak factor along reattachment lines. An assessment of the standard deviation of the prediction error for the quasi-static and peak-factor methods is given in Table 4.

324

5 CONCLUDING REMARKS

This calibration exercise rests on the premise that the C-M method correctly assesses the actual risk of a design load. The return period used throughout has been 50 years, the standard value found in the majority of the world's codes. The C-M method is unique in that it includes the contribution from the second, third, etc. highest hourly-mean wind speeds in the return period.

The quasi-static, peak-factor (once calibrated) and the C-M method all give closely comparable results on average for the longest duration Class C pressures. However, the random error of the quasi-static method is more than twice that of the peak-factor method. For the shorter duration pressures, the response is clearly not quasi-static, and only the peak-factor and C-M methods remain adequate.

The use of the calibrated values of peak factor in Table 3 effectively removes the element of arbitrary selection previously present in the method.

6 ACKNOWLEDGEMENT

The work described in this paper has been carried out as part of the research programme of the Building Research Station of the Department of the Environment and is published by permission of the Director.

7 REFERENCES

1   J A Peterka & J E Cermak, Wind pressures on buildings - probability densities, J Struct. Divn. ASCE, 1975, 101, 1255-1267.

2   K J Eaton, J R Mayne & N J Cook, Wind loads on low-rise buildings - effects of roof geometry. Proc 4th Internat. Conf. on Wind Effects on Buildings and Structures, Heathrow, 1975, Cambridge Univ. Press, 1977.

3   T V Lawson, Wind Effects on Buildings, Volume 1 : Design Applications, Applied Science Publishers, London, 1980.

4   J R Mayne & N J Cook, On design procedures for wind loading. Current Paper CP25/78, Building Research Establishment, 1978.

5   N J Cook & J R Mayne, A new approach to the assessment of wind loads for the design of buildings, Proc. 3rd Colloq. on Industrial Aerodyn., Aachen, June 1978, Fachhochshule Aachen, 1978.

6   N J Cook & J R Mayne, A novel working approach to the assessment of wind loads for equivalent static design, J. Indust. Aerodyn., 1979, 4, 149-164.

7   J R Mayne & N J Cook, Acquisition, analysis and application of wind loading data, Proc 5th Internat. Conf. on Wind Eng., Colorado, 1979, Pergamon Press, 1980.

8   N J Cook & J R Mayne, A refined working approach to the assessment of wind loads for equivalent static design, J. Indust. Aerodyn., 1980, 6, 125-137.

9   N J Cook & J R Mayne, Design methods for Class A structures, Proc. CIRIA seminar 'Wind Engineering in the Eighties', London, 1980, Construction Industry Research and Information Assoc., 6 Storey's Gate, London, 1981.

10   N J Cook, Wind tunnel simulation of the adiabatic atmospheric boundary layer by roughness, barrier and mixing-device methods, J. Indust. Aerodyn., 1978, 3, 157-176.

11   N J Cook, Determination of the model scale factor in wind-tunnel simulations of the adiabatic atmospheric boundary layer, J. Indust. Aerodyn., 1978, 2, 311-321.

12   Code of Basic Data for the Design of Buildings, CP3, Chapter 5, part 2, British Standards Institution, 1972.

13   ESDU, Characteristics of wind speed in the lower layers of the atmosphere near the ground : strong winds (neutral atmosphere), Data Item 72026, Engineering Sciences Data Unit, London, 1972.

14   C W Newberry, K J Eaton & J R Mayne, Wind loading on tall buildings – further results from Royex House, Current Paper CP29/73, Building Research Establishment, 1973.

15   W A Dalgliesh, J T Templin & K R Cooper, Comparisons of wind tunnel and full-scale building surface pressures with emphasis on peaks, Proc 5th Internat. Conf. on Wind Eng., Colorado, 1979, Pergamon Press, 1980.

TABLE 1   NOMINAL PARAMETERS OF MODELS USED FOR CALIBRATION

Model	Scale factors Length	Time	Terrain Type	z_o	h
1 UWO low-rise building	1:500	1:200	Open country	0.1m	10m
2 BACO grandstand	1:200	1:100	Open country	0.1m	24m
3 Cube	1:250	1:100	Suburbs	0.5m	25m
4 Cuboid 3:1:1	1:250	1:100	Suburbs	0.5m	25m
5 Tower 1:1:3	1:250	1:100	Suburbs	0.5m	75m
6 Hipped 45 degree roof	1:250	1:100	Suburbs	0.5m	32.5m

TABLE 2   CALIBRATION OF QUASI-STATIC METHOD

Model	1	2	3	4	5	6
$Cp^*\{A\}$ / $\overline{Cp}$	2.56	3.04	3.67	3.72	2.60	3.74
$\hat{q}\{A\}$ / $\overline{q}$	2.88	2.46	2.79	2.79	2.36	2.74
std. error$\{A\}$	0.34	0.41	0.88	0.94	0.60	0.65
$Cp^*\{B\}$ / $\overline{Cp}$	2.16	2.39	2.96	2.96	2.20	2.91
$\hat{q}\{B\}$ / $\overline{q}$	2.38	2.09	2.33	2.33	2.02	2.33
std. error$\{B\}$	0.24	0.28	0.65	0.65	0.41	0.45
$Cp^*\{C\}$ / $\overline{Cp}$	1.75	1.83	2.13	2.13	1.71	1.92
$\hat{q}\{C\}$ / $\overline{q}$	1.94	1.75	1.92	1.92	1.71	1.92
std. error$\{C\}$	0.14	0.15	0.38	0.37	0.24	0.24

TABLE 3   CALIBRATION OF PEAK-FACTOR METHOD

Model	1	2	3	4	5	6	All
$\overline{g}\{A\}$	5.42	5.22	5.82	5.78	5.17	5.63	5.51
$g'\{A\}$	1.13	1.12	1.38	1.38	1.37	1.15	1.29
$\overline{g}\{B\}$	4.74	4.66	5.07	4.84	4.51	4.82	4.77
$g'\{B\}$	0.83	0.82	1.15	0.91	1.09	1.12	1.01
$\overline{g}\{C\}$	4.31	4.09	4.19	4.08	3.83	4.02	4.09
$g'\{C\}$	0.70	0.58	0.83	0.60	0.62	0.57	0.69

TABLE 4   COMPARISON OF STANDARD DEVIATION OF PREDICTION ERROR
OF QUASI-STATIC AND PEAK-FACTOR METHODS

Method	Class A	Class B	Class C
Quasi-static	0.34	0.24	0.14
Peak-factor	0.16	0.10	0.06

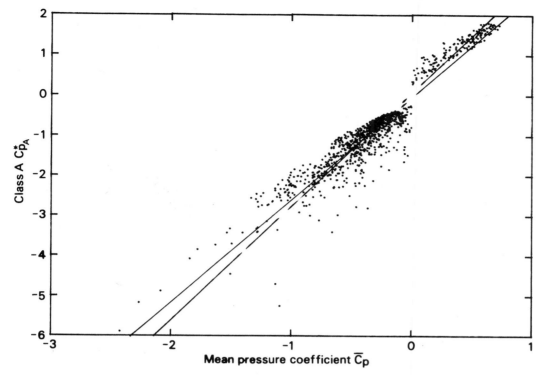

$$C^*_{p_A} = -0.03534 + 2.557 \times \overline{C}_p \pm 0.3411$$

$$\overline{C}_p = -0.01677 + 0.3532 \times C^*_{p_A} \pm 0.1269$$

R = 0.9505

Figure 1 Calibration of quasi-static approach for Class A pressures

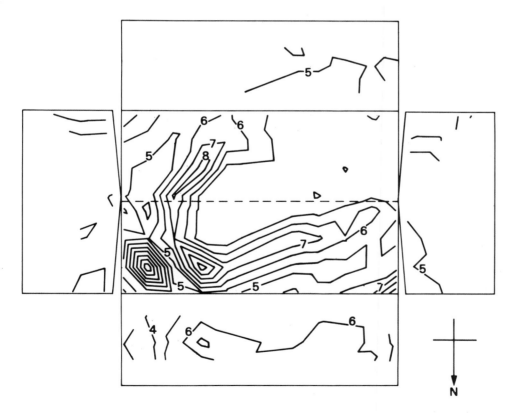

Figure 2 Class A equivalent peak factors, azimuth = 60 degrees. Mean = 5.26 SD = 1.38

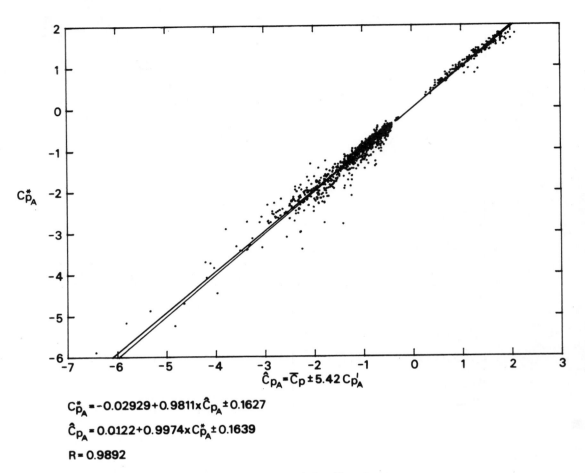

$C_{p_A}^* = -0.02929 + 0.9811 \times \hat{C}_{p_A} \pm 0.1627$

$\hat{C}_{p_A} = 0.0122 + 0.9974 \times C_{p_A}^* \pm 0.1639$

$R = 0.9892$

Figure 3 Calibration of peak-factor approach for Class A pressures

INSTRUMENTATION REQUIREMENTS FOR AERODYNAMIC PRESSURE AND FORCE MEASUREMENTS ON BUILDINGS AND STRUCTURES

Frank H. Durgin, Associate Director
Wright Brothers Memorial Wind Tunnel
Massachusetts Institute of Technology

1. INTRODUCTION

The instrumentation requirements when measuring some quantity can be subdivided into those which affect the static responses of a transducer and those which affect its dynamic response. Overall one wants to be certain that a force or pressure is measured with sufficient accuracy. Further since the objective most frequently involves measurement of extreme pressures or loads, it is those extreme pressures and loads that design the systems. The next section of this paper will consider the static requirements of the transducer. The spectral content of the flow is described in section 3. An estimate of the required instrumentation characteristics when measuring forces is obtained in section 4, and a similar analysis is applied to define the required pressure measuring systems response characteristics in section 5. The pressure measuring system used at the Wright Brothers Memorial Wind Tunnel (WBWT) will be described in section 6. The experimental procedures used at WBWT to determine the necessary frequency response of the pressure system are given in section 7. Finally some photographic evidence supporting the need for the frequency response requirements is found in section 7.

2. STATIC CHARACTERISTICS OF FORCE AND PRESSURE TRANSDUCERS

The static characteristics of transducers involve such items as linearity, hysteresis, repeatability, effects of temperature on each of the above as well as the sensitivity and zero of the transducer, long term stability, etc. Most manufacturers specify each of the above either individually or collectively or both so that in most cases one can obtain a fairly good idea of the static characteristics of a transducer from the manufacturer's specifications. Unfortunately not all manufacturers express such specifications in the same form. One must read the specifications carefully and ask the appropriate questions.

Therefore, considerable caution is needed in choosing a transducer or system for making measurements of forces or pressures due to the natural wind. The problem arises because of one's interest in measurements of the average, rms, and peak quantities. The peaks can be several times the average or rms levels; thus a large dynamic range is required if the average or rms is to be measured with reasonable accuracy. Typically the situation is made even more critical for pressure measurements since the same transducer measures both the pressure peak (at the port causing the most negative peak) and the rms and average values (at a port with the smallest peak). Further transducers are available only within certain ranges and one must be certain that the peak pressure or force lies well within that range. Thus it is not unusual to find the maximum negative peak represents only 40 to 60% of full scale.

As a result zero shifts due to changes in temperature frequently reduce the possible accuracy of pressure and force measurements. At the WBWT all systems are arranged such that the electrical output for zero force or pressure can be recorded at the beginning and end of each test. To illustrate the problem, the Setra model 237 pressure transducer used at the WBWT has a range of ±72 psf. Typically pressure tests are conducted at tunnel speeds of 60 mph (dynamic pressure, $q \simeq 9$ psf), thus allowing a measurement of $-C_p \simeq 8$. Such C_p's have been measured on occasion. Data obtained with the transducer mounted in a model D scanivalve (48 taps) typically takes about one-half hour. Zero shifts as large as $C_p = 0.3$ have been noted during a run if the transducer is turned on just prior to the test. If the transducer remains on all night and the tunnel achieves thermal equilibrium, shifts of less than 0.01 are usual. Similar zero shifts have been observed in strain gage balances used in the wind tunnel and in pressure transducers used on site.

3. SPECTRAL CONTENT OF THE WIND

An evaluation of the dynamic response requirement for force and pressure transducers requires knowledge of the spectral content of the wind and how a local structure may modify that spectral content. First we shall discuss that spectral content and later show how it relates to the required frequency response characteristics of the force and pressure transducers. The paper by Greenway [1] presents a very good review of many of the ideas presented here.

The von Karman [2] analytical description of the spectrum of longitudinal velocity (see Fig. 1a) is

$$\frac{nS(n)}{\tilde{u}^2} = 4 \tilde{n} / (1 + 70.8 \tilde{n}^2)^{5/6} \tag{1}$$

where $\tilde{n} - n\ell_x/\bar{U}$, ℓ_x is the longitudinal integral length scale (i.e. a measure of the gust size near the peak in the spectrum), n the frequency in hertz, $\bar{U}$ the average velocity, $\tilde{u}^2$ is the variance of the velocity and $nS(n)/\tilde{u}^2$ is the normalized energy at each frequency n.

An important point concerning this spectrum is that given a set of upstream roughness causing the boundary layer in the wind tunnel or full scale, ℓ_x will only be a weak function of velocity, that is, ℓ_x is determined by the upstream roughness elements and any change in $\bar{U}$ results in a change in n, not ℓ_x given the same upstream roughness.

The von Karman spectrum is applicable to flows sufficiently far behind a grid or in a fully developed turbulent boundary layer. For instance it is applicable to the flow upstream of the city model in the wind tunnel. If, however, one places an object in the flow field and then measures the downstream spectrum, the modified spectrum might appear as shown in Fig. 1b. The additional bump in the spectrum will appear at the Strouhal shedding frequency of the three-dimensional object (i.e. at $0.20 \leq nD/\bar{U} \leq 0.35$, where D is a cross-stream dimension of the object). The height and width of the new bump will depend on the size of the object, the turbulence in the flow, any motion of the object, etc.

It must be noted that the spectra shown in Fig. 1 can be experimentally verified. Only after averaging many spectra or when using a sampling time many times the longest period considered will the spectra become smooth as shown (i.e. instantaneously the spectra can vary widely). The significance is that in general the pressure, force, or moment being measured is the extreme one, and on the rare occasions that the extremes occur there can be much more energy at a given frequency which momentarily causes the phenomenon than one would expect from the spectrum.

Finally the integral scale ℓ_x is a measure of the gust sizes containing the most energy, i.e., it is a measure of the size only near the peak in the spectrum. Then given a spectrum and noting that the dimensions of $\bar{U}/n$ are length, by Taylor's hypothesis [4] that that length is a measure of gust size, then for $\tilde{n} = n\ell_x/\bar{U}$ a measure of the size of a gust at any $\tilde{n}$ in terms of ℓ_x will be

$$\ell_n = \ell_x \frac{\tilde{n}_{pk}}{\tilde{n}} \qquad (2)$$

$\tilde{n}_{pk}$ = the value of $\tilde{n}$ at the maximum value of $nS(n)/\tilde{u}^2$.

4. RELATION BETWEEN SPECTRAL CONTENT OF WIND AND FORCES TO BE MEASURED

First let us consider the forces on an object in the gusting flow. Bearman [5] has investigated the aerodynamic admittance function between the forces on a square flat plate perpendicular to the wind and the gusts in the flow. When he plotted his results versus $nD/\bar{U}$, all the data collapsed into a single admittance curve (D is the dimension of one side of the square). Thus as Davenport [6] originally pointed out and Greenway [1] reaffirmed, the ratio of D/ℓ_x becomes an important parameter. This fact is illustrated in Fig. 2 from Greenway's paper [1] in which the aerodynamic admittance function is plotted versus $n\ell_x/\bar{U}$ for different values of D/ℓ_x.

Vickery [7] performed a similar experiment and measured the admittance function for plates of various shapes (circular, rectangular, etc.). He suggests the following universal admittance function $\alpha^2(n)$ for flat plates, where A is the frontal area of the plate.

$$\alpha^2(n) = \frac{1}{\left[1 + \left(\frac{2n\sqrt{A}}{\bar{U}}\right)^{4/3}\right]^2} \qquad (3)$$

The explanation given for the behavior of the admittance function is as follows: The flow approaching the plates contains gusts of all sizes. However, only those large enough to totally engulf the plate, load it completely. Those smaller than the plate only load part of the plate. Further it is well known (see Bearman [8], Hunt [9], Durgin and Lazar [10]) that gusts smaller than the plates are distorted and attenuated as they approach the plate, making the drop-off in loading even greater than might otherwise be expected.

Since a structure is not a flat plate, but in fact is a three-dimensional object, it might be expected to change the value of $nD/\bar{U}$ at which the admittance function drops off while not affecting its shape (i.e. gusts smaller than the largest dimension of the object perpendicular to the wind stream will still not load the object over its whole frontal area).

In order to measure forces or moments on a building, one must mount the model of the building on a balance. The balance-model combination will in general have a first mode frequency for each component to be measured. Further, unless the damping is deliberately made at least 50% of critical, one can expect some over-response due to gusts with frequency components near those natural frequencies. It should be noted that the inverse Fourier transform methods proposed by Erwin et al [11] for pressure could be used to remove the over-response. Whether one damps the model or uses the inverse Fourier transform method to correct the data, there will be some fixed upper limit in frequency (n_{cutoff}) beyond which the balance no longer provides valid data.

Combining this upper limit in frequency with the admittance function given by Vickery [7], one obtains for a flat plate perpendicular to the flow:

$$\bar{U} \simeq 4n_{cutoff}\sqrt{A} \qquad (4)$$

where A is the frontal area of the structure or building. Here we have used Vickery's [7] expression for the admittance function and a 3 db cutoff. The $\bar{U}$ so calculated is an estimate of the maximum average velocity at the top of the building which one can use and still expect to obtain valid data from the balance; or, given the frontal area of a model (A) and the average velocity at the top of the building, one may estimate the required n_{cutoff}, etc.

In the above it has been tacitly assumed that the relationship between force and velocity squared is linear. This is not necessarily true, as may be seen by considering the classic case of organized vortex shedding behind a structure. Normally at high Reynolds numbers this phenomenon only occurs in uniform flow, in conjunction with model motion, but given a gust with the correct combination of change in velocity and flow direction with time one can imagine a single vortex being shed causing large forces along and across the wind. For two-dimensional blunt bodies and cylinders, however, Hoerner [3] gives data showing the $nD/\bar{U}$ for vortex shedding to vary between 0.20 and 0.35. This is very nearly the admittance function cutoff frequency found by Vickery [7]. In consideration of the above it would seem prudent to construct the balance-model combination such that the $\bar{U}$ as given by equation (4) is twice the $\bar{U}$ one proposes to test at.

It should be noted that the above assumes measurements are those of maximum aerodynamic forces on the model, assuming no aeroelastic or dynamic effects due to the model mount. This kind of test can also be used to measure the aerodynamic admittance function of a building, so that once the dynamic properties of a building are known one can estimate its response ([12], [13], [14]).

5. RELATION BETWEEN SPECTRAL CONTENT OF WIND AND MEASURED PRESSURES

The procedure is quite similar to that used for forces. Marshall [15], Bearman [8], Hunt [9], and Durgin and Lazar [10] have all measured the admittance function between the oncoming turbulence and the pressures on the faces of two or three dimensional objects. The results found are identical with those for the forces, that is:

$$\bar{U} \simeq 4n_{cutoff} D \tag{5}$$

This may be used to estimate the required frequency response of the pressure transducer that is used to measure pressures on the upstream face of a model. Note that $\bar{U}$ is the velocity at the top of the model.

Unfortunately, while the above is valid for the upstream face, it is not true for the sides of the building. More important, since in most cases the object of the facade pressure test is to find those areas on the building surface at which the loads predicted by the locally used building code are too low, the most significant measurements prove to be at the pressure levels separating the extremes.

There are many combinations of shapes which can lead to local extreme pressures, but typically they occur near three-dimensional corners, particularly near the base of the building. Fig. 3 illustrates this situation. It shows the classic vortex just upstream of the building and trailing around the corner and downstream. Frequently, especially when the building corner contains an angle greater than 90°, the vortex, after being stretched and intensified going around the corner, will end up against the side of the building roughly as shown. However, the unsteady wind implies that the strength of the vortex and its position vary with time. We have found on full-scale buildings that when trying to measure extreme peaks the true peak is not obtained unless the transducer system has a frequency response of at least 10 Hz.

Using fixed reduced frequencies implies

$$\left(\frac{nD}{V}\right)_{\text{full scale}} = \left(\frac{nD}{V}\right)_{\text{wind tunnel model}}$$

For full scale n ≃ 10 Hz and V ≃ 150 ft/sec (100 mph)

D full scale/D model = 200 to 600

V wind tunnel at WBWT = 60 ft/sec

Thus for a model n = 800 to 2400 Hz, which is 8 to 24 times the response required for measuring pressures on the upstream face.

Finally, it must be remembered that an 1/8" to 1/4" hole in a real building is a point measurement compared to a window or panel, whereas a 1/16" diameter hole in a 1 to 400 scale model building is equivalent to a 2 foot hole full scale. Thus the model measurement is in fact a measurement averaged over an "appreciable" area and the required frequency response constraint may be less than indicated above.

6. DESCRIPTION OF PRESSURE MEASURING SYSTEM USED AT WBWT

For the system in use at WBWT every effort has been made to make the frequency response as high as possible. The system that has evolved over a 10-year period consists of a type D, solenoid operated, 48 tap Scanivalve in combination with a model 237, ±0.5 psi, Setra pressure transducer (Fig. 4). Only 10 inches of 0.040 inch I.D. tubing are used between the Scanivalve and model taps. A small amount of yarn is placed near the entrance to each tube to damp out the tendency of the 10 inch tube connecting the Scanivalve to the model to resonate like an organ pipe. The frequency response [amplitude ratio (A) vs frequency (n)] of a typical port is shown in Fig. 5. It is essentially flat (±10%) up to 800 Hz.

It is important that all Fourier components of a pressure pulse arrive at the transducer simultaneously so that a pressure pulse consisting of several components is not distorted. Fig. 6 shows the delay of a sinusoidal pressure as it travels from the port to the pressure transducer as a function of frequency. The measured delay is 1.2 $\pm$0.1 ms up to 800 Hz.

Fig. 7 shows the amplitude response of the system without yarn as a function of frequency for three pressure levels of the sine wave input. At the highest amplitude ($\pm$30 psf) the resonant peak is somewhat reduced, which is of some importance. At very low pressure amplitudes a pressure pulse behaves like a sound wave and the resonance shown in Fig. 7 is due to the length of tubing (10") acting like an organ pipe. In this case very little flow along the tube is required, but as the pressure difference increases additional flow is needed and is eventually governed by Poiseuille flow for which the resonance will disappear. The data in Fig. 7 indicate that at the highest pressure difference (30 psf) the transition from sound wave to Poiseuille flow is starting for the particular system used at WBWT. The transition point is a function of the geometry of the system. It is important to know this transition point if the inverse Fourier transform method described by Irwin et al [11] is used since the transform used would have to be a function of pressure amplitude for pressure amplitudes above it. Although not shown, the delay time without yarn varies from 0.3 to 1.6 ms from 100 to 1000 Hz.

7. DETERMINATION OF REQUIRED FREQUENCY RESPONSE

A semi-empirical approach to determine the required frequency response has already been discussed in section 5. However, depending on the accuracy of the simulation in the wind tunnel and the size of the pressure port opening, that frequency response in fact may or may not be necessary. Both Benko and Gartshore [16] and Durgin [17] have investigated what frequency response is necessary. Much of what is given below is taken directly from Durgin's paper [17].

Early in 1976 a building test at the WBWT indicated some very extreme pressure coefficients. Two taps from the building were selected for analysis, one in the middle of a wall facing upstream (tap 20), and another in the area where the most extreme negative pressures occurred (tap 47). The upstream facing point was considered a reference tap. All measurements of peak pressure consisted of recording of the analog signal for 2 to 25 seconds on a Tektronix memory oscilloscope.

The first step involved eighty 10 sec samples from each tap and then an extreme value analysis on the data for each tap. The Fisher Tippett Type 1 extreme value analysis was used [7]. Figs. 8 and 9 show the results for the two taps. The ordinate is the pressure coefficient and the abscissa is the y variate defined as $y = -\ln[-\ln(M/N+1)]$. Here N is the total number of readings and M is the number of the nth ordered (by magnitude) reading. The fact that wind tunnel measured peaks allow a Type 1 extreme value analysis was recently confirmed by Mayne and Cook [10].

Next the interval of measurement was varied and the measured average peak C_p for 10 or more such intervals was plotted versus the time interval. A semi-log plot indicated a linear dependence (Figs. 10 and 11). The results are of interest for two reasons: (a) They allow correction of all data to a fixed time interval, and (b) they show that a factor of two in the time interval alters the peak values by less than 5%. Finally, a low-pass electronic filter was inserted into the output line from each transducer and the average of 10 or more peaks was measured at several cutoff frequencies. These average measured peaks are plotted as a function of the filter cutoff frequency in Figs. 12 and 13. In both cases the horizontal line is based on all data taken without a filter. Clearly for the tap with the positive C_p (tap 20) the cutoff frequency has no effect above 100 Hz as might be expected from the arguments given in section 5.

On the other hand, the data for negative C_p (tap 47) show a cutoff at 800 to 1000 Hz, which is the cutoff frequency of the pressure system. Thus, while it was clear for this tap that an 800 Hz response was required, it was not clear that a larger value was unnecessary. In order to confirm this the test was rerun, but at a 30 mph gradient velocity instead of the previous 60 mph (Fig. 14). The unfiltered expected peak in this case had increased significantly. However, there is evidence that the model was not at the same angle relative to the wind direction. Clearly, the measured peaks do reach a plateau after about 400 Hz, which is exactly what one would expect on the basis of 800 Hz cutoff at 60 mph.

The tests and results just described are from only one tap for a wind condition which caused an extremely high negative C_p. The pressure coefficients reported are based on a dynamic pressure at the top of the building. Clearly more data would be helpful, but for the model and conditions tested, the 800 Hz response was both necessary and sufficient. Benko and Gartshore [16] report a similar experiment and also found that at least 800 Hz was required. The velocity at the top of their building appears to be the same as for the 60 mph case reported here.

8. FLOW VISUALIZATION OF VORTEX CAUSING EXTREME SUCTION PRESSURE

In a recent facade pressure test carried out at WBWT an attempt was made to visualize the vortex that is responsible for a very low pressure. While the results are still very tentative because of the insight they appear to give to the way the high suctions are created, it was felt appropriate to discuss them at this workshop. The building model is shown in Fig. 15a with the wind from the left and roughly parallel to the facade. The tap used is at the center of the white area at the lower left. The standard pressure WBWT system was used.

A special electronic circuit was assembled that would trigger a flash when the input voltage reached a set level. The level was set at the equivalent of $C_p = -2$ or more negative. Smoke was inserted in the flow upstream of the model, the tunnel darkened, a camera left open until the flash occurred. One of the resulting pictures is shown in Fig. 15b.

Two things should be noted: (1) The time interval between achieving $C_p = -2$ and the flash is 20-30 microseconds, and (2) because of 10 inches of tubing there is a 1.2 ms delay between the actual pressure rise and the flash (Fig. 6). Thus the diagonal white puff shown above the smoke is believed to be the vortex originally formed at the measuring tap and now about 1 3/4 inches up the side of the building. Returning to Fig. 15a, there is also a puff but no smoke. Fig. 15a was taken the same way as Fig. 15b but without smoke. Careful examination shows the two puffs to be displaced. It is believed the puff is not smoke, but condensation due to high humidity and a very strong vortex which will have a low pressure at its core. This visual data seems to confirm the necessity of the 800 Hz frequency response and in fact may imply that an even higher response is needed for the case shown. It also gives considerable insight into the mechanisms causing these extremely low pressures.

REFERENCES

[1]   Greenway, M.E.:   An Analytical Approach to Wind Velocity Gust Factors. J. of Ind. Aero., 5, 1979, pp 61-91.

[2]   von Karman, T.:   Progress in the Statistical Theory of Turbulence. Proc.Nat.Acad.Sci., IS 34, 1928, p 530.

[3]   Hoerner, S.F.:   FLUID-DYNAMIC DRAG.   Published by author, 1958.

[4]   Hinze, O.J.:   TURBULENCE, 2nd edition, McGraw-Hill, 1979.

[5]   Bearman, P.W.:   An Investigation of the Forces on Flat Plates Normal and Turbulent Flow.   J.Fl.Mech., 46, 1971, part 1, pp 177-198.

[6]   Davenport, A.G.:   The Buffeting of Structures by Gusts.   Proc.Conf. on Wind Effects on Bldg. and Structures, NPL Toddington, 1963.

[7]   Vickery, B.J.:   On the Flow Behind a Coarse Grid and Its Use as a Model of Atmospheric Turbulence Studies Related to Wind Loads on Buildings. NPL Aero report 1143, 1965.

[8]   Bearman, P.W.:   Some Measurements of the Distortion of Turbulence Approaching a Two Dimensional Bluff Body.   J.Fl.Mech., 53, 1972, p 451.

[9]   Hunt, J.C.R.:   A Theory for Fluctuating Pressures on Bluff Bodies in Turbulent Flows.   IUTAM-IAHR Symp. on Flow Induced Vibrations, Karlsruhe, Aug. 1972.   Springer

[10]   Lazar, D., Durgin, F.H.:   Some Measurements of Pressure Fluctuations on Two and Three Dimensional Bodies Due to Longitudinal and Lateral Velocity Fluctuations.   MIT Wright Brothers Memorial Wind Tunnel report WBWT-TR-1119, 1979.

[11]   Irwin, H.P.A.H., Cooper, R.R. and Girad, R.:   Correction of Distortion Effects Caused by Tubing Systems in Measurements of Fluctuating Pressures.   J.Fl.Mech., 5, 1979, pp 93-107.

[12]   Saunders, J.W., Melbourn, W.H.:   Tall Rectangular Building Response to Cross-Wind Excitation.   Proc. 4th Int. Conf. on Wind Effects on Buildings and Structures, Heathrow 1975, Cambridge U Press, London, pp 369-379.

[13]   Whitbread, R.E.:   The Measurement of Nonsteady Wind Forces on Small-Scale Building Models.   Proc. 4th Int. Conf. on Wind Effects on Buildings and Structures, Heathrow 1975, Cambridge U Press, London, pp 567-574.

[14]   Davenport, A.G., Tschanz, T.:   The Response of Tall Buildings to Wind Effects of Wind Direction and the Direct Measurement of Dynamic Force. Proc. 4th US Nat. Conf. on Wind Eng. Rsch., U of Wash., July 1981.

[15]   Marshall, R.D.:   Surface Pressure Fluctuations near an Axisymmetric Stagnation Point.   NBS NT 563, Aug. 1971, pp 3-17.

340

[16] Benko, I., Gartshore, I.G.: The Measurement of High Frequency Pressure Fluctuations on a Building in a Wind Tunnel Test.   Proceedings 3rd Colloquium on Industrial Aerodynamics, Aachen, 1981.

[17] Durgin, F.H.: Measuring Facade Pressures at the Wright Brothers Memorial Wind Tunnel. Proc. 4th US Nat. Conf. on Wind Eng. Rsch., Seattle, Washington, July 1981.

[18] Mayne, J.R. and Cook, N.J.: Acquisition, Analysis and Application of Wind Loading Data. Proc. 5th Int. Conf. on Wind Eng. Rsch., Fort Collins, Colorado, July 1979.

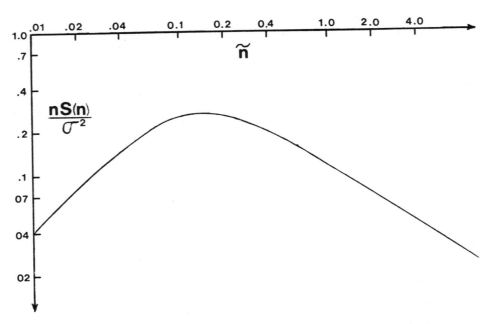

Figure 1a.   Von Karman Spectrum of Horizontal Turbulence.

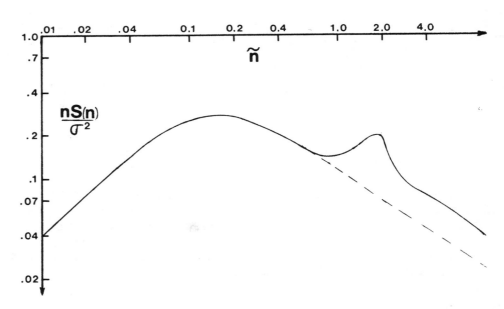

Figure 1b.   Effect of Upstream Object on
Von Karman Spectrum.

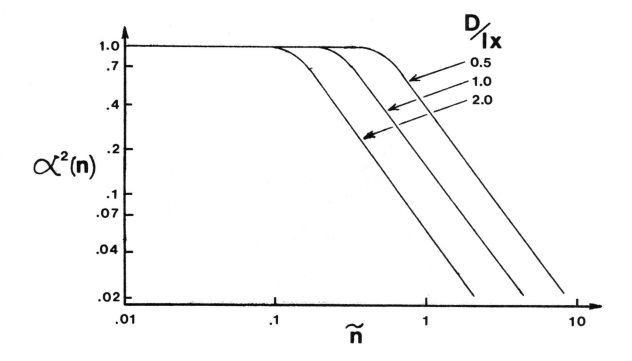

Figure 2.   Admittance Function.

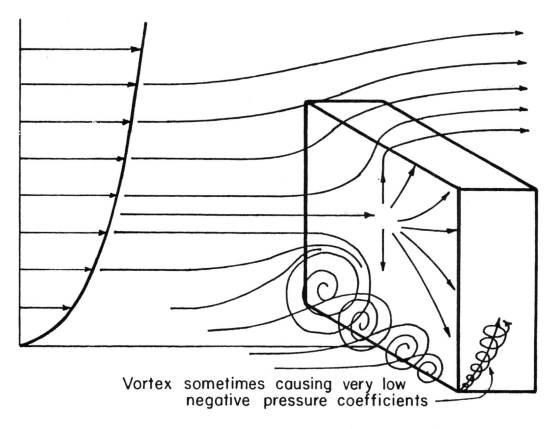

Vortex sometimes causing very low
negative pressure coefficients

Figure 3.   Typical Situation Causing Extreme
Negative Pressure Coefficients.

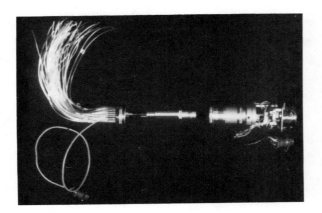

Figure 4. Setra Pressure Transducer.

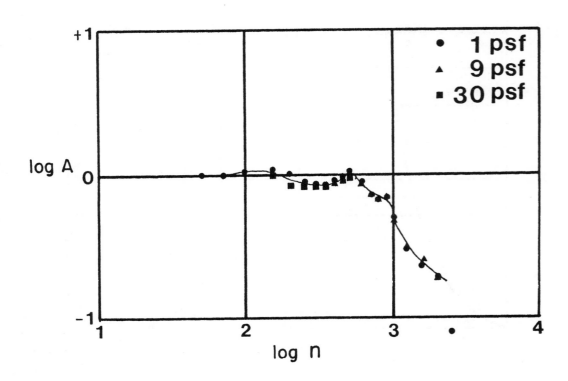

Figure 5. Frequency Response of System with Yarn.

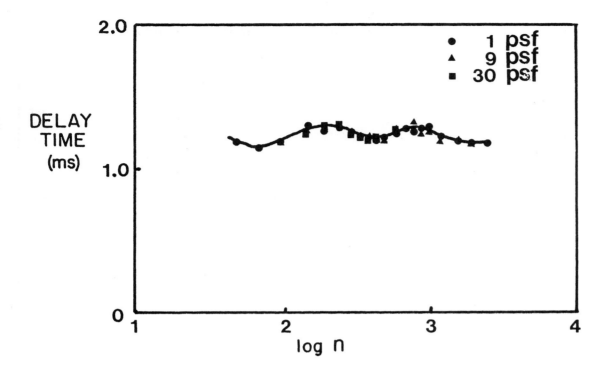

Figure 6. Delay Times of System with Yarn.

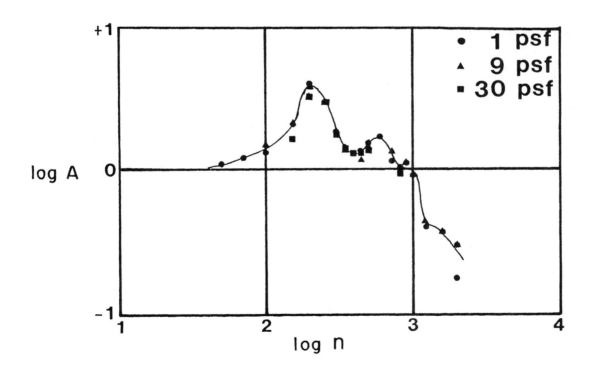

Figure 7. Frequency Response of System Without Yarn.

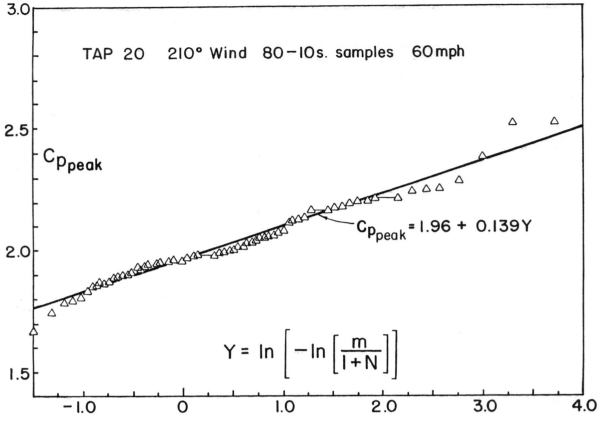

Figure 8. Extreme Value Data Analysis.

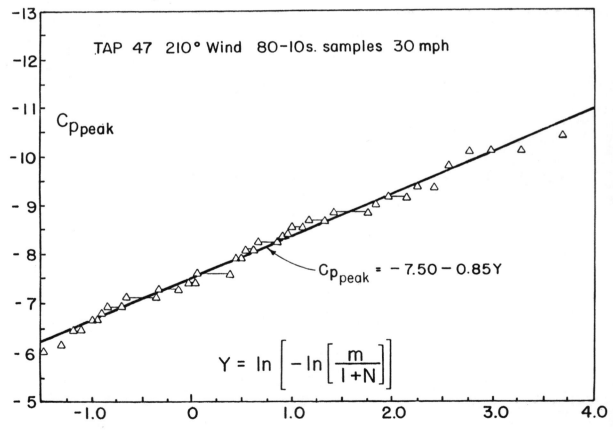

Figure 9. Extreme Value Data Analysis.

345

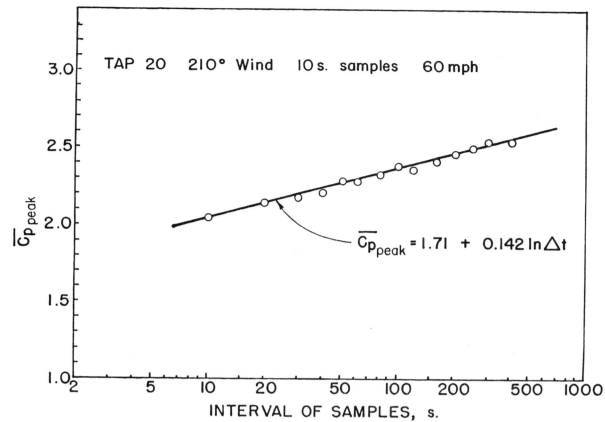

Figure 10. Effect of Sample Interval on Average
Peak Pressure Coefficient.

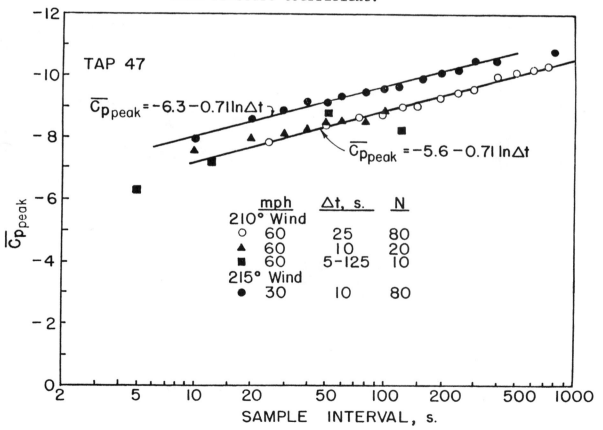

Figure 11. Effect of Sample Interval on Average
Peak Pressure Coefficient.

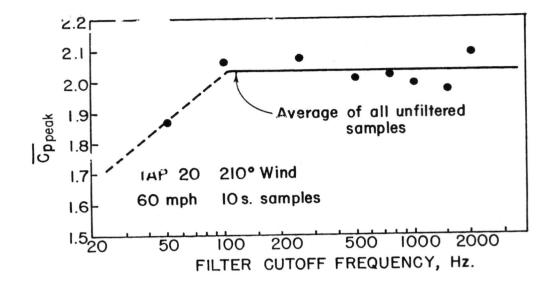

Figure 12. Effect of Filter Cutoff Frequency on Average
of 10 Peak Pressure Coefficients.

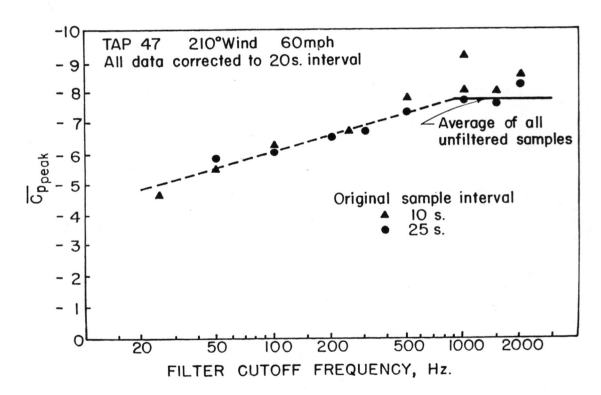

Figure 13. Effect of Filter Cutoff Frequency on Average
of 10 Peak Pressure Coefficients.

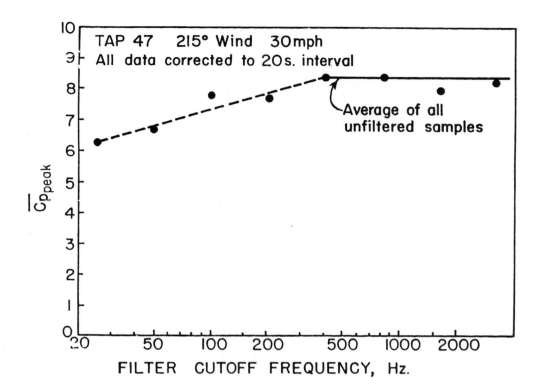

Figure 14. Effect of Filter Cutoff Frequency on Average
of 10 Peak Pressure Coefficients.

348

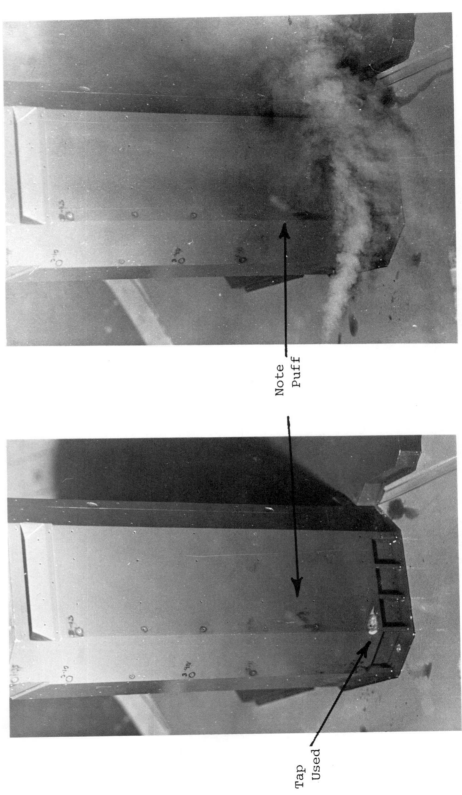

15b. Building Model with Smoke

15a. Building Model

A NEW TECHNIQUE FOR THE DETERMINATION OF STRUCTURAL LOADS AND EFFECTS ON LOW-RISE BUILDINGS

J D Holmes, CSIRO Division of Building Research (Australia)

1. INTRODUCTION

This paper describes a technique which makes use of wind tunnel pressure measurements to predict wind-induced structural loads and effects on low-rise buildings and other stiff (i.e. non-resonant) structures. The technique described here follows the development of the covariance integration method (Holmes and Best[1], Holmes and Rains[2]), but the new method differs significantly in principle, and significant improvements in both computational efficiency and statistical basis are claimed.

Other techniques, that have been used or proposed, for measuring structural loads and effects are described briefly in Section 2. The basis for the 'coincident peak' method, which is introduced in this paper, is discussed in Section 3, and an example of the application to the calculation of loads in an arch roof is given in Section 4.

2. OTHER TECHNIQUES

2.1 'On-line Weighting'

The well-known University of Western Ontario (U.W.O.) study on gable roof industrial buildings (Davenport, Surry and Stathopoulos[3]) was a first attempt to predict mean and fluctuating values of such loads as total uplift and frame bending moments, using wind tunnel pressure measurements. Following primary averaging using the 'pneumatic averaging' technique (described by Stathopoulos in this Workshop), the digitised panel pressure fluctuations were weighted using appropriate influence coefficients derived from structural analysis of 2- and 3-pin frames, to produce 'on-line' estimates of the required structural parameters, to which standard statistical processing was applied to produce mean, root-mean-square fluctuating, and expected peak values.

There are significant disadvantages in this approach, for a less well-endowed wind tunnel facility:

i) A large number (at least six) pressure transducers and data channels are required to be operating simultaneously.

ii) With many minicomputer systems associated with wind tunnels, it is not possible to process more than about three input channels 'on-line' at the required rates. Without a high-speed disc storage, rapid 'dumping' of the data and later 'off-line' processing may also not be possible.

iii) If it is required to estimate additional structural loads and effects, further wind tunnel tests are required, unless the unweighted pressure records are stored.

The recent availability of some cheaper pressure transducers, and the use of analog weighting techniques could partially alleviate (i) and (ii), but (iii) remains.

2.2 Direct Measurement

Direct measurement using force/displacement transducers, or strain gauges, is another possible approach to the measurement of overall forces and moments on a building model or a section of it (e.g. Roy and Holmes[4], Roy[5]). It can also be adapted to some simple cases of internal forces and moments, e.g. the axial force in a column of a building frame (Melbourne[6]), or the root bending moment for a cantilevered roof, but the determination of a complete internal stress distribution in indeterminate structures would be extremely difficult with this approach.

2.3 Non-Uniform Pneumatic Averaging

At present the pneumatic averaging technique has been used only for the measurement of locally averaged 'panel' loads with a uniform distribution of pressure taps[1,2,3]. It has been suggested (e.g Surry and Stathopoulos[7], Gumley[8]) that the method be extended to cover non-uniform weighting, either by varying the distribution of pressure taps per unit area, or by varying the geometry of the connecting tubes. Such approaches have had limited application to date (e.g. for the measurement of cross-wind forces on circular cylinders) but, it does not seem to be a practical proposition to use such methods in the determination of a large number of structural effects (such as those described in Section 4 of this paper).

2.4 Covariance Integration

A method, described by Holmes and Best[1], and Holmes and Rains[2], attempted to alleviate the disadvantages of the other techniques as described previously. The method made use of the pneumatic averaging method for primary averaging to produce panel loads in a similar way to the U.W.O. tests[3], but only two pressure transducers and computer channels were required to obtain the required aerodynamic information for subsequent calculations.

Standard random process averaging relationships can be employed to determine the mean and mean square fluctuating values of any structural parameter for which the structural influence coefficients are known. In the case of the mean square values, the covariance for the fluctuating pressures on each pair of panels is involved in the integration. A similar relationship can be derived for the mean square time derivative of the fluctuating structural parameter, involving the covariance matrix for the time derivatives of the panel pressures, which also must be obtained from the wind tunnel[1,2]. From these mean square values, the 'cycling rate' for the process can be calculated.

Up to this point no assumptions on the probability distribution of the structural effect, or the individual panel pressures are required. However, if the statistics of the peak value of the structural effect, such as the expected value or standard deviation, are required, some difficulties occur. If the integrated structural effect can be assumed to have a Gaussian probability distribution, then a simple relationship, derived by Davenport[9] and others, can be used to predict the expected peak value in a given time period, from the cycling rate. Note that Gaussian assumption does not apply to the individual panel pressures. However, there is now strong evidence from direct measurements (e.g. Roy and Holmes[4], Roy[5]) that the probability distributions for many structural loads and effects, influenced by even large areas on low-rise buildings, are non-Gaussian to a significant degree. This appears to be attributable, at least partly, to the high turbulence intensities occurring at the heights affecting low-rise structures and the non-linear velocity-pressure relationships (Holmes[10]).

Holmes and Rains[3] attempted to avoid this problem by using an approximation. This involved computing a peak factor for the structural effect as a weighted average of the peak factors for all the panel pressures. The weighting formula involves the correlation coefficients for the panel pressures. The approximation was based on the 'Pham-Leicester Rule', used by workers in structural reliability[11], for combining stochastic structural loads (i.e. live, snow, wind, earthquake) on structures, although the original form was used only for uncorrelated or independent loads. The expected peaks (maxima and minima) of the individual panel pressures were required as additional aerodynamic input for these calculations.

Intuitively this approximation is likely to be conservative in it simplest form. (Leicester, Pham and Hawkins at CSIRO are presently checking the validity of the approximation using wind pressure measurements and computer simulations; some modifications to the method to improve the accuracy are under consideration).

3.   THE 'COINCIDENT PEAK' METHOD

The approximation of Holmes and Rains[2] described in Section 2.4 has a disadvantage that the weighting formula used to compute the peak factor for the desired structural effect, makes use of the correlation coefficients which relate the <u>total</u> fluctuating pressures between pairs of panels. The following approach makes use of the statistical properties of the panel pressures when a <u>peak</u> value occurs on at least one panel. This has a more clearly defined relationship with the peak value of the structural effect.

In place of the correlation coefficients, the following additional values are recorded in the wind tunnel tests:

$$(C_{pj})_{\hat{\imath}} \quad \text{and} \quad (C_{pj})_{\check{\imath}}$$

$$i = 1, N$$
$$j = 1, N \qquad (N = \text{number of panels})$$

$(C_{pj})_{\hat{\imath}}$ is the expected pressure coefficient occurring on panel j, when the pressure on panel i is the maximum value recorded in the length of the record. Similarly, $(C_{pj})_{\check{\imath}}$ is the value of the pressure coefficient on panel j when panel i experiences its minimum value. Note that all the required values can be obtained using no more than two pressure transducers at a time, although clearly if more transducers are available, the values can be obtained more quickly. To improve statistical accuracy, averages over a number of repeated runs should be taken.

Figure 1 illustrates the values obtained from a single wind tunnel test run.

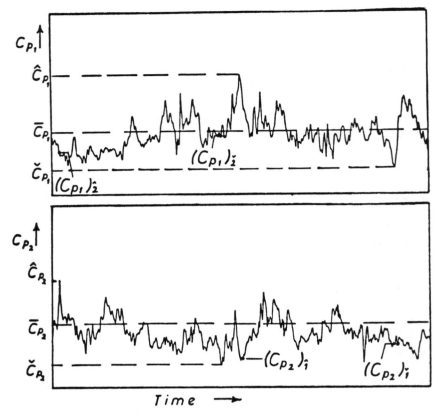

Figure 1.   'Coincident peaks' obtained in a wind tunnel test run

Thus, two N x N matrices of 'coincident peak' values are obtained; one is associated with the maxima, and one with the minima. Note that neither matrix is symmetric, in general. The diagonal values are the conventional maximum and minimum values, $\hat{C}_{pi}$ or $\check{C}_{pi}$.

The coincident peak values tend to reflect the correlation between the panel pressures. For high correlations, the coincident peaks approach the normal maxima or minima; at low correlations, they tend to approach the mean values.

The rows of each matrix form a series of coincident values of pressure coefficients occurring simultaneously on each panel. These can be used to give 2N different values of the structural parameter, η, i.e.

$$(\eta)_{\hat{i}} = [\hat{C}_{pi} \ \beta_i \ \Delta A_i + \sum_{\substack{j=1 \\ (j \neq i)}}^{N} (C_{pj})_{\hat{i}} \ \beta_j \ \Delta A_j] \ \rho \bar{u}^2/2$$

$$i = 1, N$$

and

$$(\eta)_{\check{i}} = [\check{C}_{pi} \ \beta_i \ \Delta A_i + \sum_{\substack{j=1 \\ (j \neq i)}}^{N} (C_{pj})_{\check{i}} \ \beta_j \ \Delta A_j] \ \rho \bar{u}^2/2$$

$$i = 1, N$$

where β_i is the influence coefficient for the structural effect, η, i.e. the value of η when a unit static load is applied at the centre of panel i (towards the panel). ΔA_i is the area of the panel i.

The algebraic largest and smallest values of η from the 2N values computed, represent the estimated limits of variation of η. Note that since the pressure coefficients and the influence coefficients can both vary in sign, the maximum and minimum values of η could come from either matrix of coincident peaks.

This estimation procedure is a variation of 'Turkstra's Rule' used in structural reliability studies to estimate peaks of combined stochastic loads from separate sources (as for the Pham-Leicester Rule referred to in Section 2.4). It can be seen that this procedure will always underestimate the maximum (and overestimate the minimum) value of the structural parameter, since it is possible for the expected peak value of η to occur when no individual panel is making its peak contribution. However, the errors are likely to be very small and well within the order of accuracy required in structural design calculations. The errors are probably of the same order as that caused by representing the continuous panel load by the summation of discrete point pressures in the pneumatic averaging procedure. Of course, these errors would tend to cancel each other when both techniques were used together.

4. APPLICATION EXAMPLE

Using the method described in Section 3, wind loads for central and end bays of a circular arch roof on a hangar building at Woomera, South Australia, have been calculated. As an example of the approach, representative results will be presented here.

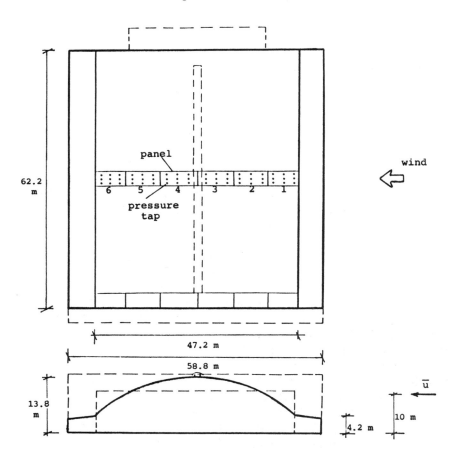

Figure 2. Panel pressure measurements on an arch roof

The building with the centre bay divided into six panels is shown in Figure 2. The fluctuating pressures on each panel were obtained by using a nine-hole manifold system and mean, maximum, minimum and 'coincident peak' values were obtained for all the panels. The sampling period was equivalent to about 10 minutes in full scale. Five separate runs were used in each case and average values computed. The coefficients obtained for the wind blowing normal to the axis of the building are shown in Table I following.

Table I. Mean, peak and 'coincident peak' pressure coefficients

Panel i	$(C_{p1})_{\hat{i}}$	$(C_{p2})_{\hat{i}}$	$(C_{p3})_{\hat{i}}$	$(C_{p4})_{\hat{i}}$	$(C_{p5})_{\hat{i}}$	$(C_{p6})_{\hat{i}}$
1	+.88	+.29	−.31	−.60	−.64	−.45
2	+.43	.00	−.13	−.44	−.48	−.38
3	+.39	−.04	−.04	−.39	−.30	−.29
4	+.08	−.16	−.09	−.22	−.27	−.35
5	+.20	−.30	−.26	−.46	−.15	−.39
6	+.40	−.30	−.46	−.75	−.61	.00

Panel	$(C_{p1})_{\check{i}}$	$(C_{p2})_{\check{i}}$	$(C_{p3})_{\check{i}}$	$(C_{p4})_{\check{i}}$	$(C_{p5})_{\check{i}}$	$(C_{p6})_{\check{i}}$
1	−.09	−.38	−.41	−.53	−.33	−.39
2	+.31	−.63	−.63	−.65	−.71	−.69
3	+.32	−.56	−.64	−.98	−.71	−.57
4	+.18	−.46	−.54	−1.10	−.74	−.52
5	+.33	−.56	−.42	−.85	−1.07	−.42
6	+.39	−.38	−.49	−.76	−.76	−1.09

	$\bar{C}_{p1}$	$\bar{C}_{p2}$	$\bar{C}_{p3}$	$\bar{C}_{p4}$	$\bar{C}_{p5}$	$\bar{C}_{p6}$
	+.27	−.31	−.31	−.52	−.48	−.46

The wind tunnel data was obtained from a 1/200 scale model of the building. The surface of the model was roughened to attempt to simulate higher Reynolds Number flow conditions, although in this case flow separations were controlled largely by the ventilator along the apex of the roof. The boundary layer simulation method and flow properties are described in another paper in this Volume[12].

The roof was supported by trussed arches in the shape of quarter-circles. The supports at the ends of the arch were intermediate between fixed and pinned conditions. Influence coefficients for the end reactions, bending moments and axial thrusts for both conditions were computed. Using the approach described in Section 3, the estimated extreme values of all these parameters were computed. The bending moments and reactions are shown for the central bay and normal wind direction in Figure 3. The values shown are the required forces or moments divided by the dynamic pressure $q = \rho \, \bar{u}_{10}^2/2$ where $\bar{u}_{10}$ is the mean wind speed at 10 m height. (Consequently the units are m^2 for forces and m^3 for moments).

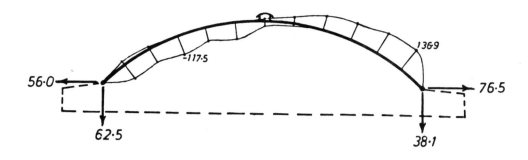

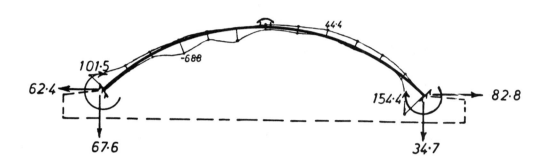

Figure 3. Predicted peak reactions and bending moment envelope for
a circular arch (pinned ends above, fixed ends below)

Calculations such as those shown in Figure 3 were repeated for an end
bay, and for four other wind directions. it is not feasible to use any
other of the other methods described in Section 2 for such a large number
of parameters at the present time.

5.   CONCLUSIONS

A new approach to the determination of peak values of structural loads
and effects for low-rise buildings, or other rigid structures, due to wind
action, has been described. All the aerodynamic data can be obtained with
two pressure transducers and data channels. The subsequent calculations are
not lengthy, enabling a large number of structural effects to be
calculated.

As an example, the envelope of bending moments and peak reactions for
a circular arch roof, calculated by this method, are shown. Determination
of such a large number of parameters is not at present feasible using any
other method.

6.   ACKNOWLEDGEMENTS

The permission of the Australian Department of Housing and
Construction to use results from the Woomera hangar study in this paper is
gratefully acknowledged by the author.

7. REFERENCES

1. J.D. Holmes and R.J. Best, An approach to the determination of wind load effects on low-rise buildings, J. Wind Engg. and Ind. Aerodyn., 7(1981) 273-287.

2. J.D. Holmes and G.J. Rains, Wind loads on flat and curved roof low-rise buildings – application of the covariance integration approach, Proc. Colloque 'Construire avec le Vent', Nantes, 1981, pp. V-1-1 – V-1-18, (also James Cook University, Wind Engineering Report 2/81, June 1981.

3. A.G. Davenport, D. Surry and T. Stathopoulos, Wind loads on low rise buildings : Final report of Phases I and II, University of Western Ontario, Report BLWT-SS8-1977, August 1977.

4. R.J. Roy and J.D. Holmes, Total force and moment measurement on wind tunnel models of low rise buildings, Proc. Colloque 'Construire avec le Vent', Nantes, 1981, pp. IX-2-1 – IX-2-15, (also James Cook University Wind Engineering Report 3/81, June 1981).

5. R.J. Roy, Total force and moment measurement on wind tunnel models of low rise buildings, James Cook University, M.Eng.Sc. thesis, 1982.

6. W.H. Melbourne, unpublished Bellman Hangar studies for the Australian Dept. of Housing and Construction.

7. D. Surry and T. Stathopoulos, An experimental approach to the economical measurement of spatially-averaged wind loads, J. Ind. Aerodyn., 2 (1977/8) 385-397.

8. S.J. Gumley, Tubing systems for the measurement of fluctuating pressures in wind engineering, University of Oxford, Report OUEL 1370/81, 1981.

9. A.G. Davenport, Note on the distribution of the largest value of a random function with application to gust loading, Proc. I.C.E., 28 (1964) 187-196.

10. J.D. Holmes, Non-Gaussian characteristics of wind pressure fluctuations, J. Wind Engg. and Ind. Aerodyn., 7 (1981) 103-108.

11. L. Pham and R.H. Leicester, Combination of stochastic loads, Proc. 7th Australasian Conference on the Mechanics of Structures and Materials, University of Western Australia, Perth, 1980 pp.154-158.

12. J.D. Holmes, Techniques and modelling criteria for the measurement of external and internal pressures, Workshop on Wind Tunnel Modelling Criteria and Techniques, National Bureau of Standards, April 1982.

A PREPARED CRITIQUE ON PRESSURE AND
FORCE MEASUREMENTS IN WIND TUNNELS.

Brian E. Lee
Department of Building Science,
University of Sheffield,
Sheffield.   S10 2TN
U.K.

1.   INTRODUCTION

The advent of our advanced knowledge of the structure of the wind together with our ability to be able to produce an accurate simulation of atmospheric boundary layers in the wind tunnel has led to the creation of increasingly realistic model analogues of the full scale wind – building interaction problem. These analogues taken together with instruments, transducers and data analysis systems based on the latest electronic and computer innovations have opened the door for wind engineers to produce reliable design information now for almost any problem. Within this relatively ideal set of circumstances what then are the origins of the problems remaining for us to solve.

These problems came about from the necessary approximations we still have to make in order to create the model analogue where the implications of reduced size will always impose a limiting condition. Here the problems of finite accommodation for transducers, distance between transducer and physical sensation, calibration for low input sensation levels and the joint problems of sensitivity and signal noise remain. These conditions are particularly true in the circumstances where individual models are dwarfed by the incorporation of large site features within the wind tunnel working section.

The seven papers which go to make up this session on force and pressure measurements present a thorough and comprehensive view of the different aspects of the field which they have each been selected to tackle. The comparisons presented, which include discussions of the different methods of tributary area load derivation and total dynamic load determination, are excellently treated and may benefit little from the addition of a reviewer. The comments offered in this critique paper may play a secondary role within the session's contribution to the modelling workshop, particularly since it has been the intention of this author to avoid the repetition of non contentious material.

2. DIRECT MEASUREMENT OF TOTAL DYNAMIC LOADS

2.1 Balance Operation at Resonance

The account given by Tschanz [1] of the measurement of total dynamic loads using elastic models with high natural frequencies contains a very thorough discussion of the advantages of this technique and of the problems he has faced and overcome in order to make it work. However he has perhaps been a little too brief in his discussion of the value of other work in this field, particularly where the use of the resonant test method is concerned.

The resonant test method, Evans and Lee [2], for the measurement of total dynamic loads is a particular case of the more general high frequency method described by Tschanz. In this method the frequency of the wind tunnel model - balance transducer combination is deliberately chosen to have a low value at frequencies excitable by the turbulence field of the simulated boundary layer. The frequency of the model, f_m may be related to the frequency of a particular building f_f by the relationship

$$\frac{f_m}{f_f} = \frac{V_m}{V_f} \cdot \frac{D_f}{D_m} \tag{1}$$

where D represents linear dimension and V represents velocity. The operation of the model - balance transducer combination at resonance instead of below resonance only represents a full aeroelastic test where the corresponding values of frequency and damping have been deliberately chosen to coincide with a particular full scale building. Elsewhere the resonance method simply reflects the utility of the below resonance method but with the advantage of the resonance phenomena, provided that scaling laws can be invoked to permit the normalization of data with respect to the values of damping and frequency at which the model - balance transducer operates. The advantages of resonance testing are that the model - balance transducer acts as a well characterized mechanical amplifier in which the useful signal is amplified by the mechanical resonance factor but in which the system noise remains unaffected, thereby increasing the signal to noise ratio by up to 2 orders of magnitude. This signal magnification is most significant if the order of size of the test results is considered. Since the purpose of tests such as these is to produce the mode generalized power spectral density function, Sn, and

$$\sigma^2 = Sn \frac{\pi f}{4\zeta} \tag{2}$$

we can write

$$\frac{Sn_m}{Sn_f} = \frac{\sigma^2_m}{\sigma^2_f} \cdot \frac{f_f}{f_m} \tag{3}$$

in order to relate model to full scale values. Since σ^2 represents a mean square response term in units of force squared, and force scaling can be represented by

$$\frac{\text{Force m}}{\text{Force f}} = (\frac{V_m}{V_f})^2 \cdot (\frac{D_m}{D_f})^2 \qquad (4)$$

the modal P.S.D. scaling relationship becomes

$$\frac{Sn_m}{Sn_f} = (\frac{V_m}{V_f})^3 (\frac{D_m}{D_f})^5 \qquad (5)$$

Using typical values of wind tunnel to full scale velocity scaling ratios and length scaling ratios of

$$\frac{V_f}{V_m} = 3 \text{ and } \frac{D_f}{D_m} = 500$$

the PSD scaling relationship is

$$Sn_f = Sn_m \times 8 \times 10^{14}$$

Hence if a typical full scale value of Sn_f for a tall building is of the order of 10^8 N^2/Hz then the equivalent model value is 10^{-7} N^2/Hz. This may serve to illustrate the cause of the measurement difficulties referred to previously in the literature as well as to indicate the advantages of a 10^2 amplifier with low noise characteristics. Additionally the resonance test method ensures that the model balance transducer acts as a narrow band filter restricting the information signal to frequencies of interest.

The resonance test method for the determination of total dynamic loads may be applied in two ways. The first method in its simplest form involves determining the model response from an r.m.s. meter reading. The r.m.s. level of the resonance peak, suitably filtered to exclude the background response contribution, can then be converted to a modal force PSD using equation 2, assuming the value of the damping, ζ, to be known from an impulse decay test. This method has been shown to be a most precise way of determining the mode generalized PSD function but is available for only one value of V_{nD} at a time, tests at different tunnel speeds being necessary to establish a useful range of V_{nD}. The second method of using the resonance test method is to employ a digital spectrum analyser. Here the full response spectrum is measured and subsequently divided by the mechanical transfer functions of the model - balance transducer system in order to yield the full spectrum of the total dynamic load. These two methods have been employed on the same model and have been shown to produce identical results. The two methods could be regarded as being complementary where the r.m.s. level is used as a precise spot check on the full spectrum method.

The resonance conditions of the model - balance configuration are dependant on the geometry and material of the balance and the mass of the model. For the single component balance described in [2] the thickness of the balance web may be determined by an equation which relates it to the desired resonance frequency and the effective mass of the model, treated as a lumped mass cantilever system. Thus, balance transducers may be designed to give a particular resonance frequency condition once the size and mass of the model are known. However the model mass is only important in determining the operational frequency of any model - balance transducer combination, and is not otherwise, significant in determining the balance transducer output, except as a second order effect on sensitivity, or in the scaling up procedure to full scale values. This may be illustrated by Figure 1 in which a light model, 600 grammes, has been mounted on a nominal 100 Hz balance and the combination found to have an exact frequency of 97 Hz. The output of the light model is contrasted with that of a heavy model, 1050 grammes, mounted on a nominal 125 Hz balance where the combination has an exact frequency of 95 Hz. The nominal balance frequencies have assumed the use of a light balance. Figure 1 demonstrates that the r.m.s. resonant response as a function of wind tunnel speed is unaffected by model mass.

Multiple graphs such as Figure 1 may be obtained for given conditions of building shape and wind regime, for all values of model - balance transducer frequency and damping level. The frequency dependant data may be collapsed by expressing the r.m.s. response as a function of V_{nD}. The damping level, independently controllable in the method described in [2] by a viscoelastic compression ring, controls the dynamic resonance magnification which in turn governs the conversion of r.m.s. peak level resonance responces to modal P.S.D.'s. An extensive series of tests has been carried out, using the resonance test method in both modes, for a wide range of values of frequency, damping and bandwidth which indicate methods of non-dimensional data collapse. It is the intention of the authors of [2], Evans and Lee, that the results of these tests be fully published in due course.

The number of force and moment components to be resolved by high frequency dynamic force balances may be worth comment. The full six components of the traditional aeronautical balance may appear at first sight to be a laudable aim and the attainment of 5 components, as described by Tschanz, and also obtained by Sauders [3] for work on vehicle aerodynamics, a creditable achievement. However the production of such a balance system may produce quantity of results at the expense of the simplicity and signal to noise ratio of the one or two component balances as used by Evans and Lee [2] and Cermak et-al [4]. Such considerations prompt the question, how many components does the structural engineer need for design purposes. If the design problem can be solved by the specification of a maximum tip deflection in two dimensions for structural purposes and a maximum resultant acceleration for human comfort purposes, then as few as three components may be adequate. Since it seems that little success is to be had currently with the problem of modelling the fundamental torsional mode of oscillation, Tschanz reporting the same difficulties as those described by Ruscheweyh [5] at the Colorado wind engineering conference in 1979, then it seems that only the two common translational modes about the vertical major and minor axes are required to produce large amounts of useful design data. Such a philosophy may lead to the design and operation of considerably less complex balances than that described by Tschanz, which in turn may have the benefit of encouraging a wider sphere of potential users.

2.2 Alternative Balance Design Methods

Tschanz has already given a useful and comprehensive review of other approaches to the problem of high frequency balance design for the direct determination of total dynamic loads. However there are a few points worth adding to his commentary.

The early work of Whitbread [6] in this field is specifically referred to by Tschanz who notes that difficulties with temperature drift and sensitivity, were encountered during the use of the balance. A careful examination of the balance from the figures in reference [6] shows that the balance depends on the operation of upper and lower strain gauged links, designed to measure tensile and compressive forces in various combinations, to determine base bending moment and base shear force for one translational mode of vibration only.

The difficulty with a design such as this is that in machining the balance structure as a whole from the solid, residual stresses are set up within the material which may account for the temperature dependant calibration draft problems reported by Whitbread.

The additional factor which may influence Whitbreads type of approach and which may also apply to the work of English and Durgin [7], is that in order for the model frequency to be significantly greater than the modelled full scale building frequency, the necessary condition for high frequency balance operation, the modelled full scale wind speed has had to be given an unrealistically high value. In Whitbreads case this value is a mean hourly wind speed of 60 m/s.

The only significant omission in the review by Tschanz of attempts to produce a balance capable of total dynamic load measurement is the work of Roy and Holmes [8]. Roy and Holmes used a 3 component dynamic force balance to measure the forces on low rise buildings and compared them with the results of integrated single point pressure measurements. Their balance consisted of three force transducers fitted to a simple framework requiring a 3 x 3 calibration matrix, since any one of the three output components registered on all three force transducers. Holmes reported difficulties in applying the calibration forces as well as some trouble with the on line data analysis procedure involving the calibration matrix. In general the frequencies of his model - balance combinations were low, like those of English and Durgin, having values of 70 Hz and 110 Hz. At these frequencies the models are quite possibly within the range liable to direct excitation by the high frequency end of the spectrum of the simulated atmospheric boundary layer wind structure. Indeed, applying the normal model scale factors to a building whose natural frequency is 0.5 Hz and which is scaled at linear and velocity ratios of 300:1 and 3:1 respectively would result in a model frequency equivalent of 50 Hz. In these circumstances the requirement for a high frequency total dynamic load balance would be a frequency of the order of 300 to 400 Hz.

3.   PRESSURE MEASUREMENTS

3.1   The Covariance Integration Approach to the Measurement of Dynamic Loads

The paper by Kareem [9] presents a thorough treatment of the covariance integration approach to the measurement of dynamic loads incorporating both the mathematical background to the process as well as a consideration of some of the practical details.   It may, however, be worth adding a little to Kareem's work since this technique was the method employed by the present author in his study of the surface pressure field of a two dimensional square prism, Lee [10].   In this paper Lee demonstrates how the total dynamic forces acting at a single cross section of the body can be evaluated from the covariance matrix formed by pre- and post-multiplying the correlation matrix of circumferential pressures by the diagonal matrices of the r.m.s. pressure fluctuations as described by Kareem in his equation (9). For a two dimensional square prism the coefficient of acrosswind forces is flow to 0.59 in a turbulence flow of 12.5% intensity, compared with the values of 0.34 and 0.27 given by Kareem for a 4:1:1 building in simulated atmospheric boundary layers which match flows over suburban and urban terrain respectively.

As Lee [10] has shown it is possible to extend the usefulness of the covariance matrix beyond the evaluation of the total dynamic load to an understanding of its physical origins at least for the problem in two dimensions, where this application is based on an original idea by Armitt [11].

The representation of the surface pressure field of a bluff body as a random function of space and time in terms of a matrix of covariance is often inconvenient, for an adequate definition of the function usually requires a high-order matrix.   It is difficult to get an understanding of these fluctuations from a large covariance matrix since many different physical causes may contribute to them.   It is often more convenient to represent such a random function as a series of modes.   Such a series of modes can be obtained from the covariance matrix by eigenvector analysis using suitable algorithms, Wilkinson [12], Barth, Martin and Wilkinson[13]. An important property of this form of modal representation is that the set of orthonormal modes produced will give the optimal representation of the random function for a given number of modes.   Explained more exactly, this means that the mean square error in the representation of the function is least for this set of modes.   Lumley[14] gives one form of this derivation.   It can be shown that the sum of the eigenvalues is equal to the integral of the mean-square pressure fluctuations over the cicumference of the body, and that each eigenvalue represents the energy contained in its associated spatial eigenvector distribution.   One disadvantage of the orthogonality condition imposed by this form of analysis is that there is no reason to suppose that the spatial variation of pressure fluctuations due to one physical cause are necessarily orthogonal with respect to that due to another cause.   Hence, whilst it is suggested that it is possible to assign a physical significance to the dominant eigenvector distributions, particularly for symmetrical structures, the mathematical constraints of the orthogonality condition mean that in some cases a unique physical cause cannot be identified for every eigenvector distribution.

The application of this concept to the surface pressure field of a
two dimensional square prism shows that the dominant eigenvector distribution is associated with a vortex shedding mode and is responsible for as
much as 80% of the surface pressure energy in a uniform flow, whilst the
second and third most significant eigenvector distributions are associated
with wake and shear layer buffeting on the sides and rear of the body.
These second the third eigenvalues can be seen to contain between 10% and
25% of the total surface pressure energy depending on the incident flow
turbulence conditions. Although this technique has been demonstrated only
for a two dimensional problem considerable scope may exist for its extension to three dimensional building aerodynamics problems in explaining the
physical origins of the total dynamic loads.

3.2 The Influence of Acoustic Pressure Fluctuations

The influence of acoustic pressure fluctuations referred to by Holmes
[5], which originate principly from the wind tunnel fan can impose a major
restraint on the success of dynamic pressure measurements carried out in
the working section. This important aspect of wind tunnel operation is
too frequently omitted from our considerations at the design stage. The
effect of acoustic pressure fluctuations within the working section of
the wind tunnel is two fold in its impact on total dynamic load assessment.
Firstly the acoustic pressure fluctuations will be superimposed on the
surface pressure fluctuations present due to aerodynamic phenomena and will
result in errors dependant on both their amplitude and phase. Secondly the
acoustic pressure fluctuations will be spatially well correlated over the
distances at least as large as typical model sizes. Thus the spatial
correlations of surface pressure fields will appear better than the aerodynamic phenomena would indicate and hence tend to produce overestimates
of the true values of total dynamic loads. The authors experience of
acoustic pressure fluctuation problems in a 0.45 x 0.45m wind tunnel at
CERL, Leatherhead indicates that in some conditions of acoustic disturbance
the spatial correlation coefficient may not fall below 0.10 to 0.20 over
comparatively large distances.

Since the existance of acoustic pressure fluctuations within the
working section of a wind tunnel tends to be classed as a fault rather
than a desireable, or even neutral, characteristic of wind tunnel design
it is difficult to obtain data on its magnitude in different wind tunnels
since such information is rarely displayed openly by the particular wind
tunnel owners. However the author is aware of the existence of certain
data which may be worth noting. Assuming the acoustic pressure fluctuations
to be measured by a wall pressure tap and expressed as an r.m.s. value
normalized by the value of the dynamic head, q, of the airstream in the
working section, then the coefficient so defined will tend to be large
for low values of q, getting progressively smaller as q increases, though
the relationship may be expected to differ between fixed speed fans with
variable pitch and variable speed fans. If the coefficient is referred to
as $C_p'_A$ then for the CERL 1.53 x 4.59m tunnel for q = 15mm water $C_p'_A \sim 0.05$.
For the 2.4 x 9.2m M.E.L. wind tunnel of the CEGB at Marchwood, $C_p'_A$ has
significantly larger values reaching as much as 0.50 at very low q values.
The original design of the BRE wind loading wind tunnel had a value of $C_p'_A$
of the order of twice that of the CERL tunnel at operational q levels
though the re-location and re-design of the inlet and frontworking section
of this tunnel has now significantly reduced the problem.

Aware of this source of error from his experience with the CEGB (1971-1974) the present author made an attempt to reduce the significance of acoustic pressure fluctuations in the design of the Sheffield University 1.2 x 1.2m boundary layer wind tunnel. Two large air conditioning system duct silencers were installed immediately upstream and downstream of the fixed speed, variable pitch fan. The result of this incorporation into the wind tunnel layout is that at q = 25mm water the value of $C_p'_A$ is 0.005 and is considered insignificant at that level.

4. THE CONSEQUENCES OF SCALING MISMATCH BETWEEN FLOW STRUCTURE AND MODEL BUILDINGS

The question of the consequences of failing to match the linear scale factor of the building model with that which describes the linear scale of the simulated atmospheric boundary layer is a taxing problem to which few reliable guidelines currently exist. The fact that few comments on the subject are available from the authors of the papers in this session bears witness to the unknown character of the problem and its solution. Yet deliberate scaling mismatch is a common feature both of practical test methods as well as an implication of some wind loading codes of practice. Within the UK Code, BSI [16] the sizes of buildings feature in the determination of the design wind speed but the use of load and pressure coefficients, or shape factors, relates only to geometrical proportion where similarly shaped buildings of different sizes are presumed to bear the same non-dimensionalized coefficients. Though a few wind tunnel experiments, notably Jenson [17], may indicate a source of error associated with this philosophy, it remains nevertheless.

As most of the participants of this workshop are aware the practice of scaling mismatch is a widespread fault and in many respects, an unavoidable one. Most wind tunnel owners and operators have by now produced for themselves a set of devices which, when fitted into the forward part of the working section of their wind tunnels will produce a simulated atmospheric boundary layer over the turntable and where a different set of devices is required for each simulated terrain type. Having produced several of these sets of devices for different terrain types, each with its unique linear flow scaling factor, few operators will see any justification in either time or expense for the reproduction of a particular terrain type simulation at a different linear scale factor, particularly if it is to suit an ad-hoc test. Furthermore the conditions of working section length and simulated flow equilibrium over the turntable will impose limits on the degree to which the linear scale can be varied for a given terrain classification. Thus circumstances arise in which the wind tunnel operator knowingly puts building models into a simulated flow where the linear scales do not agree. As an example of this in the authors experience, the 1:500 atmospheric boundary layer flow simulation in the 1.53 x 4.59m wind tunnel at CERL was used to test cooling towers at 1:500 and 1:250 and chimneys at scales as large as 1:180. Though such scaling mismatches were considered at the time liable to introduce only negligible errors into the results, few data existed to verify this decision. A decade later the situation has changed very little, a few recommendations having been suggested, i.e. Reinhold [18], but again with little convincing back-up. A few experiments of the kind shown in Figure 2 have been performed which may demonstrate little more than the fact that no scaling mismatch effects appear to be present where modelling scale changes are of a relatively small order. The results presented in Figure 2 have been performed for total dynamic load measurement using the method described in [2] for a building measureing 45 x 17 x 15m modelled at both 1:270 and 1:350 linear scales.

Before looking further into some of the more recent data which have come to light on this subject it may be worth a brief consideration of the definition of the simulated flow scaling factor itself, though the author appreciates that this is the subject of another session here within the workshop. As the author has explained elsewhere, Lee [19], it has been considered satisfactory to derive a linear flow scale factor based both on a mean velocity characteristic of the flow, the roughness length Z_o, and a turbulence characteristic of the flow, the integral scale length, L_x. However it has been suggested, Tielman [20] that for flow-building inter-action problems dependant mainly on flow separation and shear layer behaviour that perhaps neither of these lengths provides a suitable parameter, and that possibly a length parameter based on the small eddy size which affects shear layer behaviour, and in turn wake formation and body forces, might be more appropriate. Such ideas clearly deserve further consideration. However even if we remain content to assess the flow scaling factor on the basis of Z_o and L_x these atmospheric boundary layer parameters are themselves known only approximately in full scale and then with increasing uncertainty with rougher terrain conditions. If we were to take for example the size of the roughness length, Z_o, appropriate to terrain con-ditions known as Category 3 in the UK Code, where this is defined verbally as "surfaces covered with numerous large obstructions, well wooded parkland, towns and their suburbs and the outskirts of large cities where the general obstruction height is 10m", the values for Z_o appropriate to this description given by ESDU 72026 [21] range from 0.40m to 1.0m. Thus the linear scale of the simulated flow may itself be subject to a level of uncertainty of a factor greater than 2 at the outset. This situation may be improved by the joint consideration of Z_o and L_x, though to what degree is uncertain, but since many wind tunnel operators appear, from their publications, to scale on Z_o alone they must contend with the potential error as estimated above.

Specific data sources on which to base recommendations on the effect of scaling mismatch are few and some of those which purport to throw some light on the problem may themselves embody the effects of other variables. A case in point here is the work of Hunt [22] quoted by other authors in this session. Hunt's work was a specific study of the scaling mismatch problem in which he scaled simulated flows at two scale sizes and measured the surface pressure fields on isolated cube models of 4 different sizes thus providing a wide range of model scale to boundary layer scale. His conclusions were:

(i)   the pressure field on the front face is determined by the local velocity profile and the immediate upstream roughness

(ii)   the pressure field on the sides, roof and rear of the body are determined by the turbulence intensities, the integral scale of turbulence and local shear

(iii)   and, if the model scale is greater than the flow scale (i.e. the model is too big) the pressures are underestimated and vice versa.

These conclusions and the data which support them look very encouraging but a closer examination of the work reveals other factors which may have influenced Hunt's findings. Both the boundary layer simulations used by Hunt had physical heights of approximately 800mm, the measurements being performed in a 1m high working section and the four sizes of models used ranged from 50mm up to 400mm. Thus the comparison of flow scale to model scale also contained a comparison of the effect of building height to boundary layer height where this ratio varied from 1:2 to 1:16. Since the modelled partial boundary layers were of fixed actual thickness this height ratio effect may have been responsible for some of the observed variations in the pressure fields. Clearly a modelled building - wind interaction study can only ignore the importance of building height to boundary layer height within certain limits, a point made by Gartshore [23] commenting on a model study of wind effects on low rise building groups.

Whilst not wishing to simply repeat the words of other authors in this session it may be useful to highlight some of the comments of both Holmes [15] and Stathopoulos [24] with respect to the scaling mismatch problem. Holmes indicates that a mismatch of linear scales based on integral scale size, L_x, by as much as a factor of 2 is probably not significant in the derivation of local pressure data for cladding design but that more exact agreement is likely to be necessary in the case of area averaged loads. This same point regarding area averaged loads has been made by Simiu [25] who shows that increasing model sizes beyond the values given by the scaling ratio will lead to non-conservative estimates of the wind loads since the spatial correlation of the measured pressure field will be poorer over a large model than a smaller one for the same incident flow. Strathopoulos presents the results of some pressure tests on low rise industrial buildings in which buildings modelled at 1:100, 1:250 and 1:500 have been placed in a flow simulation modelled at 1:500. In these tests the building to boundary layer height ratio appeared to vary in the range 1:10 to 1:100 and so the comments offerred on the similar work of Hunt are not appropriate. Stathopoulos experiments show that a building model scale mismatch of 2, i.e. 1:250 model compared with the 1:500 model results in no significant error level in the measurement of either instantaneous area averaged peak pressures or in the measurement of generalized peak force coefficient. However he notes that in comparing the 1:100 and 1:500 model scales that a scale mismatch of this order results in the underestimate of the loads, the conclusion also reached by Hunt and indicated by Simiu. Stathopoulos concludes that the size of the tributory area and its location with respect to the separation position are likely to be important in assessing the effects of a scale mismatch.

5. CONCLUSIONS

The principle conclusion which can be drawn from this paper and from the papers in the session on pressure and force measurements is that a wealth of expertise exists within the field of instrumentation and measurement which can solve the short term problems of anyone who takes the trouble to seek it out. The longer term problems are those which involve the physical limitations which our analogue modelling philosophy imposes on us and it may be to this end that we need to devote a greater part of our research budget in the future.

ACKNOWLEDGEMENTS

The author would like to thank those other participants of this session for their co-operation and willingness to be the subject of his critique. In addition the useful discussions with Nick Cook of BRE and John Armitt of CERL held over some years are appreciated. Finally the author would like to express his gratitude to Bob Evans of Sheffield University for his help and advice on the presentation of the work on dynamic balance design and operation, and for his responsibility in conducting the experiments described by Figures 1 and 2.

REFERENCES

1. Tschanz, A. 'Measurement of total dynamic loads using elastic models with high natural frequencies.' Workshop on Wind Tunnel Modelling Criteria and Effects. N.B.S. Gaithersburg, Md., 1982.

2. Evans, R.A. and Lee, B.E. 'The assessment of dynamic wind loads on a tall building : a comparison of model and full scale results.' Proc. 4th US National Conf. Wind Engineering Research, Seattle, 1981.

3. Saunders, J. Private communication 1982.

4. Cermak, J., Sadeh, W. and His, G. 'Fluctuating moments on tall buildings produced by wind loading.' Proc. Meeting on Wind Loads on Buildings and Structures. N.B.S. Building Science Series 30, 1970.

5. Rusheweyh, H. 'Dynamic response of high rise buildings under wind action.' Proc. 5th Int. Conf. on Wind Engineering, Colorado, 1979.

6. Whitbread, R.E. 'The measurement of non steady wind forces on small scale building models.' Proc. 4th Int. Conf. on Wind Effects on Buildings and Structures, Heathrow, London, 1975.

7. English, E. and Durgin, F. 'A wind tunnel study of shielding effects on rectangular structures. Proc. 4th US Nat. Conf. on Wind Engineering Research, Seattle, 1981.

8. Roy, R. and Holmes, J. 'Total force and moment measurement on wind tunnel models of low rise buildings.' Proc. Colloque Construire avec le Vent, Nantes, France, 1981.

9. Kareem, A. 'Measurement of total loads using surface pressures.' Workshop on Wind Tunnel Modelling Criteria and Effects. N.B.S., Gaithersburg, Md., 1982.

10. Lee, B.E. 'The effect of turbulence on the surface pressure field of a square prism.' Journal Fluid Mech., Vol. 69, pp. 263–282, 1975.

11. Armitt, J.A. 'Eigenvector analysis of pressure fluctuations on the West Burton instrumented cooling tower.' CEGB Report RD/L/N - 114/68, 1968.

12. Wilkinson, J.H. Numerical Mathematics, Vol. 4, p.354, 1962.

13. Barth, W., Martin, R. and Wilkinson, J.H. Numerical Mathematics, Vol. 9, p. 886, 1967.

14. Lumley, J.L. 'The structure of inhomogenous turbulent flows.' Proc. Intern. Coll. "On the fine scale structure of the atmosphere", Doklady Academia Nauk SSSR, Moscow, 1965.

15. Holmes, J.D. 'Techniques and modelling criteria for the measurement of external and internal pressures.' Workshop on Wind Tunnel Modelling Criteria and Effects, N.B.S., Gaithersburg, Md., 1982.

368

16. British Standards Institution 'Code of Practice for Wind Loads'.
    CP3, Ch. 5, Pt. II, 1972.

17. Jenson, M. and Franck, N. 'Model scale tests in turbulent wind.'
    Part II. Danish Technical Press, Copenhagen, 1965.

18. Reinhold, T.A. 'Wind loads on solar collectors : development of
    design guidelines.' 4th US Nat. Conf. Wind Engineering Research,
    Seattle, 1981.

19. Lee, B.E. 'Model and full scale tests on the Arts Tower and
    Sheffield University.' Workshop on Wind Tunnel Modelling Criteria
    and Effects. NBS, Gaithersburg, Md., 1982.

20. Tielman, H. Private Communication, 1981.

21. Engineering Sciences Data Unit 'Characteristics of wind speed in
    the lower layers of the atmosphere.' ESDU Data Item 72026, 1972.

22. Hunt, A. 'Scale effects on wind tunnel measurements on surface
    pressure on model buildings.' Proc. Colloque Construire avec le vent,
    Nantes, France, 1981.

23. Gartshore, I. Comment on 'An investigation of the forces on three
    dimensional bluff bodies in rough wall turbulent boundary layers'
    by Lee, B.E. and Soliman, B.S. ASME Journal Fluids Engineering,
    Vol. 99, pp. 503-510, Sept. 1977.

24 Stathopoulos, T. 'Techniques and modelling criteria for measureing
    area averaged pressures.' Workshop on Wind Tunnel Modelling Criteria
    and effects, N.B.S., Gaithersburg, Md., 1982.

25. Simiu, E., 'Modern developments in wind engineering : Part II
    Engineering Structures.' Vol. 3., pp. 242-248, 1981.

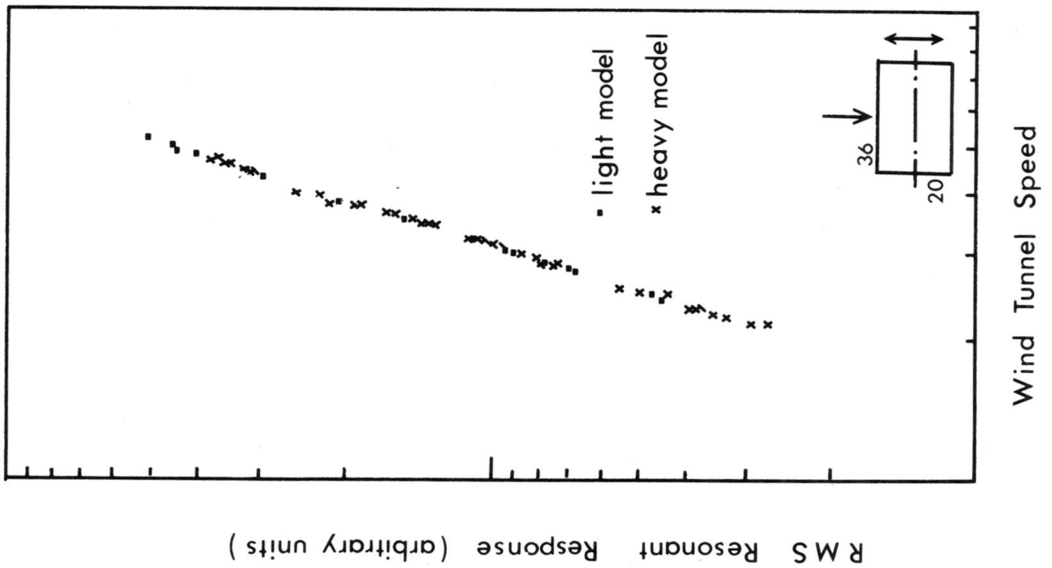

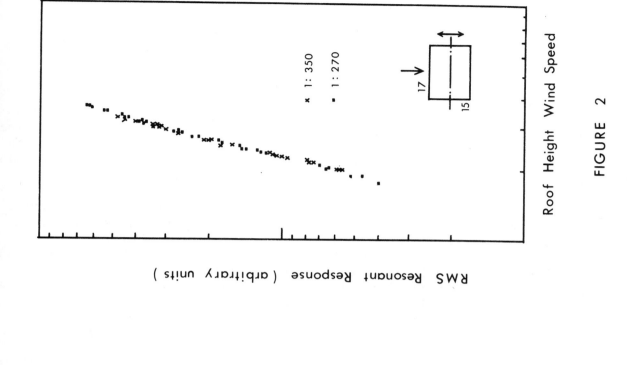

RMS Resonant Response (arbitrary units)

Roof Height Wind Speed

× 1:350

▪ 1:270

FIGURE 2

RMS Resonant Response (arbitrary units)

Wind Tunnel Speed

▪ light model

× heavy model

FIGURE 1

NEW GOALS FOR PRESSURE AND FORCE MEASUREMENTS --
DISCUSSION FOR SESSION III

W. A. Dalgliesh, National Research Council of Canada

Kareem and Tschanz each described new and effective strategies for obtaining dynamic forcing functions using pressure or force measurements on rigid models. Stathopoulos and Holmes discussed efficient techniques for sampling and averaging pressures to account for tributary areas and influence coefficients, and Cook has drawn attention to an arbitrary element in the selection of peak factors for cladding design. In each case, considerable skill has been demonstrated in the partial analysis of a larger problem; that is, certain components have been identified and solutions have been proposed.

One such problem is the prediction of net pressures on cladding elements. Great strides have been made in reducing the uncertainty associated with estimates of external pressure or suction on the outside layer of the building enclosure, but far less is known about the pressure or suction on the other side of the element. We must investigate air leakage through and between building enclosure components and develop modeling techhniques and measurement strategies for specifying these all-important internal pressure coefficients.

Another aspect of the same problem concerns the duration of peak events. As one example, the resistance of glass cladding to brittle fracture is sensitive to rate and duration of loading by peak gusts. Other mechanisms of failure may also be critically affected by the rise time and duration of individual peaks, not only by their magnitudes. Once again, field measurements will be required to guide the development of appropriate strategies for generation of design data, probably in the wind tunnel and perhaps eventually also by empirically based calculation methods.

Session IV

Aeroelastic Modeling

Chairman

Ahsan Kareem

University of Houston
Houston, Texas
USA

THE AEROELASTIC MODELLING OF TALL BUILDINGS

N. Isyumov, Manager and Associate Research Director, Boundary Layer Wind Tunnel Laboratory, University of Western Ontario, Faculty of Engineering Science, London, Ontario, Canada N6A 5B9

1. INTRODUCTION

Aeroelastic modelling is of principal value in studies of structures which are sensitive to wind induced dynamic effects. Sensitivity to dynamic action depends both on the type of structure and the nature of the potential excitation. It may range from a concern for the occurrence of violent and even catastrophic oscillations, as in the case of an aerodynamic instability, to situations where the dynamic response is strongly influenced by the dynamic properties of the structure and forms a major component of the total response. While body motion induced or "galloping" instabilities and narrow-band vortex shedding are potential concerns in some limiting situations, wind induced dynamic loads on buildings are largely due to the buffeting action of atmospheric turbulence. In the case of tall, slender and flexible buildings this excitation can be significantly amplified by the structure and the dynamic response can become resonant or narrow-band in character. In such situations, the dynamic component of the response can represent typically ½ to 2/3 of the total response and plays an important role in considerations of structural integrity and drift control. A further and often principal consideration is the need to limit wind induced horizontal accelerations to levels acceptable on the basis of occupant perception and comfort.

As in the case of most physical model studies the following two approaches are common in aeroelastic wind tunnel studies. First, such tests are carried out to provide particular empirical data for analytical formulations of wind induced effects. Second, aeroelastic studies are carried out to simulate the "entire" process, including the salient characteristics of the structure and its specific setting. The latter approach provides "direct" estimates of the full scale wind induced response and does not rely on analytical methods. Aeroelastic studies of tall buildings generally fall into this second category. In addition to the similarity of the flow, the exterior geometry and the aerodynamic forces, such studies require a representative similarity of the inertia, stiffness and damping characteristics of the building. Fortunately, the action of wind on buildings is confined primarily to the lower modes of vibration and as a result the dominant aspects of wind action can be studied with relatively simple dynamic models which provide information that can be scaled up to the full scale process without major corrections.

The following are some of the reasons for the emergence of currently used "direct" modelling approaches in aeroelastic studies of tall buildings:

i) Buildings generally have relatively low aspect ratios; typically less than 6 and seldom above 10. Their aerodynamics, therefore tend to be more 3 rather than 2 dimensional and many of the simplifying assumptions, possible in the formulation of wind loads on "line-like" structures, such as tall slender chimneys and towers, are not valid. This significantly reduces the feasibility of analytical methods.

ii) Some analytical methods, commonly referred to as "gust factor" procedures (1,2,3,4,5) are available to describe the along the wind or drag response of tall buildings due to the buffeting action of turbulence. In contrast, with some exception (6,7,8) the definition of the across the wind or lift excitation remains relatively less defined. Finally, no analytical models are currently available for describing the unbalanced or torsional excitation even for relatively simple aerodynamic shapes.

iii) It is common for buildings to be located in aerodynamically complex settings. As a result the influence of the surroundings plays a major role in modifying or determining the action of the wind. In some situations the presence of surrounding buildings has a beneficial sheltering effect. On the other hand, aerodynamic interference effects can result in significant increases in the dynamic response due to wake buffeting, and wake interaction effects (9,10,11,12,13).

iv) The dynamic action of some buildings can be highly complex with mixed degrees of freedom or three dimensional mode shapes. This compounds the difficulty of the analytical estimates and favours physical modelling which directly reproduces the main features of the wind induced response.

v) The wind induced response of tall buildings in most situations is phenomenologically understood and there is seldom a need to demonstrate the aerodynamic stability of a particular tall building design as, for example, is commonly done in section model tests for bridges and slender tower structures. Instead, the most common requirement in aeroelastic studies for tall buildings is to provide a detailed description of the response in order to insure a safe and cost effective design.

vi) Most tall buildings are shape-edged and the information obtained in small scale wind tunnel studies is representative of full scale data. As a result, direct translations of the model findings to full scale are possible without major corrections.

Most of the above reasons, alone and in combination, favour the development of "direct" aeroelastic simulations in which buildings are tested in appropriately scaled boundary layer flows which properly simulate the characteristics of the "ambient" flow, as well as the influence of the immediate surroundings. An added motivation is that aeroelastic studies for building are often carried out as part of broader investigations in which the companion phases, such as studies of curtain wall loads and pedestrian level wind conditions, have generally similar requirements for the modelling of the flow and immediate surroundings.

The objectives of the remainder of this paper are as follows:

i)   to review the scaling requirements for aeroelastic studies of tall buildings;

ii)   to provide an overview of the type of aeroelastic models commonly used; and

iii) to illustrate particular aspects of wind action on tall building examined in such studies.

2.   AEROELASTIC MODEL REQUIREMENTS

Extensive treatments of similarity requirements for aeroelastic wind tunnel studies can be found in literature (14 to 20). The principal of these are as follows:

i)   The similarity of the mean and turbulent characteristics of the flow;

ii)   geometric similarity of the structure to a scale consistent with the length scaling of the natural wind, including the depth of the boundary layer and the integral scales of turbulence;

iii) similarity of aerodynamic forces achieved by attaining a minimum body Reynolds number for sharp-edged structures and allowing for Reynolds number related differences for rounded shapes;

iv)   similarity of inertia forces and stiffness characteristics; and

v)   similarity of damping forces.

2.1 Modelling Natural Wind

It is generally accepted that models of natural wind required for aeroelastic simulations must representatively match the vertical variation of the mean wind speed, and the spectral characteristics of atmospheric turbulence including intensities and scales. In such simulations, the atmospheric boundary layer is usually taken to be locally stationary and wind effects are determined for particular conditions of mean wind speed and wind direction. Meso-scale and synoptic variations are subsequently allowed for analytically using statistical models of the wind speed and wind direction developed from available meteorological records.

The degree of similarity required in aeroelastic studies of tall buildings is generally somewhat more stringent and some comments are appropriate at this time. As in the case of the measurements of aerodynamic forces on bodies of finite size, it is particularly important to simulate the entire wind loading process at a consistent geometric scale. In other words, the geometric scale used for modelling the overall building should closely match the length scales of the simulation of natural wind. This includes the depth of the boundary layer and the scales of turbulence (integral scales or scales based on the peak wavelength). Consistent geometric scaling is important in order to achieve similarity of the spatial and temporal variations of the fluctuating wind induced pressures acting on exposed surfaces of the building. While stressing the importance of representative modelling of natural wind, it must also be recognized that the properties of natural wind vary markedly and that only approximate estimates of anticipated turbulence characteristics are generally possible. With the consequences of incomplete models of natural wind largely undefined, it is important to avoid excessive disparities particular in aeroelastic simulations of structural behaviour. Current attempts to provide specifications or guidelines for wind tunnel testing of building and structures (current ASCE Task Committee on Wind Tunnel Modelling) are therefore welcome. Of the existing code related specifications for wind tunnel model studies of buildings and structures, only the new ANSI specification offers a definite criteria. However, the permissible distortion of the scale of turbulence by a factor of 3, suggested by the current draft ANSI, appears to be excessive for aeroelastic studies of tall buildings.

2.2 Geometric Modelling

The geometric scale is chosen to maintain equality of the ratio of overall building dimensions to the inherent lengths of the generated model of the natural wind. The more important lengths of the flow for aeroelastic simulations of tall buildings are the boundary layer depth and the scales of turbulence. The matching of these lengths maintains similarity of the mean velocity variation over the height of the structure and the temporal and spatial characteristics of the aerodynamic forces. Generally, the length scaling is limited by the size of the wind tunnel facilities and should be chosen with due regard to wind tunnel blockage. Corrections are generally required if the blockage exceeds 5% of the wind tunnel cross section.

While only the roughness Reynolds number $(u_* z_0/\nu)$ usually enters the modelling of the overall flow, the attainment of a minimum body Reynolds number becomes an important consideration in the selection of the geometric scale. Geometric scales of the order of 1:300 to 1:500, commonly used in aeroelastic simulations, provide a representative simulation of the overall aerodynamic forces for sharp-edged bodies. The mismatching of Reynolds number, however, does limit the degree of detail which can be modelled. Severe distortions of the aerodynamic forces on exposed members or local detail result for local Re values below 200 to 500.

2.3 Modelling of Structural Behaviour

Having selected the geometric scale for the flow-structure simulation and having achieved representative modelling of the aerodynamic forces (with due regard to Reynolds number disparities, the additional modelling requirements relate to the similarity of the dynamic behaviour of the model and prototype structures. In the case of buildings, these additional modelling criteria require equality of the following ratios in model and in full scale;

$$\textit{Mass Scaling:} \qquad \frac{\rho_b}{\rho} \;=\; \frac{\textit{inertia forces of building}}{\textit{inertia forces of flow}} \qquad (1)$$

$$\textit{Stiffness Properties:} \qquad \frac{E}{\rho V^2} \;=\; \frac{\textit{elastic forces}}{\textit{inertia forces of flow}} \qquad (2)$$

$$\textit{Damping:} \qquad \zeta_s \;=\; \frac{\textit{dissipative structural forces}}{\textit{inertia forces of flow}} \qquad (3)$$

where ρ_b, ρ, E, V and ζ_s are respectively the bulk building and air densities; the elastic modulus or equivalent; the wind speed and the damping expressed as a proportion of critical damping.

Froude number scaling is of negligible consideration for tall buildings and free standing structures in general where the stiffness depends predominantly on elastic forces. In practical situations, the elastic stiffness in equation (2) can be taken as an effective stiffness reflecting the entire structural action including $P-\Delta$ effects.

Discussions of the above similarity requirements and the consequences of their relaxation can be found in the literature (15 to 22). For aeroelastic wind tunnel model simulations, the model air density is usually equal to that of the prototype and the similarity or inertia forces requires that the bulk density of the structure be maintained constant in both model and in full scale. Care must be taken to model this requirement correctly in situations where there are significant differences between the model and prototype air densities. Examples are differences due to significant changes in elevation or temperature. Strict modelling of the elastic properties is not necessary for buildings and a sufficient requirement is to achieve similarity of dynamic behaviour in modes of vibration for which wind action is important. Most aeroelastic studies are based on a discrete representation of the building and the modelling of inertia and elastic effects requires similarity of the mass $[M]$ and stiffness $[K]$ matrices of the model and prototype buildings; namely

$$[M]_m \;=\; \lambda_M \, [M]_p \qquad (4)$$

$$[K]_m = \lambda_K [K]_p \tag{5}$$

where λ_M and λ_K are the mass and stiffness scaling parameter.

The mass scaling parameter λ_M is determined by equation (1) and becomes

Translation

$$\lambda_M = \frac{L_m^3 \rho_{b_m}}{L_p^3 \rho_{b_p}} = \lambda_L^3 \lambda_{\rho_b} \tag{6}$$

Rotation

$$\lambda_{I_M} = \frac{L_m^5 \rho_{b_m}}{L_p^5 \rho_{b_p}} = \lambda_L^5 \lambda_{\rho_b} \tag{7}$$

In most situations $\rho_{b_m} \approx \rho_{b_p}$ and λ_M and λ_I become λ_L^3 and λ_L^5 respectively.

The stiffness scaling parameter λ_K is determined from equation (2) or its equivalent for discrete or lumped parameter systems. In the absence of Froude number scaling requirements, the velocity scale is constrained only by practical considerations and by the need to achieve a mini-mum body Reynolds number. The equivalent of equation (2) for the response of a tall building in a particular mode of vibration with frequency n becomes,

$$\left(\frac{nD}{V}\right)_m = \left(\frac{nD}{V}\right)_p \tag{8}$$

For a consistent scaling of all relevant modes of vibration, the velocity scale becomes,

$$\lambda_V = \frac{V_m}{V_p} = \lambda_L \lambda_n = \lambda_L/\lambda_T \tag{9}$$

where T is the time or period of vibration.

Having chosen the velocity scale for the simulation, the stiffness scaling becomes,

Translation

$$\lambda_K = \frac{\lambda_M \lambda_V}{\lambda_L \lambda_T} = \lambda_L \lambda_V^2 \lambda_{\rho_b} \tag{10}$$

Rotation

$$\lambda_K = \frac{\lambda_M \, \lambda_V}{\lambda_T} \qquad \lambda_L = \lambda_L^3 \, \lambda_V^2 \, \lambda_{\rho_b} \qquad\qquad (11)$$

in most wind tunnel situations $\lambda_{\rho_b} \simeq 1$ in the above expressions.

The degree of aeroelastic similarity achieved is usally dictated by the intended function of the model and the limitations imposed by model size, available model materials and economic considerations. Often it is sufficient to only model the more salient dynamic properties and it is possible to use simplified equivalent model structural systems to simulate the building. Some of the more common techniques are mentioned below. Finally, it is importnat to stress that aeroelastic wind tunnel studies to-date have been mainly confined to the use of elastic models. Simulations of non-linear behaviour present practical modelling difficulties and it is more convenient to base ultimate load considerations on theoretical arguments using elastic wind tunnel response data. For example, having defined the variation of the aeroelastic response with changes in effective stiffness and structural damping, it is possible to provide estimates of ultimate load behaviour including allowances for the "softening" of the structure and/or the increase of structural damping with increased stress level. Hybrid modelling techniques may provide alternatives. For example, measurements of instantaneous aerodynamic forces obtained from a force balance can be passed through a non-linear mechanical admittance function to determine the actual dynamic behaviour.

3.   AEROELASTIC MODEL TESTS FOR TALL BUILDINGS

Common types of aeroelastic model simulations for tall buildings are based on equivalent discrete representations which are designed to simulate the dynamic characteristics of the building only in selected modes of vibration. The aeroelastic modelling process thus becomes one of matching, at a reduced scale, the dynamic properties of the more important modes of vibration. The wind induced response is primarily in the two orthogonal fundamental sway modes of vibration for most tall buildings of compact cross-section. Of the higher modes of vibration, only the fundamental torsional mode requires consideration in certain situations. Wind induced torsional effects tend to become more important as,

i)   the building cross-section becomes less compact and there is a significant separation between the elastic centre and the line of action of the aerodynamic force;

ii)   the torsional stiffness is relatively low and the fundamental torsional period is of the same order or longer than the fundamental sway mode; and

iii)   in situations where there are significant asymmetrics in the structural system and the distribution of mass. Eccentricities between the co-ordinate and the elastic centres result in coupling terms in the stiffness matrix. Eccentricities of the centre of gravity lead to off diagonal terms in the mass matrix and inertial coupling. Both forms of coupling between the different degrees of freedom lead to more complex 3 dimensional modes of vibration in which torsional motions play a more significant part.

With the above comments it is apparent that the degree of aeroelastic modelling necessary depend upon the complexity of the dynamic and aerodynamic characteristics of the building. A number of commonly used types of aeroelastic models of tall buildings are described below.

3.1 Conventional "Stick" Aeroelastic Models

The earliest type of aeroelastic model used in wind tunnel studies of tall buildings is schematically shown in Fig. 1. This simulation comprises a rigid model of the building spring mounted near the base or some other location to provide a simulation of 1 or 2 orthogonal fundamental sway modes of vibration. In the particular example shown in Fig. 1, the mounting hardware consist of a set of gimbals located at a pivot point selected to provide a best fit between the model and prototype mode shapes; and a rigid rod extending below the model and restrained by appropriately selected springs. The other components are force transducers, which provide an indication of the wind induced overturning moment; a ballast weight adjustable to achieve correct inertial scaling and an electro magnet which provides eddy-current damping to the model.

This type of model, with variations in the supporting hardware, has traditonally in studies of slender tall buildings for which the wind induced response is principally in the two fundamental sway modes and were the higher modes of vibration and wind induced torsional effects are judged to be secondary considerations. In addition to studies of relatively compact and torsionally stiff buildings, this type of model is commonly used in exploratory studies of more complex buildings and/or situations where the across wind response is expected to dominate.

The requirements for modelling the flow for this type of study have already been discussed in Section 2 above. Fig. 3 provides a view of a typical aeroelastic model in the wind tunnel. In addition to modelling the approaching flow, it is important to include the influence of the immediate surroundings. This is normally achieved by constructing a "proximity" model which reproduces in block outline form all major buildings and structures within about five to eight building blocks of the project site. Of particular importance is the inclusion of any major nearby buildings which could lead to aerodynamic interference effects.

3.1.1 Modelling of Dynamic Behaviour

The requirements for this type of model include similarity of geometry and the modelling of the mass and stiffness properties as outlined in equations (1) and (4), and (2) and (5) respectively. The mass and stiffness matrices both are (2 x 2) diagonal matrices and it is most convenient to use rotational degrees of freedom to describe the model properties. In case of linear full scale mode shapes, namely $\mu_x (z) = \mu_y (z) = (z - h)/(H-h)$ where the point of rotation is at $z = h$, the model I_M from equation (7) becomes $I_{M_m} = \lambda_{I_m} I_{M_p}$. Normally, the full scale mode shapes are only approximately linear and the generalized mass of the prototype building in each of the two fundamental modes of vibration is used to determine the model mass moment inertia about the pivot point $z = h$. Based on the mode shape in the x-direction,

$$I_{M_m} = \lambda_{I_m} \left[(H-h)^2 \int_0^H m (z) \, \mu_x^2 (z) \, dz \right]_p \tag{12}$$

where $m (z)$ is the mass per unit length along the height of the building. Another, similar expression for I_{M_m} is obtained using $\mu_y (z)$ the mode shape in the orthogonal y direction. The pivot point $z = h$ is selected to provide the best fit for these two estimates.

In situation where the prototype mode shapes $\mu_x (z)$ and $\mu_y (z)$ depart significant from a straight line variation a correction may be necessary to allow for differences in the model and prototype generalized forces. This correction is more difficult as it requires information on the cross-spectra of the time-varying component of the wind load per unit height at various heights along the building. Providing that the mass is modelled to allow for mode shape differences, the relationship between the RMS full scale and model tip deflections of the building becomes

$$\sigma_{x_{H_p}} = \sigma_{x_{H_m}} \frac{C_x}{\lambda_L} \tag{13}$$

where C_x is a correction factor of the form

$$C_x = \left[\frac{\int_0^\infty |X_m (n)|^2 \int_0^{'} \int_0^{'} S_{F_1 F_2} (n) \, \eta_1^a \, \eta_2^a \, \mu_x (\eta_1) \, \mu_x (\eta_2) \, d\eta_1 \, d\eta_2 \, dn}{\int_0^\infty |X_m (n)|^2 \int_0^{'} \int_0^{'} S_{F_1 F_2} (n) \, \eta_1^{(1+a)} \, \eta_2^{(1+a)} \, d\eta_1 \, d\eta_2 \, dn} \right]^{1/2} \tag{14}$$

where $|X_m (n)|$ in the mechanical admittance function, $S_{F F} (n)$ is the cross spectrum of the force at the normalized heights $\eta_1 = z_1/H$ and $\eta_2 = z_2/H$; n is the frequency; and a the power law exponent.

In situations where the conventional quasi-static theory for line-like structures applies, the double integral with respect to η_1 and η_2 can be replaced by the joint acceptance function for that mode of vibration. Corrections for mode shape are discussed in the literature, for example Reference 15. Properties of joint acceptance function are discussed in References 1,

20 and 23. In view of the uncertainties in the cross-spectra of the wind forces, alternative modelling procedures should be explored in situations where C is significantly different from 1. In many cases C is slightly less than 1 and direct estimating using a linear mode model are conservative.

Consistent scaling of the elastic forces is achieved by maintaining equation (8) for both x and y directions. The estimate of the appropriate full scale damping usually represent the greatest uncertainty. Nominal values, for example $\zeta_{st} \approx 1.0$ and 2.0 for primarily steel and reinforced concrete structures, are sometimes used. In situations where ζ_{st} in difficult to estimate, conservative (lower) are used or ζ_{st} is treated as a test parameter to indicate sensitivity.

3.1.2 Experimental Technique and Instrumentation

Since the model is essentially rigid, it is common to measure the wind induced bending moment about the point of rotation and use these data to estimate wind induced deflections and accelerations. Redundancy in instrumentation, for example, using a strain-gauge balance to measure bending moments and accelerometers to obtain a simultaneous indication of the acceleration near the top of the model, provides valuable cross checks. It is important for the moment or force measuring transducers to have a flat frequency response from D.C. to well in excess of the frequency of the model. The response of this type of model is a "narrow-band" random signal and the measuring equipment must be capable of providing statistically stable estimates of the mean response, the root-mean-square value and the wind induced maximum and minimum values during a model time interval which relates to a full scale period over which wind is locally stationary (typically 20 minutes to 1 hour in full scale). While traditionally these type of measurements were carried out with various analogue meters, the current approach is to digitize the analogue signal and to analyzse the digitized data with a computer either directly or off-line.

The basic objective of the wind tunnel measurements is to define the wind induced dynamic response over a representative range of wind direction and wind speed. A 10° azimuthal increment is normally sufficient for such studies. In addition, various other types of information including power spectra of the response, estimates of aerodynamic damping, probability distribution of the parent and the peak responses, and real-time information on the interaction of wind loads acting along orthogonal directions and indications of combined load effects are commonly obtained in such studies.

3.1.3 Advantages and Disadvantages

The principal advantage of the "stick" aeroelastic model is its simplicity. The model design and construction requirement are modest and consequently there are both timing and economic advantages. Another important advantage of this type of modelling technique is the ability of readily changing the mass, stiffness, damping and even the geometric properties and to provide information on the sensitivity of the wind induced response to changes in the building configuration

The principal disadvantage of this type of simulation is that only the 2 fundamental sway modes are included. The model assumes that the mode shapes are linear and adjustments may be necessary to allow for differences between model and prototype mode shapes. The accuracy of these adjustments becomes questionable in situations of large differences. Another principal disadvantage is that the torsion and the coupling between the translational and torsional degrees of freedom (due to both inertial and elastic effects) cannot be studied with this type of model. Finally, this model simulation alone does not provide information on the distribution of the mean wind loads with height. This information, needed for estimates of effective static wind loads for structural design must therefore come from companion pressure model studies of the building.

3.2 Multi-Degree of Freedom Aeroelastic Models

Discrete mechanical analogues of buildings comprising several elastically interconnected "lumped" masses each having two translational and one rotational (rotation about a vertical axis) degrees of freedom are used to simulate a number of translational and torsional modes of vibration of tall buildings. This type of model is suited for aeroelastic studies of more complex building stuctures were torsional effects are judged to be important and/or in situations where the modes of vibration are highly 3-dimensional due to inertial and/or elastic coupling between various degrees of freedom. A schematic representation of a typical multi-degree of freedom aeroelastic model is presented in Figure 3. In this case the building has been divided into four zones each represented by a lumped mass. The masses are primarily concentrated at the rigid or flexible floor diaphragms. These diaphragms are connected by flexible columns and the entire mechanical system is enclosed by a non-structural skin which reproduces the exterior geometry. This skin is discontinuous with slits separating the various zones of the model. The mass of the flexible columns, the exterior skin and any added instrumentation are included in the total mass budget of the model.

A four lumped mass, 12 degree of freedom representation, as shown in Figure 3, is typically found to be sufficient in studies of most tall buildings. Up to 7 lumped masses have been used in unusually complex buildings (24) whereas only 3 lumped masses have been found to be sufficient in a study of an intermediate height building (25). It is generally found that the sway response of tall buildings is prodominantly in the two fundamental sway modes of vibration and the inclusion of higher sway modes in multi-degree of freedom simulations results only in marginal further improvements. A comparison of the dynamic drag and lift responses of a tall building obtained with a "stick" and 7 lumped mass model simulations is presented in Figure 4. The agreement between the estimates of the rms drag response is seen to be good. This supports the comments about the predominance of the fundamental modes of vibration and provides a useful cross-check between the two types of model simulations. The indicated structural damping of about 1% is an nominal value and the differences in the lift response around the vortex shedding peak are mainly attributed to slight differences in the actual damping values.

3.2.1 Modelling of Dynamic Properties

The modelling requirements for mass and stiffness similarity are given in equations (4) and (5) respectively. The mechanical analogues required to achieve elastic similarity vary in complexity from relatively simple model structures in which the "lumped" masses are represented by rigid plates interconnected by elastic columns to more complex systems were column shortening effects and/or the bending of the floors have to be included in order to achieve a consistent simulation of the translational and torsional characteristics of the building. Schematic representations of stiffness elements used in such models are given in given in Figure 5. The flexible columns are "built-in" at the floor diaphragms in all cases. While the floors remain essentially horizontal in the type 1 element (there is some rotation due to the axial extension of the columns), floor rotation is achieved in the type 2 and 3 elements by the addition of extension elements at the inflexion points of the columns, as in the case of type 2, or the addition of bending elements in the floors, as in type 3. In the latter case, the floor is represented by 2 rigid diaphragms connected by leaf-springs. Photographs of two aeroelastic models with the skin partially removed to show different types of structural systems are given in Figures 6a and 6b.

The addition of extension elements to the columns and/or the bending of the floors offers important additional degrees of freedom needed in order to achieve a proper modelling of the translational mode shapes and or to achieve a consistent modelling of the translation and rototional stiffnesses of the prototype building. The effect of adding extension elements on the sway stiffness of an aeroelastic model is demonstrated in Figure 7. While increasing the flexibility of the model in translation, neither the extension elements nor the bending of the floors significantly influence the torsional properties of the building (rotation about a vertical axis). The principal application of the simple rigid floor − flexible columns (typically 3 or 4 columns) element, see type 1 in Figure 5, is for shear buildings and situations where the fundamental torsional and sway periods are of similar magnitude.

An indication of the suitability of the simple type 1 element can be obtained from a consideration of its relative translational and rotational stiffnesses. Consider a symmetrically positioned set of 4 columns of equal length l and with cross-sectional dimensions of d_x and d_y in the x and y directions. The ratio of the torsional to the x-translational frequency of vibration for this type of element is of the form,

$$\left(\frac{n_\theta}{n_x}\right)_m = \left[\frac{r_y^2}{r_m^2} + \frac{r_x^2}{r_m^2}\frac{K_y}{K_x} + \frac{k_\theta}{r_m^2 K_x}\right]^{1/2} \qquad (15)$$

where r_y and r_x are the y and x co-ordinates of the column locations; r_m is the radius of gyration of mass; K_y and K_x are the effective translational stiffnesses of the set of columns in the x and y directions (typically in this case $K_y = 4\,(12\,E\,I_{xx}/l^3\,)$); and k_θ is the St. Venant torsional stiffness of column set $(k_\theta = 4\,G\,C/l$, where G is the shear modulus and C is the torsional constant). A similar expression can be written for n_θ/n_y.

An initial indication of the suitability of this type of model is obtained by comparing $(n_\theta/n_x)_p$ to the maximum possible value of $(n_\theta/n_x)_m$ from equation (15), or a similar expression for a different number of columns, for which the columns are maintained within the envelope of the building geometry. For typical model configurations the last term, containing the St. Venant torsional stiffness, is of the order of 1/8 to 1/4 while each of the other two terms is of the order of 1 to 1.4. Consequently the simple type 1 mechanical analogue is satisfactory providing that n_θ/n_x and n_θ/n_y are each less than about 1.5 to 1.7. A more complex mechanical analogue is likely needed when the fundamental torsion to the fundamental sway frequency ratios of the building exceed 1.5 to 1.7.

Standard computerized methods have been developed at the Boundary Layer Wind Tunnel Laboratory to analyse and design aeroelastic models using mechanical systems as shown in Figure 5. The input required for this analysis are the mass and stiffness matrices of the building. This information is used to determine the dynamic properties of the prototype building and to design the aeroelastic model to a particular length and velocity scale. This includes the selection and positioning of the columns, the latter is based on the required location of the centre of twist, and determining the mass and mass moment of inertia properties of the model. Nevertheless, even with computer aid, the design of a multi-degree of freedom aeroelastic model is a relatively major task particularly if more complex mechanical analogues. For example, if types 2 and 3 elements, see Figure 5, are required. Another major effort is the fabrication of the model which usually involves high precision machining of the structural components. This can be seen from the photographs given in Figures 6a and 6 b.

The structural damping in multi-degree of freedom aeroelastic models is usually obtained by adding inter-floor dampers ranging in complexity from visco-elastic columns to damping tapes attached between adjacent floors. Other possibilities are the use of miniature dash-pots and eddie-current devices. Generally it is a significant effort to attain the required damping.

3.2.2 Experimental Technique and Instrumentation

The overall experimental procedures are similar to those described in connection with the conventional "stick" aeroelastic model, however, the instrumentation requirements are generally more complex. Typically, a multi-degree of freedom aeroelastic model is mounted on a balance capable of measuring the mean and dynamic bending moments and torque near the base of the building and is fitted with accelerometers and/or displacement transducers to provide measurements of the sway and rotational motions near the top of the building. While this is sufficient to provide a complete description of the dynamic response once the model is calibrated, additional instrumentation is needed in order to determine the variation of the mean wind induced forces. In most practical situations the mean pressure data, obtained in the companion study of the curtain wall, are used to determine the distribution of the mean or static wind loads.

3.2.3 Advantages and Disadvantages

The main advantage of the multi-degree of freedom model is its ability to simulate complex structural behaviour including both sway and torsional degrees of freedom. Its main disadvantages are its greater complexity of design and fabrication and the somewhat greater instrumentation and calibration effort. Also it is generally more difficult to vary such structural properties as stiffness, mass and damping. As in the case of the "stick" model, this type of simulation does not conveniently provide information on the distribution of the mean wind load.

3.3 Other Types of Models

Other types of aeroelastic models, although less common, warrant mention at this time. As discussed above, an adequate simulation approach for most tall buildings would be a model which simulates the two fundamental sway and the fundamental torsional modes of vibration. While this approach is frequently practiced with conventional multi-degree of freedom models in which the modelling is relaxed for higher modes of vibration, special mechanical analogues have been devised to provide partial multi-mode representations. In particular, one such approach is to modify the support hardware of a "stick" model to permit torsional motion. One such approach in which the entire model was able to rotate as a rigid body has been reported by Durgin and Tong (26). While there was a clear mis-match between the model and prototype torsional mode shape, the model has a unit mode shape while the prototype varies with height, the study indicated the importance of torsional effects and the sensitivity of the dynamic response to eccentricities in the centre of gravity and the location of the elastic axis.

The development of a simple multi-mode representation of a tall building has continued at MIT and Figure 8 provides a schematic representation of a more recent version of a model which simulates the two fundamental sway modes of vibration and permits the modelling of the fundamental and possibly higher torsional modes (27). A similar model concept is being used in aeroelastic studies at MHTR (28). While this modelling approach is an important extension of the two-degree of freedom stick model, the multi-degree of freedom modelling approaches, as discussed in Section 3.2, remain important for complex structural systems particularly in situations where the degrees of freedom are strongly coupled due to structural eccentricities. It is relatively common for the effective shear centre to vary with height and the stiffness matrix of tall buildings for which torsional effects are important is usually highly three-dimensional with significant cross terms between the translational and torsional degrees of freedom. The composition of the wind excitation in such situations is largely undefined and the omission or simplification of the interaction between the various degrees of freedom, which occurs in highly coupled lateral and torsional vibrations, may lead to major differences in structural action.

Another modelling approach reported in the literature is the use of continuous rubber models or combinations of rubber models with plexiglas or aluminium cores to study the coupled and torsional lateral response of tall buildings (29,30).

3.4 Practical Considerations

The value of aeroelastic wind tunnel tests in the design of tall buildings for wind action has become well recognized. While traditionally resorted to only for unusual buildings, such tests have become acceptable alternative design methods. This change in the role of aeroelastic as well as other wind tunnel tests have received attention in the literature (31,32). It is important to stress that not all tall buildings warrant wind tunnel study and it is important to identify situations were aeroelastic tests would improve reliability and/or economy. Some guidelines for identifying "candidates" for wind tunnel study are provided by building codes. Unusual aerodynamic feature and concerns for wind induced dynamic effects are the two most common reasons for aeroelastic studies for tall buildings. Potential candidates for wind tunnel study either fall outside existing experience or differ from the norm in there sensitivity to dynamic effects, perceived importance and economic penalty imposed by wind loading considerations. The following questions identify some of the possible concerns:

— is the building unusually light, slender and/or flexible;

— are the fundamental periods of vibration unusually long for the height of the building;

— is the anticipated structural damping likely to be below normally accepted values;

— is the structural action expected to be highly complex with a possible interaction between translation and rotational motions of the building. Significant eccentricities between the centres of geometry, gravity and twist are indications of likely coupled dynamic behaviour.

— is the shape of the building unusual resulting in concerns for aeroelastic stability and/or vortex shedding;

— are the immediate surroundings likely to lead to aerodynamic interference effects and accentuate the dynamic response.

— are initial estimates using codes and/or analytic methods indicating unusually large loads and responses.

The choice of the modelling approach depends on the particular situation and is often influenced by other factors such as timing, cost and the ability of the design process to make full use of the findings. Also, to be effective improvements in the definition of the dynamic response for particular wind conditions, must be accompanied by an up-grading of other components of the wind loading process. Principal of these, is the statistical description of the speed and directional characteristics of the wind in the project area. As a result aeroelastic wind tunnel tests of tall buildings are usually accompanied by studies of meterological data available for the project area. Predictions of behaviour of the full scale building are subsequently made by viewing the dynamic behaviour of the building through the "filter" of the local wind climate. One approach is to base the predictions of the full scale process on the worst aerodynamic response within a design range of wind speed, determined on the basis of an acceptable probability level. A more effective

388

approach is to allow for the directional characteristics of the local wind climate is assessing the importance of particular wind induced effects. This is achieved by weighting the building response for a particular wind speed and wind direction by the probability of realizing that combination in full scale. The methodology for this approach has been described in the literature (24,33,34). In the case of tall buildings, the synthesis of the meteorological and the wind tunnel derived data permits probabilistic predictions of wind induced forces required for strength design, building motions or "drift" using considerations of servicability and the performance of the curtain wall system; and wind induced accelerations which

4. IMPACT ON DESIGN

4.1 General Trends

With some exceptions, aeroelastic studies tend to lead to somewhat smaller wind loads and responses in comparison to code predicted values. The following are some of the reasons for this trend:

i) Aeroelastic studies for a major building generally include improvements in the regional wind climate through more detailed analysis of available meteorological data. A 10% change in the magnitude of the wind speed for a particular probability level corresponds to approximately a 25% change in the effective wind load and a 30 to 35% change in the wind induced accelerations.

ii) A principal factor is the inclusion of the local surroundings which generally produce a sheltering effect and thus reduce wind loads. There can be exceptions to this trend in situations of aerodynamic interference effects for certain juxtapositions of isolated taller buildings, particularly in otherwise a low-rise surroundings.

iii) The recognition of the directional characteristics of the local wind climate and the wind induced response typically results in about a 20% reduction in the magnitude of the wind induced effect for a particular return period.

iv) While aeroelastic simulations improve estimates of the dynamic load and the building response, the effective shape factors, are not necessarily always below code prescribed values. Also the dynamic amplification of the response for many buildings is consistent with theoretical gust effect factor estimates. Exceptions are buildings for which the dynamic across the wind response dominates.

Generally reductions in wind induced effects, below values obtained by code or analytical methods, tend to be more pronounced for buildings of intermediate height.

An important further impact of aeroelastic studies is their contribution to a more complete understanding of the action of wind on tall buildings. This includes the recognition that wind loads generally tend to be asymmetric or unbalanced and significant torsional moments are experienced even for symmetric building shapes. Another consequence is the recognition that the design of a building cannot be based solely on the independent action of wind forces along particular building directions but must also consider situations of combined loading. The influence of aeroelastic studies on the design of tall buildings is illustrated by some examples below.

4.2 Horizontal Wind Loads

Aeroelastic studies, as described in Section 3, usually provide information on the wind induced bending moment and sometimes shears near the base of the building. While this information is sufficient to estimate a gust effect factor to be used in the context of an analytical or code method or analysis, a more tailored evaluation of the wind load is usually possible. Knowing the mass distribution of the building, and the magnitude of the dynamic response in particular modes of vibration provides a description of the dynamic part of the effective wind load at different heights along the building.

For example, in the case of the stick model, see Fig. 1, which simulates the dynamic action in a linear mode of vibration, the RMS bending moment at $z = h$ becomes

$$M_{x_{RMS}} (z = h) = \int_h^H (2\pi n_{o_x})^2 \, m(x) \, a_x \, \mu_x(z) \, (z-h) \, dz \qquad (16)$$

where a_x is the RMS modal coefficient and where in this case the mode shape would be $\mu_x(z) = (z-h)/(H-h)$.

For a uniform mass distribution the dynamic load follows the variation of $\mu(z)$ and is thus quite different in shape from the static component of the wind load which, due to the more or less constant base pressure, tends to vary more gradually with height. The variation of the static component of the wind load is usually obtained from a companion pressure study.

Knowing the variation of the mean and dynamic components of the wind induced force at various heights along the building permits the evaluation of an effective static horizontal load along a particular axis of the building and/or an effective static torque. Procedures for estimating this effective static load, which represents the combined action of wind induced static and dynamic effects, are discussed elsewhere (24).

Examples of effective statically acting wind pressures obtained in this way for 3 tall buildings are given in Figure 9. Also provided for comparison are code loads. In the case of the Calgary building the code wind load is based on the National Building Code of Canada and estimates are made for both urban and suburban terrains. Consistent with previous comments the effective wind loads estimated from the aeroelastic studies are significantly lower over most parts of the building. This is due to a more accurate estimate of the distribution of the dynamic component of the load as discussed above. The dynamic component for the two New York office towers is dominate and the differences between the code and wind tunnel derived mode shapes are more pronounced. Other comparisons of aeroelastic and code data can be found in References 35 and 36.

4.3 Torsional Effects

Most code approaches do not consider wind induced torsional loads apart from situations where there is an eccentricity between the centres of twist and building geometry. Recent trend towards more complex building shapes and structural systems have resulted in more unbalanced wind loads and larger torsional forces and motions. A particular important consequence of the latter is the increase in the wind induced accelerations near the perimeter of a building.

Current code approaches for estimating torsional forces are generally found to be inadequate. This is clearly seen from the comparison of peak base torques obtained from wind tunnel model studies and corresponding values estimated from the National Building Code of Canada for 4 tall buildings.

While there are still no reliable theoretical estimates of torsional effects, some progress in estimating the torsional response have been made using available wind tunnel data. An empirical model for estimating the wind induced dynamic torque has been developed at the Boundary Layer Wind Tunnel Laboratory. The initial formulation, based on the results of ten aeroelastic studies, has been reported in detail in Reference 37. Since then the data base has been expanded to 14 aeroelastic studies. The empirical model of the RMS torque, presented in Figure 11, provides a good representation of this data base. The length perimeter L in this formulation is a measure of the effective eccentricity of the aerodynamic force. In the empirical model used, L is the line integral of the product of the increment of the perimeter ds and the distance from its normal to the centre of rotation of the building. Further improvements in the estimate of the dynamic torque are obtained if the mean base torque coefficient of the building cross-section is known. Procedures for this improved estimate are reported elsewhere (37).

4.4 Combined Wind Loading

It is common practice in engineering to base the design of a tall building on the independent action of wind loads estimated along particular directions of the building. Unfortunately, this convenient assumption is an over simplification as for any particular wind direction the building is expected to experience simultaneously acting forces in the along and across-wind directions, as well as a torsional moment.

For members governed only by wind induced forces in one particular direction, extreme member forces or load effects can be obtained using the respective independently acting x and y lateral forces and the torque about the vertical axis. Examples of such independently acting effective static loads have been given in Figure 9. Generally, the wind induced stresses or loads effects for a particular member result from a combination of the wind forces in the two sway directions and the wind induced torque. Typically the maximum load effect for member i can be expressed as:

$$F_{i_{max}} = (C_{y_i} M_y + C_{x_i} M_x + C_{T_i} M_T)_{max} \qquad (17)$$

where $F_{i_{max}}$ = the maximum load effect or stress for member i due to wind action.

M_x, M_y, M_T = the wind induced moments due to x, y and torsional forces.

$C_{x_i}, C_{y_i}, C_{T_i}$ = the influence coefficients which relate the stress or force effect for member "i" to the wind induced moments M_x, M_y and M_T respectively.

Using available aeroelastic test data it is still convenient to evaluate $F_{i_{max}}$ from the maximum values of the independently acting wind forces. The peak value of the stress or load effect in such a case can be expressed as:

$$F_{i_{max}} = \phi \left[C_{y_i} \hat{M}_y + C_{x_i} \hat{M}_x + C_{T_i} \hat{M}_T \right] \qquad (18)$$

where ϕ is a load combination or "joint action factor" which allow for the reduced likelihood of the simultaneous occurrence of maximum x, y and torsional effects. The magnitude of this joint action factor can be estimated from the aeroelastic data using the following expression:

$$\phi = \frac{\hat{F}_i(R)|}{|C_{y_i}||\hat{M}_y(R)| + |C_{x_i}||\hat{M}_x(R)| + |C_{T_i}|\hat{M}_T(R)|} \qquad (19)$$

where $\hat{F}_i(R)$ is the actual member force or stress for member i for return period R and where the denominator is the value of F_i estimated from the combination of the independently acting maximum values of $\hat{M}_y$, $\hat{M}_x$ and $\hat{M}_T$ for the same return period.

Joint action factors for a tall building, estimated on the basis of equation (19), are shown in Figure 12. In this representation ϕ is given for different values of C_x/C_y and C_T/C_y. For both of these factors equal to 1, the wind induced load or stress within a particular member is equally sensitive to wind induced x, y and torsional forces. For a small C_T/C_y (torsional loads have little effect) the load combination factor ϕ approaches unity for both very small and very large values of C_x/C_y. These respectively represent situations where the member loads are dominated by y and x wind forces. It is noteworthy that the load combination effect of ϕ is approximately .7 for bi-axial loading cases eqully affected by x and y forces and is well below .7 when there is a significant additional influence of torque. Namely when C_T is significant in comparison to C_y or C_x. The load combination factor of .75 introduced in the 1980 edition of the National Building of Canada is shown for comparison.

The introduction of the load combination factor ϕ results in signficant simplifications in the load cases which must be considered for design. Typically, the design would be based on the independent action of effective static wind loads, as shown in Figure 9, acting in particular directions of the building. In view of the importance of the dynamic response these directions are taken to coincide with the orthogonal principal axes of the building. In addition to considering the independently acting wind loads in each of these principal directions. A combined load case with an appropriately selected joint action factor would be examined. Based on existing information, joint action factors of about .8 and about .7 would be appropriate for load cases involving the action of two or three independent wind loads respectively.

4.5 Wind Induced Accelerations

Estimates of sway induced accelerations provided by aeroelastic studies generally tend to be below suggested code values. This is illustrated in Figure 13 which provides a comparison of peak sway accelerations obtained in a number of aeroelastic studies of tall buildings with estimates based on the National Building Code of Canada. Of the 9 buildings included in Figure 13 only one was found to exceed the NBCC values. The ratio of the experimentally determined accelerations to the corresponding code predicted values for this group of 9 buildings was on average about .7. A significant portion of this difference is due to the directional effects of the wind climate and the aeroelastic response which is not allowed for in code approaches.

Wind induced torsional motions can significantly increase the effective horizontal acceleration near the building perimeter. In such situations code approaches may significantly underestimate the resultant accelerations. The effect of torsion on the horizontal acceleration near the top of a tall building is indicated in Figure 14 (39). Results are shown at locations C, which i near the centroid of the building, and at location A, which is near the acute corner. As seen from the provided predictions, significantly larger accelerations are experienced near location A. As a result, while the accelerations near the centre of the building may be within acceptable limits, increases in the acceleration due to torsional vibrations may lead to unacceptable conditions near the building perimeter. Furthermore there are suggestions that torsional motions are more visually discernible due to the rotation of the horizon and consequently may cause a greater psychological discomfort or annoyance.

As in considerations of wind induced member loads or stresses, estimates of the wind induced accelerations must be based on the simultaneous motion of the building in all three degrees of freedom. Often the resultant acceleration must be formed from information on the independently acting x, y and torsional motions. Figure 15 provides information on the relationship between the actual maximum resultant acceleration at the center of twist of a tall building and the resultant of the two independent maximum values. As expected, this ratio is 1.0 for situations were either the x or y acceleration are 0. The other expected bound of this ratio is $1/\sqrt{2}$ for the case when the x and y accelerations are equal. Although there is some experimental scatter, the data tend to follow an interaction ellipse through the two bounds. Current code approaches do not considered the resultant acceleration. A reasonable practical estimate would be to base the effective resultant acceleration on the resultant of the independently estimated accelerations along particular directions and to use a joint action factor of about 0.9.

5.   CONCLUDING REMARKS

Aeroelastic model studies are expected to continue to provide a means for the study of wind action on buildings which are sensitive to wind induced dynamic effects. In view of the greater inherent difficulties of the process — buildings have low aspect ratios and are located in highly complex flows — analytical methods are relatively more difficult and a greater reliance must be placed on direct estimates of the wind induced response obtained from representatively scaled model studies. In addition to providing particular information, collectively such studies provide an important basis for furthering the overall understanding of wind action on tall buildings and improving existing code approaches. Some examples and trends of this have been presented in this paper.

6.   ACKNOWLEDGEMENTS

This paper draws from the findings of a number of studies of wind action on tall buildings carried out at the Boundary Layer Wind Tunnel Laboratory, The University of Western Ontario. The author would like to acknowledge the contributions of Dr. A. G. Davenport, Director of the Laboratory and other colleagues in particular Drs. D. Surry and B. J. Vickery to the general area of aeroelastic modelling of tall buildings and some of the material presented in this paper. The author would like to acknowledge the contributions of various other members of the Laboratory staff; in particular Mr. B. Allison who provided some of the illustrations and Ms. K. J. Norman who typed the manuscript.

REFERENCES

1. Davenport, A. G., "Gust Loading Factors", J. Struct. Div. Proc. ASCE, Vol. 93, No. ST3, 1967.

2. Vickery, B. J., "On the Assessment of Wind Effects on Elastic Structures", C. E. Trans., Inst. Aust., pp. 183–192.

3. Vellozzi, J., and Cohen, E., "Gust Response Factors", J. Struct. Div., Proc. A.S.C.E., Vol. 94, No. ST6, June 1968.

4. Etkin, B., "Theory of the Response of a Slender Vertical Structure to a Turbulent Wind With Shear", Meeting on Ground Wind-Load Problems, NASA., Langley.

5. Simiu, E., "Gust Factors and Along-Wind Pressure Correlations", J. Struct. Div. Proc., ASCE., Vol. 99, No. ST4, 1973.

6. Saunders, J. W. and Melbourne, W. H., "Tall Rectangular Building Response to Cross-Wind Excitation", Proc. 4th Int. Conf. on Wind Effects on Buildings and Structures, Cambridge, Univ. Press, 1975.

7. Melbourne, W. H., "Cross-Wind Response of Structures to Wind Action", Proc. 4th Int. Conf. on Wind Effects on Buildings and Structures, Cambridge Univ. Press, 1975.

8. Kareem, A., Cermak, J. E. and Peterka, J. A., "Crosswind Response of High-Rise Buildings", Proc. 5th Int. Conf. on Wind Engineering, Fort Collins, Colorado, U.S.A., 1979.

9. Rusheweyh, H., "Dynamic Response of High Rise Buildings Under Wind Action With Interference Effects From Surrounding Buildings of Similar Size", Proc. 5th Int. Conf. on Wind Engineering, Fort Collins, Colorado, U.S.A., 1979.

10. Melbourne, W. H. and Sharp, D. B., "Effects of Upwind Buildings on the Response of Tall Buildings", Proc. Regional Conf. on Tall Buildings, Hong Kong, Sept. 1976.

11. Blessmann, J. and Riera, J. D., "Interaction Effects in Neighbouring Tall Buildings", Proc. 5th Int. Conf. on Wind Engineering, Fort Collins, Colorado, U.S.A., 1979.

12. Saunders, J. W. and Melbourne, W. H., "Buffeting Effects of Upstream Buildings", Proc. 5th Int. Conf. on Wind Engineering, Fort Collins, Colorado, U.S.A., 1979.

13. Cooper, K. R., "Wake Galloping, an Aeroelastic Instability", IUTAM–IAHR Symposium on Flow-Induced Structural Vibrations, Karlsruhe, Germany, 1972.

14. Whitbread, R. E., "Model Simulations of Wind Effects on Structures", Proc. Conf. on Wind Effects on Buildings and Structures, NPL Teddington, Great Britain, 1963.

15. Vickery, B. J., "On the Aeroelastic Modelling of Structures in Wind", Conf. on Struct. Models, Univ. of Sydney, Sydney Australia, May 1972.

16. Isyumov, N., "Wind Tunnel Methods for Evaluating Wind Effects on Buildings and Structures", Int. Symp. on Exp. Mech., Univ. of Waterloo, June 1972.

17. Whitbread, R. E., "The Use of Scale Models for the Determination of Wind Effects on Buildings and Structures", Course on Wind Effects on Buildings and Structures, von Karman Institute, 1972.

18. Scanlan, R. H., "Scale Models and Modelling Laws in Fluidelasticity", ASCE Nat. Struct. Eng. Meeting, April 1974, Cincinnati, Ohio.

19. Surry, D. and Isyumov, N., "Model Studies of Wind Effects — A Perspective on the Problems of Experimental Technique and Instrumentation", Int. Congress on Instrumentation in Aerospace Simulation Facilities, 1975, CHO 993−6, AES, Ottawa 1975.

20. Simiu, E. and Scanlan, R. H., "Wind Effects on Structures: An Introduction to Wind Engineering", John Wiley & Sons, New York, 1978.

21. Cermak, J. E., "Application of Fluid Mechanics to Wind Engineering", Freeman Scholar Lecture, J. Fluid Engineering ASME, Vol. 97, No. 1, March 1975.

22. Committee 7 (Wind Loading and Wind Effects) Council on Tall Buildings and Urban Habitat Chapter CL−3, "Wind Loading and Wind Effects".

23. Davenport, A. G., "The Prediction of the Response of Structures to Gusty Wind", Proc. 3rd Int. Conf. on Structural Safety and Reliability, ICOSSAR, Trondheim, Norway, June 23-25, 1981.

24. Davenport, A. G., Isyumov, N. and Jandali, T., "A Study of Wind Effects for the Sears Project", Univ. of Western Ontario, Fac. of Eng. Science Research Report, BLWT−5−1971, London, Ont.

25. Surry, D., Kitchen, R., and Davenport, A. G., "Design Effectiveness of Wind Tunnel Studies for Buildings of Intermediate Height", Canadian Journal of Civil Engineering, Vol. 4, No. 1, 1977, pp. 96−116.

26. Durgin, F. H. and Tong, P., "The Effect of Twist Motion on the Dynamic Multimode Response of a Building", IUTAM—IAHR Symposium on Flow-Induced Structural Vibrations, Karlsruhe, Germany, 1972.

27. Durgin, F. H., Private Communication Massachusetts Institute of Technology, Cambridge, Mass.

28. Irwin, P. A., Private Communication M.H.T.R. Ltd., Guelph, Ontario.

29. Friedman, P., "Design, Testing and Correlation of Aeroelastic Models", Proc. 3rd U.S. Nat. Conf. on Wind Engineering, Univ. of Florida, Gainesville, Florida, 1978.

30. Friedman, P., Goetschel, D. and Torkamani, M., "Theory and Design of Aeroelastic Models Intended for Simulating Tall Building Response to Wind Loads", U.C.L.A., Report in publication.

31. Davenport, A. G., Isyumov, N. and Surry, D., "The Role of Wind Tunnel Studies in Design Against Wind", ASCE Spring Convention, Boston, Mass., April 1979.

32. Isyumov, N., "The Role of Physical Models in Design Against Wind", 3rd Canadian Workshop on Wind Engineering, Vancouver & Toronto, 1981.

33. Davenport, A. G., "On the Statistical Prediction of Structural Performance in the Wind Environment", ASCE, Nat. Struct. Eng. Meeting, Baltimore, Maryland, 1971.

34. Davenport, A. G., "The Dependence of Wind Action on Wind Direction", ASCE—EMD Conference, Univ. of Waterloo, 1976.

35. Chen, P. W., Davenport, A. G., Isyumov, N. and Robertson, L. E., "Response of Triangular Prism in Turbulent Winds", Jour. of Struct. Div., ASCE Proc. Vol. 97, No. ST11, November, 1971.

36. Culham, G. A., "The Application of Model Test Results to Design: The Nova Headquarters Office Building", 3rd Can. Workshop on Wind Engineering, Vancouver & Toronto, 1981.

37. Greig, G. L., "Towards an Estimate of Wind-Induced Dynamic Torque on Tall Buildings", Univ. of Western Ontario, Engineering Science Master's Thesis, 1980.

38. Surry. D. and Lythe, G., "Mean Torsional Loads on Tall Buildings", Proc. 4th U.S. Nat. Conf. on Wind Engineering Research, Univ. of Washington, July 1981.

39. Greig, G. L., Isyumov, N. and Surry, D., "A Study of Wind Induced Forces and Responses for the Alberta Gas Trunkline Headquarters Building, Calgary, Alberta", Univ. of Western Ontario, Faculty of Engineering Science, Research Report BLWT—SS8—1980.

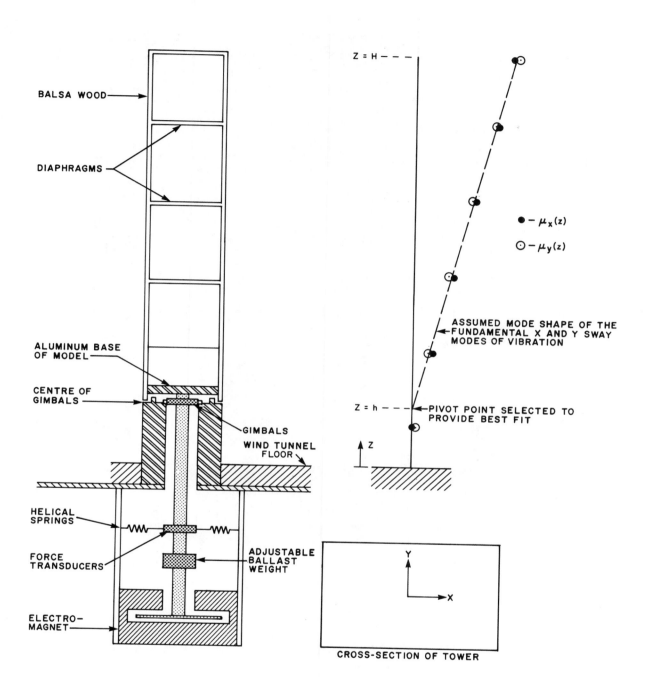

FIG. 1   SCHEMATIC REPRESENTATION OF CONVENTIONAL 2 DEGREE OF FREEDOM OR "STICK" AEROELASTIC MODEL OF A TALL BUILDING

398

FIG. 2   VIEW OF AEROELASTIC MODEL OF TALL BUILDINGS AND SURROUNDINGS
IN BOUNDARY LAYER WIND TUNNEL LABORATORY

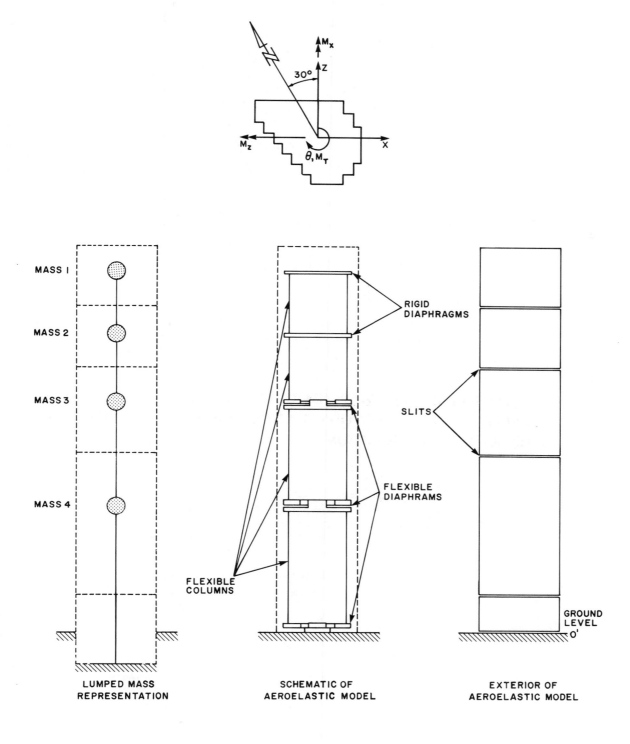

FIG. 3   MULTI-DEGREE OF FREEDOM MODEL OF TALL BUILDINGS

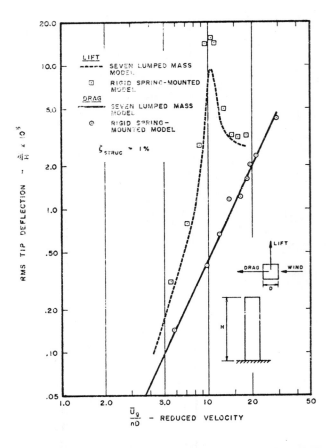

FIG. 4 AEROELASTIC SWAY RESPONSE OBTAINED WITH MULTI-DEGREE OF FREEDOM AND CONVENTIONAL STICK MODELS OF A TALL BUILDING OF SQUARE CROSS-SECTION

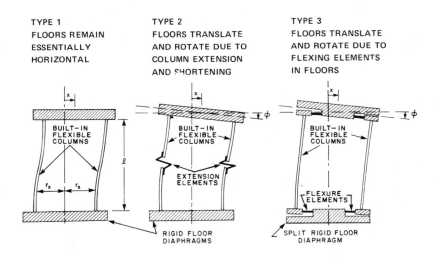

FIG. 5 TYPICAL STIFFNESS ELEMENTS USED IN MULTI-DEGREE OF FREEDOM MODELS

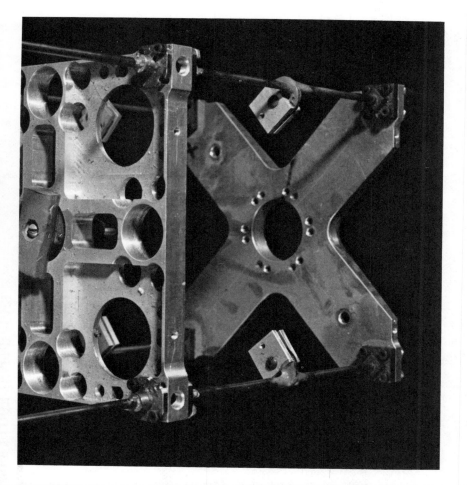

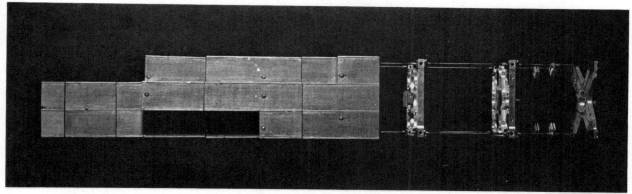

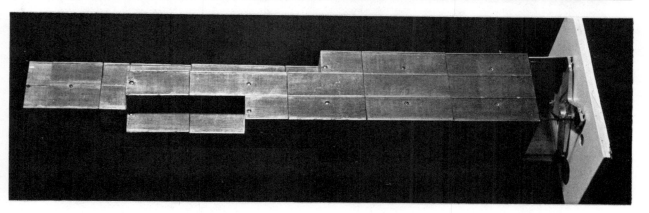

FIG. 6a   VIEW OF EQUIVALENT STRUCTURAL SYSTEM

OF 1:500 SCALE AEROELASTIC   MODEL

FIG. 6b PHOTOGRAPHS OF A MULTI-DEGREE OF FREEDOM AEROELASTIC MODEL OF
A TALL BUILDING USING FLEXIBLE FLOOR DIAPHRAGMS

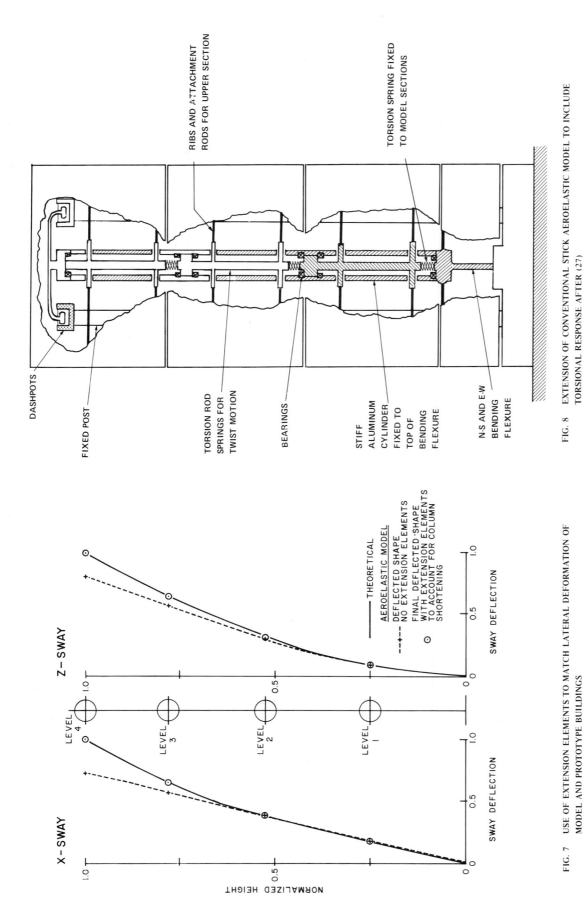

FIG. 7 USE OF EXTENSION ELEMENTS TO MATCH LATERAL DEFORMATION OF
MODEL AND PROTOTYPE BUILDINGS

FIG. 8 EXTENSION OF CONVENTIONAL STICK AEROELASTIC MODEL TO INCLUDE
TORSIONAL RESPONSE AFTER (27)

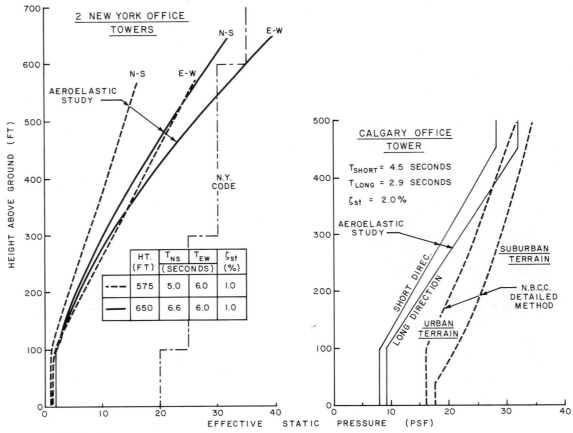

FIG. 9   EFFECTIVE STATICALLY ACTING WIND PRESSURE FOR THE STRUCTURAL DESIGN OF TALL BUILDINGS

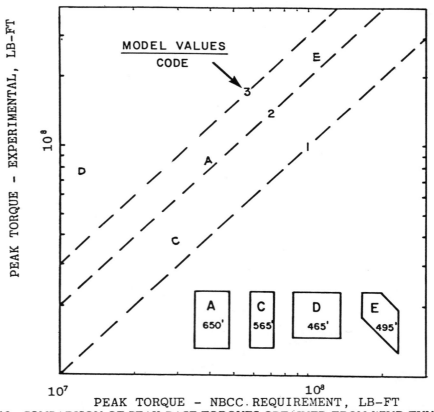

FIG. 10   COMPARISON OF PEAK BASE TORQUES OBTAINED FROM WIND TUNNEL
MODEL STUDIES AND THE NATIONAL BUILDING CODE OF CANADA FOR
4 TALL BUILDINGS

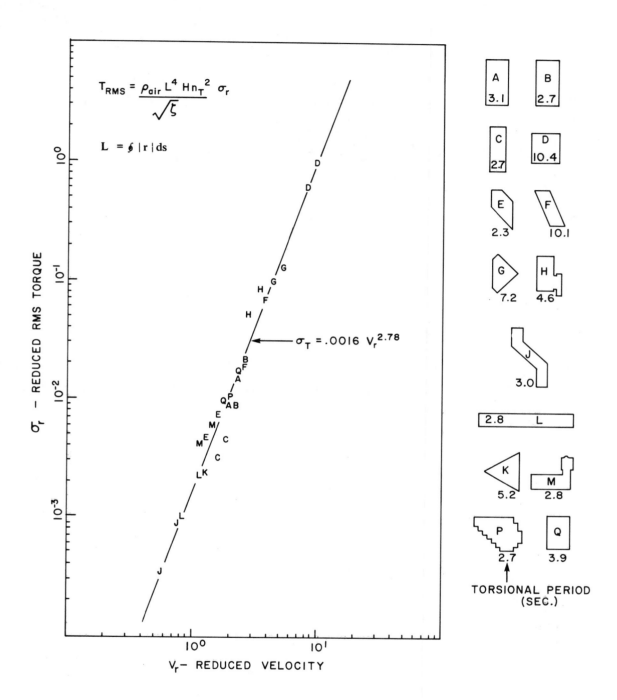

FIG. 11 DYNAMIC TORSIONAL RESPONSE OF TALL BUILDING ESTIMATED FROM
AEROELASTIC WIND TUNNEL MODEL STUDIES

406

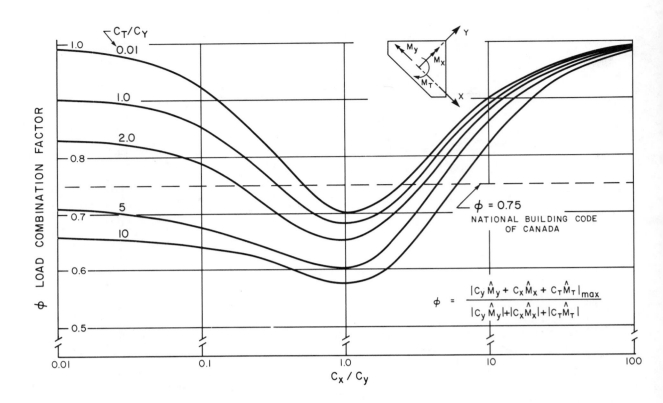

FIG. 12   JOINT ACTION OR COMBINATION FACTORS INDICATING THE SIMULTANEOUS EFFECT OF SWAY AND TORSIONAL WIND LOADS ON MEMBER STRESSES

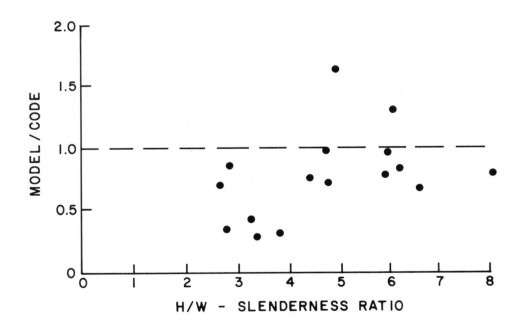

FIG. 13   COMPARISON OF PEAK SWAY ACCELERATIONS OBTAINED FROM WIND TUNNEL MODEL STUDIES WITH CODE VALUES FOR A NUMBER OF TALL BUILDINGS

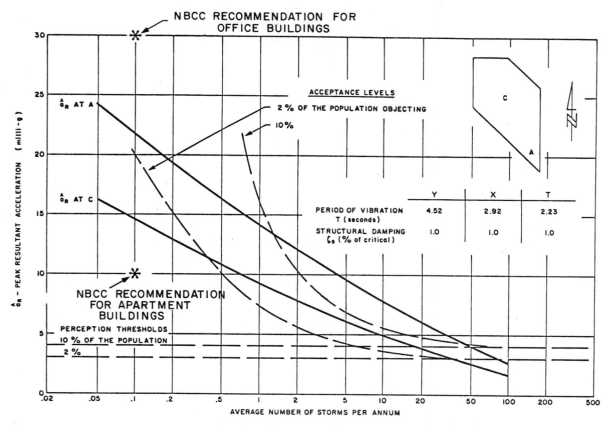

FIG. 14   PREDICTIONS OF WIND INDUCED HORIZONTAL ACCELERATIONS NEAR THE TOP OF A TALL BUILDING

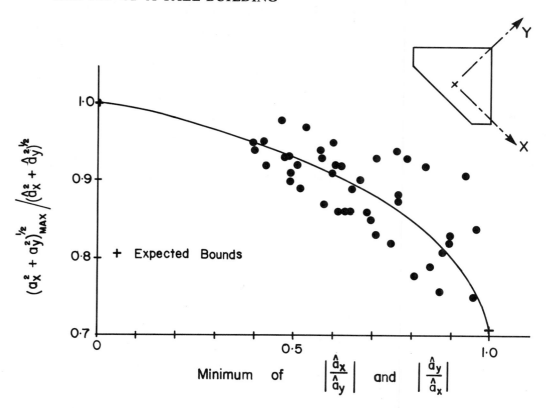

FIG. 15   MAXIMUM RESULTANT SWAY ACCELERATIONS FOR A TALL BUILDING

THE AEROELASTIC MODELLING OF CHIMNEYS AND TOWERS

B.J. Vickery, Boundary Layer Wind Tunnel Laboratory, Faculty of Engineering Science,
The University of Western Ontario, London, Ontario, Canada, N6A 5B9

1. INTRODUCTION

The general principles of aeroelastic modelling of structures subject to wind are well established (see, for example, Refs. 1 to 3) and will not be discussed here. Attention will be concentrated on factors which are peculiar to slender structures or at least more significant for slender tower-like structures and chimneys than for buildings. There are two important factors that play highly significant roles in the modelling process and/or the interpretation of the test results. These are:

    i)   the slenderness of the structure; and

    ii)   the common occurrence of circular cross-sections.

The slenderness results in flow conditions which are two-dimensional in character and, provided basic aerodynamic data are available, a theoretical solution becomes feasible. The availability of a theoretical approach permits the use of sectional tests either alone or in conjunction with a full aeroelastic model. The slenderness commonly introduces scaling problems because of the wide separation between the magnitude of the vertical and horizontal dimensions. This problem is most pronounced in modelling the individual members in a framed tower or the wall thicknesses in a tube-like structures.

The occurrence of circular cross-sections introduces severe problems because of the inability of most if not all boundary layer tunnels to achieve Reynolds Numbers comparable with full scale values. It is generally accepted that this inability is of comparatively minor consequence for sharp edged shapes, but it is well known that Reynolds Number effects for the circular cylinder are far from negligble.

2. LATTICE OR FRAMED TOWERS

The major difficulty in the aeroelastic modelling of framed towers arises because of the large separation in size between the overall height and the widths of individual members. For a typical length scale of 1:500 the required width of a 6″ x 6″ angle section is 0.012″ x 0.012″. Apart from the problem of actually constructing such a member, the local Reynolds Number is reduced to roughly 200 and the aerodynamic characteristics will not match those of the prototype and the variation of load with wind speed may differ significantly from the square-law relationship at prototype scale. In such circumstances aero-elastic testing is unlikely to lead to meaningfull results.

An approach which is more likely to lead to reliable predictions is a theoretical one with the necessary aerodynamic data being gathered from sectional model tests conducted at scale sufficient to reproduce the necessary detail and achieve the local Reynolds Numbers at which there will be little or no scale effect. Data on sharp edged shapes suggests that local Reynolds Numbers in excess of 1000 will be sufficient. Sectional tests conducted in turbulent flow can provide the data to define both the steady flow aerodynamic characteristics, the derivatives (for aerodynamic damping estimates) and the sectional aerodynamic admittance. The data gathered can then be incorporated into a theoretical model to yield predictions. The adequacy of the theoretical treatment can be tested by conducting wind tunnel tests on a properly scaled model which will, by virtue of the small scale, have different aerodynamic characteristics which, in turn, must be evaluated from a second set of sectional tests.

An approach along the lines outlined above was employed by Davenport and Vickery[4] in a study of a guyed tower (Figs. 1 & 2) and selected results have been published[5,6]. Similar techniques have been used in bridge studies[7] where the problems are very similar to those arising in studies of framed towers.

Lattice towers involving members of circular cross-section lead to additional problems since, even in a sectional test, the drag coefficient will commonly be far removed from the prototype value. In such circumstances the problem can be overcome by underscaling the model members such that;

$$\frac{(C_d \cdot D)_M}{(C_d \cdot D)_P} = \frac{L_M}{L_P} \qquad (1)$$

Provided that the response of individual members is of no concern this technique is adequate for frames of high porosity (low solidity). At moderate solidities the technique is questionable and is clearly inadequate at very high solidities. The technique of under-scaling circular members was employed in the sub-structure and in the flare boom of the model off-shore platform shown in Fig. 3. In this instance the members were underscaled but because of the moderate solidity, not quite to the extent dictated by Equation 1.

3. CHIMNEYS OF CIRCULAR CROSS–SECTION

The use of the wind tunnel to predict the behaviour of large chimneys is greatly hampered by the inability to achieve sufficiently large values of Reynolds Number. The model value of Re is normally in the subcritical regime while the prototype value is commonly in the

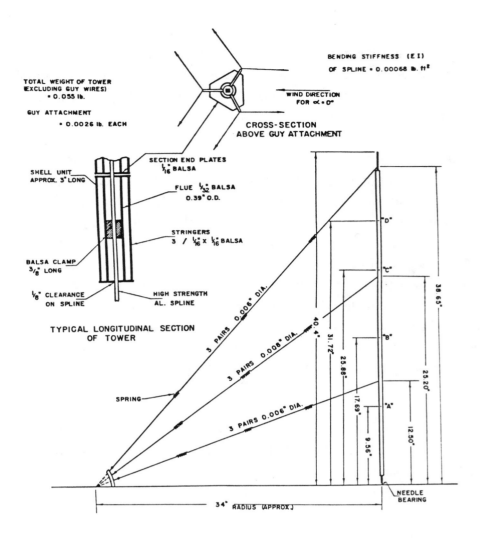

Fig. 1.   Major Dimensions of Aeroelastic Model of Guyed Stack

**Fig. 2   View of Aeroelastic Model of Stack Section Mounted in the
Test Rig**

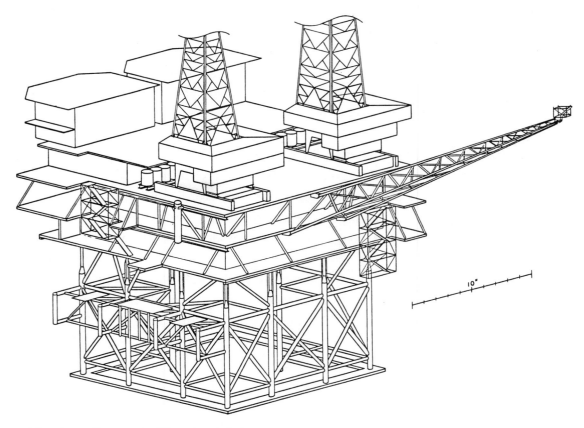

Fig. 3   Isometric Sketch of 1:120 Model

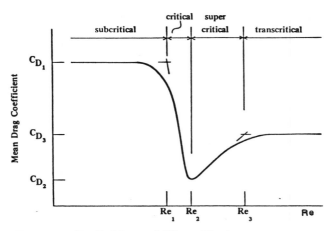

Fig. 4a   Definition of Flow Regimes

transcritical regime. The aerodynamic characteristics of the circular cylinder differ markedly between these regimes as is indicated by the variation of the drag coefficient (C_d), with Reynolds Number (R_e) and surface roughness (k/D) depicted in Figs. 4a, 4b and 4c. Other aerodynami parameters such as the coefficient of fluctuating left $(\widetilde{C}_L)$, the Strouhal Number (S) and the damping parameter (K_{a_O}) are also dependent on R_e and k/D but, unfortunately, the data base for these parameters is somewhat more limited.

In view of the obvious differences tht exist between behaviour in the sub-critical regime and behaviour in the trans-critical regime, results from aeroelastic model studies cannot be directly applied in the prediction of the prototype response. In order to apply the results of wind tunnel investigations to the prototype an adjustment must be made on the basis of some theoretical model or, alternately, a theoretical model employed directly.

A somewhat simplisitic adjustment of wind tunnel results was employed by Davenport and Vickery[8,6] in a study related to a tall reinforced concrete chimney. In this instance the measured along-wind response was multiplied by the ratio, $\dfrac{C_{d_P}}{C_{d_M}}$ where

$$C_{d_P} \quad = \quad \text{two dimensional drag coefficient at prototype scale;}$$

$$C_{d_M} \quad = \quad \text{two dimensional drag coefficient at model scale.}$$

A similar adjustment was made to the across-wind response but in this instance it was necessary to recognise differences in both $\widetilde{C}_L$ and S between model and prototype. The wind tunnel results were presented not in a conventional non-dimensional form but in a form incorporating the relevant values of C_d, $\widetilde{C}_L$ and S as shown in Fig. 5.

The above adjustment is not at all unreasonable and is based upon the assumptions that;

i)   changes (from two dimensional flow conditions) in C_d, $\widetilde{C}_L$ and S due to aspect ratio, turbulence and shear are, on a percentage basis, the same in model and prototype;

ii)   turbulence, shear and the structural properties are correctly scaled; and

iii) true aeroelastic effects are not significant or, at least, are the same in model and prototype. A theoretical model developed by Vickery & Basu[9, 10, 11, 12] permits an extension of the approach outlined above to include aeroelastic effects, i.e. the dependence of the aerodynamic forces on the motion of structure.

3.1   A Theoretical Framework for the Interpretation of Motions Induced By Vortex Shedding

The essential feature of the mathematical model referred to in the previous section is the representation of all motion induced phenomena (e.g. lock-in, improved correlation) by a

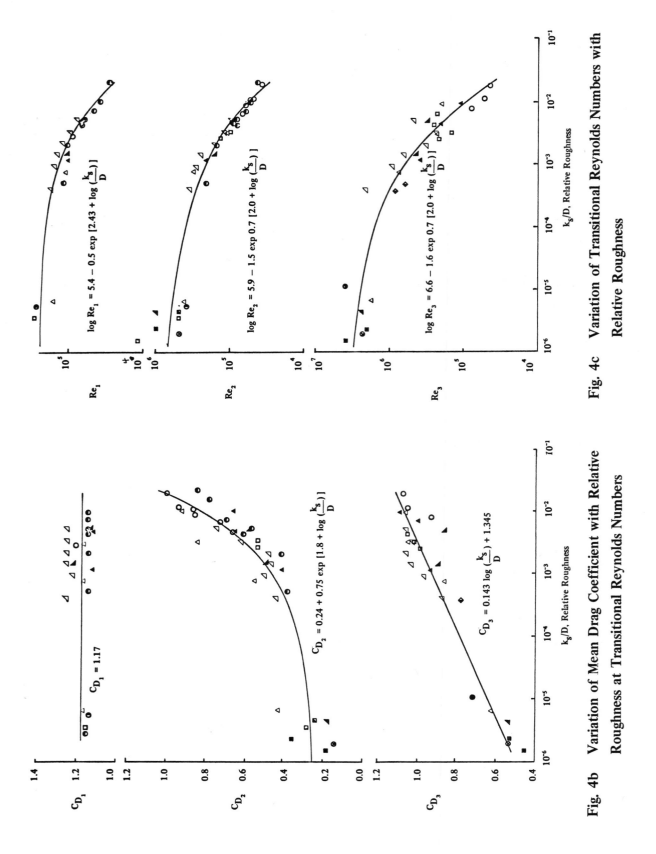

Fig. 4b Variation of Mean Drag Coefficient with Relative Roughness at Transitional Reynolds Numbers

Fig. 4c Variation of Transitional Reynolds Numbers with Relative Roughness

non-linear damping force. The aerodynamic damping, β_a, as a fraction of cirtical is given by;

$$- \beta_a = \frac{K_{a_O}}{M_e} [1 - (\frac{\sigma}{\sigma_L})^2] \quad ; \quad M_e = \frac{M_e}{\rho D^2}$$

where; K_{a_O} = a mass/damping parameter (at small amplitudes) which is dependent upon Reynolds Number, the turbulence level and the reduced velocity $V_r = \bar{V}/f_O D$.

σ_L = a limiting rms displacement.

σ = the rms displacement

The lift forces acting a stationary cylinder are characterized by a narrow-band random force given by;

$$f(t) = C_L(t) \cdot \frac{\rho \bar{V}^2}{2} \cdot D$$

where the spectrum of $C_L(t)$ is defined by the r.m.s. value, $\tilde{C}_L$, and a spectral form;

$$\frac{f S_{C_L}(f)}{\tilde{C}_L{}^2} = \frac{1}{B\sqrt{\pi}} \frac{f}{f_S} exp [- (\frac{f - f_S}{B f_S})^2]$$

For a cylinder in uniform two-dimensional flow this model leads to the following expression for the r.m.s. tip amplitude;

$$\frac{\sigma^1}{D} = \frac{C}{[K_S - K_{a_O}(1-(\frac{\sigma}{\sigma_L})^2)]^{\frac{1}{2}}} \qquad (2)$$

where $K_S = \frac{m_e \beta_s}{\rho D^2}$

$m_e = \frac{\int m \phi^2 (z) dz}{\int \phi^2 (z) dz}$

β_s = structural damping as a fraction of critical

$\phi(z)$ = mode shape

$$\sigma \; = \; \text{r.m.s. amplitude at } z = h; \text{ and}$$

$$C \; = \; \frac{\widetilde{C}_L}{8\pi^2 S^2} \; [\frac{\rho D^2}{m_e} \cdot \frac{\sqrt{\pi} \, L}{2\lambda}]^{\frac{1}{2}} \; \frac{\phi(h)}{[\frac{1}{h} \int \phi^2(h) \, dz]^{\frac{1}{2}}} \; \psi(B, V_r)$$

where S = Strouhal Number

λ = Aspect ratio (h/D)

L = Correlation length in diameters

$$\psi(B, V_r) \; = \; \frac{1}{\sqrt{B}} \; (V_r \, S)^{3/2} \; exp \, [- \frac{1}{2}(\frac{1 - \frac{1}{V_r S}}{B})^2 \,]$$

The success of an expression of the form of Equation (2) in representing the response of model chimneys is demonstrated in Figs. 6 & 7 which present results obtained by Wooton[13] from tests on a surface roughened model chimney in comparatively smooth flow at a Reynolds Number of 6×10^5 and results by Stevens[14] from a model of similar proportions but tested at a Reynolds Number of 1.1×10^4 in two turbulent boundary layers. Equation (2) provides a close fit to the experimental data for an amplitude range extending over more than two orders of magnitude. Similar agreement has also been reported by Vickery[15] for both an isolated model chimney in turbulent shear flow and a model subject to buffetting from similar models upstream.

The value of fitting experimental data by Equation (2) lies in the fact that the required coefficients $(K_{a_o}, \, C \text{ and } \sigma_L)$ are physically identifiable and data is available at both sub-critical and trans-critical Reynolds Numbers. The experimental results can be transferred to prototype scale if it assumed that;

$$C \; \propto \; \widetilde{C}_L/S^2 \qquad \text{for a cylinder in uniform two-dimensional flow}$$

$$K_{a_o} \; \propto \; K_{a_o} \qquad \text{for a cylinder in uniform two-dimensional flow}$$

$$\sigma_L \; \propto \; \sigma_L \qquad \text{for a cylinder in uniform two-dimensional flow}$$

That is, it is assumed that the modifying influence of turbulence, shear and model geometry on the behaviour in uniform two-dimensional flow is, proportionally, the same for model and prototype. There are clearly some difficulties in accepting this assumption without question but, in the absence of a body of data at prototype scale in real atmospheric flows, there appears to be no better alternative at the present time.

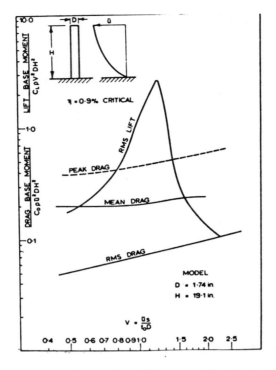

Fig. 5 Base Moments for 650' Stack

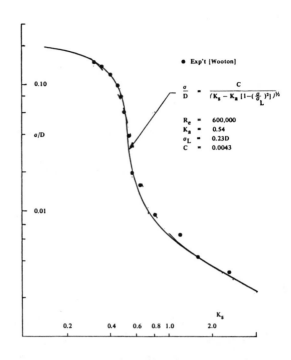

Fig. 6 Maximum Tip Response for a Rough Surfaced Model Stack (H/D = 11.5) in Smooth Uniform Flow

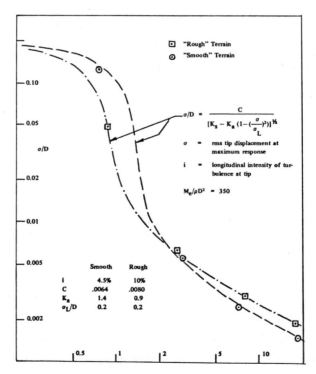

Fig. 7 Maximum Tip Response for a Model Stack (H/D = 11) in Turbulent Shear Flow

3.3 Buffetting in Grouped Chimneys

If the interpretation of wind tunnel studies of isolated chimneys presents some difficult problems then those associated with the interpretation of experimental results on grouped chimneys are almost insurmountable. Buffetting is dependent not only on the aerodynamics of the downstream cylinder but on the wake characteristics of the upstream structure and both are Reynolds Number dependent. An attempt to develop a theoretical model to deal with the buffetting problem has been made by Vickery[15] but further work is required and, more significantly, the data needed to apply the theoretical approach is meagre in the sub-critical regime and non-existent in the trans-critical regime. On a more positive note, the progress that has been made suggests very strongly that tests conducted at sub-critical values of Re will provide an upper bound to the behaviour of the prototype. A downward adjustment of wind tunnel results is in order but the degree of this adjustment cannot be quantitatively assessed and, at this stage, must be based upon "engineering judgement".

The significance of buffetting from an upstream structure is demonstrated by the results shown in Fig. 8. Fig. 8 shows the magnitude of $\widetilde{C}_L$ for a cylinder $4\frac{1}{2}''$ in diameter positioned in the wake of a square bar. The bar width, B, was varied with widths of $\frac{1}{2}''$, $1''$, $2\frac{1}{2}''$, $4''$ and $8''$ being employed and observations of $\widetilde{C}_L$ made with the bar positioned upstream at distances of 10B, 20B, 40B, 80B and 160B. At spacings of 10B and 20B the strong shedding from the bar induced substantial lift forces on the cylinder at the shedding frequency of the bar and the spectrum of the cylinder lift force indicated two peaks, one at the bar shedding frequency and one corresponding to shedding from the cylinder. The values of $\widetilde{C}_L$ given in Fig. 8 refer only to that part of the lift force near the shedding frequency of the cylinder itself. At spacings of 40B and greater the incoming flow did not have a strong spectral peak and the lift spectrum of the cylinder exhibited a single peak at the cylinder shedding frequency. The values of $\widetilde{C}_L$ in Fig. 8 are defined as;

$$\widetilde{C}_L = \frac{\widetilde{f}_L}{\frac{1}{2}\rho V_e^2\, D} \qquad \text{where} \qquad V_e = \frac{f_s\, D}{S_s}$$

$$S_s = \text{smooth flow value of } S$$

In the case of the $4''$ bar at 10B the shedding frequency of the cylinder almost coincides with the frequency of shedding from the bar and, as might be expected, the lift coefficient is greatly enhanced and reaches a value equal to more than twice that for the cylinder in a smooth stream. A more puzzling feature is that the presence of the $\frac{1}{2}''$ bar positioned 80B upstream also increases the lift coefficient markedly, in this case by a factor of 1.8 . This is clearly not a buffetting phenomenum but is probably due to interaction between the comparatively small scale turbulence from the bar and the separating shear layers of the cylinder. The above results were obtained at a Reynolds Number of 3×10^4 but even greater effects due to turbulence have been noted[16] at lower values of R_e. An increase in response with increasing turbulence may

418

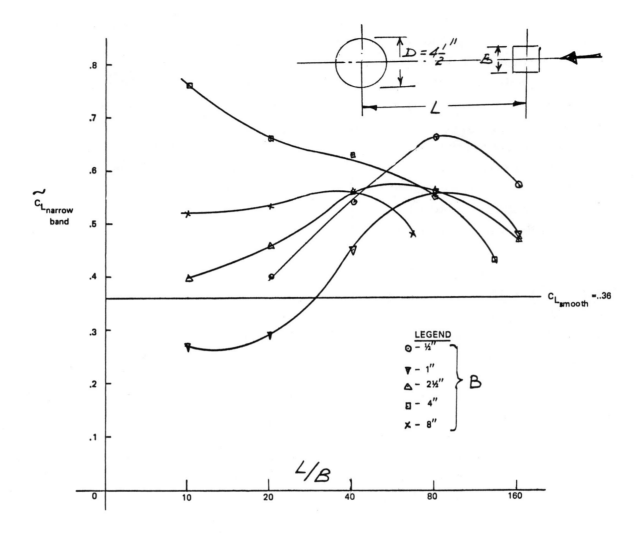

Fig. 8: Values of $\widetilde{C}_L$ for a Circular Cylinder in the Wake of a Square

also be noted in Fig. 7. The presence of turbulence certainly broadens the lift spectrum and, in this sense at least, tends to reduce the response but this reduction is masked by the substantial increase in $\tilde{C}_L$. Whether this phenomena is present at trans-critical values of R_e is questionable and this introduces another source of uncertainty into the interpretation of wind tunnel tests.

Measurements by Tunstall on a roughened cylinder ($k_s/D \approx 1/1200$) at a Reynolds Number of 7.1×10^5, which is almost in the transcritical range (see Fig. 4), showed an increase in $\tilde{C}_L$ from 0.29 to 0.38 with the addition of small scale turbulence with an intensity of 6%. It would appear that both turbulence and roughness are highly significant parameters in the transcritical range, the effect of roughness on C_D is well documented but this is not the case for $\tilde{C}_L$. The very limited data which are available suggest that for roughnesses typical of reinforced concrete chimneys ($k/D = 10^{-4}$ to 10^{-5}), $\tilde{C}_L$ is about 0.10 to 0.20 but for $k/D \approx 10^{-3}$, the work of Tunstall and others suggests values from 0.20 to 0.30 with further increases in the presence of small scale turbulence.

4.   THE SIGNIFICANCE OF DETAIL

The difficulties of modelling detail and the problems associated with very low local Reynolds Numbers has been referred to in the discussion of framed towers. An example of the significance of detail is now presented. Possible configurations of the partially open cross-section of the stem of the tower shown in Fig. 9 are presented in Fig. 10. Each of these cross-sections exhibit markedly different aerodynamic characteristics as is demonstrated in Figs. 11 to 15. Fig. 11 shows the variation in the drag coefficient with wind direction for each configuration. Figs. 12 & 13 show the lift coefficient, $\tilde{C}_L$, and Strouhal Number, S. Fig. 14 shows the maximum values of the aerodynamic mass/damping parameter,

$$K_a = -\frac{\beta_a m_e}{\rho D^2}$$

[i.e. ; positive values of K_a indicate negative aerodynamic damping]. Fig. 15 shows a modified Strouhal Number, S^*, defined such that;

$$S^* = \frac{f_o D}{\bar{V}_c}$$

where $\bar{V}_c$ is the speed at which K_a was observed to reach a maximum.

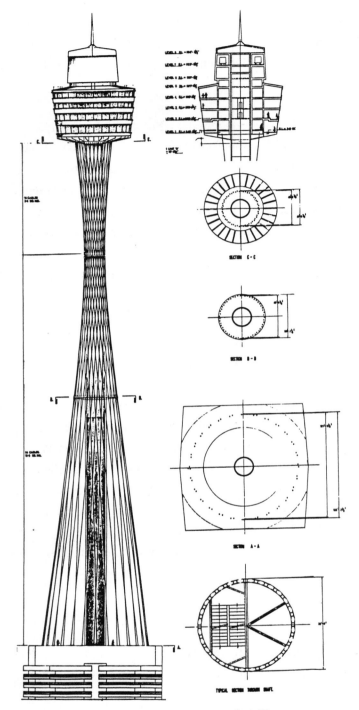

Fig. 9   Proposed Centrepoint Tower

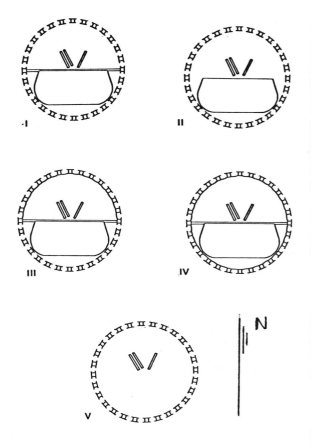

Fig. 10   Test Configurations

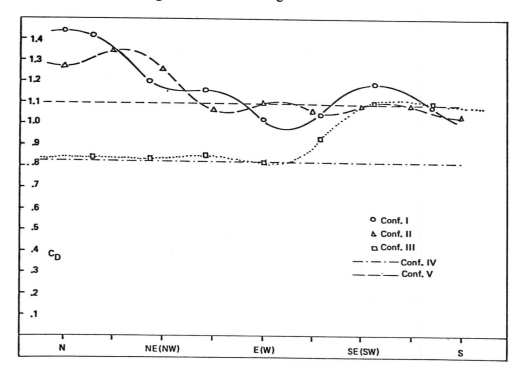

Fig. 11 C_D vs Wind Direction

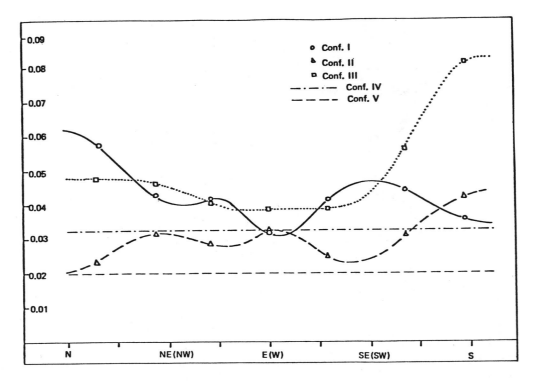

Fig. 12 $\widetilde{C}_L$ vs Wind Direction

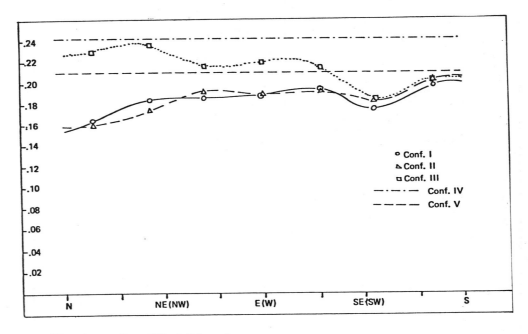

Fig. 13 S vs Wind Direction

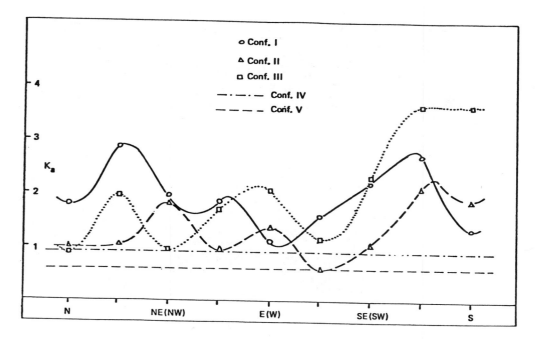

Fig. 14 $K_{a_{max}}$ **vs Wind Direction** ($\frac{a}{d} \approx 2 - 3\%$)

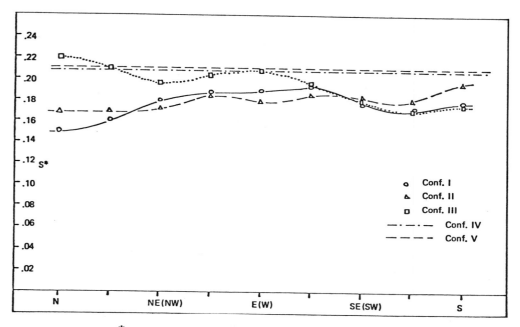

Fig. 15 S^* **vs Wind Direction**

424

Fig. 16   Sectional Model in Wind Tunnel

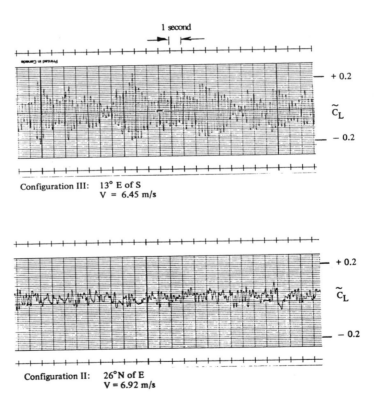

Fig. 17   Sample Traces of Lift Pressures

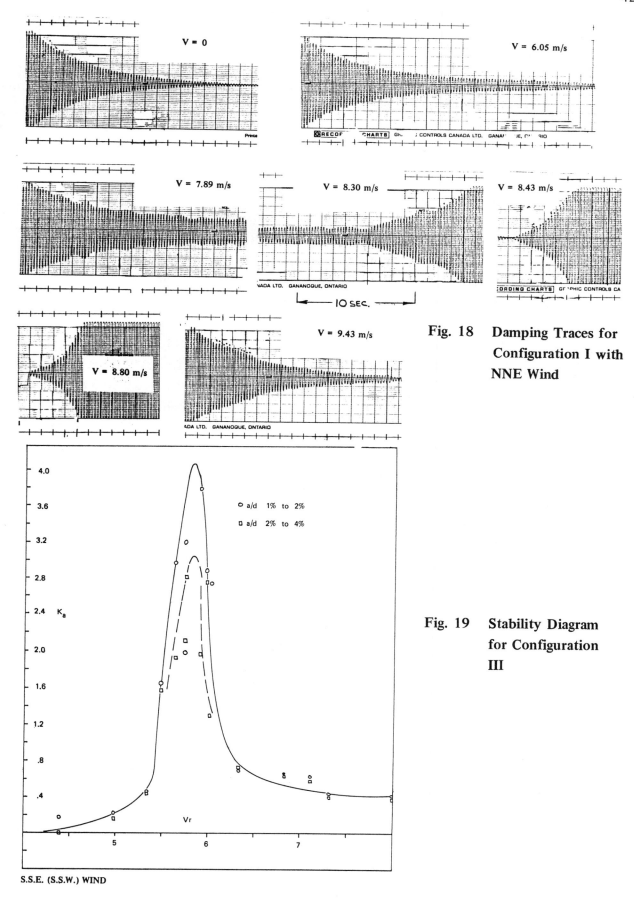

Fig. 18 Damping Traces for Configuration I with NNE Wind

Fig. 19 Stability Diagram for Configuration III

S.S.E. (S.S.W.) WIND

All the results presented in Figs. 11 to 15 were derived from comparatively the large scale (1:20) sectional model shown in Fig. 16. Sample traces of the lift force are shown in Fig. 17 and sample damping traces from which values of K_a were evaluated are shown in Fig. 18; the variation of K_a with velocity is shown in Fig. 19. None of the results presented have been corrected for wall interference effects; these corrections are substantial since the blockage ratio was about 14%

The question of Reynolds Number is still significant since the sectional tests were conducted at value of R_e of about 3×10^5 as compared to the full-scale value of roughly 10^7. The sharp edges of the individual members comprising the model and the porosity would both tend to reduce Reynolds Number effects but some doubts remain. Tests by Alridge Piper & Hunt[18] on a perforated cylinder showed only a 2.5% variation in C_d in the range $4.3 \times 10^4 < R_e < 2.3 \times 10^5$, while tests by Buresti[19] on roughened cylinders indicate independence of R_e for $R_e < 2 \times 10^5$ and a roughness $k/D > 3 \times 10^{-3}$.

The significant point arising from the results shown in Figs. 11 to 15 is that internal detail produces changes in C_D from 0.8 to 1.4, changes in $\widetilde{C}_L$ from 0.02 to 0.08, changes in S from 0.16 to 0.24 and, most significantly, changes in K_a from 0.6 to 3.8. The details producing these very marked changes in behaviour could not possibly be incorporated into a full aeroelastic model which, in order to achieve correct scaling in relation to atmospheric turbulence, would have to be constructed at a scale of, typically, 1:500. In these circumstances the predictions of full scale behaviour must be arrived at using a theoretical framework to adjust the test results obtained in small-scale aeroelastic tests.

5. CONCLUSIONS

The problems associated with the aeroelastic modelling of towers and chimneys are severe and a direct transfer (using the usual scale relationships) of data from model to prototype is rarely possible. Only in the case of sharp edged structures, such as the CN Tower, is a direct transfer of results likely to lead to adequate predictions. In the case of framed towers and, more significantly, towers and chimneys of circular cross-section, it is necessary to interpret the results in the light of a theoretical framework and to use data gathered from full scale observations and/or sectional model tests at a comparatively large scale.

6. ACKNOWLEDGEMENTS

Funding for studies referred to herein was provided by the Australian Mutual Provident Society, The Natural Sciences and Engineering Research Council of Canada, Ontario Hydro, and the DuPont Company. Collected data showing the dependence of the drag coefficient of the circular cylinder on R_e and k/D (Fig. 4) was provided by Mr. R. Basu, The University of Western Ontario.

REFERENCES

1. W.H. Melbourne, Modelling of structures to measure wind effects, Proceedings Structural Models Conference, School of Arch. Sc., University of Sydney, 1972.

2. B.J. Vickery, On the aeroelastic modelling of structures in wind, Proceedings Structural Models Conference, School of Arch. Sc., University of Sydney, 1972.

3. R.E. Whitbread, Model simulation of wind effects on structures, Proceedings 1st Conference on Wind Effects on Bldgs. & Structures, Teddington, H.M.S.O., London, 1963.

4. A.G. Davenport and B.J. Vickery, The response of the Savannah R. Guyed Stack under wind and earthquake action, Eng. Science Research Report, BLWT−5−68, The University of Western Ontario, London, Ontario, Canada, 1968.

5. B.J. Vickery and A.G. Davenport, A comparison of theoretical and experimental determination of the response of elastic structures in turbulent flow, Proceedings 2nd Int. Conference on Wind Effects on Bldgs. & Structures, Ottawa, 1967, pp. 705−738.

6. B.J. Vickery, Wind induced vibrations of towers, stacks and masts, Proceedings 3rd Int. Conference on Wind Effects on Bldgs. & Structures, Tokyo, 1971, pp. 657−665.

7. A.G. Davenport et al, A study of wind action on a suspension bridge during erection and on completion, Faculty of Eng. Science Research Report, BLWT−3−1969, The University of Western Ontario, London, Ontario, Canada, 1969.

8. B.J. Vickery, An investigation of the response to wind loads of a proposed multi-flue for the Nanticoke power station, Preliminary Report, Faculty of Eng. Science, The University of Western Ontario, London, Ontario, Canada, 1969.

9. B.J. Vickery, A model for the prediction of the response of chimneys to vortex shedding, Proc. 3rd Int. Symposium on Design of Industrial Chimneys, Munich, 1978, pp. 157−162.

10. R.I. Basu and B.J. Vickery, The across-wind response of tall slender structures of circular cross-section to atmospheric turbulence, Part I; Mathematical Model, Eng. Science Research Report, BLWT−1−1981, The University of Western Ontario, London, Ontario, Canada, 1981.

11. B.J. Vickery and R.I. Basu, Across-wind vibrations of structures of circular cross-section, Part I; Development of a mathematical model for two-dimensional conditions, submitted for publication, J. Wind Engg & Industrial Aero, 1982.

12. R.I. Basu and B.J. Vickery, Across-wind vibration of structures of circular cross-section, Part 2; Development of a mathematical model for full-scale application, submitted for publication, J. Wind Engg & Industrial Aero, 1982.

13. L.R. Wooton, The oscillation of large circular stacks in wind, Proc. Inst. Civil Eng., Vol. 43, 1968, pp. 573−598.

14. K.V. Stevens, Response of tall chimneys to turbulent wind, M.E.Sc. Thesis, Faculty of Grad Studies, The University of Western Ontario, London, Ontario, Canada, Sept. 1971.

15. B.J. Vickery, Across-wind buffetting in a group of four-in-line model chimneys, J. Wind Eng. & Indust. Aero., July 1981, Vol. 1, No. 8, pp. 177–194.

16. B.J. Vickery, Wind effects on buildings and structures — critical unresolved problems, Proc. IAHR/IUTAM Symp. on Practical Experiences with Flow-Induced Vibrations, Ed. E. Naudascher & D. Rockwell, Springer -Verlag, Heidelberg, 1980, pp. 823–828.

17. M.J. Tunstall, Fluctuating pressures on circular cylinders in uniform and turbulent flows, Central Elec. Res. Labs., Lab. Note RD/L/N, 45/70, March 1970.

18. T. Alridge, B. Piper and J.C.R. Hunt, The drag coefficient of finite-aspect-ratio perforated cylinders, J.I.A., Vol. 3, 1978, pp. 251–257.

19. G. Buresti, The effect of surface roughness of the flow regime around circular cylinders, Proc. 4th Colloquium on Industrial Aerodynamics, Aachen, June 1980, pp. 13–28.

AEROELASTIC MODELING OF MEMBRANE STRUCTURES

Richard J. Kind, Department of Mechanical and Aeronautical Engineering,
Carleton University, Ottawa, Canada.

SUMMARY

Similarity requirements for the title problem are reviewed. The
complete requirements are quite extensive and virtually impossible to fully
implement in practice. However many simplifications are usually possible
without substantial loss of accuracy. Fronde scaling, elastic scaling and
folding-stiffness scaling can usually be relaxed. Even mass scaling can
sometimes be relaxed somewhat. A new physicial model is proposed for
pneumatic effects. This gives pneumatic damping but no pneumatic stiffness.
The associated modeling requirements are outlined.

LIST OF SYMBOLS

b a reference length dimension

C_F, C_L local inverse-slope of fan and leakage pressure-vs.-flow characteristics, respectively (see eq.(2) and Fig. 6)

D aerodynamic drag force per unit length on cables

g gravitational acceleration

h fold height, see Fig. 2

K_C elastic stiffness of cable (tensile force/strain)

K_M elastic stiffness of membrane (tensile force per unit width/strain)

M_M membrane mass per unit area

M_C cable mass per unit length

P gage pressure inside building enclosure

ΔP change in P

Q_F instantaneous volume flow rate from fans into building enclosure

Q_L instantaneous leakage volume flow rate from building enclosure

T tension

U wind speed

$\dot{V}$ instantaneous rate of increase of enclosure volume due to membrane
deflection

ρ air density

ζ mechanical damping ratio

Subscripts

C   cables

M   membrane

o   static equilibrium condition (Fig. 6)

1.0   INTRODUCTION

Membrane structures have long been in use, in the form of tents.
In recent years a number of large engineered buildings have made use
of membrane structures, usually for the roof of the building.
Membrane structures are presumably attractive to architects because they
allow great scope for innovative design, can be aesthetically pleasing
and can sometimes be cost effective. Because these structures are
usually unique, costly and intended to accomodate large numbers of people,
their safe and effective behaviour in wind must be established at the
design stage. At present the only reliable and effective method for
assessing the dynamic behaviour of such structures in wind is by
aeroelastic model tests in wind tunnels. This paper reviews the
requirements for aeroelastic modeling of membrane structures.

As in other modeling situations, the mean flow and turbulence
structure of the natural wind at the building site should be correctly
simulated and the exterior of the model should be geometrically similar
to that of the prototype, with the length scale ratio corresponding to
that of the wind simulation. Wind simulation techniques have been
discussed in Session II of this conference. As in other model tests,
the prototype Reynolds number normally cannot be matched and this
usually does not constitute a serious problem. However membrane
structures often have large gently curved surfaces and flow separation
lines are not always well fixed by sharp corners. Membrane structures
can therefore be expected to be somewhat prone to Reynolds-number
effects and this possibility should be carefully considered when planning
a test program.

Membrane structures can be sub-divided into two categories: hanging
structures and air-supported structures. The remaining requirements for
aeroelastic modeling will be considered separately for each of these
categories.

2.   HANGING MEMBRANE STRUCTURES

2.1   Preliminary Remarks

Hanging membrane structures are ones in which the membrane is
hung from a number of attachment points on the rigid portion of the
building; the membrane may be attached to these points either directly
or by means of cables. Wind loads and the dead weight of the system are
reacted by tension in the cables and in the membrane itself. The cable/
membrane system may be pre-tensioned; that is the system may be drawn

taut so that internal tensions are much larger than those required merely to support the weight of the system. This of course tends to reduce displacement response to wind loads and other perturbations.

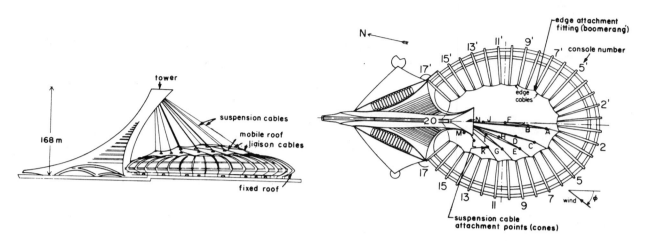

FIGURE 1: MONTREAL OLYMPIC STADIUM

Figure 1 shows the Montreal Olympic Stadium with a proposed pre-tensioned hanging membrane roof structure. The roof is retractable. This building will serve as an example in the ensuing discussion. It was the subject of two distinct aeroelastic model studies. In the first study [1], dynamic deflections of the membrane and dynamic loads at the edge attachment points were measured with the roof in its fully-deployed pre-tensioned condition. The second study [2] examined the dynamic behaviour of the roof during retraction or deployment in wind. Reference [3] describes another comprehensive aeroelastic study of a pre-tensioned hanging membrane structure.

2.2   Similarity Analysis

Aerodynamic lift forces on cables are assumed to be negligible. Pneumatic effects are also neglected in this Section on the assumption that the volume enclosed by the building is relatively leaky so that forces due to pneumatic effects are small compared with aerodynamic and other forces. Pneumatic effects are discussed in Section 3 and that discussion can be applied to hanging membrane structures if warranted by low building leakiness. For similar reasons, acoustic damping is neglected: it will be small for highly leaky buildings and it will be dominated by pneumatic damping, as discussed in Section 3, for relatively airtight buildings.

With the above assumptions, a similarity analysis indicates that the similarity requirements mentioned in the Introduction should be supplemented by matching of the following non-dimensional parameters for the model and prototype:

(a) $\dfrac{U^2}{bg}$ Froude Number - ratio of inertia or aerodynamic forces to gravity forces

432

(b) $\dfrac{M_M}{\rho b}$ ratio of membrane mass to air mass

(c) $\dfrac{M_c}{b M_M}$ ratio of cable mass to membrane mass

(d) $\dfrac{K_M}{\rho U^2 b}$ ratio of elastic tensile stiffness of membrane to inertia or aerodynamic forces

(e) $\dfrac{K_c}{\rho U^2 b^2}$ ratio of elastic tensile stiffness of cables to inertia or aerodynamic forces

(f) $\dfrac{D}{\rho U^2 b}$ drag coefficient of cables

(g) $\dfrac{h_M}{b}$ ratio of membrane fold height (see Fig. 2) to reference length

(h) $\dfrac{h_c}{b}$ ratio of cable fold height to reference length

(i) ζ mechanical damping ratio.

All symbols are defined in the List of Symbols at the beginning of this paper. The fold height h of parameters (g) and (h), above, is determined by a simple folding test as illustrated in Fig. 2; as shown in Ref. [2], it is a measure of the bending or folding stiffness of the membrane or cable material and matching of the non-dimensional parameters (g) and (h) ensures that the ratios of bending or folding forces to gravity forces are matched.

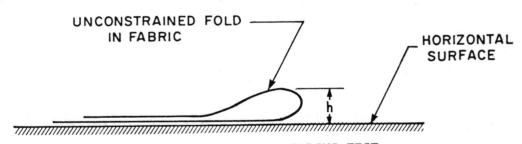

FIGURE 2: MEMBRANE FOLDING TEST

2.3   Difficulties and Simplifications

In practice it is almost never possible to match all the similarity parameters listed in the preceding Section. For example, the parameters (b), (d) and (g) all relate to membrane properties and it is seldom possible to find a material for the model membrane which allows all three parameters to be matched simultaneously. It must be kept in mind that values of wind speed U and reference length b in the model test are constrained by wind tunnel size and perhaps by Froude-number scaling requirements or other considerations.

Fortunately, in any particular case, some physical effects are normally absent or relatively unimportant. It is then possible to achieve dynamic similarity by matching only some of the above parameters or perhaps some combinations of them. Some commonly occurring simpliciations are discussed in the following paragraphs.

Usually membrane and cable radii of curvature remain very large compared with their fold heights h. Bending or folding stiffness effects are then negligibly small and the parameters (h_M/b) and (h_c/b) need not be matched. In the tests of Ref. [2] this was not always the case but it was still found impossible to match (h_M/b). The prototype membrane fabric had a fold height h = 15 cm (6 in.); even though the model-to-prototype length scale ratio was quite large, at 1:50, this required a model fold height h = 3 mm (0.1 in.) and no material, either woven or monofilm, could be found with such a small value, even ignoring the constraint on membrane mass imposed by parameter (b) of Section 2.2. This problem is discussed in more detail in Ref. [2].

In the case of pre-tensioned membrane structures, the tensile forces in the structure are typically much larger than those which arise only from the weight of the structure. In such cases static and dynamic motions will tend to remain negligible until aerodynamic or inertia forces become much larger than gravity forces. In other words, gravity forces are negligible and Froude number scaling is unnecessary. This is an important simplification since it eliminates the need to use very low wind speeds U as would be required to match the parameter (U^2/bg) in model-scale tests. The ability to use reasonably high wind speeds improves measurement accuracy and eases the difficulties of simultaneously matching the mass and stiffness parameters (b) to (e) of Section 2.2. Unfortunately if a hanging membrane structure is not pre-tensioned, it is essential to match (U^2/bg); a large wind tunnel then becomes extremely desirable, to avoid unworkably low values of U. The tests of Ref. [2] provide an example of such a situation; even with a 9 m X 9 m tunnel cross section, wind speeds were below 2.4 m/s (8 ft/sec).

The parameters (d) and (e) of Section 2.2 represent the elastic tensile stiffness properties of the membrane and cable materials. If these parameters are matched for the model and prototype and if all cable and other length dimensions are correctly scaled, the tension forces in the model will in theory be correctly scaled (tension forces proportional to $\rho U^2 b^2$; in practice the tensions in the model cables etc. would be adjusted to the correct values. The model would then have the correct tensioned shape. It can however be quite difficult to achieve accurate elastic scaling (matching of parameters (d) and (e)) while simultaneously achieving correct mass scaling (matching of parameters (b) and (c) of Section 2.2). For example, in the tests of Ref. [1], the model membrane consisted of a hand-woven mesh made of 100 denier Kevlar yarn with a mesh size of 0.5 cm X 0.5 cm, covered with 0.005 mm (0.0002 in.) polyethylene film.

Fortunately it is becoming clear from experience and analysis that it is often unnecessary to accurately scale both elastic and mass properties. Three different arguments can be adopted, as circumstances permit, in favor of relaxing these scaling requirements.

434

The first possibility is that loads are sufficiently small that elastic strains are negligible so that the cables and membrane can be considered as inextensible. Elastic scaling is then unnecessary. This, for example, was the case in the tests of Ref. [2]. The membrane and cables in this roof system are sized to carry the loads prevailing when the system is fully deployed and pre-tensioned. During retraction or deployment of the roof, loads are much smaller and elastic strains are negligible.

Another possibility is that the added or virtual mass of air moving with the vibrating body is much larger than the mass of the body itself. In that case mass scaling can be relaxed somewhat (for example the parameter $(M_M/\rho b)$ can be made somewhat larger than allowed by strict scaling) and correct elastic scaling then becomes easier. Irwin et al. [1] give theoretical and experimental evidence that this is permissible for membrane structures similar to that shown in Fig. 1. This simpliciation is clearly not permissible when the dynamics of a relatively compact system are of interest or when static or nearly static deflections of untensioned systems are of interest. For example, mass scaling could not be significantly relaxed in the tests of Ref. [2], because when partially retracted, the roof system of Fig. 1 becomes more compact and is also susceptible to large nearly-static deflections.

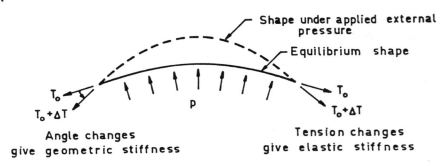

FIGURE 3:   REPRESENTATION OF MECHANISMS THAT GIVE RISE TO GEOMETRIC AND ELASTIC STIFFNESSES IN AIR-SUPPORTED OR PRE-TENSIONED MEMBRANE STRUCTURES.

The most general argument in favor of neglecting elastic scaling is that of Tryggvason [4]. He shows that for pre-tensioned or air-supported membrane structures, two restoring forces arise when a portion of membrane or cable is displaced from its static equilibrium position. The first is due to the change in geometry under the applied load and the second due to increased tension arising from strain of the material. This is illustrated in Fig. 3, taken from Ref. [4]. Tryggvason [4] shows that the geometric stiffnesses are much more significant than the elastic stiffnesses for three particular cases [3,5,6[and argues that this is typically true for air-supported and pre-tensioned membrane structures. He concludes that large errors in elastic scaling can be tolerated without incurring significant errors in the simulated response.

It must be noted that tensions in the model must be correctly scaled (tensions proportional to $\rho U^2 b^2$) even when accurate elastic scaling (matching of the parameters $K_M/\rho U^2 b$ and $Kc/\rho U^2 b^2$) is unnecessary. Furthermore, the model should have the same static shape as the tensioned prototype.

Cable drag is usually relatively unimportant and the drag coefficient $D/\rho U^2 b$ need not be accurately matched. Mechanical damping of the model is usually assumed to be similar to the prototype, and aerodynamic damping is usually dominant anyhow. Therefore no particular pains need normally be taken to match the mechanical damping parameter ζ.

To summarize, in many, but not all, aeroelastic model tests of hanging membrane structures it is sufficient to simulate the wind, use a geometrically scaled model with correctly scaled tensions (tensions proportioned to $\rho U^2 b^2$) and match the parameters $(M_M/\rho b)$ and (M_C/bM_M). Deflections, forces and frequencies are proportional to b, $\rho U^2 b^2$ and U/b respectively; prototype values can be determined from measurements on the model with the help of these relations.

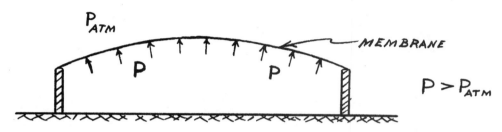

FIGURE 4 - SCHEMATIC DIAGRAM OF AIR-SUPPORTED MEMBRANE STRUCTURE

3.0   AIR-SUPPORTED MEMBRANE STRUCTURES

3.1   Preliminary Remarks

In air-supported membrane structures, the membrane is supported and stiffened by excess air pressure in the building enclosure. The excess pressure is typically serveral cm. of water gage and is several times larger than the weight per unit area of the membrane structure. The structure is, in effect, pre-tensioned by the air pressure. Figure 4 is a schematic diagram of such a structure.

3.2   Pneumatic Effects

The mechanics of these structures are much the same as for hanging membrane structures, except that pneumatic effects are obviously important and must be considered. In previous studies [4,6] the enclosed volume has been assumed to be reasonably airtight when considering dynamic volume fluctuations due to membrane motion. The ideal gas law is then used to relate pressure and volume fluctuations. However, at least in buildings occupied by people, the air is continually being changed (about 5 times per hour [7]) and the assumption of airtightness is open to challenge, even for dynamic volume fluctuations. An alternate model for building pneumatics is proposed in Fig. 5. The leakiness of the building is represented by an opening or orifice through which the volume flow at any instant is Q_L; fans supply a volume flow rate Q_F to the enclosure at any instant. The volume flow rates Q_F and Q_L are related to the gage pressure, P, in the building by the fan and building-leakage pressure/flow characteristics, as shown in Fig. 5.Quasi-steady behaviour can be assumed for the fans and for the leakage flow because full-scale membrane

436

frequencies are typically of order 1 Hz while turbomachinery exhibits
quasi-steady behaviour up to frequencies of at least 10Hz [8] and
orifice flow should remain quasi-steady up to similar frequencies.
The static pressure, P, can be assumed to be uniform over the building
enclosure at any instant because the time for propagation of an acoustic
wave across the enclosure is typically several times less than the period
of any significant membrane vibrations.

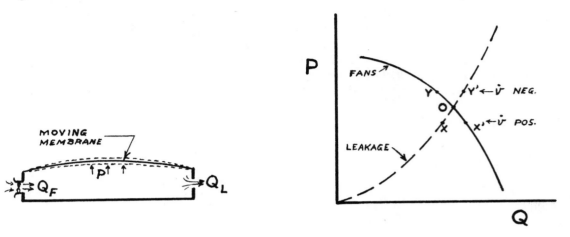

FIGURE 5 - PNEUMATIC MODEL OF BUILDING
WITH AIR-SUPPORTED MEMBRANE
STRUCTURES

FIGURE 6 - FAN AND BUILDING-
LEAKAGE PRESSURE/
FLOW CHARACTERISTICS

At static equilibrium conditions (no membrane motion), $Q_F = Q_L$ and
the system operates at point 0 on Fig. 6. If the membrane moves in a
volume-displacing mode such that the enclosure volume is increasing at
the rate $\dot{v}$, mass conservation requires that

$$Q_{Fo} + \left(\frac{dQ_F}{dP}\right)_o \Delta P = Q_{Lo} + \left(\frac{dQ_L}{dP}\right)_o \Delta P + \dot{v} \tag{1}$$

Equation (1) assumes constant density and local linearity of the
characteristic curves of Fig. 6. Since $Q_{Fo} = Q_{Lo}$, eq. (1) can be re-written
as

$$\frac{\Delta P}{\dot{v}} = -\frac{1}{C_L + C_F} \tag{2}$$

where

$$C_L = \left(\frac{dQ_L}{dP}\right)_o$$

$$C_F = -\left(\frac{dQ_F}{dP}\right)_o$$

C_F must be positive for stable operation of the fans [8]; therefore both
C_L and C_F are positive constants. Eq. (2) implies that pneumatic effects
provide damping but not stiffness; this differs from the conclusions
reached if an airtight building enclosure is assumed. Points x, x' and
y, y' on Fig. 6 illustrate the system operating condition at instants

when $\dot{v}$ is positive and negative respectively, due to membrane motion.

Acoustic damping would be somewhat difficult to assess for the conditions assumed in Fig. 5, but it is most probably small compared with the pneumatic damping of Eq. (2).

3.3 Similarity Analysis

The similarity analysis is much the same as for hanging membrane structures except that pneumatic effects must be included, and cables may be absent. Two non-dimensional parameters are required to characterize pneumatic effects; suitable choices are:

(j) $\dfrac{P}{\frac{1}{2}\rho U^2}$ internal pressure coefficient

(k) $\dfrac{b^2}{\rho U(C_L + C_F)}$ non-dimensional pneumatic damping; C_L and C_F are defined at Eq. (2).

In the most complex situations, dynamic similarity would require that these two parameters plus parameters (a) to (i) of Section 2.2 be matched for model and prototype. However many simplifications are normally possible, as outlined below.

3.4 Simplifications

It is assumed that there are no cables. If necessary, cables can be dealt with as outlined in Section 2 of this paper.

For air-supported membrane structures in which the pressure forces on the membrane act approximately vertically upwards (as they would, for example in Figure 4), similarity of both gravity and internal pressure-force effects can be achieved by use of an effective-pressure coefficient,

(1) $\dfrac{P-M_M g}{\frac{1}{2}\rho U^2}$

Provided that all the membrane structure's weight is more or less directly supported by the internal pressure, matching of this parameter replaces the requirement to match parameter (j), the pressure coefficient and parameter (a), the Froude number. This is so even if P is not much larger than $M_M g$.

If the internal pressure P is much larger than $M_M g$, the internal pressure forces dominate gravity forces and it is not necessary to match the Froude number in any case.

As discussed in Section 2.3, geometric stiffness effects in air-supported structures normally dominate elastic stiffness effects and elastic scaling is not required. Tensions in the model are correctly scaled if the effective-pressure coefficient is matched.

It may often be possible to relax mass scaling somewhat on the grounds that added mass is large. However this would seldom be necessary if accurate elastic scaling is not required.

Radii of curvature typically remain large at all times so it is unnecessary to match the folding stiffness parameter (g) of Section 2.2.

In summary, for most aeroelastic model tests of air-supported membrane structures, satisfactory dynamic similarity will be achieved by simulating the wind, using a geometrically scaled model, and matching the parameters $(M_M/\rho b)$, $((P-M_M g)/\tfrac{1}{2}\rho U^2)$ and $(b^2/\rho U(C_L + C_F)$.

The required value of $(C_L + C_F)$ can be obtained by using a constant-volume air supply (for example, a throttled high-pressure supply) for the model $(C_F = 0)$ and adjusting the leakage constant C_L to the required value. No exaggeration of enclosed volume is necessary. Prototype results are deduced from the model-test results as outlined for hanging membrane structures.

4. CONCLUSIONS

The requirements for aeroelastic modeling of membrane structures have been reviewed.

The complete similarity requirements are very difficult to implement, but extensive simplifications are usually possible without substantial loss of accuracy. Froude scaling, elastic scaling and folding-stiffness scaling can usually be relaxed. Even mass scaling can be somewhat relaxed in certain situations.

A new physical model is proposed for pneumatic effects. It leads to the conclusion that pneumatic damping is important and pneumatic stiffness is not. No exaggeration of internal building volume is required to model the revised pneumatic effects.

ACKNOWLEDGEMENTS

This review has drawn heavily on work conducted at the Boundary Layer Wind Tunnel Laboratory of the University of Western Ontario and at the Low Speed Aerodynamics Laboratory of the National Research Council of Canada (NRCC). These sources are gratefully acknowledged. The author is particularly grateful to R.L. Wardlaw of NRCC for providing the opportunity for his own modest involvement with membrane structures.

REFERENCES

1. H.P.A.H. Irwin, R.L. Wardlaw, K.N. Wood and K.W. Bateman, A Wind Tunnel Investigation of the Montreal Olympic Stadium Roof, National Research Council of Canada, LTR-LA-228, May 1979.

2. R.J. Kind and R.L. Wardlaw, Wind Tunnel Tests on the Behaviour of the Proposed Mobile Roof for the Montreal Olympic Stadium During Retraction and Deployment in Wind, National Research Council of Canada, LTR-LA-254, April 1981.

3. B.V. Tryggvason, D. Surry and A.G. Davenport, A Study of the Wind-Induced Response of the Haj Terminal Roof in Jeddah, Saudia Arabia - Phase II - An Aeroelastic Investigation, University of Western Ontario, Boundary Layer Wind Tunnel Laboratory Report, Nov. 1978.

4. B.V. Tryggvason, Aeroelastic Modelling of Pneumatic and Tensioned Fabric Structures, proc. of the Fifth International Conference on Wind Engineering, Fort Collins, Colorado, 1979, pp. 1061-1072.

5. N. Isyumov and B.V. Tryggvason, A Study of the Wind-Induced Behaviour of the Air-Supported Microwave Radome on the CN Tower, University of Western Ontario, Boundary Layer Wind Tunnel Report, 1979 (approx.).

6. B.V. Tryggvason, and N. Isyumov, A Study of the Wind-Induced Response of the Air-Supported Roof of the Dalhousie University Sports Complex, University of Western Ontario, Boundary Layer Wind Tunnel Laboratory Report - SS7-1977.

7. CRC Handbook of Tables for Applied Engineering Science, The Chemical Rubber Co., Cleveland, Ohio, 1970.

8. E.M. Greitzer, Review-Axial Compressor Stall Phenomena, ASME Journal of Fluids Engineering, Vol. 102, June 1980, pp. 134-151.

AEROELASTIC MODELING OF BRIDGES

Robert H. Scanlan, Princeton University

1.   INTRODUCTION

This paper first discusses, in some detail, the section model and its developments. Thereafter some consideration is given to full-bridge models. In the last decade and a half there has been heightened interest in the expanded role of the bridge deck aeroelastic section model. Originally conceived of as a direct representative of the dynamics of the prototype bridge, today it is increasingly viewed also as an analog source of force and moment data, both static and dynamic. The data so obtained are then incorporated into various dynamic analyses of the full-bridge prototype. Ref. 1 presents details of this approach.

Bridge deck designs have appeared in a wide variety of forms. Up to the present, at least, a comprehensive cataloguing of the design information produced by wind tunnel tests on these various forms has not been accomplished. Ref. 2 offers qualitative comments on the stability characteristics of some 40 deck sections. However, new designs, for a variety of reasons, still continue to be candidates for renewed wind tunnel testing. Unfortunately too, the techniques of mathematical and computational modeling of flows over such bluff bodies have not in general advanced to the state of offering truly viable alternatives to the more traditional direct wind tunnel testing of models. Thus the section model retains its currency as a design tool. When funds permit, it is effectively complemented by the full-bridge aeroelastic model.

The recent technical literature records the inherent tendency toward dynamic instability under wind of bluff bodies of many sorts, the long-span bridge deck being a prime example, whether it be a typical truss- or girder-stiffened form, a box or some variant thereof, or another alternative, with or without aerodynamic contour treatment. The focus of the present paper is the state of the art now centered, first, on the deck section model, and second on the degree of confirmation of its effectiveness that is offered by full-span models.

2.   PRIMARY PURPOSE OF THE SECTION MODEL

The first use of the bridge deck section model is to permit economical wind tunnel studies that will bring its eventual aerodynamic contour to a satisfactory condition from a dynamic stability standpoint. It is commonly considered minimally necessary to endow the section model with two elastic degrees of freedom -- vertical and twist, plus scaled mass similarity to a corresponding part of the prototype, in such a way that -- at least in some limited sense -- the section model becomes "representative" of prototype action. Typically, it is mounted on springs that permit scaled representation of first natural vertical and torsional mode actions at the right frequencies.

While this setup is vastly oversimplified relative to the prototype, the model thus equipped nevertheless becomes a useful instrument for the basic study of contour. In brief, used in this way the section model serves as a convenient test bench and analog computer of the stability tendencies inherent in various trial shapes. It is in this role that the section model still serves its primary function: it permits convergence upon a final external bridge deck configuration that, while meeting other design requirements, provides stability under cross winds.

3. SECONDARY PURPOSES OF THE SECTION MODEL

While formerly, considerations of the above sort used to end the wind tunnel concerns about long-span bridges, the problem has by now been examined in greater depth. Of typical present day concern for the prototype are not only insurance of its freedom from single-degree and coupled flutter under expected winds, but at least acceptable performance under vortex shedding and buffeting by the turbulence in those winds. For these problems the section model has been the analog provider of additional data which, subsequently used in mathematical analyses of the whole bridge, help to predict its general performance under wind.

Thus, once the deck configuration has been chosen, usually on the basis of acceptable vortex-influenced and flutter stabilities, the section model thereof serves as a focus of investigation of the forces and moments, both static and dynamic, expected on the prototype. Investigations of these effects raise a number of new questions: the choices among testing methods for force measurements, the effects of turbulent flow as against laminar flow, the proper modeling of wind tunnel turbulence, etc.

Perhaps the first new focus beyond the more traditional one on the section model was the concern to extract motion-linked forces (the so-called "flutter forces") therefrom. In the earliest stages this was done in oncoming laminar flow only. This approach, with strong evidence of being the most conservative, is still a basic necessity. In certain cases observed in the field the wind has in fact been of surprisingly low turbulence. This emphasizes the fact that any improvements to bridge stability that may be provided by wind turbulence may not always be counted on in nature.

Three basic methods have evolved to accomplish the task of measuring motion-induced forces: 1) driving a model sinusoidally and measuring the developed forces through strain-gage linkages; 2) driving a model equipped with direct sensors of the fluctuating aerodynamic pressures over it; 3) freely oscillating a spring-mounted model and inferring force and moment data therefrom by system identification techniques.

The first method was used in the 1950's by Halfman [3] at MIT to extract the so-called flutter derivatives of airfoils. It has been much used since the mid-1960's for bridge section models in Japan. It requires a heavy mechanical drive-linkage system, free of extraneous vibration, with means for compensation of the model inertia forces (typically, a countervibrating inertia placed outside the wind stream), since these tend to be appreciable relative to the desired motion-induced aerodynamic forces. The method has the advantage of providing accurate sinusoidal input to the model with closely controlled amplitude. Ref. 4 describes this technique.

The second technique, apparently developed at ONERA, France, for studies on turbine blade models, uses a series of small pressure taps in the model. The model being driven through prescribed vertical or torsional sinusoidal motions, the oscillating pressures at representative points on its surface are recorded. Subsequent calculational approaches resolve these oscillating pressure results into overall motion-related forces and moments. With this technique there necessarily must be some close attention paid to the frequency-response characteristics of the pressure-measurement system. In addition, there may be portions of a given structural contour -- such as roadway barriers or fences on bridge decks -- that produce local pressure anomalies or concentrated moments but may perhaps not be convenient locations for the mounting of pressure taps that completely define these local effects. Such questions require careful resolution prior to data collection. The basic pressure-tap technique for oscillating bridge deck section models has been described in Ref. 5.

An interesting variant on the pressure-tap technique is that in which several pressures are pneumatically "averaged" by feeding several tubes from individual taps to a plenum chamber whose net fluctuating pressure is recorded [6]. The wind tunnel group at the University of Western Ontario has further extended the averaging chamber pressure-tap technique beyond the simple summation for lift forces to a configuration in which the geometric arrangement of the tap holes is designed in a special way that permits the direct reading of torsional moment, from the average plenum-chamber pressure. This is accomplished by spacing the pressure taps so that the pressure contribution of each hole to the plenum chamber is proportional to the afferent area of the hole times the pertinent lever arm to the moment axis. With a direct pressure-measurement system the model can be either driven or spring-suspended. It can, for example, be driven by an electrodynamic shaker in place of more cumbersome machinery.

The third type of dynamic force measurement was developed in the U.S. [7] and consists of freely oscillating the deck section model under cross wind, measurement of its response, and inference therefrom, through an analytical model plus identification techniques, of the appropriate flutter derivatives consonant with the model's performance. This technique requires the least special laboratory equipment. An interesting feature attendant on the extraction of flutter derivatives from section models is the insight offered by such derivatives for comparing the inherent qualities of different deck shapes. The derivatives portray exclusively the <u>aerodynamic</u> stability tendencies of a given shape irrespective of what mechanical or dynamic properties the deck bearing that shape might have. This offers valuable criteria for comparisons among shapes alone, divorced from all parameters but their geometry and the way this influences their aerodynamics.

In addition to the three dynamic methods most prominent in the measurement of flutter derivatives, it is usually necessary to obtain static force and moment values (in the form of coefficients) from a section model. This is done typically through sensitive high-frequency, low-displacement strain gage linkages or the more standard type of wind tunnel balance, when this is available.

4. SECTION MODEL TESTING UNDER TURBULENCE

One may assume that the experimentation discussed to this point has been concerned with laminar approach flow. Bridge deck section models being bluff bodies, they tend to break up the laminar oncoming flow into a pattern of locally-induced "signature" turbulence that is characteristic of the deck shape itself, particularly of its windward side. This will result in practically all bridge-deck model measurements being influenced, to a greater or lesser extent, by fluctuating flow. Hence, even laminar-flow testing results in some randomly unsteady pressures over the model.

The natural wind is very often turbulent, and more recent considerations have made it imperative that attention be paid to this fact when assessing the wind performance of deck section models. Both vortex-induced oscillation and the onset of flutter instability are strongly influenced by the presence of turbulence in the oncoming flow. Vortex lock-in may be reduced or suppressed by the lack of lateral coherence under turbulent flow, and the onset of flutter instability -- often extremely abrupt under laminar flow -- may be replaced by a more gradual increase in rms response amplitude with increasing mean wind speed. Furthermore, the flutter derivatives measured under laminar oncoming flow may be altered by the presence of turbulence in this flow. In more recent years, therefore, it has become a part of section model testing to include turbulence in the oncoming flow.

While this is generally a desirable trend from the standpoint of the additional insights to be gained, it raises some questions that require careful answers. Firstly, turbulence is in general a three-dimensional flow phenomenon; therefore, its presence in the flow over a section model -- heretofore conceived of in essentially two-dimensional terms -- requires explanation and interpretation. In short, it requires that full knowledge of the three-dimensional characteristics of the test flow be acquired and that the section model be viewed as a three-dimensional object both in interpreting its response and in projecting to the prototype the characteristics inferred from that response. In short, a conceptual theory of the action of the turbulent flow on any bluff surface is required as a background for interpretation of results. This is typically offered by the existing theories dealing with effects that are randomly distributed in both space and time. A particularly outstanding and important characteristic of section model response under laminar flow has been the evolution of aerodynamic damping with mean wind velocity. How this is affected by turbulence in specific cases needs close delineation.

When steady-state aerodynamic force and moment derivatives are used to describe the effect of turbulent flow upon a bluff structure, the frequency dependence of such coefficients becomes a question of importance. This is summed up in another study, namely, the focus upon the so-called aerodynamic admittance, or modification factor that must be determined to describe the force or moment on a model under turbulent flow as against its counterpart developed under laminar flow.

5. APPROPRIATENESS OF THE TEST TURBULENCE

Finally, the nature, quality, and appropriateness of the turbulence provided by the wind tunnel itself are pertinent questions. In testing section models of bridges over navigation channels it is usually possible to assume that the prototype bridge deck is at a height of some 60 to 70 meters, which location is mainly affected, not by the entire changing profile of the turbulent wind (as a tall building or stack would be), but by only a single horizontal "slice" of it at one elevation. Under these circumstances a single state, rather than a whole profile, of turbulent approach flow, properly modeled, suffices for the test.

Aside from questions arising because of its three-dimensional nature in the presence of a two-dimensional model, such turbulence can appropriately be set up by a grid, or across-flow lattice of ordinary slats. This setup however, in most available tunnels, typically can produce a turbulence scale length of the order of only some 10 to 30 cm. Although there is even some doubt in meteorology as to what lengths to assign to the turbulence scales that develop in the actual atmosphere under the mechanical stirring of high winds, values of the order of 60 to 180 meters may be cited as typical. This suggests model scale factors of the order of 1:600 when wind tunnel turbulence is developed passively by a grid. Since actual models, for reasons of realizing structural detail, etc., have a preferred scale of between 1:20 and 1:100, it is clear that strictly correct turbulence scale over a section model can seldom be duplicated by means of a grid.

The idea has occurred to some, therefore, to set up wind tunnel turbulence by means of an active device that produces the correct velocity spectra and scales. This, to date, has resulted in at least one device -- that developed by Bienkiewicz, Peterka, and Cermak in collaboration with the author [8]. It consists of two sets of flapping airfoils that regulate, respectively, the horizontal and vertical components of turbulent flow past a deck section model. The device has the added feature that the turbulence produced is largely two-dimensional in character as it passes over the deck model, consistent with the originally conceived two-dimensional character of that model. Experimentation with flapping airfoils has also been carried out at Kyoto University by Shiraishi.

It may not in fact prove, in the long run, to be critically important -- as some already maintain -- to establish the fully correct turbulence environment around the bridge deck section model. It is the view of some that only the small-scale turbulence produces the important flow-modification effects. This may explain why truss-section models, which produce their own small-scale signature turbulence, produce relatively reliable stability results under laminar flow. Answers to this question will be a prime objective of future bridge aerodynamic research, including that involving the actual production of larger-scale turbulence.

In general, there is continued need to improve many aspects of bridge deck section model studies, but most especially those associated with model motion and proper turbulence. Mechanical devices and procedural simplification are needed in extracting motional force derivatives and aerodynamic admittances. In this connection, much future use can be made of the numerous electronic devices now available for rapid signal processing and display.

The number of parameters involved in approaching reasonable flow and dynamic similitude and projecting test results over to the prototype are great enough that the enterprise of bridge deck section model testing is one requiring the very careful attention of a professional aerodynamicist.

6.   TESTING OF FULL-BRIDGE MODELS

The full-bridge model must be a faithfully scaled aeroelastic representation of the whole prototype. Its principal advantage when surrounding terrain is also properly modeled, together with the natural wind itself, is clearly that it is an analog computer of the _entire_ response of the bridge to expected environmental winds acting on the prototype. In principal it requires no focus upon force coefficients, motional or otherwise, as in the section model case, and requires no ensuing concerns over dynamic response analysis methods, either for extracting and interpreting data or for projecting results over to the prototype.

As will be familiar, there is, however, some price to be paid for the convenience. First, in all likelihood a section model study will have been necessary in any case for the same project. This will have been necessitated by prior emphasis on determining a stable deck-section shape. With this already in hand, much parallel data will be made available. Second, the full model will be generally a more costly and time-consuming enterprise. It will require separate "engineering" in most cases since aeroelastic similitude will require replacement of prototype strength elements (such as trusses or boxes) by hidden splines and the like, developed to provide the required stiffness properties in a manner much simpler to realize than via a true replica model. Third, the scale of the full-bridge model will necessarily be small, exacerbating certain problems of model making on the one hand and the already grossly violated Reynolds number similitude on the other. Ref. 9 clearly reviews the issues of full-bridge versus section model testing.

7.   APPROPRIATENESS OF TURBULENCE FOR FULL-MODEL TESTS

A number of atmospheric flow simulation techniques have been developed to date and are by now well known. In "long" tunnels, a full atmospheric boundary layer simulation is accomplished by proper choice of surface roughness and a sufficiently long fetch preceding the test area. In converted aeronautical-type tunnels or others with relatively short sections available for flow modification, large spires followed by a short section of roughness elements suffices to produce a good distribution of velocity and turbulence intensity. A third simulation device for the shorter tunnel is the so-called counter-jet technique, in which the desired turbulence is tripped by jets entering from the floor and directed against the mean flow direction.

In all cases of full-bridge modeling to date the model scale has, of necessity, matched the range of achievable turbulence scale relative to atmospheric. This has necessarily been small, ranging from about 1/100 to 1/600, in all cases of turbulence modeled by passive flow-impeding devices or by the counter-jet technique. There is little present prognosis for full model testing at greatly increased scales if proper atmospheric simulation is desired.

In almost all model testing it has been correctly assumed that the critical component of mean wind is the one normal to the centerline of the bridge deck. However, full-bridge model testing offers the added possibility of directly studying effects of winds from any angle. This can become useful for studies concerned with the construction stages of bridges or for unusual bridge configurations such as high, strongly arched spans that may pick up wind forces below the deck under quartering winds.

Notable full-bridge model studies have been conducted by Farquharson [10], Scruton [11], Irwin and Schuyler [12], Melbourne [13], Olivari [14], Davenport and his co-workers [15], and researchers in Japan, of which Refs. 16 and 17 are representative publications among a number.

8. THE TAUT-STRIP MODEL

An abridged version of the full model has been introduced by Davenport and Tanaka [18]. This is the so-called "taut strip," consisting of two parallel wires pulled to an appropriate tension across a wind tunnel to set up scaled vertical frequencies and separated by appropriate distances to permit achievement of correct torsional frequencies when "dressed" with appropriately weighted deck cross-section elements. This model form achieves, in a rather economical fashion, certain of the objectives of a full model without comparable cost. The model and turbulence scales necessarily remain small, but three-dimensional coherence aspects of turbulence are well duplicated. Correct prediction of prototype behavior from taut strip performance, strictly speaking, requires careful, theoretically-based interpretation of the model action and its reinterpretation for the prototype. In a few cases, however, action of the model can be qualitatively interpreted as giving a good direct impression of prototype performance.

9. CONCLUDING REMARKS

At the present juncture a very reasonable hope is that, within a short future period, developments that are now proceeding apace will enable all necessary information and data for bridge aerodynamic design to be obtained directly from very economical basic model tests, particularly section model tests. A few cases have already been explored wherein full-bridge model behavior has been compared to theoretical predictions based on section model data. Refs. 17 and 19-22 exemplify work of this kind. At any rate, this appears to be a present trend to which it is hoped that corroborative information can be added from the several present or recent ongoing full-scale field studies such as those on the Luling, Pasco-Kennewick, Golden Gate, and Lions' Gate Bridges.

Some very recent section and full-bridge model studies representing current state of the art have been reported [23-26]. Ref. 27 provides a summary as of 1979 of a number of important issues in bridge stability under wind. Finally, a recent conference in England on Bridge Aerodynamics should be cited [28].

10. ACKNOWLEDGMENT

In preparation of this paper the author solicited photographs of current wind tunnel research from several individuals active in this work. A large number of excellent pictures were graciously offered and are hereby acknowledged. In particular, the author would like to express his thanks to the following individuals:

B. Bienkiewicz, Colorado State University, Ft. Collins, Colorado; H. Bosch, Federal Highway Administration, Washington, DC; J. E. Cermak, Colorado State University, Ft. Collins, Colorado; A. G. Davenport, University of Western Ontario, London, Canada; E. Hjorth-Hansen, Norwegian Institute of Technology, Trondheim, Norway; M. Ito, University of Tokyo, Japan; W. H. Melbourne, Monash University, Australia; D. Olivari, Von Karman

Institute, Rhode-St. Genese, Belgium; N. Shiraishi, Kyoto University, Japan; D. E. Walshe, National Maritime Institute, United Kingdom; R. L. Wardlaw, National Aeronautical Establishment, Ottawa, Canada.

The photographs were to numerous to be included in their entirety with this paper. However, selected examples, with captions, have been appended to the present text. Many more of those received by the author were shown as slides with the oral presentation of this paper at the NBS/NSF Workshop.

11. REFERENCES

1. Scanlan, R. H., "State-of-the-Art Methods for Calculating Flutter, Vortex-Induced, and Buffeting Response of Bridge Structures," Federal Highway Administration Rept. FHWA/RD-80/050, Washington, D.C. (April 1981), Nat'l. Tech. Info. Service, Springfield, Virginia 22161.

2. Scanlan, R. H. and Wardlaw, R. L., "Aerodynamic Stability of Bridge Decks and Structural Members" from Cable-Stayed Bridges, Struct. Engrg. Series No. 4, U.S. Dept. of Transportation, FHWA (June 1978).

3. Halfman, R. L., "Experimental Aerodynamic Derivatives of a Sinusiodally Oscillating Airfoil in Two-Dimensional Flow," NACA Tech. Rept. 1108 (1952).

4. Ukeguchi, N., Sakata, H., and Nishitani, H., "An Investigation of Aeroelastic Instability of Suspension Bridges," Proc. Int'l Sympos. on Suspension Bridges, Lab. Nac. de Engen. Libson, pp. 273-284 (1966).

5. Loiseau, H., and Szechenyi, E., "Etude du comportement aéroélastique du tablier d'un pont à haubans," C. R. des journées AFPC, ONERA, Chatillon, France (April 1974).

6. Surry, D. and Stathopoulos, T., "An Experimental Approach to the Economical Measurement of Spatially Averaged Wind Loads," Jnl. Indust. Aerod., Vol. 2, pp. 385-397 (1977-78).

7. Scanlan, R. H. and Tomko, J. J., "Airfoil and Bridge Deck Flutter Derivatives," J. Eng. Mech. Div., ASCE, Vol. 97, No. EM6, pp. 1717-1737 (Dec. 1971).

8. Cermak, J. E., Bienkiewicz, B., Peterka, J., and Scanlan, R. H., "Active Turbulence Generator for Study of Bridge Aerodynamics; EMD Specialty Conf., ASCE, Austin, Texas (Sept. 1979).

9. Wardlaw, R. L., "Sectional Versus Full Model Wind Tunnel Testing of Bridge Road Decks," DME/NAE Quarterly Bulletin No. 1978(4), National Res. Council, Ottawa, Canada, pp. 25-47, (Reprint Jan. 1979).

10. Farquharson, F. B. (ed.), "Aerodynamic Stability of Suspenson Bridges," Univ. of Washington Engrg. Expt. Sta. Bulletin No. 116, Parts I-V, (June 1949-June 1954).

11. Scruton, C., "Experimental Investigation of Aerodynamic Stability of Suspension Bridges with Special Reference to the Proposed Severn Bridge," Proc. Inst. of Civil Engrs, London, Vol. 1, No. 2, pp. 189-222 (March 1952).

448

12. Irwin, H. P. A. H. and Schuyler, G. D., "Experiments on a Full
     Aeroelastic Model of Lions' Gate Bridge in Smooth and Turbulent Flow,"
     NRC Canada, NAE Rpt. LTR-LA-206 (Oct. 1977).

13. Melbourne, W. H., "Model and Full Scale Response to Wind Action of the
     Cable-Stayed Box Girder West Gate Bridge," Proc. Sympos. on Practical
     Experiences with Flow-Induced Vibrations, Karlsruhe, W. Germany
     (Sept. 1979).

14. Olivari, D. and Thiry, F., "Wind Tunnel Tests of the Aeroelastic
     Stability of the Heer-Agimont Bridge," Tech Note No. 113, Von Karman
     Inst. for Fl. Dyn., Rhode-St. Génesè, Belgium (1975).

15. Davenport, A. G., Isyumov, N., Fader, D. J., and Bowen, C. F. P.,
     "The Study of Wind Action on a Suspenion Bridge--The Narrows Bridge,
     Halifax," BLWT Report 3-69, University of Western Ontario, London,
     Canada (May 1969).

16. Hirai, A., "Wind Tunnel for a Full Model of Suspension Bridges," (in
     Japanese), Univ. of Tokyo (1969).

17. Miyata, T., Kubo, Y., and Ito, M., "analysis of Aeroelastic
     Oscillations of Long-Span Structures by Nonlinear Multidimensional
     Procedures," Proc. 4th Int'l Conf. on Wind Effects on Bldgs and
     Structures, Heathrow, 1975, Cambridge Univ. Press, pp. 215-225 (1977).

18. Davenport, A. G. and Tanaka, H., "Aerodynamic Study of Taut Strip
     Models in Turbulent Boundary Layer," BLWT Rept. No. 1-74, Univ. of
     Western Ontario, London, Canada (Jan. 1974).

19. Irwin, H. P. A. H., "Wind Tunnel and Analytical Investigations of the
     Response of Lions' Gate Bridge to a Turbulent Wind," NAE, Nat'l Res.
     Council, Rept. No. LTR-LA-210, Ottawa, Canada (June 1977).

20. Holmes, J. D., "Prediction of the Response of a Cable-Stayed Bridge
     to Turbulence," Proc. 4th Int'l Conf. on Wind Effects on Bldgs. and
     Structures, Heathrow, U.K., Sept. 1975, Cambridge Univ. Press,
     pp. 187-197 (1977).

21. Konishi, I., Shiraishi, N., and Matsumoto, M., "Aerodynamic Response
     Characteristics of Bridge Structures," Proc. 4th Int'l Conf. on Wind
     Effects of Bldgs. and Structures, Heathrow, U.K., Sept. 1975, Cambridge
     Univ. Press, pp. 199-208 (1977).

22. Soo, H. S.-W., "Comparison Between Theory and Experiment in the Flutter
     and Buffeting of Long-Span Suspension Bridges," Doctoral Dissertation,
     Princeton University (Jan. 1982).

23. Hjorth-Hansen, E. and Freriksen, R., "A Comparison of Two Wedge-Shaped
     Box Girder Bridge Decks," Rept. on Sectional Model Tests in Wind Tunnel,
     Norwegian Institute of Technology, Trondheim (Sept. 1980).

24. Hjorth-Hansen, E., "Aerodynamic Aspects of Box-Girder Decks with
     Wedge-Shaped Fairings for Suspension Bridges," from Norwegian Bridge
     Building (honoring Arne Selberg): Tapir Publ. Trondheim, 1981.

25. Irwin, H. P. A. H., Duffy, K., and Savage, M.-G., "Wind Tunnel Investigation of the Palmerston Bridge, Pugwash, Nova Scotia," Rept. No. LTR-LA-244, NAE, NRC, Ottawa, Canada (July 1980).

26. Isyumov, N., Tschanz, T., and Davenport, A. G., "A Study of Wind Action for the Weirton-Steubenville Bridge," U.W.O. Engr. Rpt. BLWT-SS1-77, London, Ontario (1977).

27. Scanlan, R.H., "On the State of Stability Considerations for Suspended-Span Bridges Under Wind," from Proc. Symposium on Practical Experiences with Flow-Induced Vibrations, Karlsruhe, W. Germany, pp. 595-618 (Sept. 1979).

28. "Bridge Aerodynamics," Proc. of a Conference, Inst. of Civil Engrgs., London, U.K. (March 1981).

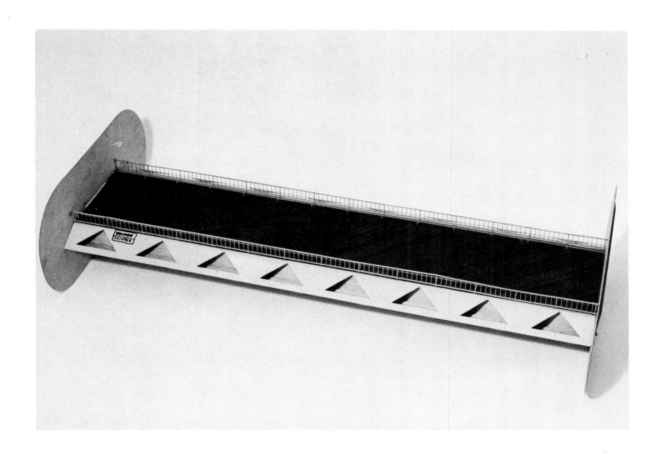

Fig. 1 Section model of box girder deck with wedge-shaped fairings. Project for Norwegian Public Roads Administration. (Courtesy of E. Hjorth-Hansen, Norwegian Institute of Technology, Trondheim).

Fig. 2 1:110 full-bridge model of Lions' Gate Bridge,
Vancouver, in 9m x 9m wind tunnel, National Aero-
nautical Establishment, Ottawa Canada (Courtesy of
R. L. Wardlaw).

Fig. 3 Full model of Heer-Agimont Bridge in VonKarman Institute wind tunnel, Rhode-St. Génèse, Belgium (Courtesy of D. Olivari).

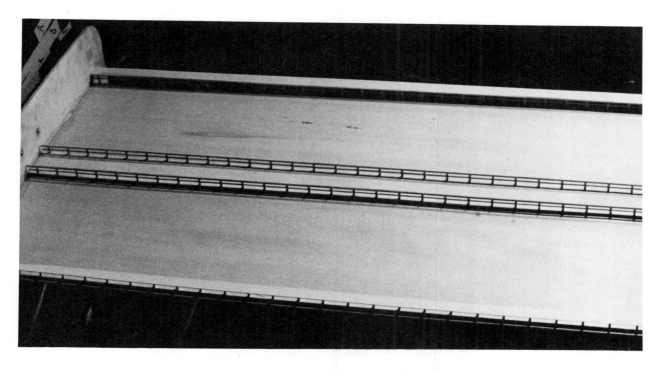

Fig. 4 Section Model of Yamato River Bridge, Osaka (Courtesy of N. Shiraishi, Dept. of Civil Engineering, Kyoto University, Japan).

Fig. 5. Full bridge model of suspension bridge with island beneath.
University of Tokyo Full-bridge tunnel (Courtesy of M. Ito, Dept. of Civil
Engineering, University of Tokyo, Japan).

Fig. 6. Array of airfoils for active turbulence generator. (Courtesy
of J.E. Cermak and B. Bienkiewicz, Colorado State
University, Ft. Collins, Colorado).

453

Fig. 7. Full model of the Murray Mackay Bridge, Halifax (Courtesy
of A.G. Davenport, Boundary Layer Wind Tunnel, University
of Western Ontario, London, Canada).

454

Fig. 8 Test model suspended in open jet of George Vincent wind
tunnel, F.H.W.A., U.S.D.O.T., McLean, Virginia (Courtesy of
H. Bosch, FHWA, Fairbank Res. Lab.).

Fig. 9 Full model of West Gate Bridge in wind tunnel at Monash University (Courtesy of W. H. Melbourne).

Fig. 10 1:60 scale model, final form, Tsing-Ma Bridge, Hong Kong
(Courtesy of Mott, Hay and Anderson, Far East, Hong Kong, and
D. E. Walshe, National Maritime Institute, Feltham, Middlesex, U.K.).

PREPARED CRITIQUE OF PAPERS ON AEROELASTIC MODELING

R.L. Wardlaw, Low Speed Aerodynamics Laboratory, National Research Council
of Canada, Ottawa K1A 0R6

INTRODUCTION

The term "aeroelastic" is used to describe the behaviour of structures
when both aerodynamic and elastic forces interact; but other forces may
also be involved and in wind engineering problems the effects of gravitational, inertial and damping forces must frequently be considered. The
need for aeroelastic modeling and wind tunnel testing of structures arises
entirely as a result of our inability to analytically predict the aerodynamic forces that affect the structure. The aerodynamic problem is particularly intractable because we are dealing with non-streamlined bluff
shapes with complex wakes and because the structures are located, not in a
smooth uniform flow but in the highly turbulent surface wind layer. Without the aerodynamic forces the interaction of the other forces could be
readily resolved analytically.

The essential requirement of aeroelastic modeling is that the forces
that influence behaviour be in the same proportion to each other in the
model structure as they are in the prototype structure. At reduced scale
it is not possible to correctly model all the forces at the same time and
it becomes necessary to make a judgement as to which forces are most
important to the particular problem and must be correctly modeled and which
forces can be safely disregarded.

The four papers of this session differ from each other in that they
require different modeling compromises. For tall buildings the gravitational forces are unimportant and typical shapes are insensitive to
Reynolds Number. However, for chimneys and towers, Reynolds Number sensitive characteristics are of considerable concern. From the discussion of
membrane modeling it appears that correct scaling of both the mass and
elastic parameters may be relaxed, although material and cable tensions
will have to be correct. Gravitational forces may or may not be important
depending on tension levels. In the case of bridges, Froude Number modeling is essential for full models, and, at the small scales that are required,
care must be taken to avoid problems arising from very low component
Reynolds Numbers such as those for cables.

THE AEROELASTIC MODELING OF TALL BUILDINGS

At first glance, the modeling of tall buildings may appear as the simplest conceptually of the four topics to be discussed; however, painstaking care in all aspects of the investigation are essential. Accurate simulation of the surface wind is required, yet the properties of the full scale wind are frequently in doubt and they must be determined by separate full scale and model studies before testing of the structure itself can proceed. The elastic design of the model can be very difficult and highly developed data acquisition and processing systems must be available for obtaining precise results at low cost.

Attention has been properly focussed on the limitations in model scale associated with blockage and Reynolds Number effects. As pointed out by Isyumov, the job is not finished when building motion has been measured. The designer must be able to determine from the results both building accelerations and loads in members. He also draws attention to the need to allow for variations in air density between model and full scale, that may be of the order of 10 percent (Ref. 1).

By comparing wind tunnel measurements of loads to design values obtained from codes, Isyumov has demonstrated that substantial savings can be realized by following the wind tunnel procedure. Clearly, on-going examination of our codes must continue in order to ensure that they take advantage of experience gained with model and full scale investigations.

THE AEROELASTIC MODELING OF CHIMNEYS AND TOWERS

This paper deals primarily with adjusting or interpreting results to accommodate Reynolds Number and surface roughness effects that cannot be correctly simulated at model scale. One tool involves the comparison of sectional model tests done at the same Reynolds Number as the full model and also at a much higher Reynolds Number. The difference in the results can then be used to adjust the full model results. The question arises of the acceptability of modeling the turbulence for the sectional model tests using grids.

The paper discusses wake interference problems, but overlooks the type of wake instability that can occur with closely spaced towers such as have been observed at heavy water plants (Ref. 2) or the problems at larger spacing such as occur with power cables (Ref. 3). For the latter problem a theoretical base exists for predicting behaviour.

After considering the techniques described, particularly the use of sectional model testing, it is tempting to conclude that the time has come when research funds should be made available for measuring and cataloguing the aerodynamic properties, both static and dynamic, of structural members along the lines promoted by Scanlan for bridge road decks. Such information would go a long way towards the prediction of loads and the avoidance of instabilities such as those that have been encountered with structural components of bridges and other structures (Ref. 4). The measurements of Delany and Sorensen (Ref. 5) of the static forces on a variety of prismatic shapes are a demonstration of the usefulness of systematically recorded data - much more along these lines, preferably including dynamic forces, would be invaluable.

AEROELASTIC MODELING OF MEMBRANE STRUCTURES

Although this is a relatively new area for wind tunnel aeroelastic modeling, much insight has been gained by the work of Tryggvason (Ref. 6), Irwin (Ref. 7) and Kind (Ref. 8).

In his paper for this session Kind has introduced a novel and simple physical model for pneumatic effects that accounts for building leakiness. An interesting conclusion from analysis using this model is that the exaggeration of internal building volume that is required for the non-leaky model as described by Tryggvason is no longer necessary. The limitations of Kind's useful hypothesis should be investigated.

Much was learned in the modeling of the Montreal Olympic Stadium that was done by Irwin (Ref. 6). Great care had been taken to closely match the shear and tensile stiffness and the mass of the material. In hindsight it became clear that these modeling requirements could have been relaxed.

The reason mass scaling can be relaxed is that the added mass for such a large membrane is much greater than the mass of the membrane itself. Irwin has estimated using a much simplified model that, for the Stadium, the added mass would be about twenty times the mass of the material. These calculations have been recently refined by Campbell (Ref. 9) who has treated several cases including circular and square plates with both sealed and unsealed edges. For the lowest mode, he has estimated the following ratios of added mass to membrane mass where the membrane mass is 2.2 kg/m^3

Circular plate	– unsealed edge	84
(diameter = 150 m)	– sealed edge	131
Square plate	– unsealed edge	50
(width = 134 m)	– sealed edge	89

The ratios are proportional to the size of the membrane so that for small membranes the added mass will be much less. Similar reductions will apply for higher modes of deformation. The high added mass will have the effect of reducing system damping and may partly negate the windfall of high acoustic damping that Irwin has postulated.

AEROELASTIC MODELING OF BRIDGES

The concept of the sectional model for predicting road deck behaviour is attractive because of relatively large model scale, flexibility in configuration changes, low cost and short lead time. The ability to experimentally separate the aerodynamic forces from the other forces allows for the cataloguing of aerodynamic data for families of aerodynamic shapes. With continued research into analytical techniques, behaviour prediction methods using these force data may be developed to a stage where design without testing is realized for most configurations.

At the present, full model testing remains the most reliable technique for forecasting prototype behaviour, particularly the response to turbulence; although the sectional model in smooth flow continues to play a useful role in assessing section stability. Promising experiments at the University of Western Ontario (Ref. 10) suggest that useful information may be obtained from sectional model testing in turbulent flow. Further

460

research in this area is being undertaken by Utsunomiya at the National Research Council of Canada.

Observations of low turbulence intensity over water (Ref. 11,12) is of critical importance to the wind tunnel investigation. This has been demonstrated by Irwin's studies of the Lions' Gate Bridge where instabilities appeared at low intensities that were not present in high intensities. It is interesting to observe that two identical bridges in eastern Canada over the same river about 10 miles apart had different response to wind. In one case the upstream fetch was clearly over water whereas in the second case the fetch was over land as a result of the bridge being located on a bend in the river. Serious vortex shedding excited motion was observed in the former but none in the latter (Ref. 13).

Finally it should be remembered that bridge components can have serious motion as a result of wind action, in particular stay cables, long truss members and deck hangers. Aeroelastic modeling would normally be done using sectional models.

REFERENCES

1. K.J. Eaton, Buildings and Tropical Windstorms, Overseas Building Note No. 188, Building Research Establishment, UK, April 1981
2. K.R. Cooper, H.P.A.H. Irwin, R.L. Wardlaw, Aerodynamic Investigations of In-Line Slender Towers for Heavy Water Plants; Proc. ASCE Specialty Conference, Methods of Structural Analysis, Univ of Wisconsin, Madison, Aug. 22-25, 1976, pp 286-307
3. R.L. Wardlaw, K.R. Cooper, R.G. Ko, J.A. Watts, Wind Tunnel and Analytical Investigations into the Aeroelastic Behaviour of Bundled Power Conductors, IEEE Trans. on Power Apparatus and Systems, Vol PAS-94, No 2, March/April 1975 pp 642-654
4. R.L. Wardlaw, Approaches to the Suppression of Wind-Induced Vibrations of Structures, IAHR/IUTAM Symposium, Practical Experiences with Flow-Induced Vibrations, Karlsruhe, Sept 3-6, 1979 pp 650-670
5. N.K. Delany, N.E. Sorenson, Low Speed Drag of Cylinders of Various Shapes, NACA TN 3038, November 1953
6. B.V. Tryggvason, Aeroelastic Modeling of Pneumatic and Tensional Fabric Structures, Proc., Fifth International Conference on Wind Engineering, Fort Collins, Colorado, 1979, pp 1061-1072 Vol 2
7. H.P.A.H. Irwin, R.L. Wardlaw, A Wind Tunnel Investigation of a Retractable Fabric Roof for the Montreal Olympic Stadium, Proc., Fifth International Conference on Wind Engineering, Fort Collins, Colorado, 1979, pp 925-938, Vol 2
8. R.J. Kind, R.L. Wardlaw, Wind Tunnel Tests on the Behaviour of the Proposed Mobile Roof for the Montreal Olympic Stadium During Retraction and Deployment in Wind, National Research Council of Canada, Laboratory Technical Report LTR-LA-254, April 1981
9. W.F. Campbell, The Added Mass of Some Rectangular and Circular Flat Plates and Diaphragms, National Research Council of Canada, Laboratory Technical Report LTR-LA-223, 1982
10. A.G. Davenport, N. Isyumov, H. Rothman, H. Tanaka, Wind Induced Response of Suspension Bridges - Wind Tunnel Model and Full Scale Observations, Proc. Fifth International Conference on Wind Engineering, Fort Collins, Colorado, 1979, pp 807-824, Vol 2
11. M. Shiotani, H. Arai, Lateral Structures of Gusts in High Winds, Proc., Second Int. Conf. on Wind Effects on Buildings and Structures, University of Toronto, 1968

12. H.W. Teunissen, Validation of Boundary-Layer Simulation; Some Comparisons Between Model and Full-Scale Flows, Proc. International Workshop on Wind Tunnel Modeling Criteria and Techniques, National Bureau of Standards, Gaithersberg, Maryland, April 14-16, 1982

Local Wind Environment and Snow Transport

Chairman

Richard D. Marshall

National Bureau of Standards
Washington, D.C.
USA

NATURAL VENTILATION MODEL STUDIES

Richard M.Aynsley, Department of Architecture and Building, Papua New
Guinea University of Technology, Lae, P.N.G.

1.INTRODUCTION

Although natural ventilation has been utilised since antiquity to
provide fresh air and indoor air movement for thermal comfort,(Vitruvius,
Ref.1), it was not until the 1950s that wind tunnels were used to make
quantitative estimates of natural ventilation. Dick (Ref.2) used wind
pressure distributions from wind tunnel studies in conjunction with discharge coefficients to estimate natural ventilation through small wall
vents and cracks around doors and windows. Smith (Ref.3) measured windspeeds through window openings in a model in a wind tunnel and expressed
them as windspeed coefficients referenced to a windspeed outside the
model. Both studies incorporated limited full-scale correlation to verify
their method.

Later Van Straaten (Ref.4) performed similar studies to those of Dick
on windspeeds through large window openings in models of school classrooms.
Many of these early studies were performed in uniform flow conditions with
no attempt to model boundary layer flow or control blockage effects. Limitations of such studies are discussed by Aynsley (Ref.5) while McKeon (Ref.
6) has studied pressure changes due to high tunnel blockage. Data from
these early studies were used to formulate guidelines for design of naturally ventilated houses (Claudill,Ref.7), classrooms (Claudill,Ref.8) and
factories (Weston,Ref.9). Little further development took place in techniques for determining quantitative estimates of natural ventilation until
the 1970s when rapid advances in boundary layer wind tunnel techniques and
full-scale studies of wind characteristics provided opportunities for more
detailed design data.

2.ADVANTAGES OF WIND TUNNEL MODELLING FOR ESTIMATING NATURAL VENTILATION OVER THE TRADITIONAL APPROACH.

By taking advantage of developments in boundary layer wind tunnel
techniques (Surry, Ref.10), together with refinements in airflow criteria
and long term climatic data in probabilistic form (Vickery,Ref.11), a designer can vastly improve on the traditional methods used in estimating
natural ventilation.

The traditional methods are described in mechanical engineering reference sources such as A.S.H.R.A.E. (Ref.12) and I.H.V.E. (Ref.13) handbooks. Typically they rely on a simple equation such as:

$$Q = E.A.V \qquad \text{(Equation 1)}$$

where Q is the airflow in cubic metres per second,
     E is the efficiency of an opening, ranging from 0.25 to 0.60,
     A is the free area of an opening in square metres,
and   V is the mean external windspeed in metres per second.

A range of correction factors are often applied for conditions which
are known to grossly effect the estimate. Such an approach cannot

effectively account for:

1. Physical structure of wind at a particular site in terms of mean windspeed profile or turbulence characteristics.

2. Influence on local wind due to nearby obstructions, topographic features or vegetation.

3. Influence of architectural features such as extended eaves, sunscreen devices or wall projections on airflow around a building.

4. Influence of size, proportion and location of openings in a building on airflow pattern.

Model studies in boundary layer wind tunnels provide a means of investigating all these influences for a particular site and building design. Most boundary layer wind tunnel techniques have been shown to be sufficiently accurate for most design purposes (Vickery,Ref.11, Dalgleish,Ref.14 and Isyumov,Ref.15) and are continually being refined. Some wind tunnel techniques for determining influences such as vegetation on local winds (White,Ref.16) need further correlation of model and full-scale studies to determine their order of accuracy.

3. STRATEGIES FOR UTILISING NATURAL VENTILATION IN BUILDINGS

There are two strategies for utilising natural ventilation. One assumes primary reliance on natural ventilation with use of fans only during periods of calm (Aynsley,Ref.5). The other strategy assumes reliance on a mechanical ventilation or air conditioning system, primarily, with the possibility for natural ventilation by providing openable windows and doors in the event of systems breakdowns or interruptions to power supplies. This is being promoted by the governments of Singapore and Papua New Guinea (Ref.17 and 18). In this case natural ventilation is a substitute for an on-site generator system. In either case power failures do occur, standby generators can fail and winds can cease to blow.

Any rational system for assessing these strategies should involve the determination of "acceptable risk" in terms of loss of acceptable indoor environmental conditions. This amounts to estimating the probability of the occurrence of an "unacceptable event" and the consequences of that event. Obviously to make such an evaluation, long term climatic data on temperature, humidity and wind is needed in probabilistic form so that an estimate of the likely frequency and duration of unacceptable environmental conditions can be made.

"Unacceptable events" in terms of indoor environmental conditions may be determined from thermal comfort criteria (Givoni,Ref.19), temperature or humidity limits for storage or processing of goods, or operation of equipment, or fresh air requirements for each indoor space.

"Consequences of events" can be evaluated in terms such as; loss of thermal comfort, monetary cost, loss of production, ongoing viability of a business enterprise, or the need to relocate site of operations.

4. CURRENT APPROACHES TO ESTIMATING NATURAL VENTILATION

In recent years studies of natural ventilation in buildings have taken two distinct paths. One path pursuing "infiltration" through small cracks and openings associated with heat loss problems or build-up of indoor pollutants in cold climates. This path has concentrated on full-scale measurements using pressurised buildings and tracer gas decay or replenishment techniques, together with the development of mathematical models for computing infiltration rates for typical building types. Research in this area is being co-ordinated by the Air Infiltration Centre, Bracknell in Great Britain.

Natural ventilation studies of flow through larger openings for indoor thermal comfort in warm climates have taken two paths. One approach uses windspeed coefficients measured in boundary layer wind tunnels in conjunction with thermal comfort criteria and climatic and wind frequency data.

The other approach makes use of existing published wind pressure distribution and discharge coefficient data in conjunction with thermal comfort criteria and climatic and wind frequency data.

5. ESTIMATING NATURAL VENTILATION USING MEAN WINDSPEED COEFFICIENT METHODS.

A useful method of quantifying natural airflow inside buildings suggested by the writer, Aynsley (Ref.20) for the design of buildings in hot humid climates involves the use of windspeed coefficients. These coefficients, C_v, are the ratio of the mean windspeed, $\overline{V}_1$, at a point of interest indoors, usually one metre above floor level, to the mean windspeed, $\overline{V}_z$, at a specified reference height, z, usually 10 m, upstream from the building in the airflow undisturbed by the building (Fig.1) so that:

$$C_{\overline{V}_1} = \frac{\overline{V}_1}{\overline{V}_z} \qquad \text{(Equation 2)}$$

Given the variation in individual's response associated with thermal comfort criteria and the limited availability of detailed temperature, humidity and windspeed data, a one hour meaning period for windspeed data seems reasonable to the writer for thermal comfort estimates.

Figure 1:
Wind Tunnel Technique
for Determining Mean
Windspeed Coefficients

$$C_{\overline{V}_1} = \frac{\overline{V}_1}{\overline{V}_{10}}$$

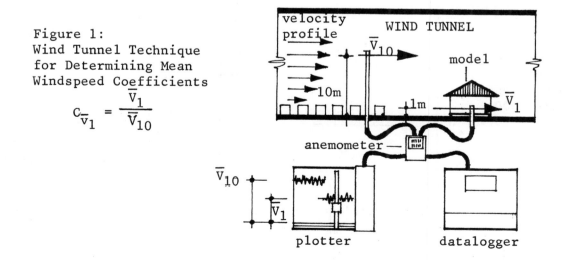

468

By adopting a common reference windspeed in similarly modelled winds, ventilation windspeeds can be directly compared from a range of different building models. When wind frequency data for mean 10 metre windspeeds ($\overline{V}_{10}$) and wind directions near the building site are available for a particular month and time of the day, airspeed $\overline{V}_1$, through the full-scale building can be estimated. Using the appropriate mean windspeed coefficient $C_{\overline{V}_1}$ from a wind tunnel study, with the model orientated at the corresponding wind direction:

$$\overline{V}_1 \text{(full-scale)} = C_{\overline{V}_1} \cdot \overline{V}_{10} \text{ (full-scale)} \qquad \text{(Equation 3)}$$

Some mean windspeed coefficients determined from a simple model by the writer for windspeeds 1 m above floor level and related to 10 m reference windspeeds for buildings subdivided by solid partitions (Fig.2), are indicated in Figure 3.

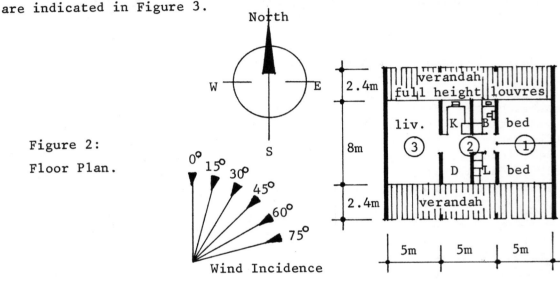

Figure 2:

Floor Plan.

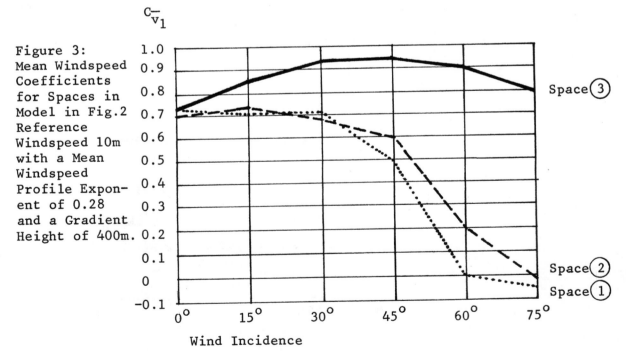

Figure 3:
Mean Windspeed
Coefficients
for Spaces in
Model in Fig.2
Reference
Windspeed 10m
with a Mean
Windspeed
Profile Exponent of 0.28
and a Gradient
Height of 400m.

Wind Incidence

Windspeed coefficients in industrial buildings were studied by
Weston (Ref.9) in the wind tunnel at the Commonwealth Experimental Building Station in Australia during 1954. Further wind tunnel studies by Van
Straaten (Ref.21) of school classrooms provided additional windspeed
coefficient data. More recent sources of windspeed coefficients for
housing come from studies by Givoni (Ref.22) and Chand (Ref.23).

These windspeed coefficients were based on low level external reference windspeeds, and tests were conducted in uniform airstreams,which
makes estimation of actual prototype velocities from the meteorological
wind data difficult. Hamilton (Ref.24) showed the effect of velocity
profile on wind pressures on low buildings. The writer's method stresses
the need for the reference windspeed to be measured in an appropriate
mean windspeed profile at a height for which detailed long term wind
data is available.

A similar technique,described by Vickery and Apperley (Ref.11) for
predicting the probable frequency of strong wind gusts near ground level
in urban terrain, was found to have an accuracy in the order of plus or
minus 10%. With careful modelling, similar accuracy should be possible in
estimates of indoor windspeeds.

The windspeed coefficient approach is ideally suited to steady airflow through and around buildings of complex or unusual shape. A discharge coefficient approach for such a building would require a detailed
wind tunnel study of wind pressure distributions at many locations over
the model surfaces. It is much simpler to measure windspeed coefficients
directly.

A typical natural ventilation study in this category was performed
by the writer in the Department of Architectural Science's wind tunnel
at the University of Sydney. The building was a large single storey
building designed for a location in the central Pacific (Fig.4). Most
external walls consisted of floor to ceiling sliding screens and extensive
use was made of quadrant-shaped ventilation hoods fixed at various locations on the roof. The design also incorporated a number of courtyard
areas within the building to which wall openings were provided. Windspeeds
inside this model at a height equivalent to 1 metre above floor and
externally at a 10 metre reference height were measured with a 'Davimeter'
thermocouple type anemometer. Output from the anemometer was recorded
using an X-Y plotter on paper calibrated to the anemometer response.

The principal advantage of the windspeed coefficient approach is that
mean windspeed estimates can be obtained for any position at which a small
anemometer can be located. A disadvantage is that larger, more detailed
models are necessary than are necessary for wind pressure distribution
studies. Components which deflect the airstream, such as louvres or
tilting sashes, need to be modelled with extreme care due to their significant influence on indoor airflow patterns.

The simplicity of the velocity coefficient method for estimating air
velocities in buildings is appealing; however the time and expense involved
in wind tunnel studies to determine windspeed coefficients restricts its
application for many small buildings. Most studies on low cost buildings
have been made in research programmes at government research stations and
at universities.

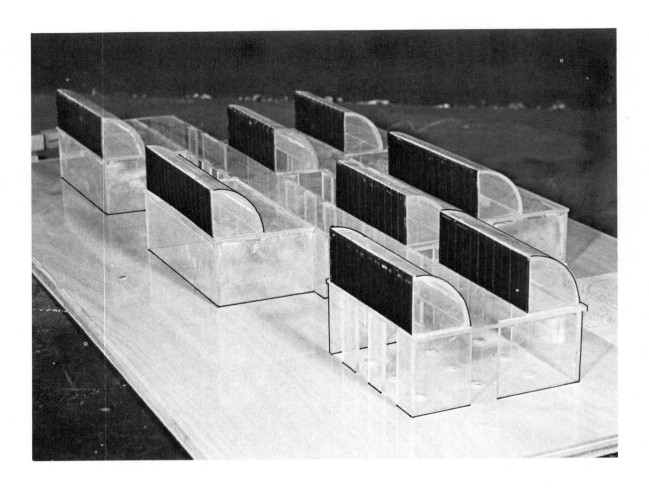

Figure 4: Wind Tunnel Model used to Study the Effectiveness of Natural
Ventilation Scoops Incorporated in Roof Form.

The full-scale effects of insect screens or other wall components
with curved or rounded sections in the airflow cannot be assessed on wind
tunnel building models. Even if insect screening could be reproduced
accurately at small scale, flow past rounded elements in insect screening,
for example, could be significantly different due to differences in
Reynolds numbers of the flow at full and model scale.

For these reasons, unless components installed in wall openings are
composed of sharp-edged, flat surfaced shapes and can be accurately
modelled at small scale, openings in models should be kept clear. A
windspeed reduction factor, established from full-scale studies of wall
components, can then be applied to the flow through clear openings in
the model to allow for the effect of inserting components such as insect
screens. Typical windspeed reductions due to insect screens are indicated
in Table 1.

Screen Material	Windspeed through clear opening	% Windspeed reduction with screen
	m/s	%
Bronze wire screen, 5.5 wires/cm, porosity 80%	2	26
	4	17
	6	11
Plastic-coated fibreglass, 7 threads/cm, porosity 60%	2	41
	4	34
	6	19

Table 1: Windspeed Reductions through Screens.

5.1 Estimating Indoor Thermal Comfort from Natural Ventilation using Mean Windspeed Coefficients

By combining estimates of airflow through openings with the frequency of occurrence and direction of winds, relative humidity, dry bulb temperature and thermal comfort criteria by Macfarlane (Ref.25), the writer estimates the frequency of occurrence of windspeeds sufficient to restore thermal comfort in buildings. Estimates and indoor comfort are generally limited to 9 am and 3 pm as these are the only times for which simultaneous long term windspeed, relative humidity and 86 percentile dry bulb air temperature data for each month are readily available in Australia (Ref.26 and 27).

Determination of windspeed necessary to restore indoor thermal comfort in well shaded insulated and naturally ventilated housing in hot humid regions, based on Macfarlane's method, can thus be quickly assessed using Figure 5 developed by the writer. An example of such a calculation is provided in Appendix 1 to this paper.

5.2 Summary of Advantages and Disadvantages of the Mean Windspeed Coefficient Method.

Advantages:

1. Simplicity of the method, only reference and indoor mean windspeeds at points of interest are measured,

2. Windspeed coefficients can be obtained at any point inside a model where an anemometer can be placed, in main airstream as well as eddy zones.

472

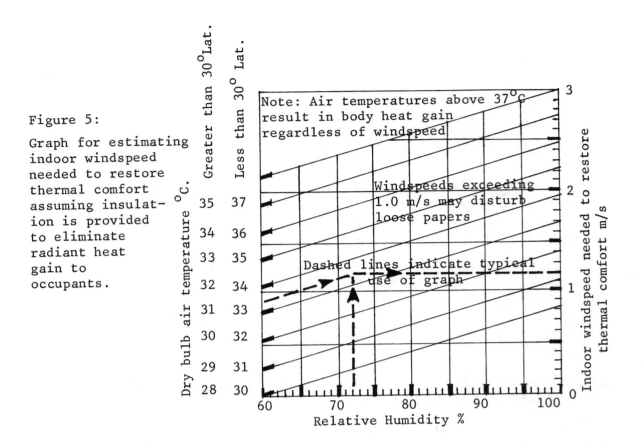

Figure 5:

Graph for estimating
indoor windspeed
needed to restore
thermal comfort
assuming insulat-
ion is provided
to eliminate
radiant heat
gain to
occupants.

Disadvantages:

1. Designers need access to boundary layer wind tunnel test
   facilities.

2. Need for extremely accurate modelling of openings where inclined
   sashes or louvres are involved.

6. WIND DISCHARGE COEFFICIENT METHOD

The attraction of a discharge coefficient method is that the increasing amounts of pressure distribution data from boundary layer wind tunnel
studies for wind loading and discharge coefficients used in design of
ventilation ducts, offer a means of estimating natural ventilation in a
design office without resorting to special wind tunnel studies.

Obviously using pressure coefficients or discharge coefficient data
derived for ventilation ducts or wind loadings to improvise estimates in
another application is not an ideal approach. However limited sources
of discharge coefficients derived from natural ventilation studies do
not account for smaller contraction of flow at entries of openings in
buildings when flow can divide around the building.

With regard to pressure distributions used in wind loading applications they are usually derived from models without openings. Typical differences between pressure differences across models with and without openings have been measured, (Aynsley, Ref.28). However if a particular building does not warrant the expense of a special wind tunnel study, the approximate methods such as discharge coefficient approach may be justified as an improvement over traditional methods for estimating natural ventilation such as Equation 1.

6.1 Estimation of Flow Through Openings using Discharge Coefficients

An equation (4) can be written for the mean windspeed $\overline{V}_0$ through an opening using appropriate discharge coefficients, C_d, from Tables 2 or 3 and pressure coefficients for total windward pressure C_{p_1}, and leeward static pressure C_{p_2}, referenced to dynamic pressure of the mean windspeed $\overline{V}_z$ m/s, z being the height in metres above ground (usually 10 m) for which long term mean windspeed data is available.

$$\overline{V}_0 = C_d \left[(C_{p_1} - C_{p_2}) \, \overline{V}_z^{\,2} \right]^{\frac{1}{2}} \quad \text{m/s} \qquad \text{(Equation 4)}$$

or in volumetric flow rate Q where A is the area of the opening (m^2)

$$Q = C_d A \left[(C_{p_1} - C_{p_2}) \, \overline{V}_z^{\,2} \right]^{\frac{1}{2}} \quad m^3/s \qquad \text{(Equation 5)}$$

These equations for airflow through a single opening in a building due to differences in wind pressure are difficult to apply, as there is little detailed data available on pressure due to wind inside buildings, with significant openings in windward and leeward walls. To overcome this lack of data, in cases where internal airflow follows a simple path through a building without branching (Figure 6), it is convenient to combine the discharge characteristics of each opening in series and use the wind pressure difference between the inlet and outlet openings on external walls. The condition of no branching in the airflow path within the

Figure 6:
Typical Wind Flow through Building Openings in Series indicating Terms for Areas of Openings, A, Local Windspeeds, V, and Pressure Coefficients C_p.

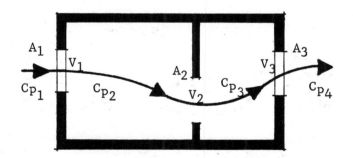

building is not a major restriction, as it is a feature of most buildings designed to encourage natural airflow.

For a number of openings in series:

$$Q = \left\{ \frac{(C_{P_1} - C_{P(n+1)}) \, V_z^2}{\dfrac{1}{C_{d_1}^2 A_1^2} + \dfrac{1}{C_{d_2}^2 A_2^2} + \dfrac{1}{C_{d_3}^2 A_3^2} + \cdots \dfrac{1}{C_{d_n}^2 A_n^2}} \right\}^{\frac{1}{2}} \quad m^3/s$$

(Equation 6)

The area of openings in windward walls should be reduced by a factor equal to the cosine of the angle of wind incidence in the case of inclined wind incidence to windward openings.

Mean windspeeds through each of the openings is found by dividing the common discharge rate Q, by each local opening area A.

6.2 Estimating Indoor Thermal Comfort from Natural Ventilation using Discharge Coefficients

Volumetric discharge necessary for thermal comfort through each opening can be calculated for each wind direction. This discharge is substituted into the discharge equation to determine the 10 m windspeed necessary for comfort at each opening. The frequency of occurrence for each 10 m windspeed interval exceeding that needed for comfort for a particular wind direction are summed. Repeating this process for each wind direction and totalling all percentages of occurrence indicates the total percentage of time when natural ventilation is likely to restore thermal comfort near the openings under study. An example calculation is provided in Appendix 2 of this paper.

6.3 Sources of Discharge Coefficients for Estimating Natural Ventilation.

Most discharge coefficients found in texts on fluid mechanics are for single, square-edged, round openings in flat plates, or a range of standard orifice designs used for flow measurement in pipes.

Full-scale studies by Dick (Ref.2) in 1950 established that discharge coefficients of approximately 0.65 were appropriate for very small square edged cracks and openings in houses. A.S.H.R.A.E. Guide (Ref.12) contains a wide range of both discharge coefficients and dynamic loss coefficients for rectangular ventilation ducts. Discharge coefficients for larger window openings in series were determined by Wannenburg and Van Straaten (Ref.4) in 1957 for estimating airflow for comfort in school classrooms. These window openings occupied approximately 10% of the wall area. Snyckers (Ref.29) extended the range of discharge coefficients for a range of rectangular openings in series up to 20% of wall area. Airflow rates through openings in series estimated using discharge coefficients from Tables 2 and 3 were found to be within 15% of measured values in a series of wind tunnel studies on a variety of opening geometries and sequences in which openings did not exceed 10% of the wall area (Aynsley,Ref.5). Cosine corrections were made to areas of windward openings in these studies for wind directions not normal to the windward wall.

Description of opening	Typical range of discharge coefficients for normal incidence	Jet characteristics
Small openings in thin walls less than 10% of wall area near the centre of the wall	0·50–0·65	Small inertia due to small mass of air in jet
Openings 10–20% near the centre of a wall with aspect ratio similar to the cross-section of the downwind space	0·65–0·70	Significant inertia due to increased mass of air in jet
Openings 10–20% of a wall with one edge common with the downwind space such as a doorway	0·70–0·80	Wall effect reduces energy losses on one side of jet
Openings similar in size to the cross-section of the downstream space	0·80–0·90	Wall effect around the perimeter of the jet significantly reduces turbulent energy losses

Table 2: Typical Discharge Coefficients for Single Inlet or Intermediate Openings in Buildings.

A_0/A_1	C_d
Approaching	
0·0	0·63
0·2	0·64
0·4	0·67
0·6	0·71
0·8	0·81
1·0	1·00

Table 3: Discharge coefficients for Outlet Openings.

6.4 Sources of Pressure Coefficient Data for Estimating Natural Ventilation

There are two major sources of pressure distribution data for building forms, those of Chien, and those of Jensen and Franck. Models used by Chien (Ref.30) in pressure distribution studies were simple block forms with none of the architectural features common to many buildings. Studies by Jensen and Franck (Ref.31) included some models with projecting eaves. Less detailed average pressure coefficients for a range of building forms are provided in wind loading codes (Ref.32). Significant eaves projections are a common sun control feature on low buildings in warm humid climates. Proportions of such architectural features become very significant on smaller buildings and a common practice of raising the building off the ground can cause appreciable changes in pressure distributions (Best, Ref.33).

Pressure distributions at mid-height of walls on recent housing forms built in Australia's hot humid tropics are provided in graphical form by Aynsley (Ref.5) for a number of wind incidences to a long wall.

Most pressure distributions on buildings are determined from wind tunnel studies of isolated models. Where a particular building has other buildings or obstructions nearby accurate pressure distributions can only be obtained by modelling such obstructions during the wind tunnel study. In the case of groups of buildings some studies were performed by Aynsley (Ref.28) and Holmes (Ref.34).

6.5 Advantages and Disadvantages of the Discharge Coefficient Method

Advantages

1. Ability to make use of increasing sources of wind pressure distribution data from wind load research.

2. Approximate estimates of windspeeds through openings can be made without resorting to specific wind tunnel tests by using existing data for building with wall openings less than 20% of wall area.

Disadvantages

1. When building openings exceed 20% of wall area wind pressure differences vary significantly from building forms without openings.

2. Estimates of mean windspeed are restricted to locations in jets close to wall openings, a significant disadvantage when the majority of occupied space is located away from openings.

7. MODELS FOR NATURAL VENTILATION STUDIES

The writer's selection of a scale for a wind tunnel model for measuring windspeed coefficients is governed by:

1. Maximum acceptable tunnel blockage usually 2% of tunnel cross-section for any single building.

2. The ability to be able to reproduce suitable wind characteristics at that scale. For mean windspeed measurements inside buildings a well modelled mean windspeed profile is the principal requirement as long as models are sharp-edged, flat surfaced bluff bodies.

3. The ability to place the anemometer probe inside the model at points of interest without causing more blockage than would be caused by the equivalent of an adult person. This is affected by the size and types of probes available.

In the case of mean pressure distributions studies on solid models, model scale is governed by the factors outlined in 1 and 2 above.

8. WINDSPEED AND WIND PRESSURE MEASUREMENTS

As indoor airflows are often turbulent and undergo frequent changes in direction it is best to use an anemometer probe with a response that is not sensitive to flow direction such as Disa's spherical film probe or Thermo Systems miniature cylindrical film probe. Alternatively a typical miniature hot wire probe can be orientated so as to ensure that the principal changes in flow direction occur around the circumference of the wire and not its length. Where airflow studies are performed to investigate thermal comfort it should be remembered that thermal comfort indices refer only to airspeed and not direction so it is not necessary to record flow direction. However flow visualisation using a short tuft on the end of a wire probe is useful for aligning directionally sensitive anemometer probes with the flow direction.

Where wind pressure tappings on surfaces of small solid models are being measured using a multitube manometer data can be rapidly recorded for each wind direction by photographing the manometer. When data logging facilities are available, measurements are made with a suitable pressure transducer via a scanning valve.

9. NEED FOR FURTHER MODEL AND FULL-SCALE CORRELATION STUDIES

While the correlation studies of natural ventilation estimates from wind tunnel models and full-scale measurements has been encouraging to date (Dick, Ref.2, and Smith, Ref.3) there will need to be many more. Techniques for modelling vegetation to reproduce full-scale flow characteristics need attention as well as proving the reliability of 10 m scaled windspeed measurements in tunnels as a reference for calculating windspeed and pressure coefficients. Where the local 10 m recording anemometer is too distant to include in the wind tunnel model, a smaller scale terrain model can be used to determine direction and windspeed corrections to existing records needed for a substitute reference anemometer location within the perimeter of the model including the building under test (Holmes, Ref.35). Correlation studies will need to be studied in a variety of terrain. The time alone needed for full-scale measurements as well as long term access to suitable test buildings prevents many studies of this nature being undertaken.

10. NEED FOR PRELIMINARY WIND TUNNEL STUDIES IN BUILDING DESIGN

As the size, shape and relative location of a building to adjacent buildings and obstructions are prime influences on the characteristics of airflow around a building, it is essential that these influences be studied in the early stages of the design of a building. It is only during these stages that fundamental changes that may be necessary can be made without suffering a severe cost penalty.

By using small scale clear plastic models together with foam plastic beads as a flow visualisation media, an experienced wind tunnel user can easily locate areas with poor natural ventilation and suggests modification for improvement. Such studies are inexpensive and can help the tunnel user select anemometer probe location points for later quantitative studies.

11. CONCLUSIONS

Use of boundary layer wind tunnel model studies to estimate natural ventilation in buildings appears very promising, particularly windspeed coefficient methods. The mean windspeed coefficient approach to estimating airflow through buildings for thermal comfort seems to offer a number of attractions to architects and other building designers.

Firstly the frequent use of models by architects during the design of buildings suggests that, if the scale was determined so as to meet blockage criteria for an available wind tunnel, a simple series of tests could provide valuable design data.

Secondly the increasing number of boundary layer wind tunnels at universities and building research establishments, suitable for studies of building models, means they are available to many designers at reasonable cost.

Thirdly, the method, when based on readily available meteorological data, provides a simple rational basis for evaluating the relative performance of alternative designs for low-energy, naturally ventilated buildings in oppressive hot humid climates.

Early correlation studies of model and full-scale data show close agreement. However the extreme complexity of wake flows from groups of bluff bodies in a turbulent layer suggests many more such studies will be needed before the reliability and order of accuracy of current modelling techniques can be confirmed.

REFERENCES

1   Vitruvius, The ten books of architecture, translated by Morris Hickey Morgan, Dover, New York(1960).
2   J.B.Dick, The fundamentals of natural ventilation of houses,J.Institution of Heating and Ventilating Engineers, 18,no.2 (1950).
3   E.G.Smith, The feasibility of using models for predetermining natural ventilation, Texas Engineering Experiment Station,Res.Rept.26 (1951).
4   J.J.Wannenburg and J.F.Van Straaten, Wind tunnel tests on scale model buildings as a means for studying ventilation and allied problems, J.Institution of Heating and Ventilating Engineers, (March 1957).
5   R.M.Aynsley,W.Melbourne and B.J.Vickery,Architectural aerodynamics, Applied Science, London (1977).
6   R.J.McKeon and W.H.Melbourne, Wind tunnel blockage effects and drag on bluff bodies in a rough wall boundary layer, Paper from Proceedings of the Third Int.Conf.on Wind Effects on Buildings and Structures (Tokyo), Saikon Shuppan Co.Ltd (1971).
7   W.Caudill, Some general considerations in the natural ventilation of buildings, Texas Engineering Experiment Station, Res.Rept.22 (1951).
8   W.Caudill and Bob H.Reed, Geometry of classrooms as related to natural lighting and natural ventilation, Texas Engineering Experiment Station Res.Rept.36 (1952).
9   E.T.Weston,Natural ventilation in industrial-type buildings, Commonwealth Experimental Building Station, Spec.Rept.14 (1954).
10 D.Surry and N.Isyumov, Model studies of wind effects - a perspective on the problems of experimental techniques and instrumentation, Int.Congress on Instrumentation in Aerospace Simulation Facilities, Record (1975) 76-91.
11 B.J.Vickery and L.W.Apperley, On the prediction and evaluation of the ground wind environment, Dept.of Civil Engineering,University of Sydney Res.Rept.R277 (1973).

12 A.S.H.R.A.E., Handbook of fundamentals, American Society for Heating, Refrigerating and Air-Conditioning Engineers (New York) (1981) Ch.14.

13 I.H.V.E.,Guide, Institute of Heating and Ventilating Engineers (Lond.) (1972).

14 W.A.Dalgleish, Comparison of model/full-scale wind pressures on a high rise building, J.Ind.Aerodynamics, 1,no.1 (1975).

15 N.Isyumov and A.G.Davenport, Comparison of full scale and wind tunnel windspeed measurements in the Commerce Court Plaza, J.Ind.Aerodynamics, 1 (1975) 201-212.

16 R.F.White,Effects of landscape development on the natural ventilation of buildings and their adjacent areas, Texas Engineering Experiment Station, Res.Rept.45 (1954).

17 T.Wheeler, An energy conservation in buildings programme for P.N.G., Energy Planning Unit,Department of Minerals and Energy (Konedobu, P.N.G.) Report 2/81 (1981).

18 R.Stedman, Kent Ridge teaching hospital,Singapore, Institution of Engineers Australia,Nat.Symposium on Energy Conservation in Buildings, (Sydney),(1978) 70-74.

19 B.Givoni,Man,climate and Architecture,2nd ed, Applied Science,London (1976).

20 R.M.Aynsley, Tropical housing comfort by natural airflow, Building Res. and Practice 8,no.4,(1980) 242-253.

21 J.F.Van Straaten,Thermal performance of buildings, Elsevier,Amsterdam, (1967).

22 B.Givoni, Study of ventilation problems in housing in hot countries, Technion,Building Research Station ,Report (Oct.1962).

23 I.Chand, Prediction of air movement in buildings,Central Building Res. Institute (Roorkee) Building Digest 100 (Sept.1972).

24 G.F.Hamilton,Effect of velocity distribution on wind loads on walls and low buildings, University of Toronto,Technical Paper 6205 (Nov. 1962).

25 W.V.Macfarlane, Thermal comfort zones, Architectural Science Review, 1,no.1 (1958) 1-14.

26 Climatic averages Australia, Metric ed., Bureau of Meteorology,Canberra (1975).

27 Wind frequency analyses: computer output for individual locations. Available for purchase from the Bureau of Meteorology,Melbourne.

28 R.M.Aynsley, Wind generated natural ventilation of housing for thermal comfort in hot humid climates, Fifth Int.Conf.on Wind Engineering, July 1979, Colorado State Univ.(Fort Collins) ,Vol.1 (1979) III-3-1 - III-3-12.

29 W.A.Snyckers, Wind tunnel studies of the flow of air through rectangular openings with application to natural ventilation in buildings, University of Pretoria, Master of Science in Engineering Thesis(1970).

30 Ning Chien et al, Wind-tunnel studies of pressure distribution on elementary building forms, Iowa Inst.of Hydraulic Res.,State Univ. of Iowa (1951).

31 M.Jensen and N.Franck,Model-scale tests in turbulent wind,Part II Phenomena dependent of the velocity pressure.Wind loads on buildings, Danish Technical Press,Copenhagen (1965).

32 S.A.A.loading code Part 2 wind forces, Australian Standard 1170,Part 2 (1975).

33 R.J.Best and J.D.Holmes, Model study of wind pressures on an isolated single-storey house, James Cook University,Dept.of Civil and Systems Engineering, Wind Engineering Rept. 3/78 (1978).

34   J.D.Holmes and R.J.Best, A wind tunnel study of wind pressures on
     grouped tropical houses, James Cook Univ.(Townsville) Dept.of Civil
     and Systems Engineering,Wind Engineering Report 5/79 (1979).

35   J.D.Holmes, G.R.Walker and W.E.Steen, The effect of an isolated hill
     on wind velocities near ground level - initial measurements, James
     Cook University (Townsville), Dept.of Civil and Systems Engineering
     Wind Engineering Report 3/79 (1979).

APPENDIX 1.

Example

Determine the likelihood of maintaining comfort by natural airflow in the living area of a low set residence with extended eaves and end walls (Fig.2) in Townsville at 3 pm during January.

Climatic Data

> 3 pm, 86 percentile daily maximum dry   = 32.9°C
>> bulb temperature for January
>
> 3 pm average relative humidity   = 62%
>
> Airflow needed to restore comfort from   = 0.83 m/s
>> Figure 5

Mean Windspeed Coefficients

Method : Using mean windspeed coefficient $C_{\overline{v_1}}$ of 0.72 for the living room indicated in Figure 3 for wind from the North, the windspeed, $\overline{V}_{10}$ required at the reference height of 10 metres to ensure thermal comfort in the living room can be calculated by substituting 0.83 m/s for the local indoor windspeed $\overline{V}_1$,

$$\overline{V}_{10}(\text{for comfort}) = \frac{0.83}{C_{\overline{v_1}}} \qquad \text{from Equation 3}$$
$$= \frac{0.83}{0.72}$$
$$= 1.15 \text{ m/s}$$

By summing the frequency of occurrence of all windspeeds 10 metres above the ground, from Table 4, which exceed 1.15 m/s for winds from the North, an estimate of the percentage of time winds from the North would provide thermal comfort in the living room, at 3 pm during January in Townsville.

| Directions | Windspeed intervals (metres per second) | | | | | |
	Calms	0.5-1.5	2.0-3.0	3.6-5.0	5.7-8.2	8.7-10.8	Total(%)
	0.6						0.6
NNE		0.6	2.6	1.9	3.2	–	8.4
NE		1.3	4.5	14.3	12.3	–	32.5
ENE		–	0.6	11.7	7.8	–	20.1
E		–	0.6	3.2	3.9	1.9	9.7
ESE		–	–	0.6	1.9	1.3	3.9
SE		–	0.6	1.9	–	–	2.6
SW		–	0.6	–	–	–	0.6
NW		–	–	1.3	–	–	1.3
NNW		–	1.3	1.9	0.6	–	3.9
N		0.6	3.9	8.4	2.6	0.6	16.2

Table 4: Percentage Distribution of 10 m Winds at 3 pm at Townsville during January.

From Table 4 percentage occurrences of northerly winds for all wind-speed intervals exceeding 1.15 m/s are 3.9% for 2.0 to 3.0 m/s; 8.4% for 3.6 to 5.0 m/s; 2.6% for 5.7 to 8.2 m/s and 0.6% for 8.7 to 10.8 m/s, giving a total of 15.5% of time when northerly winds will restore comfort. This process is repeated (Table 5) for each wind direction until one obtains a total percentage of time during which comfort is likely to be maintained.

Direction	$C_{\overline{v}_1}$	10m $\overline{V}$ to restore comfort $(0.83/C_{\overline{v}})$ m/s	Cumulative Percentage of time winds exceed %
N	0.72	1.15	15.5
NNE	0.90	0.92	7.8
NE	0.95	0.87	31.2
ENE	0.85	0.98	20.1
E	negligible	–	negligible
ESE	0.85	0.98	3.9
SE	0.95	0.87	2.6
SSE	0.90	0.92	negligible
S	0.72	1.15	"
SSW	0.70	1.19	"
SW	0.52	1.60	0.6
WSW	negligible	–	negligible
W	"	–	"
WNW	"	–	"
NW	0.52	1.60	1.3
NNW	0.70	1.19	3.9
			86.9%

Table 5:   Total Percentage of Time during which Comfort is likely to be maintained for all Wind Directions.

The actual percentage of time that thermal comfort can be maintained will be higher than the estimated 87.0 percent because all windspeeds through the living area (with a 2 m/s external reference windspeed) exceed by a significant amount the minimum internal windspeed for thermal comfort of 0.83 m/s. However the size of windspeed increments provided in the available wind data 0.5-1.5 and 2.0-3.0 m/s prevent a more accurate appraisal.

The remaining periods of time when thermal comfort cannot be achieved by natural ventilation, that is less than 13 percent, should be catered for by the occasional use of ceiling fans.

APPENDIX 2.

Example

Determination of the extent to which natural ventilation is likely to maintain comfort in the living room of the isolated high set house (Fig.7) in Townsville at 3 pm during January using the discharge equation.

Figure 7:
Plan of High Set
House Showing
Living room Wall
Openings.

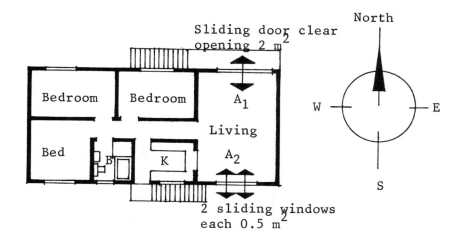

Climatic Data

    3 pm, 86 percentile average maximum dry bulb
        temperature for January   = 32.9°C

    3 pm average relative humidity   = 62 %

Assuming adequate insulation is provided in the building envelope and windows and walls are shaded to control radiation, the minimum indoor airflow needed to restore thermal comfort for the above temperature and humidity from Figure 5 is 0.83 m/s.

Windspeed frequency and directions for Townsville at 3 pm during January are indicated in Table 4. It would be reasonable to expect this data to be respresentative of most Townsville suburbs as there is no significant change in terrain over most of the suburban area.

Building Data

Orientation and layout of isolated high set house are as indicated in Figure 7.

Openings in living room walls:

	Dimensions	Clear area of Opening
Sliding door north facing wall	2 m x 1 m	2.0 m^2
2 sliding windows south facing wall	1 m x 0.5m each	1.0 m^2
Cross-section of living room	3 m x 4 m	12.0 m^2

Method

NE Winds

Wind incidence used 45°

Pressure coefficients at locations near openings from Aynsley (Ref.5):

Windward   +0.33
Leeward   -0.38

Free area of windward opening = 2 m^2

Correction factor for 45° incidence = cosine 45° = 0.707
Corrected free area = 2 x 0.707 m^2
= 1.4 m^2

Discharge coefficient from Figure 7 for windward
opening with two edges common with
downwind space   = 0.75
Free area of leeward openings   = 1 m^2

Discharge coefficient for leeward openings
from Figure 8 with an A_o/A_1 ratio of 1/12th
or 0.08   = 0.63

Discharge Q, to achieve thermal comfort at the windward opening is the product of the effective area of the opening ($A_1 \cos \theta$) and the minimum windspeed for thermal comfort 0.83 m/s, that is:

$$Q = A_1 \cos \theta \, \bar{V}_1$$
$$= 2 \times 0.707 \times 0.83$$
$$= 1.17 \text{ m}^3/\text{s}$$

Equation 6 can be rearranged to solve for the windspeed 10 m above ground needed to achieve the above discharge rate and corresponding minimum indoor windspeed for thermal comfort.

$$V_{10} = \frac{A_1 \cos \theta \, \bar{V}_1}{\left[\dfrac{(C_{p1} - C_{p2})}{\dfrac{1}{C_{d_1}^2 A_1^2} + \dfrac{1}{C_{d_2}^2 A_2^2}} \right]^{\frac{1}{2}}}$$

$$= \frac{1.17}{\left\{ \dfrac{(0.30-(-0.38))}{\dfrac{1}{0.75^2 \times 1.4^2} + \dfrac{1}{0.63^2 \times 1^2}} \right\}^{\frac{1}{2}}}$$

$$= 2.63 \text{ m/s}$$

Discharge Q to achieve thermal comfort at the leeward openings is the product of the effective area of the opening and the minimum local windspeed needed for thermal comfort.

$$A = A_2V_2$$
$$= 1 \times 0.83$$
$$= 0.83 \ m^3/s$$

Solving for the 10 m windspeed to achieve this discharge rate as before:

$$V_{10} = \frac{0.8}{\left(\frac{(0.30-(-0.38))}{\frac{1}{0.75^2 \times 1.4^2} + \frac{1}{0.63^2 \times 1^2}}\right)^{\frac{1}{2}}}$$

$$= 1.87 \ m/s$$

Clearly for winds from the North East the governing 10 m windspeed requirement is 2.63 m/s for the windward opening, which falls within the 2.0 - 3.0 m/s 10m windspeed interval on the wind frequency table (Figure 4).

Summing the percentage occurrence of higher windspeed intervals 3.6 - 5.0 m/s of 14.3% and 5.7 - 8.2 m/s of 12.3% for the North East direction gives a total of 26.6% of time.

Wind direction	Incidence θ	C_{p1}	C_{p2}	A_1	$A_1 Cos \ \theta$	A_2	$A_2 Cos \ \theta$	C_{d1}	C_{d2}	Minimum windspeed interval for thermal comfort	% Occurrence of thermal comfort
NNE	22½	0.36	0	2	1.8	1	–	0.75	0.63	5.7	3.2
NE	45	0.30	-0.38	2	1.4	1	–	0.75	0.63	3.6	26.6
ENE	67½	0.14	-0.07	2	0.77	1	–	0.75	0.63	5.7	7.8
ESE	67½	-0.07	0.14	2	–	1	0.38	0.64	0.70	14.4	–
SE	45	-0.38	0.30	2	–	1	0.71	0.64	0.70	5.7	–
SW	45	-0.16	0	2	–	1	0.71	0.64	0.70	11.3	–
NW	45	0.0	-0.16	2	1.4	1	–	0.75	0.63	5.7	–
NNW	22½	0.16	-0.07	2	1.8	1	–	0.75	0.63	5.7	0.6
N	0	0.55	-0.12	2	2.0	1	–	0.75	0.63	3.6	11.6
									TOTAL		49.8 %

Table 6: Total Percentage of Time during which Comfort is Likely to be Maintained for all Wind Directions.

When all wind directions are considered (Table 6) comfort is likely to be achieved near both windward and leeward openings for the 86 percentile dry bulb temperature of $33^{\circ}C$ and relative humidity of 62% for approximately 50% of the time at 3 pm during the month of January.

WIND TUNNEL MODELING APPLIED TO PEDESTRIAN COMFORT

Shuzo Murakami, Institute of Industrial Science
University of Tokyo
Tokyo 106, Japan

SUMMARY

Methodology for wind tunnel modeling concerning pedestrian comfort is described. As the effects of wind on the human body are so complicated and varied, the modeling of both the human body and the natural wind is particularly difficult in connection with this subject. It is thus impossible to advance research concerned with this subject solely by means of wind tunnel experiments. It is also necessary to conduct outdoor experiments in natural wind.

In a specific site study, the minimum number of wind-directions which must be tested is important. A method for reducing the number of wind direction to be tested without loss of accuracy in the assessment is also described.

1. DIFFICULTY OF WIND TUNNEL MODELING FOR WIND EFFECTS ON PEOPLE

The physical effects of wind on people may be classified into two types. The first includes the mechanical effects[1-4,6-14] and the second the thermal effects[1,5,13-16]. Stumbling due to gust is an example belonging to the former type while coldness due to long exposure at a windy outdoor space belongs to the latter. For both types of effects wind tunnel modeling is very difficult for the following reasons.

In the first place, modeling of the human body is very difficult. In the case of mechanical effects, it is almost impossible to reproduce the behavior of the human body by any mechanical model. In the case of thermal effects, modeling is also difficult because the heat loss from a human body consists of not only sensible heat but also latent heat. The most damageable points in the human body are the eyes and it is especially difficult to model these[2]. In a windy space, people are affected not only mechanically and thermally but also psychologically, especially by sound due to wind.

In the second place, modeling of the natural wind is also difficult in connection with this subject. The behavior of the human body is influenced most greatly by change in the wind speed, and especially by turbulence of 1.0 - 0.3 c/s frequency.

The human body can resist rather easily changes of wind speed which are anticipated but is vulnerable to non-anticipated ones. When holding an umbrella, a person is more vulnerable to gust than he is without one. It is difficult to produce turbulent wind of 1.0 - 0.3 c/s frequency in the usual wind tunnel. However, Dr. Hunt et al.[3] have reported that they succeeded in producing a turbulent wind in this range of frequency by means of a new system of oscillating vanes in the wind tunnel.

It is thus almost impossible to advance research of wind effects on pedestrian comfort solely by means of wind tunnel experiments and using only models of the human body. It is necessary to conduct both wind tunnel experiments with real human bodies and also outdoor experiments in natural wind. Therefore, this report describes not only wind tunnel techniques but also experimental techniques which do not use a wind tunnel.

2. VARIOUS METHODS TO STUDY WIND EFFECTS ON PEDESTRIANS

There are many considerations in the study of wind effects on pedestrians. First is the time-length of the investigation. In the case of a long term observation of wind effects the investigation may continue for a few years, but in the case of a short term experiment it may continue for only a few seconds to a few minutes. A second consideration is the number of persons who take part in the experiment as subjects. As many as over two thousand pedestrians[2] and as few as two or three subjects may be observed in various experiments. A last consideration is the type of wind flow, which includes both the flow in a wind tunnel and that in natural wind.

There are thus many considerations to investigate when studying wind effects on people. Since the experimental method of wind tunnel modelling can provide information from only a limited viewpoint, it is necessary to utilize the many methods discussed below.

Table 1 Outline of walking tests in large wind tunnel [2]

1) Walking of subjects on the wind tunnel test floor was observed. The subjects were directed to walk on lines as shown in Fig 1(1) under various conditions: namely, walking with baggage; walking with umbrella (refer to Fig 1(2));etc.

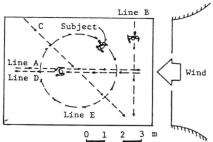

(1) Walking lines used by subjects,   (2) Walking with umbrella, U_∞=12.5m/s
    drawn on working section

Fig. 1 Working section of wind tunnel

2) Three methods were used for observation and analysis of walking:
    a) Mechanical recording and analysis of footsteps;
      (Each subject's walking was recorded by a 16 mm movie camera. The footsteps were analysed in detail by a motion-analyser.)
    b) Visual assessments by the experimenters on walking conditions, disturbance of hair and clothes, etc;
    c) Verbal assessments by the subjects on the relative ease or difficulty of walking.

3) The large wind tunnel used was of the open-jet type. The size of the working section measures 6 m in width, 13 m in length, and 5 m in height.

4) Two types of wind profile were used in the test:
    a) uniform-flow
    b) non-uniform flow with fences.
The height of the fences used was 1.8m and their lengths were 1.5m and 1.0m.

2.1. Wind Tunnel Modelling Under Condition of Constant Flow

The effect of short-term exposure to wind can be studied. Here, short-term means from a few seconds to twenty or thirty minutes.

An outline of the walking tests in a large wind tunnel as conducted by the author[2] is shown in Table 1. Table 2 shows an outline of the experiments conducted in a wind tunnel for measurement of drag force on people[4]. The experiments outlined in Tables 1 and 2 were concerned with the mechanical effects of wind. Table 3 shows an outline of the experiments concerned with the thermal effects of wind[5]. The wind tunnel used here was the silent type. The sense of coldness or warmness may be influenced by the noise from the blower set in the wind tunnel.

Table 2 Outline of experiment for measurement of drag force
on people [4]

1) The wind tunnel used was the same as that used for Table 1

2) Equipment for measuring drag force:
The wind drag was measured by means of a large balance installed in the floor of the wind tunnel. The measuring platform (3m by 4m) is shown in Fig 2. Load cells were used to measure the drag force. Fig 3 illustrates this equipment.

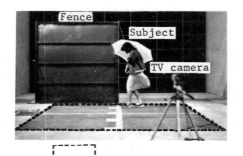

Measuring platform (3m×4m)
Fig 2 Inside view of wind tunnel

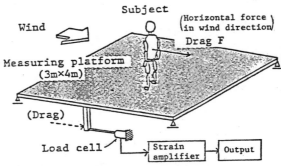

Fig 3 Measuring equipment for drag force

3) Wind profile at working section was the same as that in Table 1.

4) Test procedure:
The experiments consisted of three main tests
    a)tests when subjects were standing still;
    b)tests when walking in uniform flow;
    c)tests when walking in non-uniform flow with a fence.
Tests were conducted two times per subject, once when holding and once when not holding an umbrella.
5) After the walking tests, subjects assessed verbally their sense of difficulty or ease of walking. Walking postures were also recorded by a TV camera.
6) The projected area of each subject was measured by a planimeter from photographs taken by a 35 mm camera.

Table 3   Outline of wind tunnel experiment concerned with thermal effect of wind on human body [5]

1) Methodology of the experiment used to determine the grade of coldness:
Here the thermal effect is defined as the heat loss from a human body exposed to wind for a long time. The subjects sat still on a seat while exposed to the constant flow of the wind tunnel. Their sense of coldness became ever more severe as time elapsed. The subjects assessed changes in the grade of coldness as the experiment progressed.   The grades of coldness are shown below.   The experiment was stopped either when the subject said it was "impossible to bear" or when thirty minutes had elapsed.

Grades of coldness

1) Comfortable or not cool
2) Cool
3) Chilly
4) Cold
5) Bitter cold
6) Impossible to bear

Subject exposed to wind tunnel flow

2) Exposing of subjects to wind:
The condition under which subjects were exposed to wind is shown in Fig 4. The flow of this wind tunnel is silent. The wind tunnel belongs to Institute of Industrial Science, University of Tokyo.

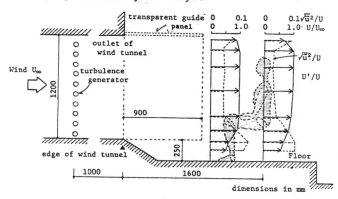

Fig 4   Velocity profile of wind exposed to subject

The experiments were conducted during various seasons, so the clothes worn by subjects changed according to air temperature.

3) Measurement of body surface temperature:
The body surface temperature of the subject was measured at various points using thermocouples.   The relationship between the grade of coldness and the surface temperature was then analyzed.

2.2. Outdooor Experiments Under Natural Wind Flow

The effect of natural wind can be observed by using as subjects either those chosen beforehand by the experimenters or ordinary pedestrians chosen at random.   An outline of the outdoor experiments conducted by the author[2] is shown in Table 4. In this experiment the subjects were chosen beforehand.   Table 5 shows an outline of observations made of a large number of ordinary pedestrians chosen at random [2].

2.3. Questioning Inhabitants Concerning Wind Effects

Long-term general impressions of wind effects can be obtained by interviewing many inhabitants of a given area[17-20].

490

Table 4 Outline of outdoor experiments of wind effects on walking, using subjects chosen beforehand [2]

1) The high-rise building and the pedestrian precinct are located in Shinjuku, Tokyo. A plan of the precinct is shown in Fig 4.

2) Methodology used for walking tests : The walking of the subjects was recorded by a 16 mm movie camera. The subjects were directed beforehand to walk on the walking line. The walking line was set at the windiest place in the precinct and was changed according to the prevailing wind direction, as shown in Fig 4.   Its length was about 7 to 8 m. One walking trial along the line took about 10 seconds per subject, and the trials were repeated many times. After each walking trial, verbal assessments by the subjects were recorded, as well as visual assessments by the experimenters. The wind speed at the walking line was measured by a sonic-anemometer set at a height of 1 m. Footsteps of the subjects were analysed by a motion-analyser using the film record of the walking tests.

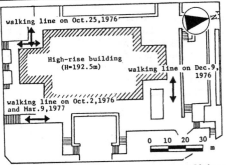

Fig 4 Plan of precinct and walking lines

3) Wind velocity on the date of tests :

Date of test	Daily maximum wind speed at Tokyo Met.Observ.	Mean wind speed $\overline{U}_{10-sec}$ at walking line	Intensity
No.1 Oct. 2, 1976	6.9 m/s	1.56 m/s	about25 %
No.2 Oct.25, 1976	23.4	3.80	53
No.3 Dec. 9, 1976	13.5	3.92	10
No.4 Mar. 9, 1977	25.4	7.62	15

Table 5 Outline of observations of wind effects on ordinary pedestrians chosen at random [2]

1) The high-rise building and the pedestrian precinct are located in Shinjuku, Tokyo. A plan of the precinct is shown in Fig 5.

2) Methodology used for walking tests : The walking conditions of pedestrians were recorded by a video TV camera set at a distance from the pedestrian passageway in a covered space. Seven serial pictures per second could be reproduced from the record of the video camera. Wind velocity at the passageway was measured by a sonic-anemometer set at a height of 1 m.

June.20,1978
Aug.3,1978
Wind direction:
SSE at Tokyo
Met.Observ.

Fig 5 Plan of precinct and wind direction during tests

3) Grading system for footstep-irregularity and walking-balance:
The serial pictures of pedestrians were analysed and grades of wind effects were assessed for each pedestrian. The grades for footstep-irregularity and walking-balance are shown below. Fig 12(1) is an example of "footstep-irregularity Grade 5" and (2) is an example of "walking-balance Grade D."

(1) Grades of footstep-irregularity

Grade 1 no effect

Grade 2 minor irregularity of footsteps

Grade 3 footsteps irregular

Grade 4 walking line greatly irregular
(walking difficult to control, body sometimes blown sideways)

Grade 5 body blown sideways or to leeward
(walking impossible to control, walking difficult, walking stops)

(2) Grades of walking-balance

Grade A no effect

Grade B sensitive to wind
(face turns sideways to avoid gust, face covered with hand to break gust)

Grade C upper-half of body bends to windward

Grade D whole body bends to windward, whole body swings

3. SCALES TO MEASURE WIND EFFECTS

As the effects of wind on pedestrians are both complicated and varied, the first step must be to find a suitable scale by which to measure the effects. Some useful scales are shown below:

3.1. Drag Force

This is the most fundamental information required to measure wind effect. While it is easy to measure exactly the drag force of wind on a person standing still[11,12], it is not easy to measure the drag force on a walking person because the response of the measuring equipment is not fast enough[4]. Fig 6 shows the drag force on a person who is standing still[6]. Fig 7 shows the fluctuation of the drag force on a walking person[4].

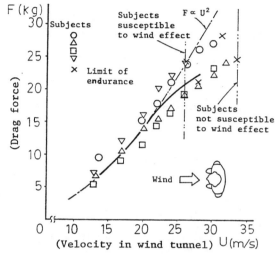

Fig. 6 Drag force on person standing still
(refer to Table 2 for techniques of measurement)

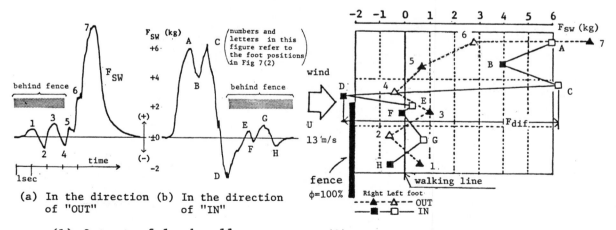

(a) In the direction (b) In the direction
of "OUT" of "IN"

(1) Output of load-cell

(2) Relationship between drag force
and position of fence

Fig. 7 Fluctuation of drag in non-uniform flow behind fence (ref. to Tab.2)

$\left(\begin{array}{l}\text{OUT : walking direction of subject from strong} \\ \qquad \text{wind area into wake area of fence} \\ \text{IN : opposite of "OUT"} \\ \text{Fsw : drag force on walking person}\end{array}\right)$

3.2. Scales Given by Analysis of Mechanical Recording of Walking

Useful scales are obtained by analyzing filmed records of various walking conditions. The scales include the irregularity of footsteps, the walking speed, the length of one step, the distance between left foot and right foot, etc. The relationships between these scales and the wind speed and direction also give useful scales. The relationship between the wind speed and footsteps[2] is shown in Fig 8. Fig 9 shows an example of analysis of footstep-traces[2]. Here X means the length of one step and Y means the distance between right and left foot. These results belong to the experiments outlined in Table 4.

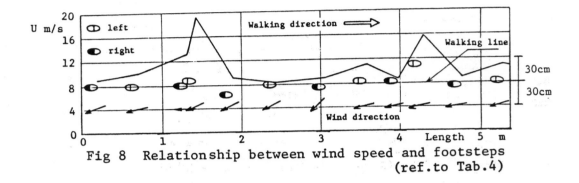

Fig 8 Relationship between wind speed and footsteps
(ref.to Tab.4)

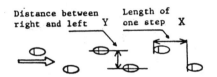

Length of one step and distance
between right and left foot

Date of tests	Oct. 2	Oct.25	Dec. 9	Dec. 9
Highly turbulent	–	✗	–	–
Non-highly, follow	●	–	■	▲
Non-highly, against	○	–	□	△

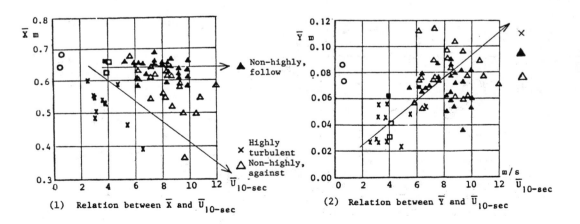

(1) Relation between $\overline{X}$ and $\overline{U}_{10-sec}$ (2) Relation between $\overline{Y}$ and $\overline{U}_{10-sec}$

Fig 9 Analysis of footsteps (ref.to Tab.4)

3.3. Verbal Assessments by Subjects

Impressions of subjects are useful scales. While they do not give quantitative information they do give very valuable qualitative information.

3.4. Visual Assessments by Experimenters

Observation by the experimenters and repeated viewings of the filmed records of pedestrians also give general information on the effects of wind on walking conditions. Table 6 shows verbal and visual assessments by subjects and experimenters based on wind tunnel experiments[2]. Table 7 shows those based on experiments in natural wind. Fig 12 shows the assessment of walking conditions by the experimenters, using serial pictures of pedestrians[2].

Table 6 Results of verbal and visual assessments of wind tunnel
experiments (ref.to Tab.1 for measuring techniques)

wind tunnel U_∞ (can be assumed, $=\hat{U}_{n\,max}$) (natural wind $\bar{U}_n$ (estimated as G.F.≒2.0)		5 2.5	10 5	15 7.5	20 m/s 15)
Uniform flow	effects on walking		walking not easy footsteps-irregular body-balance lost	walking diffi-cult to control upper-body bends to windward	walking very difficult whole body bends to windward
	on hair	minor disturb-ance	disturbed	violently dis-turbed	
	on face eyes ears	wind felt on face	frequent blinking wind noisy	impossible to open eyes con-tinuously; falling tears	impossible to face wind, facial pain ear-ache, headache breathing difficult
	on clothes and umbrella	minor disturb-ance	fluttering of clothes difficult to hold umbrella	impossible to hold umbrella	violent flutter-ing of clothes
Non-uniform flow	effects on walking		walking not easy	walking diffi-cult stumbling	dangerous possible falling

(for non-uniform flow downward revision is required)

3.5. Verbal and Visual Assessments by Inhabitants

The general impressions of inhabitants about wind effects give good scales by which to measure the long-term effects of wind in a given area[17-20].

4. SIMULATION OF WIND FLOW IN WIND TUNNEL: Consideration of Effect from Gustiness

In the outdoor experiments, pedestrians are influenced most strongly by gusts of 1.0 - 0.3 c/s frequency. If the wind velocity is constant, a person can endure very high-velocity wind--30 m/s or so--by bending his whole body to windward. Fig 10 shows a person exposed to a constant wind flow of 32 - 34 m/s[6]. It is therefore necessary to produce changes in wind velocity when experiments concerned with wind effects on pedestrians are planned for a wind tunnel.

494

While it is not easy to produce turbulent wind of 1.0 - 0.3 c/s frequency, it is possible to produce large changes in wind speed in a wind tunnel using certain types of fences.

Fig.10 Subject exposed to constant wind flow of 32-34 m/s
(ref.to Tab.2)

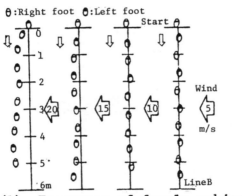

(1) Footsteps of female subject taken in uniform cross-wind
(as indicated by walking line B in Fig.1)

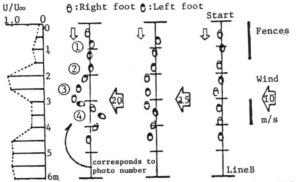

(2) As (1), but taken in non-uniform flow behind fences

Serial photos of female subject taken at 1-second intervals in non-uniform flow behind fences, with U∞ = 20m/s

Fig.11 Comparison between footsteps when walking in uniform flow and footsteps when walking behind fences (ref.to Tab.1)

Measurements taken of walking from a low-speed area protected by a fence into a high-speed area as well as the reverse are very valuable for studying the effect of velocity changes on walking. The effect of gustiness on walking is suggested by analogy with these experiments. Comparison between footsteps when walking in uniform flow and footsteps when walking behind fences[2] is shown in Fig 11. Fig 7 also shows the severe effect of velocity changes on pedestrians. Table 7 shows the difference in assessments made for the condition of highly turbulent wind and those made for non-highly turbulent wind[2].

Table 7 Difference in assessments made for condition of highly turbulent wind and for non-highly turbulent wind (ref. to Tab. 4)

$\overline{U}_{10\text{-}sec}$ (at height of 1 m, averaged over 10 sec)		1 2 3 4	5 6	7 8	9 10 m/s
non-highly turbulent wind (around 20 %)	effects on walking		little effect	footsteps irregular	walking difficult to control
	on hair		hair disturbed	hair blown strongly	hair violently disturbed
	on clothes		partial fluttering	clothes violently fluttered	
	on eyes face breathing etc.			skirt flew up / wind felt strongly	eyes felt dry / difficult to face wind / wind felt violent
highly turbulent wind (50~60 %)	effects on walking	walking difficult to control / footsteps irregular			
	on hair	disturbed	hair blown strongly		
	on clothes	partial fluttering	clothes violently fluttered		
	on eyes face breathing etc.	wind felt strong	skirt flew up / difficult to face wind		

5. CONDITION OF SUBJECTS WHEN WIND EFFECTS ARE EXAMINED

Wind effects on pedestrians are influenced greatly by the condition of the subject. Factors which contribute to subject-condition include the following.

1) Age and sex of subject:

Because an older person does not move so easily as a younger one, the effect of wind on the old is more severe than it is on the young. In general, women are more affected by the wind than are men. These points must be considered when wind criteria for walking are proposed. The filmed records for a man and a woman who lost their walking balance in natural wind are shown in Fig 12.

2) Belongings:

The effect of wind on an opened umbrella is extremely great. Criteria based on experimental data which omit such a case are limited in usefulness. Fig 13 shows the difference of drag force between the case with umbrella and that without umbrella[4].

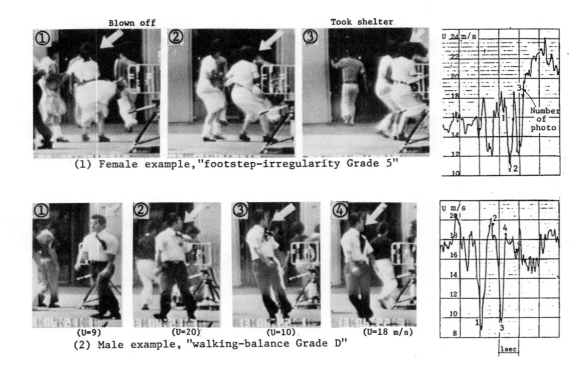

Fig.12   Examples of assessments of footstep-irregularity and
          walking-balance grades for male and female pedestrians
                                                        (ref.to Tab.5)

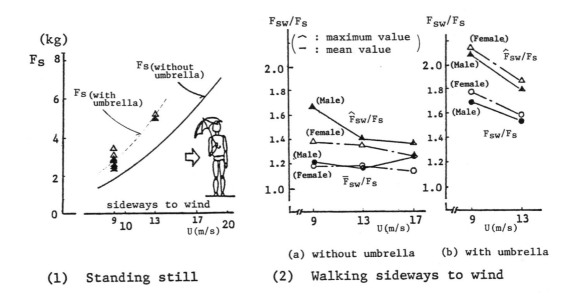

(1) Standing still   (2)   Walking sideways to wind

Fig. 13   Drag force on person with umbrella (ref. to Tab.2)

$$\left(\begin{array}{l} F_{SW} : \text{drag force, walking in uniform flow} \\ F_S \ \ : \text{drag force, standing still} \end{array}\right)$$

3) Clothes:

A person wearing a skirt is more vulnerable to wind effects than one wearing trousers.

4) Hair, eyes, etc:

Long hair is easily blown and disturbed by the wind. Women are generally more sensitive to the effect of wind on their hair. Eyes are the point most easily damaged by the wind. As soon as the eyes are damaged, it becomes impossible to continue walking. It becomes easier to endure high speed wind when spectacles of the goggle or skiing-type are worn.

5) Temperature:

The discomfort caused by wind is more severe when it is cold than when it is warm.

All these conditions should be considered when planning experiments for testing pedestrian comfort or when suggesting criteria for wind-acceptability.

6. INTERPRETATION OF WIND TUNNEL DATA

As mentioned above, wind tunnel modeling of turbulence with frequency of 1.0 - 0.3 c/s--that which most greatly affects pedestrians--is very difficult. Therefore the wind-tunnel flow with no such turbulence expressed as a mean value is less uncomfortable than is a natural wind flow of the same mean velocity. That is, when the mean velocity of both flows is the same, the wind tunnel flow has less affect on pedestrians than does the natural wind. Thus, for example, the limit of wind speed dangerous for walking as given by the wind tunnel experiment corresponds to a much lower mean value for natural wind conditons. When the air is dusty, the natural wind becomes even more uncomfortable.

In light of these points, the correspondence between wind tunnel flow and natural wind is proposed as follows.

Among the various effects of turbulent wind, that of instantaneous maximum velocity is most severe on people. We must therefore take note of the instantaneous maximum value of natural wind. From the viewpoint of wind effect, the maximum value of natural wind corresponds to the maximum value of wind tunnel flow. In addition, for the purposes of these wind-tunnel experiments, the constant velocity in the wind tunnel may be assumed to represent the continuous case of instantaneous maximum velocity.

It is then possible to convert the wind-tunnel velocity to that of natural wind by using Gust Factor. When the value of G.F. for natural wind in streets is assumed to be 2.0, half of the wind-tunnel velocity corresponds to the value of natural wind expressed by mean value. The velocity scale in Table 6 is shown in two ways in consideration of this correspondence.

7. ACCEPTABLE CRITERIA

Various acceptable criteria for wind effects on people are compared in Table 8. Most admirable are those suggested by Hunt et al. which are based on extensive wind tunnel experiments(No.2 in Table 8). In their experiments a type of gusty wind was reproduced. The criteria suggested by the author are based on wind tunnel experiments(No.3), on outdoor experiments(No.4, No.5), and on the observation of a great number of ordinary pedestrians(No.1).

Next, criteria for windiness as suggested by Davenport[7, 8] are shown in Table 9. These criteria are suggested not only for pedestrians but also for general out-door activities. It is notable and also admirable here that the wind velocity for each grade (Beaufort Number) is related to its exceedance probability. It is not clear, however, on what experiments these proposals for values of wind speed and probability are based.

Finally, the criteria suggested by the author are shown in Table 10. These criteria too are not only for pedestrians but also for such out-door activities as shopping. These proposed criteria are based on the many types of experiments shown in Table 1 - 5, and also on extensive questioning of inhabitants[17-20] and long-term wind observation near the ground[21-24].

Table 8 Various suggested criteria for wind effects on pedestrians
(based on instantaneous wind speed)   [2,3,7-10,14]

```
0          5          10          15          20    m/s 25
|----------|----------|----------|----------|----------|
```

No.1 ■ Murakami and Deguchi : on the basis of experiments in Table 4
; based on instantaneous wind speed U_{3-sec}

No effect	Minor irregularity of footsteps	Footsteps irregular Walking line greatly irregular	Body blown sideways or to leeward

2 ■ Hunt et al.[3]   ; based on gusty wind

For comfort and little effect on performance	Most performance unaffected	For control of walking	For safety of walking	

3 ■ Murakami and Deguchi (Table 1); based on uniform flow U_∞ = peak of street wind U_s

		Walking not easy	Walking difficult to control	Walking difficult or almost impossible

4 ■ Murakami and Deguchi (Table 4); based on non-highly turbulent wind $\bar{U}_{10-sec}$

	Footsteps irregular	Walking difficult to control	

5 ■ Murakami and Deguchi (Table 4) ; based on highly turbulent wind $\bar{U}_{10-sec}$

	Walking difficult to control Footsteps irregular	

6 ■ Melbourne and Joubert [9]   ; based on gust speed

	Comfortable		The limits of safety	Danger

7 ■ Gandemer [10]   ; based on $u = \bar{U} + \sqrt{u'^2}$

	Comfort	Discomfort	

8 ■ Lawson and Penwarden [14]   ; based on Beaufort Scale, U_{3-sec}

1·2	3	4	5	6	7
Wind felt on face	Hair disturbed clothing flaps etc.	Hair disarranged	Danger of stumbling when entering a windy zone etc.	Difficult to walk steadily etc.	

(difference of averaging-time in each criteria should be considerd with care)

Table 9   Tentative Comfort Criteria for Windiness
          (after Davenport [7,8])

Activity	Areas Applicable	RELATIVE COMFORT			
		Comfort	Tolerable	Unpleasant	Dangerous
1. Walking fast	Sidewalks	5	6	7	8
2. Strolling, skating	Parks, entrances skating rinks	4	5	6	8
3. Standing, sitting -short exposure	Parks, plaza areas	3	4	5	8
4. Standing, sitting -long exposure	Outdoor restaurants bandshells, theatres	2	3	4	8
Representative criteria for acceptability			<1/wk.	<1/mo.	<1/yr.

Units: Beaufort Number   Temperetures > 10°C

Table 10   Acceptable criteria for wind environment
            based on daily maximum instantaneous wind speed
            [2,4,17-24]

	Rank 1	Rank 2	Rank 3
Purpose of area	Areas used for purposes most susceptible to wind effects*1	Areas used for purposes not too susceptible to wind effects	Areas used for purposes least susceptible to wind effects*2
Limit of exceedance probability	Limit of exceedance probability of 10 m/s (daily maximum instantaneous wind speed)		
	below 10 %	below 22%	below 35%
	37 days in a year	80 days in a year	128 days in a year

-Wind speed is defined as that at a height of about 1.5m from ground.
-The instantaneous wind speed 10m/s may be converted to mean velue
    using the following Gust Factors:
        G.F.=1.5~2.0; at places where wind speed is very high,
                for example in the precincts of high-rise buildings,
            =2.0~2.5; at places where wind speed is normal,
            =2.5~3.0; at places where wind speed is low,
                for example in narrow streets in overcrowded areas.

*1 for example, outdoor shopping areas, outdoor restaurants, home gardens.
*2 for example, sidewalks.

8. NUMBER OF WIND DIRECTIONS TO BE INVESTIGATED IN STUDY OF SPECIFIC SITE

When a high-rise building is projected wind tunnel tests are conducted to assess wind-induced discomfort, using a city model of the specific site in scale about 1/200 - 1/1000. When the wind speed at one point is assessed, the value of probability of exceeding the tolerable velocity is the most preferable scale by which to evaluate the strength of the wind speed at that point averaged over one year. Therefore the result of the wind tunnel test for each point should be shown using this probability.

In order to estimate the exceedance probabiblity precisely, wind tunnel tests should be conducted for all 16 wind directions. If the values of the velocity for each wind direction are integrated for 16 wind directions, with the weight of occurrence probability given for each direction, the value of exceedance probability can be calculated for one point. The details of this method are discribed by Isyumov and Davenport[8].

However, as a matter of fact, it is not easy to carry out wind tunnel tests for 16 wind directions and for numerous measuring points because such requires too much time and work. It would thus be very useful for a researcher engaged in wind tunnel experiments to be able to reduce the number of wind directions used in the wind tunnel tests without loss of accuracy in his estimation of exceedance probability. The most reasonable method of doing this is to eliminate those directions whose occurrence probability is low.

Fig 14 shows comparisons of the values of exceedance probability of wind speed near ground level in the case of a single block model and in the case of a specific site study[25]. The estimated values, using 16, 8, 6, and 4 wind directions, are then compared. For example, in the case of 6 directions, wind tunnel experiments were conducted for only those 6 directions whose occurrence probabilities were prevailing, and for the remaining 10 directions wind tunnel tests were not conducted. But when the exceedance probability was calculated, integration of wind velocity was made for all 16 directions. The velocity values for those directions for which wind tunnel tests were not conducted were assumed to be equal to the mean value of the velocities of the 6 directions for which wind tunnel tests were conducted. The estimated values of exceedance probability for 8 directions are the same as those for 16 directions. The values for 6 directions are also rather similar to those for 16 directions.

In conclusion, wind tunnel tests should be conducted for at least 6 wind directions in order to estimate the exceedance probability. If wind tunnel tests are conducted for 8 directions, the data is sufficient to estimate the exceedance probability more exactly.

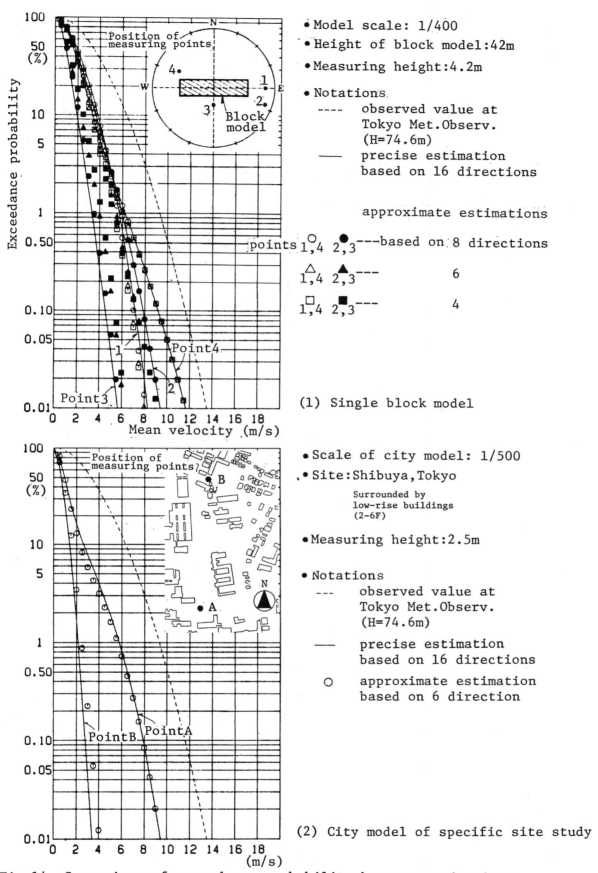

• Model scale: 1/400
• Height of block model:42m
• Measuring height:4.2m
• Notations.
  ----   observed value at
          Tokyo Met.Observ.
          (H=74.6m)
  ―――   precise estimation
          based on 16 directions

  approximate estimations

points $\overset{○}{1,4}$ $\overset{●}{2,3}$ ----based on 8 directions

$\overset{△}{1,4}$ $\overset{▲}{2,3}$ ---   6

$\overset{□}{1,4}$ $\overset{■}{2,3}$ ---   4

(1) Single block model

• Scale of city model: 1/500
• Site:Shibuya,Tokyo

  Surrounded by
  low-rise buildings
  (2-6F)

• Measuring height:2.5m

• Notations
  ---   observed value at
         Tokyo Met.Observ.
         (H=74.6m)

  ―――   precise estimation
         based on 16 directions

  ○   approximate estimation
         based on 6 direction

(2) City model of specific site study

Fig 14   Comparison of exceedance probability between estimations
         based on 16, 8, 6 and 4 wind directions [25]

502

9. CONCLUDING REMARKS

1. The effects of wind on people are so complicated that it is almost impossible to advance research concerned with this problem solely by the method of wind tunnel modeling.

2. The turbulence of wind speed is very influential on the behavior of the human body. The wind tunnel data must be interpreted and related to acceptable criteria with this point in mind.

3. In addition to wind tunnel modeling, it is also necessary to conduct outdoor experiments in natural wind before suggesting acceptable criteria.

4. The exceedance probability is the most preferable scale for assessing the wind environment in a specific site study. In order to estimate the exceedance probability, wind tunnel tests should be conducted for the 6 - 8 wind directions whose occurrence probabilities are prevailing.

ACKNOWLEDGEMENT

Acknowledgement is hereby given to the generous co-operation of K. Deguchi, Lecturer, Hosei University, and Graduate Student, University of Tokyo.

REFERENCES

[1] A.D.Penwarden, Acceptable Wind Speeds in Towns, Building Science vol.8,(1973).

[2] S.Murakami, K.Uehara and K.Deguchi, Wind Effects on Pedestrians; New Criteria based on Outdoor Observation of Over 2000 Persons, 5th International Conference on Wind Engineering, Colorado, (1979)III-6.

[3] J.C.R.Hunt, E.C.Poulton and J.C.Mumford, The Effects of Wind on People; New Criteria based on Wind Tunnel Experiments, Building and Environment, vol.11,(1976) pp15-28.

[4] S.Murakami and K.Deguchi, Measurement of Drag Force on People Walking in Wind Tunnel, Symposium Designing with the Wind, C.S.T.B., Nantes, France, (1981)II-2.

[5] S.Murakami and K.Deguchi, Thermal Effects of Wind on People, Report of the Annual Meeting of Architectural Institute of Japan,(1981).

[6] S.Murakami, K.Deguchi and T.Takahashi, Acceptable Wind Speeds for Windy Working Environment High Above Ground, J. Instute of Industrial Science, University of Tokyo vol.34, No.3,(1982).

[7] A.G.Davenport, An approach to human comfort criteria for environmental wind conditions, Colloquium on Building Climatology, Stockholm, Sept,(1972).

[8] N.Isyumov and A.G.Davenport, The Ground Level Wind Environment in Built-up Areas, Proceedings of Wind Effects on Buildings and Structures, London, (1975).

[9] W.H.Melbourne and P.N.Joubert, Problems of Wind Flow at the Base of Tall Buildings, Proceedings of Wind Effects on Buildings and Structures, Tokyo, (1971).

[10] J.Gandemer, Wind Environment around Buildings: Aerodynamic Concepts, Proceedings of Wind Effects on Buildings and Structure, London,

(1975).

[11] A.D.Penwarden, P.E.Grigg and R.Rayment, Measurements of Wind Drag on People Standing in a Wind Tunnel, Building and Environment, vol.13, (1978),pp75-84.

[12] S.Soma, Experimental Studies on High Wind around the High-rise Building, Grant-in-aid for Publication of Scientific Research Result No.A-51-3 (Head Investigator: H.Ishizaki, Grant-in-aid for Research in Natural Disaster), 1977,pp59-69.

[13] P.S.Jackson, The Evaluation of Windy Environments, Building and Environment, vol.13, (1980), pp251-260.

[14] T.V.Lawson and A.D.Penwarden, The Effects of Wind on People in the Vicinity of Building, Proceedings of Wind Effects on Buildings and Structures, London, (1975).

[15] S.Olesen, J.J.Bassing and P.O.Fanger, Physiological Comfort Conditions at Sixteen Conbinations of Activity, Clothing, Air Velocity and Ambient Temperature, ASHRAE Transaction, (1972),pp199-206.

[16] R.G.Steadman, Indices of Windchill of Clothed Persons, Journal of Applied Meteorology, vol.10, Aug.(1971).

[17] S.Murakami and K.Ikeda, Effect of High-rise Building on Air Flow at Ground Level - Part 2 Research on wind environment, based on long term recording of wind induced discomfort by inhabitants, Report of the Annual Meeting of Architectural Institute of Japan,(1978).

[18] S.Murakami, Y.Iwasa et al., Effect of High-rise Building on Air Flow at Ground Level - Part 3 Research on wind environment, based on long term recording of wind induced discomfort by inhabitants, Report of the Annual Meeting of Architectural Institute of Japan,(1979).

[19] S.Murakami,Y.Uchiumi et al., Effect of High-rise Building on Air Flow at Ground Level - Part 4 Research on wind environment, based on long term recording of wind induced discomfort by inhabitants, Report of the Annual Meeting of Architectural Institute of Japan,(1979).

[20] S.Murakami, Y.Iwasa et al., Effect of High-rise Building on Air Flow at Ground Level - Part 7 Research on wind environment, based on long term recording of wind induced discomfort by inhabitants, Report of the Annual Meeting of Architectural Institute of Japan,(1980).

[21] S.Murakami and K.Fujii, Effect of High-rise Building on Air Flow at Ground Level - Part 5 Long term wind observation near ground in built-up area, Report of the Annual Meeting of Architectural Institute of Japan,(1979).

[22] S.Murakami, A.Kawaguchi et al., Effect of High-rise Building on Air Flow at Ground Level - Part 6 Long term wind observation near ground in built-up area, Report of the Annual Meeting of Architectural Institute of Japan,(1979).

[23] S.Murakami and K.Fujii, Effect of High-rise Building on Air Flow at Ground Level - Part 8 Long term wind observation near ground in built-up area, Report of the Annual Meeting of Architectural Institute of Japan,(1980).

[24] S.Murakami, A.Kawaguchi et al., Effect of High-rise Building on Air Flow at Ground Level - Part 9 Long term wind observation near ground in built-up area, Report of the Annual Meeting of Architectural Institute of Japan,(1980).

[25] S.Murakami and K.Fujii, Method for Approximate Estimation of Exceedance Probability of Wind Velocity, using Minimum Number of Wind Directions, Report of the Annual Meeting of Architectural Institute of Japan,(1981).

SIMULATION AND MEASUREMENT OF THE LOCAL WIND ENVIRONMENT

Jacques Gandemer, Centre Scientifique et Technique du Batiment

1. FOREWORD

It is not the aim of this communication to review all the different methods (and their associated limitations) for simulating the local wind environment in wind tunnels. For this information, the synthesis published by T. V. Lawson in the Journal of Industrial Aerodynamics (July 1978) and entitled "The Wind Content of the Built Environment" can be consulted.

Therefore, this paper will include a discussion of concepts, experimental methods and perhaps "practical ideas" for simulation and measurements. These remarks or comments are the result of ten years of experimentation in the field of environmental wind engineering. The modeling approach presented can now be regarded as fully operational as far as the simulation of the local wind environment and the generation of representative atmospheric winds are concerned.

2. MARKET SITUATION

In the past, the development of urbanization often led, for space reasons, to building in areas such as plateaux which were once known as exposed. Urbanization has also often proceeded without taking into account topographical effects and without realizing that built fronts or buildings of great height and important dimensions in relation to the wind scale were being constructed. Consequently, some unexpected alterations in the ground flows have resulted due to the complex interaction of the site, the site plan and the turbulent wind. In some cases, these alterations have led to wind effects which are particularly unpleasant, "aerodynamic accidents," and which can not be easily dealt with by the inhabitants.

In spite of the complex interactions between the turbulent wind and the built environment which can lead to unusual wind effects, researchers have been able to show that these phenomena can be reproduced in an atmospheric wind tunnel and therefore ways of controlling them can be devised [1]. A strong need for integrating wind studies into the development of site plans for city centers has subsequently been recognized where the main goal of these studies has been the avoidance of wind nuisance.

Now, architects, urbanists, and planners do not intend to "put up with" climatic phenomena any more, but instead intend to play on them. The present day desire to create successful urban areas, to control the atmospheric climate according to expected activities, to use the wind in beneficial ways such as natural ventilation, or to protect oneself from wind effects such as increasing heat loss from roofs and facades, have resulted in a high density and diversity of programs. For example, we can list the following actions:

504

- The search for an urbanization (mesh, density, variation with height, profiles, etc) integrating topographical effects in a "self-protection" or "self-ventilation" perspective (at the scale of the site plan in the aggregate).

- A knowledge of the charts of wind speed fields over water or in coastal zones, taking into account the wakes of built-up areas and the wind regime.

- Optimization of exterior areas in site plans for new programs (in particular according to the activities) or when a rehabilitation of under-privileged areas is being carried out or when a district function is being modified.

- Treatment of an aerodynamic accident noticed "a posteriori" (permeability under buildings, etc).

- Search for a device or mechanism that would be efficient in an urban area for protection against the wind (urban furniture, wind-break, sand-trap, aerodynamics of gangways).

- Control of a particular micro-climate: outdoor swimming pools, tennis courts, outdoor theatres, solariums and verandas on ships, etc.

To conclude, the climatic phenomena and more particularly the wind are now usually taken into account in the conception stages of a project and the recent bioclimatic reflections only tend to foster this process. The latter goes beyond comfort near ground level and includes the control of flows on roofs and facades (ventilation, heat loss, etc). Finally, it is becoming increasingly necessary to be able to associate the pressure field with the wind speed field (wind loads and the functioning of solar collectors and positioning of windmills).

NOTE: This tendency, apart from the scientific considerations involved in producing a simulation, reinforces the idea that environmental aerodynamics studies require an elaborate simulation which reproduces the wind speed field or the pressure field induced by wind phenomena (simulation of static and dynamic properties of the wind at the model scale and with the surrounding terrain reproduced over a sufficient distance).

3. EVOLUTION OF THE DISCOMFORT CRITERION

We will not enlarge here upon the methodology adopted by the CSTB since that has already been presented in various publications [2, 3]. In brief, the adopted methodology involves integrating a study of the local climatology with systematic measurements of a nuisance parameter, ν, for each point of the site plan (in a wind tunnel simulating the wind) to express locally the notion of nuisance in terms of a yearly period during which a given level is reached or exceeded. As a complement to this, the areas associated with the iso-duration of the nuisance can be used to identify zones with strong horizontal wind speed gradients which are particularly unpleasant.

Although this method is in current use, there remain some problems which need to be addressed.

3.1 The Reference Threshold

The nuisance level which is usually taken as the beginning of wind manifestation ($\bar{U}s = 5$ m/s and $\sigma s = 1$ m/s), is not accepted unanimously.

However, since the suggested methodology is not dependent on the chosen threshold, the only rule consists of defining an event that would be clear enough for the designers and aerodynamists to identify the same phenomenon. They can then discuss (in the same language) the tolerable frequency of occurrence of that event.

Thus, in various types of programs, we are induced to express the period of the nuisance (or of the acceptable conditions) taking into account three (or even more) types of "threshold events."

3.2 Nuisance Parameter Formulation

The nuisance parameter, ν, is essentially related to the dynamic action of the wind (mean level and fluctuation). However imperfect the parameter is (since heat action, meteorological conditions, human context, etc. are not taken into account), it can be said that the parameter schematizes the micro-climates rather well for discomfort felt by human beings aside from the harsh climatic spells (cold, rain, snow, etc) which they are prepared to confront (from the point of view of their psychology, activities, clothes).

The nondimensional expression of the nuisance parameter is:

$$= \frac{\bar{U} + \gamma\sigma}{\bar{U}ref + \gamma\sigma ref}$$

where $\bar{U}ref$ and σref are, respectively, the mean speed and standard deviation of the wind at a height of 1.5 meters in an upstream zone without buildings, but where the approach flow $\bar{U}$ and σ have been simulated. According to various authors, the relative weighting of turbulence (γ) can vary from 0 (Penwarden [4]), 1.5 (Isyumov-Davenport [5]), to 3 and 4 (Hunt [6]).

The CSTB has adopted $\gamma = 1$ for nuisance due to the wind and $\gamma = 3.5$ for safety considerations (pedestrian's fall linked to the peak speed).

NOTE: In some cases, only the mean speed will interfere and $\gamma = 0$ (for example the case of stream navigation).

In order to have a better understanding of the importance of the differences in the ν formulations, we carried out the following experiment:

From the meteorological data at a given station (Fig. 1) and about 200 measurement points in a site plan (Fig. 2), orientated following the sector of the main winds, three formulations of the nuisance parameter were adopted:

$$\nu_1 = \frac{\bar{U} + \sigma}{\bar{U}\text{ref} + \sigma\text{ref}} \quad \text{(CSTB formulation)}$$

$$\nu_2 = \frac{\bar{U} + \sigma\text{filter}}{\bar{U}\text{ref} + \sigma\text{ref filter}}$$

where "σfilter" concerns the speed fluctuations within the frequency range mainly felt by human beings (roughly speaking 0.03 to 1 Hz).

$$\nu_3 = \frac{\bar{U} + 3.5\ \sigma}{\bar{U}\text{ref} + 3.5\ \sigma\text{ref}}$$

where the ν_3 formulation is based on the peak speed and was compared to the ν_4 formulation which is based on direct measurement of peak speed.

$$\nu_4 = \frac{\hat{U}}{\hat{U}\text{ref}}$$

To facilitate the experimental work, the speed of the wind tunnel experiments on the 1/200 model was set at 10 m/s (about the same value as the full-scale speeds).

Because of experimental considerations, the values have all been integrated over a period of time amounting to 3 hours in full scale (i.e., 1 minute in the wind tunnel). This period was chosen so that the variation between measurements at one point does not exceed a few percent.

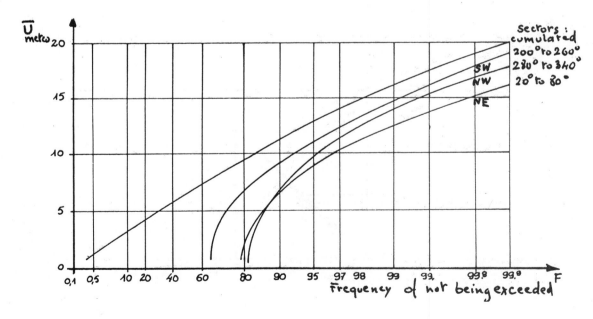

Fig. 1. Climatic data (10 m above the ground). Station: DUNKERQUE

508

Fig. 2. Site plan (wind tunnel model at 1/200).

We discovered an excellent agreement between the parameters ν_1 and ν_2 (Fig. 3). Consequently, in the chosen example, the filtering of turbulence over the supposed range of sensitivity does not change the results and therefore does not offer any interest.

NOTE: A few additional points were far from the bisector. Those were excluded from the figure because all corresponded to turbulence intensities of between 80 and 120 percent.

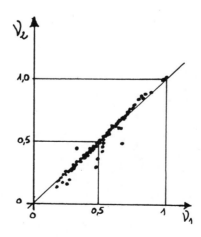

Fig. 3. Correspondence between the parameters ν_1 and ν_2.

Taking into account the statistical distribution of winds at the site which were estimated from meteorological station data and the threshold accepted as the beginning of the discomfort event, ($\overline{U}s = 5$ m/s, s = 1 m/s), the cumulative yearly nuisance periods (T1 and T3) were calculated (Fig. 4).

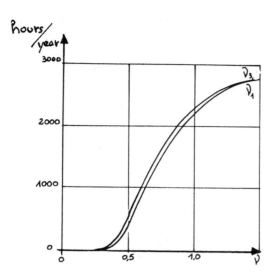

Fig. 4. Correspondence between the nuisance parameters and the cumulative period (hours/year) of nuisance. (Station: DUNKERQUE)

The comparison between T3 and T1 cumulative periods illustrates an important difference in the predicted cumulative period for sheltered zones (small and medium T1 value) (Fig. 5).

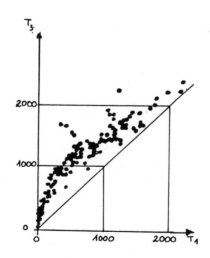

Fig. 5. Correspondence between T3 and T1 cumulative period (hours/year).

510

The two approaches become practically identical for areas with a long cumulative nuisance period. The formulation used to derive the ν_3 parameter emphasizes the contribution of turbulence in creating nuisance much more than the formulation used to derive the ν_1 parameter. It is possible to determine the multiplying coefficients T3/T1 in relation to the turbulence intensity, I, (Fig. 6). Note that there is a discontinuity in the coefficient for increasing turbulence intensity, and the cumulative nuisance period calculated using the ν_3 formulation becomes twice as long as that calculated with the $_1$ formulation for $I \geq 40$ percent.

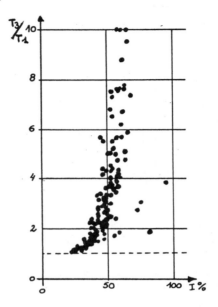

Fig. 6. Variation of the coefficient T3/T1
with the intensity of turbulence I.

NOTE: A ν_4 formulation, based on a direct peak measurement and on the peak reference speed, was compared to that of ν_3. The tendencies are the same, but the consequences are accentuated. Therefore we will not linger on the ν_4 subject.

The ν_3 formulation strongly increases the cumulative nuisance periods in areas of <u>high turbulence</u> which frequently correspond to areas with low mean speeds. Consequently, the representation of micro-climates in site plans based on $_3$ produces both a <u>regrouping</u> of areas and a general increase of the cumulative nuisance period as compared to a representation based on ν_1. The resulting representation of the micro-climates offers no advantage in studying a site plan and the general increase of the cumulative nuisance period is an obvious drawback.

In addition, measurements are obtained in a laboratory using hot wire anemometers with vertical probes. It was shown that in a nonhomogeneous and nonisotropic flow [7] the mean speed and turbulence measurements were over-estimated and this is especially true for high turbulence levels. Therefore, the ν_3 formulation which assigns more importance (compared to ν_1), to areas of strong turbulence, <u>focuses</u> more on the areas where measurements are less reliable.

Finally, from a philosophical viewpoint, the ν_3 parameter deals with the peak speed event which provides a poor representation of the general wind climate. The obvious exception is the problem of pedestrian safety or of stability considerations which are directly connected with the notion of peak speed.

Upon taking into account the fact that full-scale experiments on persons [6-8] have not really shown the prevalence of turbulence over the mean flow in defining comfort limits (as opposed to safety limits) and upon considering the reasons listed above, it can be concluded that the ν_1 format, which integrates $\bar{U}$ and σ (the whole spectrum) on an equal basis, is a useful criterion for evaluating local wind environments studied using atmospheric wind tunnels. The ν_1 format produces a representation of micro-climates which is easy to use in the comparison of different solutions for the same site plans. The ν_1 format also can be used to calculate nuisance periods which are quite reliable and well accepted in practice.

3.3 Thermal Considerations and the Nuisance Parameter

If the mean speed (which includes a rate of air exchange) locally approaches 0, heat related phenomena can represent considerable nuisances (even greater than a "dynamic" wind nuisance) which are not identified by the suggested parameter, ν_1. In other words, creation of successful local micro-climates may require a protection against the wind which simultaneously does not create scorching areas.

While it is not yet possible, using a theoretical approach and mete-orological data, to derive a nuisance parameter which balances dynamic wind and temperature considerations, a practical method resulting from full-scale real situation experiments in solariums and outdoor swimming pools [9] can be suggested. The practical method is based on defining as inadequately ventilated, all the areas for which the local speed ratio will be less than 20 percent of the reference speed. These areas will be considered alarming from a microclimatic viewpoint in spite of the fact that the nuisance parameter leads to a noncritical situation.

3.4 Improvement of Estimated Cumulative Yearly Nuisance Periods Through Consideration of Directional Variations

In order to provide a more realistic evaluation of the annual cumulative nuisance periods to be used in designing a site plan, the variation of the wind phenomenon with relatively small variations in wind angle is studied for the two or three most frequent wind directions (in France SW - NE) and an integrated estimate of the nuisance parameter is obtained.

The presentation of the micro-climate in terms of an annual cumulative nuisance period for a wind mean axis introduces the notion of cumulative meteorological data for several sectors centered around the studied direction and consequently implies that the speed and turbulence field remains almost unaltered for a wind angular variation equal to ±20° (and even ±30° in some cases), (Fig. 1 and 7).

If this approach remains acceptable in a first approximation, it is necessary to analyze (in statistical terms) the variability of the speed fields in relation to the change of the wind direction. Then, we will be able to define in scientific terms the maximum angular sector where the speed field can be regarded as invariant, taking into account a given error domain. This work is now in progress at the CSTB and it ought to soon enable us to optimize the experimental approach (productivity development) and to improve the quality of the predictions.

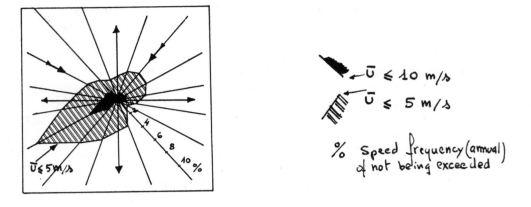

Fig. 7. Compass card. (Station: DUNKERQUE - Period 1962-1970).

As an example, Fig. 8 illustrates the variation of the   values (standard deviation of the variation equal to 0.12), obtained at identical points for an angular variation of 20° for the same statistical wind.

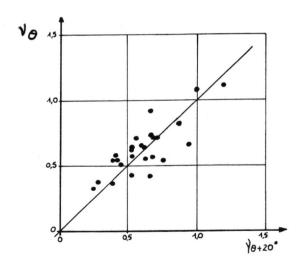

Fig. 8. Variation of the nuisance parameter (same points) for the wind direction Θ and Θ + 20.

4.   SIMULATION OF LOCAL WINDS AND ENVIRONMENT IN BUILT AREAS WITH
     PARTICULAR EMPHASIS ON MODEL SCALES

4.1 <u>Choice of Scales for Use in Studying the Local Wind Environment</u>

Comparisons of results of wind tunnel and full-scale measurements have
been obtained (Fig. 9) for several specific studies [9-11].

Fig. 9.   Full-scale measurements (reference pylone and tripod anemometer).

Taking into account the various uncertainties (full-scale and wind
tunnel), it was proved that at a 1/200 scale, the nuisance ν parameter obtained on the model (for which the wind was dynamically simulated), was in
full agreement with that measured on the site (Fig. 10).

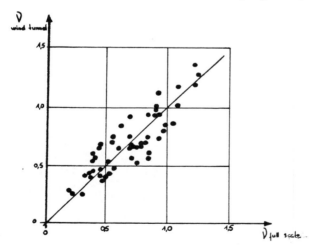

Fig. 10.   Comparison of the nuisance parameter obtained in wind
            tunnel and in full-scale.

The causes of error include on the one hand the uncertainties of
measurement at the site (propeller anemometer, wind nonstationarity, etc.)
and of measurement in the wind tunnel (model accuracy, point spotting, the
wind mean axis, hot wire anemometer techniques, etc.) and the essential
results of this work have already been published in the <u>Journal of Industrial
Aerodynamics</u> [3].

In conclusion, we can only recommend the use of the range of scales from 1/250 to 1/150 which represent an interesting compromise between competing requirements. In addition, use of these scales reduces problems associated with model construction since classical materials such as wood, dense polystyrene, balsa, plaster, cork, etc., can be used without any particular refinement of the surface.

NOTE: In the case of smooth, rounded shapes, the flow around the shape must be characteristic of super- (or hyper-) critical flow regimes. Surface roughness can be used to trigger this type of flow which is characterized by the separation zone and the value of the Strouhal number. The roughness used should be such that the Reynolds number based on the thickness is greater than 1000 but $e/D < 10^{-2}$ [20]; where "e" is the roughness size and D is the diameter of the structure. An example of the dependence of the separation point position and of the Strouhal number on the surface roughness is given in the following table:

Shape: cooling tower; Incidental turbulence intensity: 5 percent
Reynolds number of the flow 3.3 10^5

e/D	Re_e	separation angle (stagnation point 0°)	S_t
4.10^{-4}	320	130°	0.21 – 0.45
8.10^{-4}	264	90–110°	0.19 – 0.20
12.10^{-4}	396	90–100°	0.18 – 0.19
16.10^{-4}	528	90–100°	0.19 – 0.20
20.10^{-4}	660	90–100°	0.18
40.10^{-4}	1320	90–100°	0.16

At these scales (1/250 – 1/150), urban furniture and wind breaks require a metal or plexiglas treatment. Lichen are commonly used to model vegetation.

Fig. 11. Wind breaks on a model (1/200).

Fig. 12. Topographic model (1/500).

For relatively open areas, glossy paints will have to be avoided as they cannot provide an aerodynamically rough regime $\frac{U* \; Zo}{\nu} > 2.5$. In this context, it will be necessary to take the precaution to slightly sand (roughly 0.1 mm grain diameter) the surfaces.

Finally, every measurement point (hot wire anemometers) needs an environment faithfully reproduced over a radius of a minimum of 0.5 m (that is to say 100 m in full-scale). Upstream of the faithfully reproduced environment, the statistical properties of the wind must be simulated.

NOTE: This size of a faithfully reproduced environment also sets the minimum distance between the wind tunnel walls and measurement points.

4.2 Aerodynamic Study of a Specific Device

For the study and designing of particular aerodynamic devices, stream-lining, deflector, urban furniture, wind break, etc., where the methods of quantitative visualization by laser chronophotography are particularly interesting, the 1/20 scale can be used. In these conditions, only the turbulent intensity will be reproduced at this scale (Fig. 13).

The real efficiency of the device will then be quantified in its environment at the 1/200 scale.

Fig. 13. Protected gangway (visualization by laser chronophotography).

4.3 Topographical Studies or Studies at City or Partial City Scale

In this type of program, the aim is not a precise knowledge of the flows at ground level, but rather the analysis of the fluxes and their behavior in relation to the site. These studies are often commissioned to provide relief from a particular problem or to provide information on different options of urbanization. Consequently, the interesting "layer" is generally situated between 5 and 20 meters and we will then be able to use model scales like 1/500 or even 1/1000 (treatment of the model floor by sanding). Work at these scales showed, during comparisons with site studies [14-15], that the topographical phenomena were perfectly reproduced and could easily be quantified. This range of scales is used in numerous programs to acquire a better knowledge of the site, in order to prepare the simulation at the scale of ground level studies (1/200).

4.4 A Method for Obtaining Full-Scale Measurements for Comparison with Wind Tunnel Data

In order to improve the measurements of the flow field at a site and particularly to make the full-scale experimental processes easier, a technique of <u>remote transmission</u> was worked out. A portable transducer visually provides simultaneously the local speed (transmitter-tripod with propeller anemometers) and the speed and its direction* at the reference (pylone with anemometer and vane).

The power of the transmitter-receiver system allowed transmissions over a distance of about 150 m. The instantaneous and mean values (from 3 seconds to 5 minutes) are available as well as the standard deviation which can be parallely recorded.

Easy to operate, to move and largely autonomous (a day long), the device is a particularly efficient "site" instrument.

Fig. 14. Anemometric technique with remote transmission (full-scale measurements).

5. SIMULATION OF THE WIND STATIC AND DYNAMIC CHARACTERISTICS IN A RANGE FROM 1/250 TO 1/150

The CSTB has at its disposal boundary layer wind tunnels which are particularly efficient and well adapted to the simulation of the wind phenomenon in the environment and notably the "atmospheric wind tunnel" [13] (Fig. 15).

In all simulation cases, only the neutrally stable conditions are considered and the characteristic dimension is based on the turbulence scale.

---

*Now being realized.

Fig. 15.   The CSTB atmospheric wind tunnel: test chamber
20 m x 4 m x adjustable roof (2.5 m to 3.5 m)

The simulation of winds of different statistical nature is reproduced
to a height of about 1 m and is referred to full-scale data presented in
the synthesis by J. COUNIHAN [16] or to results obtained by the CSTB climatology division [17].

The simulation technologies are those used in the long test-section
wind tunnels and consist of a subtle combination of ground roughness and
vortex generators.

The essential originality of the device consists in using tridimensional
blocks of different dimensions positioned:

- on the one hand in the test chamber to provide an energetic distribution and a level similar to the spectrum proposed by A. G. DAVENPORT.

- on the other hand, in the collector to refocus the maximum of
spectral density upon low frequencies (1.2 x 1.2 x 1.2 m blocks) (Fig. 16).

Fig. 16.   Big blocks in the collector and roughness and blocks at the
beginning of the test chamber.

518

This technique enables us to use large reproduction scales (1/100 and 1/150) without blocking the test chamber or heterogeneifying it and to minimize pressure losses which would jeopardize the great blowing speeds ($\overline{U}$flux > 20 m/s).

As an example, the figures and diagrams below give the flow characteristics for the simulation of a countryside-type wind at the 1/150 scale (Fig. 17).

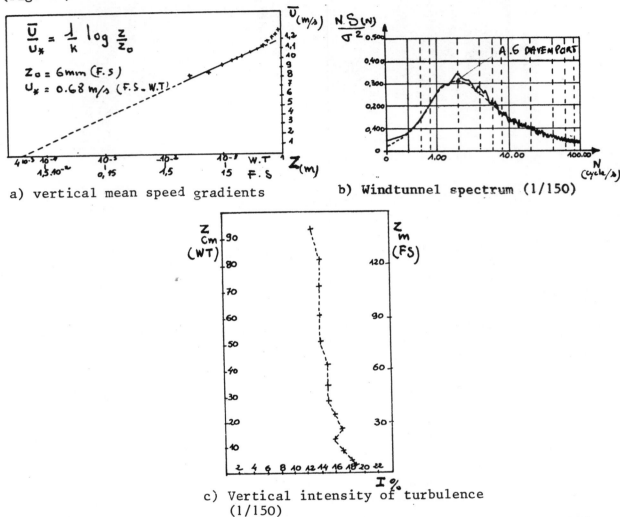

a) vertical mean speed gradients

b) Windtunnel spectrum (1/150)

c) Vertical intensity of turbulence (1/150)

Wind tunnel (WT) measurements Country wind: α = 0.12				Full-scale (FS) measurements Country wind: α = 0.12 to 0.15			
z (m)	I	Lx (m)	Ly (m)	Z (m)	I	Lx (m)	Ly (m)
30	0.17			30	0.14	100	35
45 to 100	0.14	102	36	100	0.11	200	60 to 70

Fig. 17.   Characteristics of the wind tunnel simulations.
Wind: country-type (Scale 1/150).

From the point of view of anemometric measurements, we will make the two following comments:

- The adjustable variable roof enables us to realize a 0 pressure longitudinal gradient. The optimal slope for a given model must be realized in the presence of the probe support (in relation to the explored area), which, in spite of its aerodynamic design and a measurement point extending one meter ahead of and one meter below the support beam, increases the anemometric measurements by a few percent.

- The flow thermic regulation (at $\pm 0.5°C$) greatly reduces the derivatives of the hot wires anemometer measurement system and thus, reduced the experimental variations and has led to an excellent measurement simulation.

6. REMARK ON THE SIMILITUDE CONDITIONS FOR THE COUPLED STUDY OF PRESSURE FIELDS AND SPEED FIELDS

When pressure measurements are studied simultaneously with the wind field (mean, peak, standard deviation, correlation, etc.) the following complementary similitude conditions will have to be met:

- For the models, the pressure taps must be flush with the walls of the model and metal or plexiglas are used for the construction of experimental models.

- For the measurements to be meaningful, it will be necessary to avoid all acoustical pollutions; the one linked to the measurement technology, taking into account the presence of tubes that can introduce resonance effect [19], the other linked to the wind tunnel itself, which with its geometry and the presence of rotating blades (whistle effect) can superimpose a disturbing acoustic pressure field [18].

NOTE: This last point was the object of particular research for the CSTB atmospheric wind tunnel: the rejection of the whistle effect outside the measurement range (600 rpm/minute, 12 blades, fundamental at 120 Hz > 100 Hz), a flow regulation on the blades pitch (constant fan rotation and the treatment of the walls to absorb the acoustic peak, induced a perturbation of about one Pascal for pressures greater than 250 Pascals.

7. CONCLUSION ON THE WIND TUNNEL ENVIRONMENT STUDIES

Apart from a few precautions concerning similitude and measurement technologies, it is possible to say that the reproduction of small scale wind effects is accurate and that researchers can efficiently help to control climatic atmospheres.

As it is, it is important to develop the nuisance criteria, notably from its thermic viewpoint, as well as to further develop pedagogical methods (laser chronophotographic visualization or andoscopy) which can make the designers more sensitive to the aerodynamic concepts.

Finally, researchers will have to keep in mind the limits of the application context (the climatic phenomena only represent one part of the programs) and the fact that only multidisciplinary teamwork will make aerodynamic studies more efficient.

REFERENCES

1. J. Gandemer and A. Guyot, Guide, "Integration Du Phenomene Vent Dans La Conception Du Milieu Bati," CSTB-Paris (1976).

2. J. Gandemer and A. Guyot, Guide, "Protection Contre Le Vent," Aerodynamique des brise-vent et conseils pratiques, CSTB-Paris (1981).

3. J. Gandemer, "Aerodynamic Studies of Built-Up Areas Made by CSTB at Nantes-France," Journal of Industrial Aerodynamics, 3 (1978).

4. A. D. Penwarden and Afe Wise, "Wind Environment Around Buildings," Building Research Establishment Report, HMSO (1975).

5. A. G. Davenport and N. Isyumov, "The Ground Level Wind Environment in Built-Up Areas," Fourth Intl. Conf. on Wind Effects on Buildings and Structures, Heathrow (1975).

6. J.C.R. Hunt, E.C. Poulton and J.C. Mumford, "The Effects of Wind on People, New Criterias Based on Outdoor Observation of Over 2000 Persons," Building and Environment, 11 (1976).

7. J. Gandemer and G. Barnaud, "Etude Aerodynamique Du Champ De Vitesse Dans Les Ensembles Batis," CSTB Nantes - ADYM. 12-73.

8. S. Murakami, K. Vehara, and K. Deguchi, "Wind Effects on Pedestrians: New Criteria Based on Outdoor Observation of Over 2000 Persons," Fifth International Conference on Wind Engineering, Ft. Collins (1979).

9. J. Gandemer and G. Barnaud, "Etude Aerodynamique D'Un Paquebot, Application A L'Optimisation Du Micro-Climat Des Ponts Superieurs," Colloque on Construire Avec Le Vent, Nantes (1981).

10. J. Gandemer, "Etude Aerodynamique Des Structures Brise-Vent Entrant Dans L'Amenagement De La Ville Nouvelle De Lille-Est," Synthese de l'etude en vraier grandeur, CSTB Nantes-ADYM. 81-11.L.

11. J. Gandemer, "Etude Du Micro-Climat De La Ville De Vaudreuil," Confrontation des resultats en soufflerie et sur le terrain, CSTB Nantes - ADYM. 76-6.C.

12. H. Maille, "Methode De Visualisation Quantitative Par Chronophotographie Laser," CSTB Nantes, Colloque on Construire Avec le Vent (1981).

13. C. Roulle, "Simulation De La Couche Limite Atmospherique Au 1/150 eme," CSTB Nantes - ADYM. 80-10.L.

14. G. Grillaud, "Site De Gruissan: Recalage Climatique Et Etude Topographique," CSTB Nantes - ADYM. 78-7.C.

15. G. Grillaud, "Simulation De La Couche Limite Atmospherique A L'Echelle Topographique," CSTB Nantes - ADYM. 78-9.L.

16. J. Counihan, "Adibatic Atmospheric Boundary Layers: A Review and Analysis of Data from the Period 1880-1972," Atmospheric Environment (1975).

17. R.E.E.F., Vol. II, Climatologie – Sciences du Batiment. CSTB Nantes.

18. G. Barnaud, "Comparison Des Mesures Du Champ De Pression Instationnaire Sur Un Batiment Dans Diverses Souffleries," CSTB Nantes, Colloque on Construire Avec Le Vent (1981).

19. J. Vimal du Monteil, "Optimisation Et Automatisation De La Technologie Des Mesures De Pression Instationnaire," CSTB Nantes – ADYM. 80-2.L.

20. C. Roule, "Reproduction Des Regimes Hypercritiques Sur Formes Rondes En Soufflerie Atmospherique," CSTB Nantes – ADYM. 80-20.L.

SMALL-SCALE MODELING OF SNOW-DRIFT PHENOMENA

J. D. Iversen
Department of Aerospace Engineering
Iowa State University
Ames, Iowa 50011

1. INTRODUCTION

The problem of drifting snow in the colder populated regions of the world has long been a difficult one, both in terms of predicting drift patterns and in establishing possible control by means of vegetation, snow fences, or other obstructive devices. A host of snowdrift problems exist in connection with highways, railroads, airports, parking lots, approaches to buildings, and so forth. Experimentally, it is time-consuming and often unrewarding to study snowdrift patterns in the field, since control of weather conditions is not possible and measurements are difficult.

It is less expensive, quicker, and easier to test snow-drifting phenomena with a small-scale model whether the test is of the drifting around a model building or the means of controlling snow-drifting. Control of the experiment (wind direction, duration, etc.) is a primary advantage with small-scale models, but wind tunnel models are also relatively cheap and many more experiments can be conducted in a given period of time. The important question for model experiments concerns the validity of full-scale prediction from model results. Small-scale models were first tested by Finney [1,2,3] in a small wind tunnel; many of the results he obtained are still used in snow fence placement and highway design. Other wind tunnel experiments have been performed by Anno [4], Jensen [5], Anfilofev [6], Kimura and Yoshisaka [7], Becker [8], Nøkkentved [9], Kreutz and Walter [10], Gerdel and Strom [11,12], Stehle [13], Sherwood [14], Brier [15], Kind and Murray [16], and by the writer [17-21]. Model experiments in water have been performed by Wuebben [22], Theakston [23], Isyumov [24], Norem [25], Calkins [26,27], and de Krasinski et al. [28,29]. An interesting and carefully performed set of experiments has been reported by Tabler [30,31] and Tabler and Jairell [32] using scale models placed on a frozen lake with natural snow itself used as the modeling material. Tobiasson and Reed [33] have also performed outdoor modeling.

The exact similitude requirements for modeling drifting snow problems are not met on a small scale because of the large number of modeling parameters that cannot be satisfied simultaneously. Those who have reviewed the similitude requirements include Imai [34], Gerdel and Strom [11], Odar [35,

36], Isyumov [24], Norem [25], de Krasinski et al. [28,29], Kind [37], Iversen [17,18,38], Dyunin [39], and Tabler [31]. The complexity of the similitude problem is emphasized by the fact that, in general, none of these authors agrees as to the most important or appropriate sets of parameters on which to base a similitude. Similar requirements are necessary for modeling drifting sand and dust in air, but fewer attempts have been made in this area. An early set of experiments was conducted by Woodruff and Zingg [40]. Most of the recent wind tunnel experiments in this area have been performed by the writer and his associates at the Iowa State University Wind Tunnel Laboratory and at the National Aeronautics and Space Administration (NASA) Ames Research Center (Greeley et al. [41], Iversen et al. [42,43,44]) and in Denmark [45].

2. THE SALTATION PHENOMENON

The movement of loose surface particles by wind is a complex phenomenon. Bagnold's classic work [46], based on wind tunnel studies and field observations, defines the basic parameters and relations of sand movement by wind. Recent work on threshold experiments and calculations (the threshold shear stress is defined as that minimum value of stress at which particles begin to move) is reported by Iversen et al. [47,48,49].

The dimensionless threshold friction speed, A_1, is given by

$$A_1 = \frac{u_{*t}}{\left(\frac{\rho_p g D_p}{\rho}\right)^{1/2}} \tag{1}$$

For water, ρ_p is replaced by net density $\rho_p - \rho$. Bagnold and most later researchers assumed A_1 is a unique function of particle friction Reynolds number B:

$$A_1 = A_1(B), \quad B = \frac{u_{*t} D_p}{v} \tag{2}$$

In addition, however, the threshold parameter A_1 must be a function of cohesive forces for small particles [49,50].

Loose particulate material can move in one of three ways: creep, saltation, or suspension. Creep is the motion in which large particles roll along the surface but do not become airborne. Medium-size and smaller particles will become airborne as a result of aerodynamic lift or because of a combination of lift and impact from returning particles. The saltating particle rises from the surface in a nearly vertical direction and then gradually returns via a shallow angle trajectory. Very small particles, after becoming airborne, may go into suspension and perhaps rise to great heights before gradually settling out of the atmosphere. The ratio U_f/u_{*t} (terminal speed to threshold friction speed) is

$$\frac{U_f}{u_{*t}} = \frac{2}{A} (3C_D)^{-1/2} \tag{3}$$

where the drag coefficient C_D is a function of Reynolds number

$$C_D = C_D \left(R = \frac{U_f B}{u_{*t}} \right) \tag{4}$$

$$C_D = \frac{24}{R}, \ (R \leqq 0.1); \ C_D > \frac{24}{R}, \ (R > 0.1) \tag{5}$$

The mean vertical turbulent eddy velocity in the boundary layer is of the same order of magnitude as the friction speed u_*. Thus, particles that go into suspension do so because their terminal speed, U_f, is smaller than u_* (or u_{*t}). An approximate division between dust (material in suspension) and sand (saltating particles) is therefore found by setting the ratio $U_f/u_{*t} = 1$. For particle diameters large enough so that U_f/u_{*t} is greater than 1, particles will not become suspended until that ratio u_*/u_{*t} is reached when U_f/u_* becomes approximately unity. Except for very strong winds, nearly all blowing snow particles move in the saltation mode [37].

3. SIMILITUDE CRITERIA

3.1. Basic Parameters

Jensen [51] showed that for simulation of the atmospheric boundary layer the roughness parameter in the model should be the same as that in the atmosphere, i.e.,

$$\frac{z_{o_m}}{z_o} = \frac{L_m}{L} \tag{6}$$

The variables for a boundary layer problem in a neutral atmosphere without particle motion are:

g gravitational acceleration

h reference height

L reference length

ℓ other horizontal lengths

z_o roughness height

U wind speed at reference height

η other heights

ν kinematic viscosity

The presence of both a reference height and a reference length allows the possibility of a geometrically distorted model.

Adding blowing particles to the problem requires the following additional variables:

D_p particle diameter

t time

ρ fluid density

ρ_p particle density

These basic variables result in the following set of dimensionless parameters:

1) $\frac{h}{L}, \frac{\ell}{L}, \frac{\eta}{h}$ geometric similarity

2) z_o/L Jensen's roughness criterion

3) UL/ν Reynolds number

4) U^2/gL Froude number

5) D_p/L particle diameter-length ratio

6) ρ/ρ_p density ratio

7) Ut/L time-scale

The Froude number, parameter 4, is usually not important without particle motion (or without buoyant effluent, etc.) but it is of importance in blowing snow similitude. Parameter 5 can almost never be satisfied in a small-scale model because too small a particle results in too small a value of U_f/u_{*t} and motion is by suspension rather than by saltation. The density ratio, parameter 6 cannot be satisfied in the water channel, and the Froude number cannot be satisfied in the wind tunnel because at least a portion of the model must have wind speeds above the threshold for motion, and thus there is a lower limit on wind speed (in addition to the usual Reynolds number limitation).

Thus the small-scale snowdrift model is necessarily a distorted model, and if other than unsubstantiated qualitative information is to be obtained from a model test, it is necessary to determine the effect of model distortion.

3.2. Combination of Terms Using Particle Equations of Motion

Since it is impossible to satisfy all the basic dimensionless parameters simultaneously, it is useful to combine these terms by theoretical means, such as the particle equation of motion. Which terms result depends on how the equations of motion are nondimensionalized. Two modeling parameters that result are [38]:

8) $C_D \rho L/\rho_p D_p$

9) $gL^2/u_*^2 h$

3.3. Transport Rate Similitude

As shown in [52], a similitude can be derived on the basis of mass transport rate (representing either deposition or deflation rate). The resulting parameter is

$$10) \quad [d(V/L^2 h)/d(u_* t/L)]/\left[\frac{\rho}{\rho_p}\left(\frac{u_*^2}{gh}\right)\left(1 - \frac{u_* t}{u_*}\right)\right]$$

The dimensionless drift volume $V/L^2 h$ in parameter 10 can be replaced by drift plan area A_d/L^2 or cross-section area A_c/Lh. Parameter 10 involves parameters 4, 6, and 7 explicitly and parameter 5 implicitly.

3.4. Densimetric Froude Number

The equivalent roughness height in saltation, z_0', has not generally been determined. However, Owen [52] shows that it is proportional to u_*^2/g, i.e.,

$$z_0' \sim \frac{u_*^2}{2g} \tag{7}$$

The writer has found that, by using particulate modeling materials of widely varying density, the effective mass deposition or deflation rate could be correlated by the product of density ratio ρ/ρ_p and roughness ratio z_0'/h. It was thus considered that the effective roughness height should be expressed by Owen's result times the density ratio. However, though the roughness height relationshp expressed by Eq. (7) may be too simple, the roughness height is probably at most only a weak function of density ratio.

Consider a deposition situation in which deposition is governed by a topographic feature of height h and lateral width L (such as a fence, rearward- or forward-facing step, or building). The volumetric rate at which material is deposited should be proportional to the product of wind speed (or friction speed), volumetric concentration, and frontal area. The maximum mass concentration possible in saltation is usually considered to be the fluid density ρ, so that the volumetric concentration would be ρ/ρ_p. Thus

$$dV/dt \cong (\rho/\rho_p)hLu_* \tag{8}$$

and the mass deposition rate would be

$$q_d L = \rho_p \, dV/dt \cong \rho h L u_* \tag{9}$$

Since $q \cong \rho u_*^3/g$, the ratio of deposition rate to total mass flow rate is

$$q_d/q \cong \rho h L u_*/(\rho u_*^3 L/g) = gh/u_*^2$$

or

$$q_d \cong q/(u_*^2/gh) \tag{10}$$

so that a dimensionless deposition rate would be expected to decrease with increase in Froude number.

Experiments with different values of particle density, to be discussed later, have shown the deposition rate to be a function of density ratio as well. If the particle concentration within the deposition region is volume limited rather than mass limited, then Eq. (10) becomes

$$q_d \cong q/(\rho u_*^2/\rho_p gh) \tag{11}$$

As shown later, the densimetric Froude number of Eq. (11) correlates the deposition rate data better than the Froude number alone, although the deposition rate is not as strong a function of the densimetric Froude number as predicted by Eq. (11).

3.5. The Similitude Function

The mass transport parameter 10 provides a basis for analysis of snow-drifting simulation. The similitude problem reduces to, e.g.,

$$\frac{d(A_d/L^2)}{d(u_{*t}/L)} = \frac{\rho u_*^2}{\rho_p gh} \left(1 - \frac{u_{*t}}{u_*}\right) f\left(\frac{\rho}{\rho_p}, \frac{u_*^2}{gh}, \frac{u(h)L}{\nu}, \frac{U_f}{u_{*t}}, \frac{\eta}{h}, \frac{\ell}{L}, \frac{z_o}{L}, \frac{h}{L}\right) \tag{12}$$

Equation (12) allows for the possibility of vertical geometric distortion and uses Jensen's criterion for normal roughness modeling. It does not allow for the satisfaction of many of the original dimensionless terms nor for parameters 8 and 9. Thus the snowdrift model is still highly distorted. By varying particle size and density, model speed, scale and possibly vertical distortion, a range in degree of distortion can result so that full-scale predictions can be made. The neglect of parameters 8 and 9 is probably not serious unless small-scale features such as surface ripples become large enough to obscure or interfere with gross drift geometry associated with important topographic features of the model.

4. MODELING EXPERIMENTS

4.1. Highway Grade-separation Snowdrift Control

A series of experiments was performed to determine snowdrift characteristics of, and optimum control geometry for, an interstate highway grade separation structure [17,18,21]. Two models of the highway grade-separation structure were constructed. Model 1 was built at a 1:120 scale and Model 2 was built at the same scale horizontally but with a 1:60 scale in the vertical direction. Both models were covered with cloth to simulate grass except for the highway lanes which were smooth plastic [21]. Three particles were selected for simulation of snow [18]: 269 μm walnut shell (1100 kg/m³), 101 μm glass spheres (2500 kg/m³), and 49 μm dense glass spheres (3990 kg/m³).

The models were placed at various orientations to the wind in a boundary layer wind tunnel (1.2 m × 1.2 m × 7 m test section). The boundary layer depth at the model location (5 m downwind) was increased to 0.25 m by turbulence-generating spires placed at the test section entrance. Particle material was placed to a uniform depth of 0.015 m (0.03 m for Model 2) across the test section width from 2.3 m to 3.7 m downwind of the test section entrance prior to the start of each experiment. Plan view photographs of the model were taken at recorded times during each test run in order to measure the planform drift area on each stretch of highway lane as a function of time.

Both the undistorted and vertically distorted grade-separation models were first tested without drift-control simulated vegetation not only in order to obtain comparison with control planting configurations but also to obtain appropriate similitude relationships for more exact configuration comparisons and for possible extrapolation to full scale. With the wind direction parallel to the bridge center line, a total of 13 bare (simulated grass only) model experiments were analyzed to produce the relationship desired. Ten of these experiments were with the undistorted model (Model 1) and three with the distorted model (Model 2). Photographs of one of these and a later experiment are shown in Fig. 1. The planform drift area A_d divided by L^2 (bridge length squared) is shown for the 13 experiments as a function of dimensionless time in Fig. 2. The values of area A_d correspond to the snowdrift area (in plan) covering a 1 meter length of both lanes of highway including the shoulders.

Parameter 10 (in terms of wind speed rather than friction speed, since that is easier to measure) was used to define a dimensionless drift rate. The results are shown in Fig. 3 with dimensionless drift rate plotted versus the densimetric Froude number (cf. Eq. 12). The trend of decrease in deposition rate with increase in densimetric Froude number is clear but the decrease is less than predicted by Eq. (11). The full-scale values of densimetric Froude number are just to the left of most of the model values. It could thus be expected that the full-scale dimensionless drift rate would be relatively somewhat larger than the model values. The dimensionless planform drift area is plotted against the combined mass rate-densimetric Froude number parameter in Fig. 4. The correlation is quite good with the correlation coefficient R = 0.9918 (R = ± 1 is a perfect correlation; R = slope time ratio of horizontal and vertical standard deviations).

Values of friction speed were estimated from the wind speed measurements using Owen's [52] equation for velocity profile. The data for mass deposition rate from the ten Model 1 experiments were curve fit by linear regression to compare Eqs. (10) and (11). The results are as follows:

$$q_d g/\rho u_*^3 = 1.4 \left(u_*^2/gh \right)^{-0.49}, \quad R = 0.74 \tag{13}$$

$$q_d g/\rho u_*^3 = 0.064 \left(\rho u_*^2/\rho_p gh \right)^{-0.41}, \quad R = 0.90 \tag{14}$$

It is clear that the correlation (R = correlation coefficient) with densimetric Froude number is better than with the ordinary Froude number. The dimensionless deposition rate is not as strong a function of Froude number or densimetric Froude number as predicted by Eqs. (10) or (11), however. By multiple regression, treating the Froude number and density ratio separately, the best fit was found to be

Fig. 1a. Photograph of uncontrolled (grass only) calibration highway
model snowdrift experiment, wind direction left to right
parallel to bridge center line. Note cornice formation.
Model cars added after experiment completion.

Fig. 1b. Photograph of highway model with simulated hedgerow (located
too close to roadway). Wind direction right to left at 20°
angle with bridge center line.

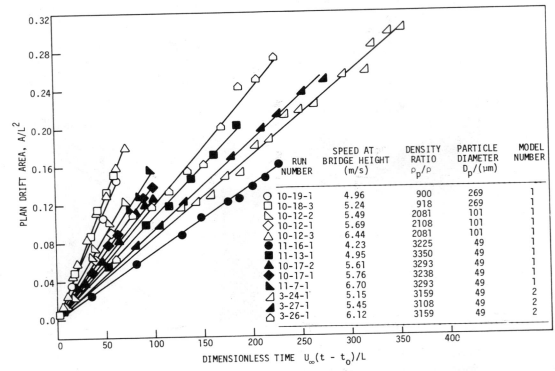

Fig. 2. Plan view drift area on interstate highway lanes as a function of time from the start of drifting. Speed ranges from 4.23 m/s to 6.70 m/s at 0° wind direction. Particle density ranges from 1100 to 3990 kg/m³ and particle diameter from 49 to 269 microns.

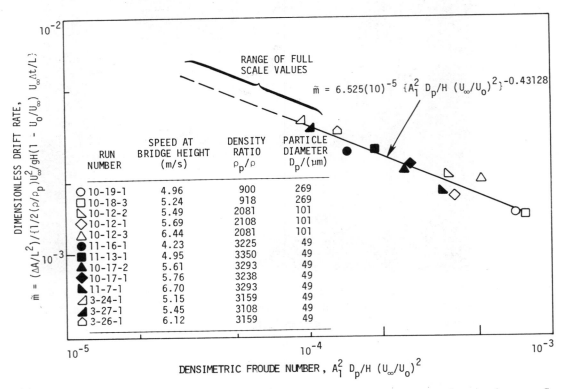

Fig. 3. Dimensionless drift rate versus densimetric Froude number. Same experiments as in Fig. 2.

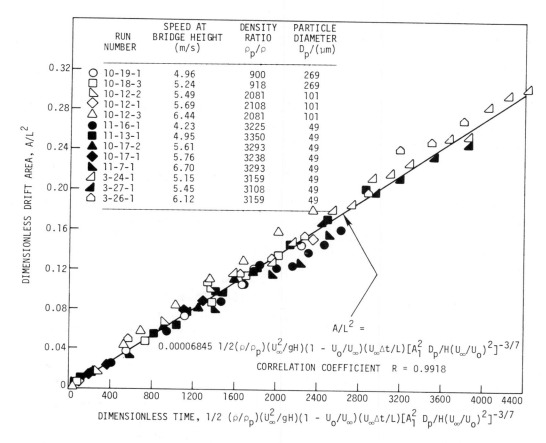

Fig. 4. Dimensionless drift area versus combined mass rate-densimetric Froude number parameter. Same experiments as Figs. 2 and 3.

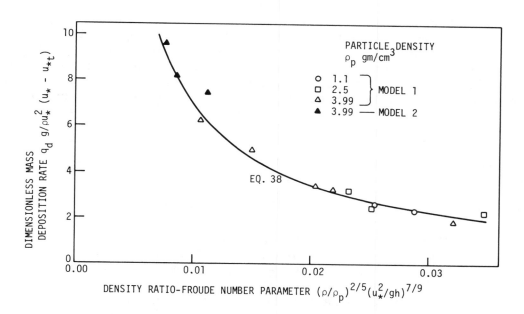

Fig. 5. Deposition rate versus density ratio-Froude number parameter. Same experiments as in Figs. 2, 3, and 4.

$$q_d g/\rho u_*^2 (u_* - u_{*t}) = 0.0696(\rho/\rho_p)^{-2/5}\left(u_*^2/gh\right)^{-7/9}, \quad R = 0.97 \quad (15)$$

The form of the best fit to any set of data is undoubtedly a strong function of the geometry of the model. Since only a very few experiments of this type have yet been performed, much work remains to establish the mass transport similitude on a firmer foundation. Equation (15) is used in plotting the data as shown in Fig. 5.

4.2. Testing of Simulated Vegetative Snowdrift Control

Simulated rows of bushes (3 m full-scale height) were placed on the grade-separation model to act as snowdrift control barriers. A total of 51 experiments were performed with a variety of control configurations and with wind speed direction angles of 0°, 20°, and 40° to the bridge center line. Typical results are shown in Fig. 6. Two control configurations, one of good performance and one of mediocre, are compared in Fig. 6 with a bare model calibration at a 20° wind direction angle. It is equally possible to create a planting configuration that can be considerably worse than no control as well as one that is a considerable improvement, such as plan L-20.

4.3. Effect of Vertical Distortion

The results with the vertically distorted model (Model 2) at zero degree wind direction for one bare model experiment and one controlled model experiment are shown in Fig. 7 along with the corresponding results for the undistorted Model 1. The results for both configurations are displaced to the left for Model 2 compared with Model 1. The reason for this is that Model 2 is effectively horizontally distorted as well as vertically distorted where there are separated flow or reduced speed regions such as the lee side of the fill slope under the bridge and downwind of the simulated plant control. Because of the greater relative height of the fill slope and plant control in the distorted Model 2, the regions of reduced speed extend farther downwind than in the undistorted Model 1 (or in full scale). Thus early drifting occurs farther downwind for Model 2 than for Model 1 and the drift area on the roadway is larger for corresponding times as shown in Fig. 7. That Model 2 still gives valid relative results at least for certain configurations is shown by the fact that the two sets of curves are parallel and displaced vertically about the same distance. Without use of the combined parameter including the effect of densimetric Froude number, the curves would not be parallel.

4.4. Extrapolation of Model Results to Full Scale

Previous investigators have usually used the Froude number U^2/gL as a means of determining the full-scale wind speed and modeled snow storm time duration. This is incorrect for a distorted model according to the preceding similitude analysis and wind tunnel test results. If, however, the Froude number is used (albeit incorrectly) as a means of extrapolation, then the full-scale-to-model wind speed ratio U/U_m would just be the square root of the length ratio L/L_m. The time ratio $\Delta t/\Delta t_m$ would be the same value as the speed ratio. Using run number 10-17-1 as an example (model speed = 5.76 m/s and time duration = 18 min), the full-scale wind speed would be 63.05 m/s and the duration would be 3 hrs and 17 min, which are not very realistic values.

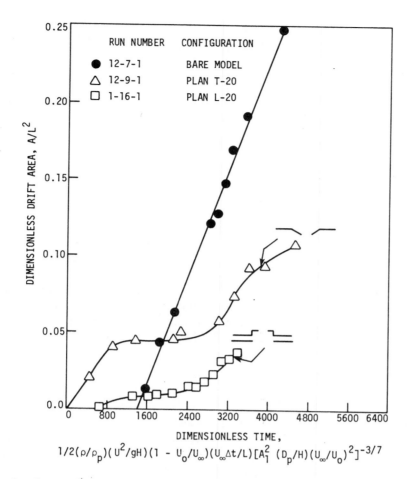

Fig. 6. Dimensionless drift area versus combined mass rate parameter. Wind direction 20° to bridge center line. Comparison of uncontrolled (bare model) and two different plant snowdrift control configurations, Model 1.

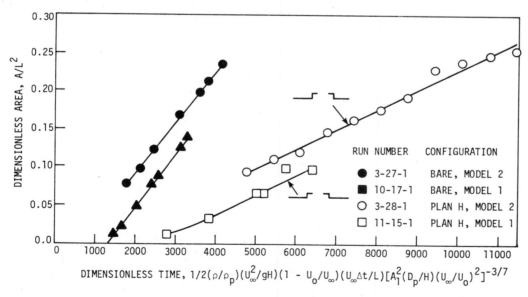

Fig. 7. Dimensionless drift area versus combined mass rate parameter. Comparison of 0° wind direction tests, controlled and uncontrolled, using Model 1 (true geometric model) and Model 2 (vertical geometric distortion factor of two).

According to the wind tunnel results of the 13 bare model experiments, the appropriate extrapolation is to equate the combined mass-transport rate parameters for model and full scale:

$$(\Delta A/L^2) \left[A_1^2 \frac{D_p}{H}\left(\frac{U}{U_o}\right)^2 \right]^{3/7} \Bigg/ \left\{ \frac{\rho}{\rho_p} \frac{U^2}{gH} \left[1 - \left(\frac{U_o}{U}\right)\right] \frac{U\Delta t}{L} \right\} \Bigg|_{model}$$

$$= (\Delta A/L^2) \left[A_1^2 \frac{D_p}{H}\left(\frac{U}{U_o}\right)^2 \right]^{3/7} \Bigg/ \left\{ \frac{\rho}{\rho_p} \frac{U^2}{gH} \left[1 - \left(\frac{U_o}{U}\right)\right] \frac{U\Delta t}{L} \right\} \Bigg|_{full\ scale} \quad (16)$$

It is assumed that everything is known in this equation except the full-scale values of U and Δt (or, the model values of U and Δt which represent a given full-scale storm). Thus another equation is needed in order to solve for the two unknowns. One equation can be obtained by assuming that the ratio of particle speed to wind speed is the same in model and full scale. The second equation would then be:

$$\frac{U\Delta t}{L}\Bigg|_{model} = \frac{U\Delta t}{L}\Bigg|_{full\ scale} \quad (17)$$

For the example run 10-17-1, the values of full-scale wind speed U and time duration Δt would be 14.68 m/s and 14 hrs, 6 min, respectively.

The ratio of particle to wind speed is not likely to be the same in full scale as in the model so it is probably more appropriate to search for an alternate second equation. This was done by equating the modified densimetric Froude numbers; i.e.,

$$\frac{\rho}{\rho_p} \frac{U^2}{gH} \left(1 - \frac{U_o}{U}\right)\Bigg|_{model} = \frac{\rho}{\rho_p} \frac{U^2}{gH} \left(1 - \frac{U_o}{U}\right)\Bigg|_{full\ scale} \quad (18)$$

Then, in order to satisfy Eq. (16),

$$\frac{\Delta A}{UL\Delta t} \left[A_1^2 \frac{D_p}{H}\left(\frac{U}{U_o}\right)^2 \right]^{3/7}\Bigg|_{model} = \frac{\Delta A}{UL\Delta t} \left[A_1^2 \frac{D_p}{H}\left(\frac{U}{U_o}\right)^2 \right]^{3/7}\Bigg|_{full\ scale} \quad (19)$$

Equations (18) and (19) result in full-scale values for run 10-17-1 of 22.16 m/s for wind speed and 5 hrs, 32 min for storm duration. If the fundamental similitude Eq. (16) is valid for all wind speeds, then either the wind speed and time set of values of 14.68 m/s and 14 hrs 6 min or 22.16 m/s and 5 hrs 32 min are appropriate since both sets satisfy Eq. (16). There probably are at least subtle differences with changes in wind speed, however, and thus one set may be more valid than the other.

5. MODEL - FULL-SCALE COMPARISON OF SNOW FENCE DRIFTS

Very little accurately measured full-scale snowdrift data are available with which to validate small-scale model measurements. A number of measurements have been made, however, of the drifts associated with the Wyoming Highway Department snow fences [30, 53, 54]. The typical snow fence is 3.66 m in height, with a 0.1 h gap at the fence bottom, is inclined 15° from the vertical, has horizontal alternate 15 cm slats and slat spacing, and a porosity of 50%.

The flat terrain wind tunnel experiments were performed with a miniature fence 2.54 cm in height and 76.2 cm in length (length-scale ratio of 144), to allow comparison with Tabler's full-scale and frozen-lake model data. The material used to simulate snow consisted of glass spheres of density 3990 kg/m^3 and average particle diameter of 49 μm. The value of the ratio U_f/u_{*t} = 1.3 and the threshold friction speed is 21.5 cm/s. To extrapolate to full scale it was assumed that a typical dry snow particle would have a diameter of 150 μm, density 700 kg/m^3, and u_{*t} = 14.0 cm/s. The large density of the spheres helps decrease the model value of the densimetric Froude number, making it easier to simulate full-scale drifts with reasonable full-scale wind speed values. The small particle diameter aids also in simulating the capability of real snow to form cornices. The friction caused by the wind blowing over the particles and then moving them results in the particles becoming electrostatically charged. This electrostatic charge results in a temporary angle of repose much greater than the usual 34°. For snow, the larger angle of respose (which can, of course, be greater than 90°) is often permanent because of interparticle sintering. For the glass spheres, after a test is completed, gradual disappearance of the charge results in a return to the normal 34° angle of repose and avalanching occurs. Cornice formation simulation is not possible with spheres that are much larger because the charge cannot be held for a sufficient length of time. Views of the development of the downwind drift as a function of time and other views involving tests of simulated hedgerows and trees have been reported previously [19,20].

At the end of the test of the model fence, a profiler was used to obtain accurate measurements of the drift profile. Also, photographs of the drift formation from directly above (plan) and from the side (profile) were taken simultaneously at regular intervals during the wind tunnel experiments. The results of measurements of the volume of the drifts taken from the photographs are shown in Fig. 8. The deposition volume increases linearly with time initially but the rate slows as equilibrium is reached. At the end of the experiment the drift volume was within 93% of Tabler's full-scale geometry. Comparisons of full-scale and model values of areas, lengths, and volumes are shown in Table 1 and in [20].

6. COMPARISONS OF WATER CHANNEL AND WIND TUNNEL SIMULATIONS

Probably the best way to compare the results of wind tunnel and water channel simulations is to compare the characteristics of two experiments [20,22] which were very nearly identical except for the fluid medium. The two experiments were simulations of the Wyoming Highway Department's 3.66 m snow fence. The characteristics of the two experiments are listed in Table 2.

536

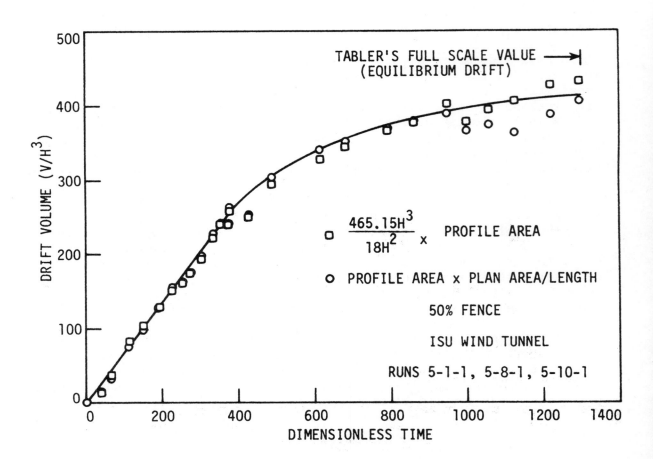

Fig. 8. Drift volume as a function of dimensionless time for the 50% porous fence.

Table 1. Comparison of full-scale and wind tunnel downwind drift measurements for the Wyoming 50% porosity snow fence.

Drift Parameters	Full-scale [55]	Model-scale [20]
Profile Area (A_{PR})	18 H^2	16.8 H^2
Plan Area (A_{PL})	693 H^2	645 H^2
Volume (465 $H^3 \times A_{PR}/18\ H^2$)	465 H^3	434 H^3
Volume ($A_{PL} \times A_{PR}/L$)	455 H^3	406 H^3
Length (maximum)	27.4 H	26.7 H
Length (average, A_{PL}/H)	23.1 H	21.5 H

The equivalent full-scale speed is predicted (for comparison purposes) by equating the model and full-scale densimetric Froude numbers. Because these numbers are larger in the water channel, the predicted full-scale values are also larger, from 43 to 158 m/s compared to 23 to 40 m/s from the wind tunnel. (If the method of [20] is used, these velocities would be smaller.)

The largest difference in the results of the two experiments is due to the differences in bed form features. The bed form features are functions of the terminal-friction speed ratio, the Froude number, and the drag factor of parameter 8. The first two terms are closer to full-scale values (much closer for Froude number) in the wind tunnel. The drag factor parameter values from the two experiments bracket the full-scale data. As a result of the differences, the bed form features in the water channel greatly affected the drift shape but are not considered important in the wind tunnel experiment. Comparisons of drift shapes are shown in Fig. 9. The ratio of ripple wavelength to fence height appears to be a factor of one to two in the water channel and only about 0.25 in the wind tunnel.

In both water channel and wind tunnel simulations, care should be taken to ensure that bed-form features are sufficiently small so that topographically controlled drift features are not obscured. This is particularly difficult to do at small scale in a water channel. Also, cornice formation caused by electrostatic effects is not possible in a water channel.

7. MODELING OF FALLING SNOW

The simulation of drifting snow in a wind tunnel or water channel can be carried out by laying an initial deposit of material on the tunnel or channel floor (usually upwind of the model), by introducing material from above the model during the experiment, or by a combination of the two. The second method is important (1) if it is necessary to investigate drift features associated with topographic obstacles when the wind speed is not sufficient to move quiescent particles (i.e., below threshold), or (2) if roof snow loads are to be simulated, or (3) if visibility during falling

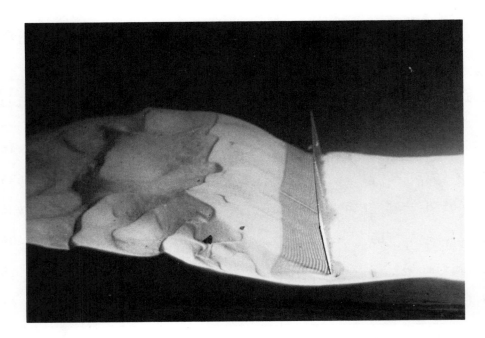

Fig. 9a.   Snow-fence drift simulation in the water-channel [22]. Flow right
            to left. Fence height ripple length ratio $\simeq$ 0.5. Photo courtesy
            James L. Wuebben of the Cold Regions Research and Engineering
            Laboratory, Hanover, New Hampshire.

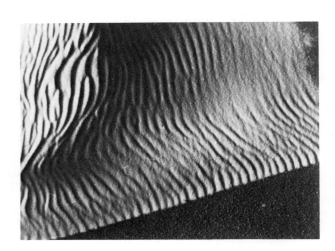

Fig. 9b.   Snow-fence drift simulation in the wind tunnel [20]. Flow left
            to right. Upper photo is close-up of cornice formation early
            in the experiment. Lower photo is near-equilibrium drift.
            Fence height-ripple length ratio $\simeq$ 4.

Table 2. Comparison of wind tunnel and water channel snow drift simulations.

Parameter	Wind Tunnel [20]	Water Channel [22]	Full Scale
Fluid density	0.001226 gm/cm^3	1	0.001226
Particle density	4 gm/cm	2.6	0.9
Geometric scale n	144	46 to 152	
Model speed U	400-700 cm/s	30-60	
Fence height H	2.5 cm	2.4 to 8	366
Terminal speed friction speed ratio U_f/u_{*t}	1.4	0.95	2.5
Froude number U^2/gh	65 to 200	0.12 to 1.5	14 to 44[†] 53 to 700[‡]
Densimetric Froude number $\rho U^2/(\rho_p-\rho)gH$	0.013-0.04	0.07-0.96	equal to model
Drag factor $C_D\rho H/\rho_p D_p$	7.3	1950-6500	660
Predicted full-scale speed U	23 to 40 m/s	43 to 158	

[†]Predicted from wind tunnel data

[‡]Predicted from water channel data

snow is to be simulated. Both methods have been used in both wind tunnels and water flumes although there have been more experiments of the second type in water. Roof snow loads have been simulated in a water channel by Isyumov [24,56] and visibility effects by Norem [25].

The modeling criteria for the falling snow problem may be different from the drifting snow problem, depending upon the desired simulation experiment. For falling particles, the particle trajectory parameters become important, i.e., the Froude number and drag factor parameters 8 and 9. The writer has not performed this type of experiment, but the procedure of varying particle density and diameter, wind speed and geometric scale should also prove quite useful in simulating falling snow as well as drifting snow.

8. CONCLUSIONS

The dimensionless parameters derived by simple dimensional analysis for a given similitude problem can often be combined for modeling purposes by theoretical means. If not all original dimensionless parameters can be satisfied at model scale, then the model is distorted and some means of interpreting the effect of distortion must be ascertained. For the snow-drifting model the effect of varying the amount of distortion can be determined by using particles of different diameter and density, by using different wind tunnel speeds and by testing models with vertical geometric distortion as well as true geometric models.

The primary modeling parameter is considered to be the mass-rate parameter which is derived from consideration of the amount of mass/unit time capable of being transported by the wind. This parameter is combined empirically with the densimetric Froude number to successfully correlate model-scale snowdrift accumulation data as well as to predict equivalent full-scale wind speeds and storm durations.

It is certainly possible to simulate snow drifting in a qualitative fashion in either water channels or wind tunnels. The writer believes that there are several advantages to the wind tunnel simulation, both qualitatively and quantitatively. The particle-to-fluid-density ratio in water is sufficiently small that it becomes difficult to satisfy the densimetric Froude number requirement. Also, the electrostatic effects for cornice formation are absent in water, and channel bed forms (dunes and ripples) will often obscure the desired drift features. The comparisons of model and full-scale simulations discussed here present evidence that successful and useful results can be obtained from careful wind tunnel snowdrift modeling experiments.

9. ACKNOWLEDGMENTS

The research summarized here was supported by the Iowa Department of Transportation and the Engineering Research Institute of Iowa State University. The author gratefully acknowledges many helpful discussions with Professor Stanley Ring, Professor James Sinatra, and Mr. Jeffrey Benson. Thanks are also due to research assistant V. Ethiraj and technical assistants Cory Wandling and Rory Deichert.

10. REFERENCES

1. E. A. Finney, Snow control on the highway, Michigan Experiment Station, East Lansing, Michigan, Bulletin No. 57 (1934).

2. E. A. Finney, Snow control by tree planting, Part VI, Wind tunnel experiments on tree plantings, Michigan Engineering Experiment Station, East Lansing, Michigan, Bulletin No. 75 (1937).

3. E. A. Finney, Snow drift control by highway design, Michigan Engineering Experiment Station, East Lansing, Michigan, Bulletin No. 86 (1939).

4. Y. Anno and T. Konishi, Modeling the effects of a snowdrift-preventing forest and a snow fence by means of activated clay particles, Cold Regions Science and Technology, 5 (1981) 43-58.

5. M. Jensen, Aerodynamik i den naturlige vind, Teknisk Forlag, København (1959), pp. 219-227.

6. B. A. Anfilofev, Construction of an aerodynamic tunnel for glaciological investigations, In Physics of Snow, Avalanches and Glaciers, USDA Technical Translation TT 69-53051 (1973).

7. K. Kimura and T. Yoshisaka, Scale model experiments on snowdrifts around buildings, Draft translation, Cold Regions Research and Engineering Laboratory, Hanover, New Hampshire (1971).

8. A. Becker, Natural snow fences along roads, Bautechnik, 22 (1944) 37-42.

9. C. Nøkkentved, Drift formation at snow fences, Stads-og Haveingeniren, Copenhagen, Denmark, 31, 9 (1940), Sept.

10. W. Kreutz and W. Walter, "Der Strömungsverlauf sowie die Erosionsvorgänge und Schneeablagerungen an künstlichen Windschirmen nach Untersuchungen un Windkanal, Berichte des Deutschen Wetterdienstes, 4, 24 (1956) 1-25.

11. R. W. Gerdel and G. H. Strom, Scale simulation of a blowing snow environment, Proceedings of the Institute of Environmental Sciences, 53 (1961) 53-63.

12. G. H. Strom, G. R. Kelly, E. L. Deitz and R. F. Weiss, Scale model studies on snow drifting, U.S. Army Snow, Ice and Permafrost Research Establishment, Hanover, New Hampshire, Research Report 73 (1962).

13. N. S. Stehle, Snow movement characteristics--The wind duct Borax tests, U.S. Naval Civil Engineering Laboratory, Port Hueneme, California, Technical Note N-682 (1964).

14. G. E. Sherwood, Preliminary scale model snowdrift studies using Borax in a wind duct, Naval Civil Engineering Laboratory, Port Hueneme, California, Technical Note N-682 (1967).

15. F. W. Brier, Snowdrift control techniques and procedures for polar facilities, Naval Civil Engineering Laboratory, Port Hueneme, California, Technical Report R767 (1972).

16. R. J. Kind and S. B. Murray, Saltation flow measurements relating to modeling of snowdrifting, unpublished manuscript (1980).

17. J. D. Iversen, Drifting snow similitude--transport rate and roughness modeling, Journal of Glaciology, Vol. 26 (1980), 393-403.

18. J. D. Iversen, Drifting snow similitude--drift deposit rate correlation, In J. Cermak (Ed.), Wind Engineering, Pergamon Press, Oxford (1980), pp. 1035-1047.

19. J. D. Iversen, Wind tunnel modeling of snow fences and natural snowdrift controls, 37th Eastern Snow Conference, Peterborough, Ontario (June, 1980), pp. 106-124.

20. J. D. Iversen, Comparison of wind tunnel model and full-scale snow fence drifts, Journal of Wind Engineering and Industrial Aerodynamics, 8 (1981), 231-249.

21. S. L. Ring, J. D. Iversen and J. B. Sinatra, Wind tunnel analysis of the effects of plantings at highway grade separation structures, Engineering Research Institute, Iowa State University, ISU-ERI-Ames 79221 (1979).

22. J. L. Wuebben, A hydraulic model investigation of drifting snow, Cold Regions Research and Engineering Laboratory, Hanover, New Hampshire, CRREL Report 78-16 (1978).

23. F. A. Theakston, Model technique for controlling snow on roads and runways, Cold Regions Research and Engineering Laboratory, Hanover, New Hampshire, Special Report 115 (1970).

24. N. Isyumov, An approach to the prediction of snow loads, PhD Dissertation, University of Western Ontario, Ontario, Canada (1971).

25. H. Norem, Designing highways situated in areas of drifting snow, Cold Regions Research and Engineering Laboratory, Hanover, New Hampshire, Draft Translation 503 (1975).

26. D. J. Calkins, Model studies of drifting snow patterns at safeguard facilities in North Dakota, Cold Regions Research and Engineering Laboratory, Hanover, New Hampshire, Technical Report 256 (1974).

27. D. J. Calkins, Simulated snowdrift patterns, Cold Regions Research and Engineering Laboratory, Hanover, New Hampshire, Special Report 219 (1975).

28. J. S. de Krasinski and W. A. Anson, A study of snow drifts around the Canada building in Calgary, University of Calgary, Department of Mechanical Engineering, Report No. 71 (1975).

29. J. de Krasinski and T. Szuster, Some fundamental aspects of laboratory simulation of snow or sand drifts near obstacles, University of Calgary, Department of Mechanical Engineering, Report No. 151 (1979).

30. R. D. Tabler, Geometry and density of drifts formed by snow fences, Journal of Glaciology, 26 (1980) 405-419.

31. R. D. Tabler, Self-similarity of wind profiles in blowing snow allows outdoor modeling, Journal of Glaciology, 26 (1980) 421-434.

32. R. D. Tabler and R. L. Jairell, Studying snowdrifting problems with small-scale models outdoors, Western Snow Conference, April 15-17, Laramie, Wyoming (1980).

33. W. Tobiasson and S. Reed, Use of field models to study snow drifting. Cold Regions Research and Engineering Laboratory, Hanover, New Hampshire, Informal memorandum (1966).

34. T. Imai, On the principle of similitude in the model test of snowstorm, Seppyo - Journal of the Japanese Society of Snow and Ice, 11 (1949) 14-16.

35. F. Odar, Scale factors for simulation of drifting snow, Journal of the Engineering Mechanics Division, ASCE, 88, EM2 (1962) 1-16.

36. F. Odar, Simulation of drifting snow, Cold Regions Research and Engineering Laboratory, Hanover, New Hampshire, Research Report 174 (1965).

37. R. J. Kind, A critical examination of the requirements for model simulation of wind-induced erosion/deposition phenomena such as snow drifting, Atmospheric Environment, 10 (1976) 219-227.

38. J. D. Iversen, Drifting snow similitude, Journal of the Hydraulics Division, Proceedings, American Society of Civil Engineering, 105, HY6 (1979) 737-753.

39. A. K. Dyunin, Basic requirements of aerodynamic models in applied glaciology, In Physics of Snow Avalanches and Glaciers, USDA Technical translation TT 69-53051 (1973), 220-229.

40. N. A. Woodruff and A. W. Zingg, Wind-tunnel studies of fundamental problems related to windbreaks, SCS-TP-112, U.S. Department of Agriculture, Soil Conservation Service, Washington, D.C. (1952).

41. R. Greeley, J. D. Iversen, J. B. Pollack, N. Udovich, and B. R. White, Wind tunnel studies of Martian eolian processes, Proceedings of the Royal Society, London, England, Series A, Vol. 341 (1974) pp. 331-360.

42. J. D. Iversen, R. Greeley, J. B. Pollack, and B. R. White, Simulation of Martian eolian phenomena in the atmospheric wind tunnel, Space Simulation, National Aeronautics and Space Administration Special Publication No. 336 (1973) pp. 191-213.

43. J. D. Iversen, R. Greeley, B. R. White, and J. B. Pollack, Eolian erosion on the Martian surface, Part 1, Erosion rate similitude, Icarus, 26, (1975) 321-331.

44. J. D. Iversen, R. Greeley, B. R. White, and J. B. Pollack, The effect of vertical distortion in the modeling of sedimentation phenomena: Martian crater wake streaks, Journal of Geophysical Research, 81, (1976) 4846-4856.

544

45. J. D. Iversen and V. Jensen, Wind transportation of dust from coal piles, Skibsteknisk Laboratorium Lyngby, Denmark (1981), SL report 81054.

46. R. A. Bagnold, The Physics of Blown Sand and Desert Dunes, Methuen, London (1941).

47. J. D. Iversen, J. B. Pollack, R. Greeley, and B. R. White, Saltation threshold on Mars; the effect of interparticle force, surface roughness, and low atmospheric density, Icarus, 29, (1976) 381-393.

48. J. D. Iversen, R. Greeley, and J. B. Pollack, Windblown dust on Earth, Mars, and Venus, Journal of Atmospheric Science, 33, (1976) pp. 2425-2429.

49. J. D. Iversen and B. R. White, Saltation threshold on Earth, Mars, and Venus, Sedimentology (1981) in press.

50. R. A. Schmidt, Threshold wind-speeds and elastic impact in snow transport, Journal of Glaciology, 26 (1980), 453-467.

51. M. Jensen, The model-law for phenomena in natural wind, Ingeniøren, 2 (1958) 121-128.

52. P. R. Owen, Saltation of uniform grains in air, Journal of Fluid Mechanics, 20, (1964) 225-242.

53. R. D. Tabler, New snow fence design controls drifts, improves visibility, reduces road ice, Proceedings of the Annual Transportation Engineering Conference, 46 (1973) 16-27.

54. R. D. Tabler, New engineering criteria for snow fence systems, Transportation Research Record, 506 (1974) 65-78.

55. R. D. Tabler, Geometry and density of drifts formed by snow fences, Abstract, Snow in Motion Symposium, Ft. Collins, Colorado (1979).

56. N. Isyumov and A. G. Davenport, A probabilistic approach to the prediction of snow loads, Canadian Journal of Engineering, 1 (1974) 28-49.

LIST OF SYMBOLS

A_d Planform drift area, m^2

A_c Longitudinal drift cross-section area, m^2

A_1 Threshold speed coefficient, $u_{*t} (\rho/\rho_p g D_p)^{1/2}$

B Friction Reynolds number, $u_{*t} D_p/\eta$

C_D Drag coefficient

D_p Particle diameter, μm

g Gravitational acceleration, m/s^2

h,H   Characteristic vertical dimension, m

η All other vertical dimensions, m

k   von Kármán's constant

L   Characteristic horizontal dimension, m

ℓ All other horizontal dimensions, m

R   Reynolds number, correlation coefficient

q   Mass-transport rate, kg/sm

q_d Mass-deposition rate, kg/sm

t   Time, s

u_* Surface friction speed, m/s

u_{*t} Threshold friction speed, m/s

U   Reference wind speed, m/s

U_∞ Wind tunnel free stream speed (above boundary layer), m/s

U_o Free stream or reference wind speed at initiation of motion (threshold), m/s

U_F Particle terminal speed, m/s

V   Drift volume, m^3

z_o Aerodynamic roughness height, m

z_o' Aerodynamic roughness height in saltation, m

ν Kinematic viscosity, m^2/s

ρ Fluid density, kg/m^3

ρ_p Particle density, kg/m^3

INSTRUMENTATION CONSIDERATIONS FOR VELOCITY MEASUREMENTS

P. A. Irwin, Director of Technical Services,
MHTR Ltd. Wind Tunnel Laboratories,
650 Woodlawn Road West,
Guelph, Ontario, Canada.

SUMMARY

The characteristics of pedestrian level winds and how these characteristics affect the choice of instrumentation for measuring wind speed in model studies are discussed. Hot-wire and hot-film anemometer methods are reviewed. This is followed by a description of a recently developed omnidirectional pressor device for measuring the speed and then by a discussion of some of the less commonly used methods. In the discussions the characteristics of the speed measuring system that are desirable for routine use are outlined. The importance is emphasized of carrying out flow visualization studies to complement the quantitative measurements of speed.

1. INTRODUCTION

To assess the local wind environment using wind tunnel models requires that the local wind speed be measured at various locations on the model for a number of different wind directions. To obtain a fuller understanding of the wind environment a flow visualization study, using smoke for example, is also very desirable. The speed measurements are typically at a height equivalent to about 2m above ground at full scale. For wind tunnel models in the range 1:500 to 1:200 this implies heights of 4mm to 10mm on the model. Figure 1 illustrates a typical model study. The flow field is usually very turbulent and unpredictable as to direction. In addition to the mean speed, the gusts associated with the turbulence are also felt by the pedestrians and must therefore be measured by the instrumentation. Reference 1 is a special issue of the Journal of Industrial Aerodynamics that contains a number of papers discussing the special characteristics of pedestrian level winds.

The lack of predictability of mean wind directions and the high turbulence at the measurement location pose problems for many types of wind velocity probes because they require alignment in highly turbulent flow. However, two important simplifications arise. The first is that at close proximity to the ground the vertical component of velocity is generally very small or at too high a frequency to be of significance to pedestrians (the term vertical is used loosely here to denote normal to the ground surface, which may be sloped). The second simplification is that provided a good flow visualization study is carried out in parallel with the quantitative measurements it is virtually always unnecessary to carry out quantitative measurements of the wind direction, only its speed. Pedestrians are primarily sensitive to the absolute value of the horizontal wind speed (more exactly the dynamic pressure resulting therefrom) which is given the symbol Q in this paper.

$$Q = \sqrt{U^2 + V^2} \tag{1}$$

Figure 1: Wind Tunnel Model for a Pedestrian Level Wind Study

548

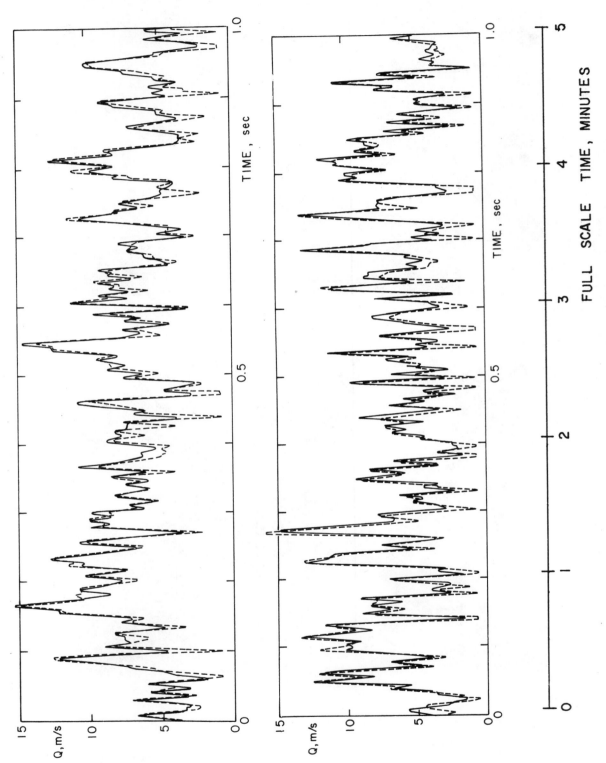

Figure 2: Time Histories of Speed Measured on a Wind Tunnel Model by a Hot-Wire Anemometer (———) and an Omnidirectional Pressure Device (- - -)

where U and V are the two horizontal velocity components. It is worth noting that whereas U and V can be either positive or negative, Q is always positive. Typical examples of time histories of Q measured by two different methods on a model are shown in Figure 2.

The characteristics of pedestrian level winds have led to the development or adoption of particular methods of measuring windspeed in wind tunnel studies. These are discussed in this paper.

2.   GENERAL CONSIDERATIONS

It is desirable for the instrumentation to be able to respond to the instantaneous speed, Q. From the time history of Q, statistical parameters such as the mean, $\bar{Q}$, the root-mean-square (r.m.s.) value, $\sqrt{q^2}$, of the fluctuating part, q, of the speed, and the peak gust speed $\hat{Q}$ can be computed. In addition, the probability distribution of Q may be obtained. These statistical parameters would normally be evaluated by the wind tunnel computer system and would be combined with long term meteorological data in order to predict the probability of various wind speeds occurring at each measurement location at full scale.

A number of investigators have used an effective gust speed, Q_{eff}, to assess the comfort of pedestrians where

$$Q_{eff} = Q + \gamma \sqrt{q^2} \tag{2}$$

The constant γ has been given different values ranging from 1 to 4 at different laboratories (see the papers in Reference 1). Other methods have been based directly on the measured gust speed, $\hat{Q}$, which does not then involve any assumption as to the relation between the r.m.s. speed and the peak gust speed. However, the duration of the peak gust warrants some discussion, e.g. whether it is a 3 second or a 10 second gust, because the longer the gust duration the lower the gust speed. A 3 second gust corresponds approximately to $\gamma = 3.5$. A 10 second gust would correspond to a lower value of γ. Where the 3 second gust speed is high the 10 second gust speed will also tend to be high and the outcome of a pedestrian level wind study will not be very sensitive to the particular choice of gust duration within the 3 to 10 second range particularly if the comfort criteria are tailored to the chosen gust duration.

From the above discussion it is concluded that the frequency response of the instrumentation should be such that it is flat from zero frequency up to a value corresponding to about 0.1 to 0.3 Hz at full scale, depending on the gust duration being considered. The relation between model and full scale frequencies is

$$n_m = n_p \left(\frac{b_p}{b_m}\right) \left(\frac{Q_m}{Q_p}\right) \tag{3}$$

where n = frequency, b = length scale, and the subscripts m and p refer to model and full scale respectively. Assuming $Q_m/Q_p \simeq 1$ then for a 1:400 scale model the instrumentation should have a flat frequency response up to a frequency, n_m, of 40 to 120 Hz, depending on the gust duration required.

The question of exactly how flat the frequency response should be is partly a question of the overall accuracy that is required to realistically assess whether a wind problem will exist or not at any given location. For this purpose an accuracy of $\pm$ 10% in the measurements of gust speed is probably sufficient so this tolerance would be appropriate for the frequency response.

Other factors to consider are the practicality of the method of measurement if it is to be used on a routine basis. It should lend itself to automation, not require lengthy or often repeated calibration procedures, and not involve too large an expenditure on equipment. The sensing devices should not be too fragile. The method should be such that the speed can be obtained at a large number of measurement locations. Poor coverage of pedestrian areas through use of too few speed sensors can result in high wind speed zones going undetected. It may be worth using a slightly less accurate measuring system if it results in improved coverage.

3. HOT-WIRE AND HOT-FILM METHODS

A hot-wire anemometer, which consists of an electrically heated wire that is cooled by the air flow, can be used to measure Q provided the wire is aligned vertically, Figure 3(a). If the wire is horizontal it measures only the component of horizontal speed normal to the wire, not the total speed. The response for the vertical wire can be expressed as

$$E = A + B \, Q^{\beta} \tag{4}$$

where E is the output voltage, A and B are calibration constants and β is an exponent usually with a value of about 0.45. Because of the non-linear calibration expression the signal is often passed through an analogue linearizing unit which then makes the output voltage proportional to Q. The linearization may alternatively be carried out digitally using the wind tunnel computer system. However it is done, the linearization is an essential step in evaluating accurate values of the mean speed, $\bar{Q}$, and the r.m.s. value $\sqrt{\overline{q^2}}$ in highly turbulent flow. The hot-wire anemometer possesses very good frequency response, usually measured in thousands or even tens of thousands of Hz. However, the wire must be held physically by two prongs and a supporting stem and these introduce undesirable disturbances to the flow for particular directions not only for the hot-wire probe in question but also for other hot-wire probes that may be present nearby.

The solution to the problem of stem or prong interference is to use a cylindrical hot-film probe. This consists of an electrically heated thin film of conducting material on the surface of a thin, vertical, nonconducting cylinder which is of sufficient diameter and strength to be self supporting. Figure 3(b) illustrates this type of probe. Its calibration curve is similar to that of the hot-wire probe and its frequency response, although inferior to the hot-wire, is still more than sufficient for pedestrian level wind studies. It can be easily mounted so that the cylinder protrudes from the model street surface with negligible effect on the flow or no more than that of a pedestrian.

Disadvantages of both hot-wire and hot-film probes are that they are easily broken, their calibrations can drift significantly during the wind tunnel tests due to temperature changes and accretion of dust part-icles, and they involve appreciable expense especially if it is desired to make measurements at many locations. The errors due to temperature

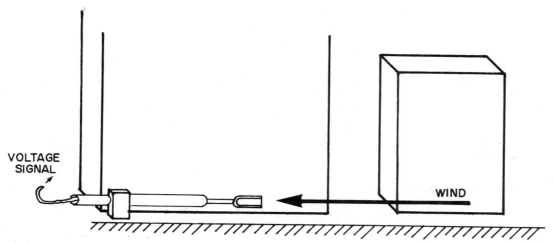

(a) Hot-Wire Probe

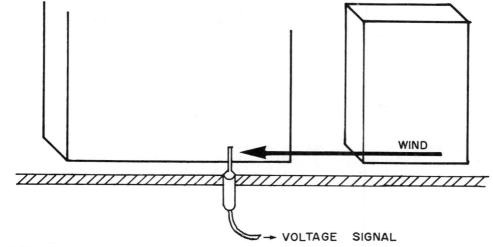

(b) Cylindrical Hot-Film Probe

FIGURE 3
HOT-WIRE AND HOT-FILM PROBES

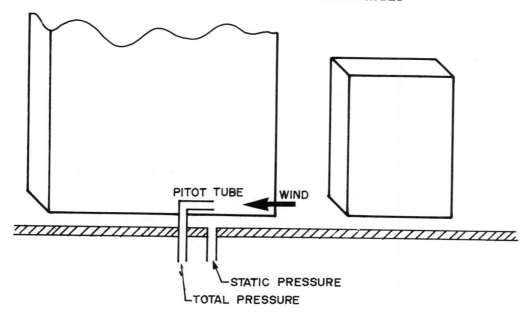

FIGURE 4
PITOT TUBE AND STATIC PRESSURE TAP

drift can be reduced by applying temperature corrections, especially when the wind tunnel has a computer system on line. In practice, the number of measurement locations is often limited to a bare minimum when the hot-wire or hot-film approach is used. Ten to twenty locations may be chosen in a situation where forty to fifty would be desirable. Nonetheless, the ready availability of off-the-shelf hot-wire and hot-film equipment and its excellent frequency response have made it a prime tool in pedestrian level wind studies.

4. PRESSURE METHODS

A pitot tube pointing into the flow and a static pressure tap under it in the model street surface have been used to measure pedestrian level winds, Figure 4. With proper attention to the details of the tubes connecting the pitot tube and static pressure tap to the pressure transducer, and by using a transducer with small internal volume it is probable that a sufficient frequency response for pedestrian level wind studies could be obtained by this method. Techniques for measuring fluctuating pressures through tubes have been developed at many laboratories. Some are described in Reference 2. However, the problem remains that the pitot tube will tend to read low because of the high intensity turbulence and that it must be aligned with the mean flow direction. For routine measurements the need to frequently realign the pitot tube is a severe impediment.

However, the principal of a pressure method is attractive because highly developed techniques and equipment now exist for rapidly measuring pressures at many locations. These techniques were developed for pressure model studies of cladding loads where the number of pressure taps is typically measured in hundreds. It was considerations such as these that led to the development of the omnidirectional wind speed sensor described in Reference 3 and shown in Figure 5. It overcomes the problem of directional sensitivity, is simple to make, requires no calibration since its calibration is fixed, and enables the mass production methods of the present day pressure study to be applied to the measurement of pedestrian level wind speeds.

The geometry of the omnidirectional sensor can be seen in Figure 5 to consist of a hole of diameter D in the street surface out of the centre of which protrudes a tube of external diameter slightly less than D. The tube protrudes to a height h above the street surface and the top of the tube is flat. Experiments with other shapes of top have indicated there is little to be gained by using more complex shapes. The instantaneous excess pressure Δp at the bottom of the sensor hole over that at the top of the sensor tube is measured and from Δp the wind speed at a chosen height, h_s, above the surface is calculated using an appropriate calibration formula. Usually h_s is chosen to be the central tube height, h. The principal of operation and calibration of the omnidirectional sensor is described in detail in Reference 3. Figure 6 shows the data from which the calibration formula was derived. This plot shows the skin friction velocity, u_τ , expressed in nondimensional form and plotted against the pressure difference Δp, also in nondimensional form. The skin friction velocity is closely related to the wind speed at pedestrian level. The calibration formula for wind speed at the level of the top of the sensor tube is determined once and for all by the geometry and may be expressed as

$$Q = 85(\nu/h) + 1.74(\Delta p/\rho)^{\frac{1}{2}} \tag{5}$$

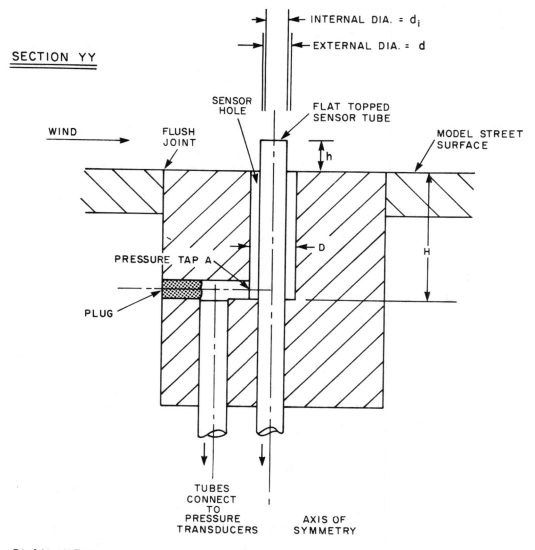

INTERNAL DIA. = d_i

EXTERNAL DIA. = d

SECTION YY

SENSOR HOLE

FLAT TOPPED SENSOR TUBE

WIND

FLUSH JOINT

MODEL STREET SURFACE

h

D

PRESSURE TAP A

PLUG

H

TUBES CONNECT TO PRESSURE TRANSDUCERS

AXIS OF SYMMETRY

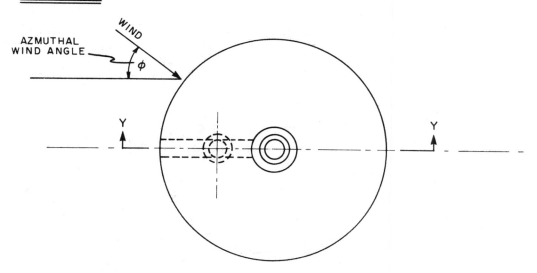

PLAN VIEW

WIND

AZMUTHAL WIND ANGLE

ϕ

Y

Y

Figure 5: Omnidirectional Pressure Device for Measuring Pedestrian Level Wind Speed.

554

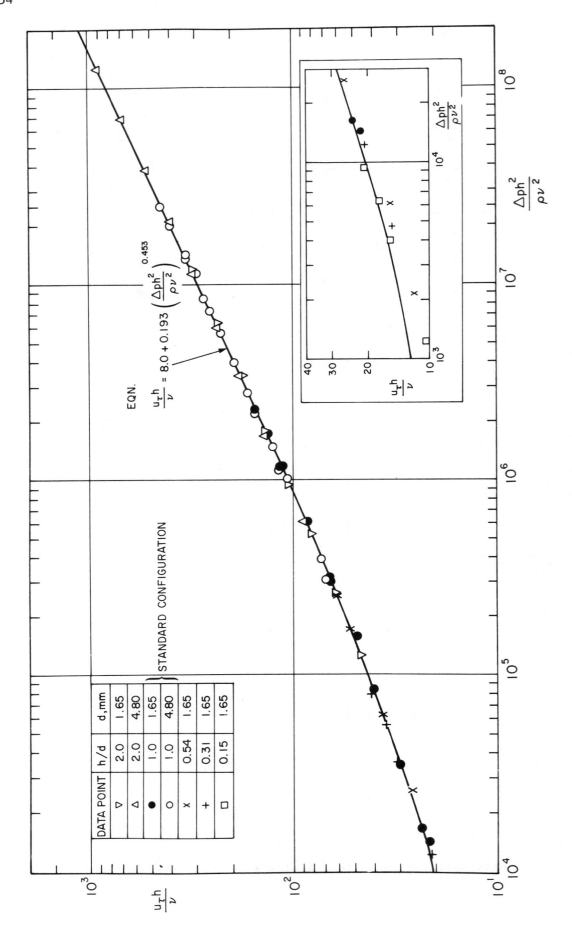

Figure 6: Calibration of Omnidirectional Wind Sensor

where ν = kinematic viscosity of air and ρ = air density. This formula is valid for $\Delta p h^2/\rho\nu^2 > 10^4$, $h/d > 0.5$, $d_i/d = 0.72$, $D/d = 1.56$ and $H/d > 1.5$ (see Figure 5 for definitions).

To measure the instantaneous pressure difference Δp with the minimum distortion of the pressure signal requires careful attention to the details of the tubing system connecting the sensors to the pressure transducers. However, the methods involved are no different from those used in pressure studies, and in practice it is not the tubing system that limits the frequency response. The limitation comes from the physical size of the sensor itself compared with the turbulence eddies. When the eddy size is less than about 70 times the sensor size the above calibration formula starts to break down. However, the frequency response does improve as the wind speed increases, since an eddy takes less time to go by, and this is actually a desirable characteristic because speed fluctuations become more noticable to pedestrians at high speeds. Figure 2, which is taken from Reference 3, shows a comparison between simultaneous measurements on a city model by an omnidirectional surface wind sensor and by a hot-wire anemometer. The agreement can be seen to be good, particularly on the peak gust speeds. It is found in practice that the omnidirectional sensor's frequency response is sufficient to measure gusts with periods as short as 5 to 10 seconds at full scale with an accuracy of $\pm$ 10% on speed.

The pressure signal Δp falls to low values that become difficult to measure at speeds below about 2m/s so a high tunnel speed is desirable. Zero drifts on pressure transducers must be carefully accounted for. Tunnel noise is not a problem because Δp is an instantaneous difference in pressure between two very closely spaced points and the pressure fluctuations due to noise, being essentially equal at the two points, get eliminated in the substraction of one pressure signal from the other.

To determine the peak gust value of Q from the peak Δp, Equation 5 can be used directly. If it is desired to obtain the mean speed, $\bar{Q}$, and the r.m.s., $\sqrt{\overline{q^2}}$, then two routes are possible. One is to convert the complete time history of Δp, point by point, to a time history of Q and then compute $\bar{Q}$ and $\sqrt{\overline{q^2}}$ in the normal manner. The second route is first to evaluate $\overline{\Delta p}$ and $\sqrt{\overline{\Delta p'^2}}$, where $\Delta p'$ = fluctuating part of Δp, and then use approximate relations between mean and r.m.s. speeds and pressures. The approximate relations are[3]

$$\bar{Q} = 85\frac{\nu}{h} + \frac{1.74}{\sqrt{\rho}}\left(\overline{\Delta p} - \tfrac{1}{4}\left(\overline{\Delta p'^2}/\overline{\Delta p}\right)\right)^{\frac{1}{2}} \tag{6}$$

$$\sqrt{\overline{q^2}} = \frac{1.74}{2\sqrt{\rho}}\left(\overline{\Delta p'^2}/\overline{\Delta p}\right)^{\frac{1}{2}} \tag{7}$$

It is worth commenting that the omnidirectional sensor gives an output that is close to proportional to the dynamic pressure of the wind. Although wind speed has become the benchmark quantity in assessing pedestrian comfort, what pedestrians actually feel is the dynamic pressure. A time history of the dynamic pressure, as opposed to speed, magnifies greatly the difference between gusts and lulls. When the output Δp of the omnidirectional sensor is observed on an oscilloscope the gusts and lulls visible in the Δp signal bring to mind strongly the experience of being outside on a windy day.

5. OTHER METHODS

Other methods of measuring wind speed include the Laser Anemometer, the inference of wind speed from the measured drag of a small object projecting into the air stream, and the sand erosion method.

At the present the Laser Anemometer is not in routine use in pedestrian level wind studies. To be of use in this application a back scattering technique would be necessary. There are potential problems caused by the interference of line of site by the model buildings. Also Laser Anemometer equipment is at present more costly than hot-wire equipment. However, it is possible that future developments will make the Laser Anemometer a more attractive method.

The measurement of speed via the drag on a small body is a possible method and instrumentation has been developed along these lines[4]. So far this method has not been widely used. If the miniature drag balances could be produced sufficiently economically this approach could conceivably compete with the hot-film or omnidirectional pressure probe methods. A variation on the drag method is reported in Reference 5 where the movement of a suspended styrofoam ball was used to measure speed and direction. However, slow data reduction was a drawback with this method.

The sand erosion method is attractive from the point of view of giving overall patterns of wind speed rather than point measurements. The extent of the high speed regions is made clearly visible. References 6 and 7 describe the application of this method. However, there are quantitative uncertainties in relating the sand patterns to the wind speed which have not yet been resolved. The sand erosion technique has not been widely employed partly because of the quantitative uncertainties but possibly also because it is not so easy to use on a routine basis as the hot-wire, hot-film or omnidirectional pressure sensor methods.

6. CONCLUDING REMARKS

In this paper the various approaches to measuring pedestrian level winds at model scale have been discussed. The hot-wire, hot-film and pressure methods have been given most space because they are at present the most practical approaches for everyday use. However, they do give only point measurements and there is much room for innovation in developing new methods that would give more complete spatial patterns of speed both accurately and economically. It is appropriate to re-emphasize a final point and that is the value of a flow visualization study to complement the wind speed measurements. The flow visualization can be accomplished by several methods, the usual being the introduction of smoke streams. One of the most effective is to take the wind tunnel model and immerse it in a water flume with an appropriately tailored water flow. Dye streams are then introduced to show the flow patterns. The prime value of flow visualization is that it enables the causes of high wind speed areas to be diagnosed and demonstrated to the client in an easily understandable way. It is also an almost essential aid in developing solutions to wind environment problems, such as landscaping or changes of building details.

REFERENCES

1) Lawson, T.V. (editor)

"The Wind Content of the Built Environment", Special issue of the Journal of Industrial Aerodynamics 3 (1978), 89-249.

2) Irwin, P.A., Cooper, K.R., and Girard, R.

"Correction of Distortion Effects Caused by Tubing Systems in Measurements of Fluctuating Pressures", Journal of Industrial Aerodynamics, 5 (1979), 93-107.

3) Irwin, P.A.

"A Simple Omnidirectional Sensor for Wind-Tunnel Studies of Pedestrian Level Winds", Journal of Wind Engineering and Industrial Aerodynamics, 7 (1981), 219-239.

4) Ettouney, S.M. and Fricke, F.R.

"Anemometer for Scale Model Environmental Wind Measurements", Building Science, 10 (1975), 17-26.

5) Isyumov, N., and Davenport, A.G.

"The Ground Level Wind Environment in Built-up Areas", Proceedings of the 4th International Conference on Wind Effects on Buildings and Structures, London, Sept. 1975.

6) Beranek, W.J.

"General Rules for the Determination of theWind Environment", Proceedings of the Fifth International Conference on Wind Engineering, Fort Collins, Colo, 1979, 225-234

7) Borges, A.R.J. and Saraiva, J.A.G.

"An Erosion Technique for Assessing Ground Level Winds", Proceedings of the Fifth International Conference on Wind Engineering, Fort Collins, Colo., 1979, 235-242.

SIMULATION OF BUOYANCY AND WIND INDUCED VENTILATION

Michael Poreh and Samuel Hassid, Technion, Haifa, Israel

1.   INTRODUCTION

Ventilation and air exchange in buildings and industrial plants can be induced by external winds, which create pressure differences between various opennings of the buildings, and by buoyancy forces created by temperature-density differences between the inner and outer air (a stack effect). When the outer air motion is very slow, the warmer air in the buildings will leave it through upper opennings and will be replaced by cooler air entering the building from lower opennings.

Assuming that the characteristic temperature rise in the building is ΔT and that the vertical distance between the upper and lower opennings is L the exit velocity of the warm air will be determined by the balance of the pressure losses at the opennings $[O(\rho U^2/2)]$ and the buoyant pressure $[O(\Delta \rho g L)]$ so that

$$u_{exit} \div ((\Delta\rho/\rho)gL)^{\frac{1}{2}} = ((\Delta T/T)gL)^{\frac{1}{2}} \tag{1}$$

The heat flux through the upper opennings, whose area is A, would be

$$H \div \rho C_p \Delta T\, u_{exit} A \div \rho C_p \Delta T^{3/2} g^{1/2} L^{1/2} A/T^{1/2} \tag{2}$$

When the outer wind speed U increases, the air exchange pattern will be changed, as pressures of the order of $\pm\ \rho U^2/2$ will be built on the envelope of the buildings. The ventilation will either increase or decrease, depending on the particular building geometry and the position of the opennings relative to the wind direction. At high wind speeds $U \gg u_{exit}$, or

$$U/((\Delta T/T)gL)^{\frac{1}{2}} \gg 1 \tag{3}$$

the effect of buoyancy will be negligible and the air exchange will be determined primarily by the wind induced pressures.

The dependence of the air exchange and heat transfer on a large number of factors, including the detailed configuration of the building and surroundings, makes an analytical or numerical analysis of practical design problems impractical, particularly when both the buoyancy and the wind induced pressures are of the same order of magnitude. It would thus be convenient if the combined effect of the wind motion and buoyancy in a particular geometry could be simulated in small scale wind tunnel models.

This paper discusses the requirements for such simulations. It is shown that in many cases only approximate simulations can be obtained in small wind tunnel models. Their scaling laws are specified and some of their limitations are discussed. The modelling of a chemical plant, which produces a considerable amount of heat and pollution, is described as an example.

2. CRITERIA FOR MODELLING

Consider a large heat source within a building. The air motion and heat transfer is determined by the following independent dimensional parameters:

H   –   the heat flux from the inner source

L   –   a characteristic length of the building, such as its height

ρ –   the density of the ambient air

T   –   the temperature of the ambient air

C_p –   the specific heat of the air (at constant pressure)

U   –   a characteristic velocity of the ambient wind field

ν –   the kinematic viscosity of the air

α –   the molecular thermal diffusivity of the air

and a large number of dimensionless parameters describing the relative geometrical configuration, the relative velocity and turbulence distribution in the wind field and the boundary conditions. These dimensional parameters should be the same in the model and the prototype.

It follows from dimensional considerations that dependent dimensionless parameters in the field, such as the relative temperature rise at a given point of the prototype and the corresponding point in the model, would be a function of the set of the independent dimensionless parameters which can be grouped from the above list of variables. Accordingly one may write that a relation exists

$$F \left(\frac{\Delta T}{T} \; ; \; \frac{H}{\rho C_p T U L^2} \; ; \; \frac{U^2}{C_p T} \; ; \; \frac{U^2}{gL} \; ; \; Re, \; Pr \right) = 0 \tag{4}$$

where Re is the Reynolds number UL/ν and Pr is the Prandtl number ν/α. To ensure a complete similarity between the model and prototype, the values of all the independent dimensionless parameters in (4) must be equal in model and prototype [1]. Obviously this requirement makes modelling impossible. Fortunately, it is recognized that under certain conditions the effect of several dimensionless parameters on the phenomenon is insignificant and they can be neglected.

It is well established [1,2] that when the Reynolds number of a flow is sufficiently large, namely

$$Re > Re_{minimum} \tag{5}$$

the flow will be turbulent and its primary features would be independent of the Reynolds number. The effect of the Prandtl number is also negligible because the flow is turbulent and its value is the same in the model and prototype, so that one may write

$$F \left(\frac{\Delta T}{T} \; ; \; \frac{H}{\rho C_p T U L^2} \; ; \; \frac{U^2}{C_p T} \; ; \; \frac{U^2}{gL} \right) = 0 \tag{6}$$

The term U^2/C_pT is recognized as the ratio of the kinetic energy to the thermal energy of the gas and is significant only when the Mach number of the flow is large, [3]. Neglecting this term in Eq. 4 gives:

$$F \left(\frac{\Delta T}{T} \; ; \; \frac{H}{\rho C_p TUL^2} \; ; \; \frac{U^2}{gL} \right) = 0 \qquad (7)$$

Denoting by x_m and x_p the value of any variable in the model and prototype respectively and defining the scaling of x as $\lambda(x) = x_m/x_p$, it follows from (7) that the necessary conditions for similarity, in addition to Eq.(5), are

$$\lambda \left(H/\rho C_p TUL^2 \right) = 1 \qquad (8)$$

$$\lambda \left(U^2/gL \right) = 1 \qquad (9)$$

The scaling of the relative temperature rise in such a model would be

$$\lambda \left(\Delta T/T \right) = 1 \qquad (10)$$

Simulations which satisfy these scalings laws are usually termed <u>exact simulations</u>.

Before proceeding with the analysis, it is worthwhile to stress that physical arguments which lead to neglection of terms in general functional forms, like Eq. (6), must be used carefully and the results must be critically examined [4]. Assume for example that the functional dependence described in Eq. 6 is expressed in a different form, such as

$$F \left(\frac{\Delta T}{T} \; ; \; \frac{H}{\rho U^3 L^2} \; ; \; \frac{U^2}{C_p T} \; ; \; \frac{U^2}{gL} \right) = 0 \qquad (11)$$

which is fully equivalent of course to (6). If the earlier arguments about the term U^2/C_pT are applied to the new equation, the resultant equation is

$$F \left(\frac{\Delta T}{T} \; ; \; \frac{H}{\rho U^3 L^2} \; ; \; \frac{U^2}{gL} \right) = 0 \qquad (12)$$

which is not equivalent to Eq. (7) and gives different scaling laws. The only way to check out the validity of the proposed approximation is to examine whether dimensional or dimensionless parameters which are physically significant in the problem have been omitted, and whether the remaining dimensionless parameters can describe the problem correctly. Such an examination will easily show that Eq. 12 is not a valid approximation. It is known that convective motion is induced by buoyancy and it is thus expected that the buoyancy flux, which is proportional to $H/\rho C_p$, should appear in one of the dimensionless parameters. Obviously any form, such as Eq. (12), which does not include the specific heat C_p

has to be rejected. Clearly the wrong result was obtained in our case by formulating the interdependence between the various variables in terms of dimensionless parameters which are not significant in this problem, as $H/\rho U^3 L^2$.

3. APPROXIMATE SIMULATIONS

The use of significant dimensionless parameters is also very helpful in drawing additional conslusions from the dimensional analysis. It is clear for example that a typical heat flux in our problem is $\rho C_p \Delta T U L^2$ rather than $\rho C_p T U L^2$.

Similarly since the body force on a heated element is proportional to $\Delta \rho g$, which for small $\Delta T/T$ is proportional to $\rho g \Delta T/T$. Thus, one expects that $U/\sqrt{(\Delta T/T)gL}$ would be a more significant dimensionless parameter than $U/\sqrt{gL}$. These arguments suggest that it is more meaningful to rewrite Eq. 12 in the form

$$F \left(\frac{H}{\rho C_p \Delta T U L^2} \; ; \; \frac{U^2}{\Delta T g L/T} \; ; \; \frac{\Delta T}{T} \right) = 0 \tag{13}$$

One further expects, by analogy to similar problems that T would not be a significant parameter in this problem, except for its effect on the body force. It thus follows that for small $\Delta T/T$ one may write that

$$F \left(\frac{H}{\rho C_p \Delta T U L^2} \; ; \; \frac{U^2}{\Delta T g L/T} \right) = 0 \tag{14}$$

or

$$F \left(\frac{H T^{1/2}}{\rho C_p g^{1/2} L^{5/2} \Delta T^{3/2}} \; ; \; \frac{U^2}{\Delta T g L/T} \right) = 0 \tag{15}$$

Based on this conclusions, the criteria for achieving simulation in small scaled models are

$$\lambda \; (H/\rho C_p \Delta T U L^2) = 1 \tag{16}$$

$$\lambda \; (U^2/(\Delta T/T)gL) = 1 \tag{17}$$

which imply of course that

$$\lambda \; (H T^{1/2}/\rho C_p g^{1/2} L^{5/2} \Delta T^{3/2}) = 1 \tag{18}$$

Simulations which satisfy these requirements are usually termed approximate simulations [3]. The scaling of $\Delta T/T$ in such models can be set arbitrarily to any desired value, provided $\Delta T/T$ is not a large number. When $\lambda(\Delta T/T) = 1$, these requirements become identical to those specified for exact simulations.

It is interesting to note that when the outer velocities are very small, Eq. 15 becomes

$$F\ (HT^{\frac{1}{2}}/(\rho C_p g^{\frac{1}{2}} L^{5/2} \Delta T^{3/2})) = 0 \qquad (19)$$

or

$$HT^{\frac{1}{2}}/(\rho C_p g^{\frac{1}{2}} L^{5/2} \Delta T^{3/2}) = \text{constant} \qquad (20)$$

which is consistent with Eq.2. When U becomes very large, on the other hand, the effect of buoyancy is negligible and the last term in Eq. 14 can be eliminate. One finds for this case that

$$H/\rho C_p \Delta T U L^2 = \text{constant} \qquad (21)$$

The relaxation of the requirement $\lambda(\Delta T/T) = 1$ in approximate simulations has two benefits. When one uses higher temperature differences in the model, the velocity scaling, which according to Eq.(17) is equal to

$$\lambda(U) = \lambda((\Delta T/T)gL)^{\frac{1}{2}},$$

becomes larger and the Reynolds number of the flow becomes larger. It is thus easier and many times the only possible way to satisfy the requirement Re > Re$_{minimum}$. The larger temperature differences in the model are are more easy to measure.

4. LIMITATIONS OF THE PHYSICAL MODELLING

The limitations of both the exact and the approximate simulations are primarily related to the accuracy of the assumption of Reynolds-number-independence. There is evidence that above Re = 4×10^4 [1,2] the effect of the Reynolds number in external flows around bluff bodies is small. Measurements of the internal velocities and turbulence in a model of a house suggest that this rule might also hold for internal flows [5]. One recalls that the flow of air through windows is basically similar to a jet flow, which becomes turbulent at even lower Reynolds numbers, and thus this conclusion is not surprising. In case of buoyant flows Reynolds-number independence is expected to start even earlier [6, p. 512].

It must be realized, however, that the appropriate Reynolds number for flows near boundaries is a local Reynolds number which is based on the distance from the wall. Close to the wall this Reynolds number is always small and one cannot expect processes which are controlled by the wall region to be Reynolds-number-independent. This is particularly true of heat transfer from the wall. Consider for example a simulation of a room in which one section of a vertical wall is heated to a temperature $T + \Delta T_s$.

The value of ΔT_s is related to the heat flux H. Assuming that the heat flux is by free convection, the Nusselt number, Nu, which is proportional to H, will be a function of the Grashof number $G = gh^3 \Delta T_s/(T\nu^2)$ where h is the height of the heated wall. When the product of the Grashof number and the Prandtl number GP < 10^8 the flow is laminar and Nu + (GP)$^{1/4}$. When GP > 10^{10}, the flow is turbulent and Nu + (GP)$^{2/5}$. Taking h = 2m, ΔT_s = 30°C one gets in the prototype for Pr = 0.7, GP > 10^{10} whereas in a 1:10 scale model GP will be reduced by $\lambda(L)^3$ giving GP < 10^8. Clearly in a 1:100 model the flow will be totally laminar. Thus for a given H, ΔT_s will be highly dependent on the viscosity, and the temperature rise ΔT_s between the wall and the environment will not be scaled as the temperature rise of the air in the room above the ambient temperature. Fortunately, the thermal boundary layer on the wall is very small, so that one could therefore build the model according to the scaling derived earlier, using as an independent parameter H, rather than ΔT_s, calculating the H_p from known correlations.

The temperature rise in the building itself in such a model would scale according to Eq. 18, except that near the wall, where the temperature rise will be larger. An estimate of the additional temperature rise near the wall can be made using heat transfer correlations. The same approach can be used in cases where the heat transfer from the internal sources, is by forced convection. Deviation from the model scaling is expected to exist only near the heat sources.

5. A SIMULATION OF HEAT DISPOSAL FROM A CHEMICAL PLANT.

A plant for manufacturing chlorine and sodium hydroxide by electrolysis is being built by Makhteshim Chemical Works Ltd. in Israel. The process, which is performed in 18 large cells located above large opennings on the first floor (40 × 40 m²) of a single building, produces a relatively high flux of heat (550 KW). The air heated by the cells is expected to rise and leave the building through a large openning along the centerline of the roof. The ground floor is partially open to the atmosphere and fresh air can easily enter the building. The designer of the plant has originally proposed to build a rather elaborate roof which would avoid the entrance of rain water and secure the disposal of the heat through the roof, independent of the ambient wind, See Fig. 1 (a). The cost of the original roof was very high and the question rose whether one can not use a much simpler and less expensive design of the type shown in Fig. 1(b).

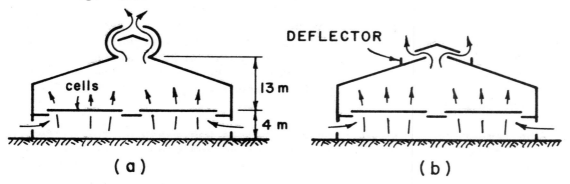

Fig. 1   Schematic description of the tested plant
(a) Original roof configuration
(b) Final roof configuration

The ventilation and heat disposal from the plant have been investigated in the M. David Lipson Environmental Wind Tunnel at the Technion – Israel Institute of Technology. The cross section of the 15m long wind tunnel is 2m × 2m.

It was decided to simulate the phenomena in a 1:83.3 scale model of the building and the surroundings.
Since T, C_p and g are equal in the model and prototype Eqs.(16)-(18) imply that:

$$\lambda(\Delta T) = \lambda(H^{2/3})/\lambda(L^{5/3})$$

and

$$\lambda(U) = \lambda(H^{1/3})/\lambda(L^{1/3})$$

Most of the tests were made using H_m = 630 W, which gives $\lambda(H)$ = 1:870. The scaling of the temperature rise and velocities were therefore

$$\lambda(\Delta T) = (870^{-2/3})(83.3)^{5/3} = 17.4:1$$

and

$$\lambda(U) = (870^{-1/3})(83.33)^{1/3} = 1:2.18$$

The temperature rise in the building was later found to be around 4°C, which corresponds to a 70°C rise in the model. The maximum value of $\Delta T/T$ in the model was therefore of the order 0.25. Note that $-\Delta\rho/\rho = T/(T + \Delta T) - 1 = 0.23$ whereas $\Delta T/T = 0.25$, suggesting a 10% error in the scaling laws due to the large value of $\Delta T/T$ in the model.

The exit velocity in the model is expected to be of the order of $\sqrt{0.23 \cdot 9.81 \cdot 10/83.3}$ = 0.525 m/sec, giving a Reynolds number of the order of 4×10^3, however, flow visualization showed that the flow was turbulent, and the effect of the Reynolds number was apparently small.

To examine the validity of the scaling laws the values of the temperature rise ΔT_m for different values of H_m were measured at different points in the building. Typical results are shown in Fig. 2 for two roof configurations. It appears that the 2/3 power dependency of ΔT_m on H_m is confirmed in the model. The points for H = 840 Watt appear to be slightly above the 2/3 lower law which fits the measurements with smaller H_m, probably due to effective lower density differences at this value. Since the values of the temperature rise in the prototype are one order of magnitude lower, the accuracy of the model appears to be satisfactory for design purposes.

The temperature rise at different points was measured for different roof configurations for different wind speeds and wind directions. The detailed results, which are of little value to the reader will not be described in this paper. These results showed, however, that when the wind blew normal to the openning in the roof, the simpler design (b) did not perform as well as the original design. However, a short vertical deflector placed along the roof, see Fig. 1, cured this deficiency completely.

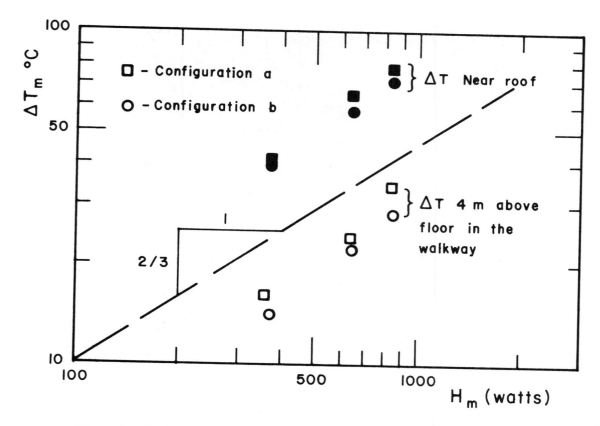

Fig. 2 The temperature rise in the model for different values
         of heat flux.

The model can also be used to estimate the concentration of pollutants
emitted during the process. Assuming that the discharge of a given pollutant is Q, then

$$Q = C \cdot A \cdot u_{exit}$$

where C is the concentration of that pollutant at the exit.
According to Eq. 2

$$H = \rho C_p \Delta T \cdot A \, u_{exit}$$

and thus it is expected that

$$\frac{C}{\Delta T} = \frac{Q}{(H/\rho C_p)}$$

The similarity between heat and mass transfer in turbulent flows suggests
that this relation holds at any point in the model, except of course close
to the cell where both the temperature rise and the concentrations can not
be accurately simulated.

The help of Mrs. Z. Vider and Mr. E. Gantz in the experimental work is
gratefully acknowledged.

566

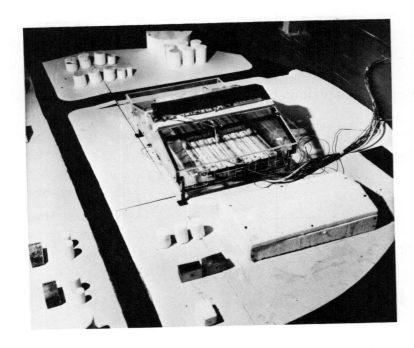

Fig. 3.   Photograph of the model in the wind tunnel

REFERENCES

1.   J.E. Cermak, Application of fluids mechanics to wind engineering, Freeman Scholar Lecture, ASME, J. Fluid Eng., 97 (1975) 9-38.

2.   W.H. Snyder, Guidline for fluid modelling of atmospheric diffusion (draft for public comment), U.S. Environmental Protection Agency, Research Triangle Park, NC 27711, EPA-450/4-79-016, June 1979.

3.   A.B. Camber, Compressible flow, in Handbook of Fluid Dynamics, Section 8-24, V.L. Streeter Editor, McGraw-Hill 1961.

4.   M. Poreh and A. Kacherginsky, Simulation of plume rise using small-tunnel models, J. of Wind Engineering and Industrial Aerodynamics, 7(1981) 1-14.

5.   J.E. Cermak, J.A. Peterka, S.S. Ayad and M. Poreh, Wind tunnel investigation of natural ventilation Fluid Dynamics and Diffusion laboratory Colorado State University Report, Oct. 1981.

6.   H. Schlichting, Boundary-layer theory, McGraw-Hill Book Co., 4th Ed., 1979, p. 318.

PREPARED CRITIQUE OF PAPERS ON THE LOCAL WIND
ENVIRONMENT AND SNOW TRANSPORT

Hatsuo Ishizaki, Kyoto University

1. INTRODUCTION

In every wind tunnel experiment, simulation criteria and measurement techniques must carefully be considered so as to obtain reliable experimental results. Since all similitude requirements cannot be satisfied, and since various measuring systems cannot be applied in each experiment, suitable simulation and measurement techniques depend upon the purpose of the experiment. In the following, a few topics on simulation and measurement in experiments concerning local wind environment and snow transport are described.

Modeling scales are of the first importance when considering simulation criteria. For the case of wind effects on buildings and other structures, model scales are normally on the order of 1:100. Models used to study wind environmental problems are often much smaller. In experiments on topographical effects, models are occasionally on the order of 1:1000. On the other hand, actual people are used in wind tunnels to study pedestrian comfort, and such experiments are thus performed in full scale. Therefore, simulation and measurement techniques in experiments on environmental problems must allow for very small and very large model scales.

As models become smaller, the similitude laws and model fabrication become more difficult and errors due to scaling become large. Measuring systems such as anemometers, probes or sensors used in the wind tunnel must also be small. In large scale experiments, anemometers and wind vanes used out of doors can be used in the wind tunnel. It is difficult to describe all modeling problems concerning environmental winds inclusively.

In order to simulate the wind in the wind tunnel, natural wind characteristics must be known. If not, rational criteria for simulation cannot be given. However, wind characteristics, particularly turbulence characteristics, have not yet been defined well enough and more investigations are needed.

An example to illustrate the difficulties in defining the turbulent quantities which should be simulated in the model experiment is described in the following. Time histories of wind speeds and reciprocals of power law indices of wind profiles during typhoons are shown in Fig. 1 [1]. As seen from the figures, both quantities similarly vary with time and the power law indices depend upon the wind speeds. Variations of the power law indices show also variations of wind turbulence. If wind effects on structures are considered, only those values of the severest wind should be taken into account to simulate the wind, but the severest winds are not always important for environmental wind problems. It is difficult to fix the values of the turbulent quantities, for instance, turbulence intensity, wind profile, etc. for the model experiment because they vary with time. Thus, prototype winds are difficult to define.

2. EFFECTIVE GUST DURATION

As the natural wind is not uniform flow, problems on effective gust durations are fundamentally important in order to consider instrumentation and wind effects. In the paper by Irwin titled "Instrumentation Considerations for Velocity Measurement," the duration of the most effective gust is shown to range from 3 to 10 sec, with corresponding frequencies of about 0.1 to 0.3 Hz in full scale. He notes that the corresponding frequency for a 1:400 scale model ranges from 40 to 120 Hz. Murakami, in his paper "Wind Tunnel Modeling Applied to Pedestrian Comfort," notes that the behavior of the human body is most influenced by the turbulence in the frequency range of 0.3 ~ 1.0 Hz. These full-scale frequencies correspond to the frequencies of 120 to 400 Hz for a 1/400 model.

Based on the above descriptions, the effective gust frequencies of 0.1 ~ 1.0 Hz in full scale and 40 Hz or more in model experiments should be considered.

3. EFFECTIVE GUST SPEED

An effective gust speed, Q_{eff}, shown by Irwin is

$$Q_{eff} = Q + \gamma \sqrt{\overline{q}^2}$$

where $\sqrt{\overline{q}^2}$ is the r.m.s. value of wind speed fluctuation and γ is a constant or a peak factor. Irwin shows that the longer the gust duration the lower the gust speed, and $\gamma = 3.5$ for the 3 sec gust. In Gandemer's paper, "Simulation and Measurement of the Local Wind Environment," nuisance parameters ν_1, ν_2, ν_3 and ν_4 are defined. Typically two of them, ν_1 and ν_3, are

$$\nu_1 = \frac{U + \sigma}{U_{ref} + \sigma_{ref}} \qquad , \qquad \nu_3 = \frac{U + 3.5\,\sigma}{U_{ref} + 3.5\,\sigma_{ref}}$$

As seen from these expressions, $\gamma = 1$ in the case of ν_1 and $\gamma = 3.5$ in the case of ν_3, but the gust duration is not shown in his paper. The relationship between gust duration and peak speed or the value of γ has not been discussed in the above papers. The relation obtained by the writer from wind data recorded for typhoons and other high winds is

$$\gamma = \frac{1}{2} \ln \frac{T_o}{T} \tag{1}$$

where T = gust duration or averaging time of peak gust
       T_o = averaging time of mean wind speed, normally 600 sec.

The values of γ, calculated from equation (1) for the various gust durations, are shown in the following table.

Table 1. Gust duration, T, and peak factor γ.

T(sec)	0.2	0.5	1	2	3	5	10	20	30	60	80
γ	4.0	3.5	3.2	2.9	2.7	2.4	2.1	1.7	1.5	1.2	1.0

γ = 2.7 for a gust duration of 3 sec and this value of γ is smaller than the value, γ = 3.5, shown by Irwin. If a gust duration of 0.5 sec is assumed, the value of γ becomes 3.5. The nuisance parameter in Gandemer's paper corresponds to a peak gust of 0.5 sec duration.

The gust factor, G, can be expressed by the turbulence intensity, I, and the peak factor, .

$$G = 1 + \gamma I \qquad (2)$$

In Murakami's paper, the gust factor, G = 2.0, is shown for the frequency range of 1.0 ~ 0.3 Hz or gust duration of 1 ~ 3 sec. From Table 1, γ = 3.2 for a gust duration of 1 sec and γ = 2.7 for a duration of 3 sec. If the turbulence intensity I = 0.2 is assumed,

$$G = 1.5 ~ 1.6$$

from equation (2). If I = 0.3,

$$G = 1.8 ~ 2.0$$

The value G = 2.0 given by Murakami is nearly appropriate. It is obvious that the value G must be determined from the turbulence characteristics at a site considering the gust duration.

4. INSTRUMENTATION

With regard to methods for measuring wind speeds in the wind tunnel, the hot-wire, hot-film and pressure methods are mainly described in Irwin's paper. These methods are practical and are widely used. However, another useful method that the writer can suggest is the thermistor-anemometer.

In the early stage in its development, the thermistor-anemometer could not follow wind speed fluctuations. However, the thermistor probe now used is very small as shown in Fig. 2 and possesses high frequency response. Although it does not have as good a response as a hot-wire anemometer, it can quantitatively analyse wind speed fluctuations with frequencies up to 40 Hz [2].

The thermistor-anemometer is much more stable than the hot-wire anemometer against temperature changes and can measure total wind speeds. The response of the thermistor is invariable with wind direction changes in turbulent wind. It is easily manipulated as an omnidirectional sensor in wind measurements.

The flow visualization is a useful complement to wind speed measurements. However, it is difficult to visualize air flow on the same model which is used for wind speed measurements. Instead of the usual flow visualization, for instance smoke streams, a number of streamers or tufts on the model can show the general tendencies of stream lines. Dandelion fuzz is a good streamer. Its movement shows not only the wind direction at a point on the model but also the extent of wind turbulence.

5.   VENTILATION MODEL STUDIES

A difficult problem in wind tunnel studies on ventilation of buildings is the modeling of screens, louvres or small openings. Even if small models are geometrically similar to full-scale structures, Reynolds numbers of the flow are different at full and model scale as described by Aynsley in his paper "Natural Ventilation Model Studies."

The other problem concerns wind properties. As shown in Table 1 in Aynsley's paper, wind speed reduction through screens depends on the wind speed itself. Moreover, the wind turbulence around buildings varies with wind speed. Therefore, mean wind speed coefficients and wind discharge coefficients must depend on the outdoor wind speed. Wind tunnel experiments on ventilation studies should therefore be performed over a range of wind speeds.

6.   SNOWDRIFT PROBLEM

Modeling of snowdrift phenomena is elaborately described in Iversen's paper "Small-Scale Modeling of Snowdraft Phenomena." In the model experiment on drifting snow, the simulations of snow and wind must be considered, and the similitude requirements become complicated. The various similitude parameters are shown by Iversen and the significant parameters suitable for the objective of the experiment are described.

One of the difficulties of snowdrift modeling is simulation of snow. Perhaps the best material for the experiment would be snow, but the snow cannot be used conveniently in the wind tunnel. Iversen uses two kinds of glass spheres and walnut shells for snow particles. Other materials, for example white acid clay, should be considered for snow.

7.   CONCLUSIONS

The subjects discussed above are some aspects of the problems on wind tunnel modeling concerning wind environment and snow transport. Many other problems are mentioned in the papers presented and these problems should be given additional consideration in order to improve wind tunnel experiments.

8.   REFERENCES

1.   H. Arai, et al., "Study on vertical distributions of wind velocities and evaluation of angles of sea wind," Railway Technical Research Report, No. 627 (in Japanese) (1968).

2.   Y. Iwasa, et al., "Experimental study on measurement of turbulence in wind tunnel," Transactions of the Architectural Institute of Japan, No. 280 (in Japanese) (1979).

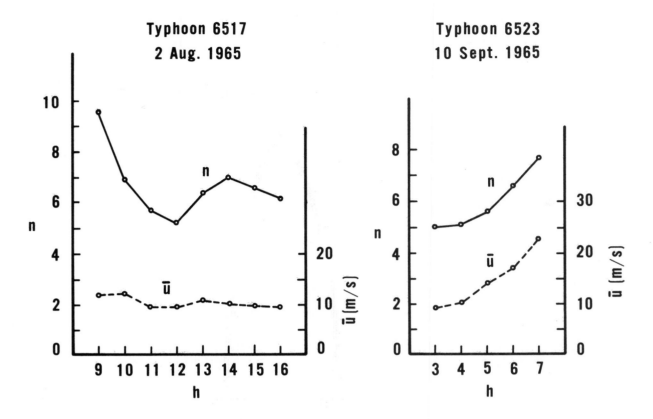

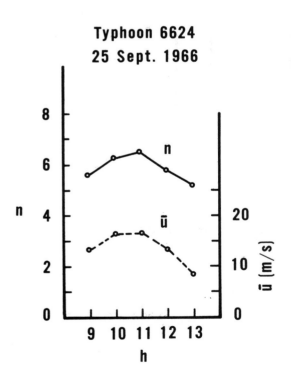

Fig. 1. The time variations of the reciprocal, n, of the power law index of wind profile and the mean wind speed, $\bar{u}$, at the height of 10 m during typhoons at Nakagawa, Shikoku.

(a) Typhoon 6517, wind direction: WNW, NW
(b) Typhoon 6523, wind direction:E, ESE
(c) Typhoon 6624, wind direction: SSE, S, SSW.

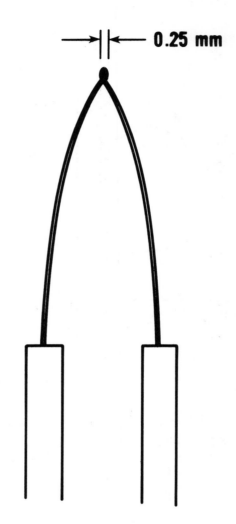

Fig. 2.   Thermistor probe.

Session VI

Validation of
Wind Tunnel Testing

Chairman

Frank H. Durgin

Massachusetts Institute of Technology

Cambridge, Massachusetts

USA

COMPARISON OF MODEL AND FULL SCALE TESTS
OF THE COMMERCE COURT BUILDING IN TORONTO

W. Alan Dalgliesh
Division of Building Research
National Research Council of Canada

1. OVERVIEW AND SCOPE

The first wind tunnel study of the Commerce Court Tower, sponsored by the structural engineers for the project in 1969, included measurements of mean and fluctuating pressures, dynamic response of a two degree-of-freedom aeroelastic model, and a synthesis with meteorological data for predicting structural loads, dynamic response, and exterior cladding and window pressures[1]. With full cooperation from the project owner, design consultants, contractors and building management staff, the National Research Council of Canada (NRCC) instrumented and monitored this 57-storey office tower continuously from first tenancy in 1973 until October 1980.

The first comparisons of mean and fluctuating pressure coefficients were presented in 1975[2] and a preliminary assessment of both wind tunnel and building code predictions of tip displacement appeared in 1978[3]. Viewed as a validation of wind tunnel testing, the comparisons were moderately successful, but they also showed that field measurement techniques and data analysis methods are themselves in need of development and validation. The National Aeronautical Establishment (NAE) of NRCC cooperated in the design of both a rigid pressure tap model and an aeroelastic model with 21 degrees-of-freedom at a length scale of 1:200 for testing in their 9 m x 9 m wind tunnel. In a sense, the parallel studies of the next few years reversed the validation process as specific situations encountered in the field were investigated in the wind tunnel.

The next comparison, presented in 1979[4], exposed some of the limitations of field studies for the evaluation of mean pressure coefficients as well as the difficulties of specifying peak pressures, whether in the field or the wind tunnel. The aeroelastic studies were instrumental in revealing how much torsion contributes to accelerations at end walls of this particular building[5,6].

It is the purpose of this report to review and illustrate the main findings of the earlier comparisons and to supplement them with new examples comparing the dynamic behaviour of the building and the aeroelastic model. The interdependence of wind tunnel and field testing in carrying out a validation process will be stressed, and suggestions will be offered for further work to develop improved modelling and consulting procedures.

2. DESCRIPTION OF BUILDING AND NAE PROXIMITY MODEL

The Commerce Court Tower, at 239 m the second tallest building in downtown Toronto, is partially sheltered from southwest to northwest by buildings from 175 to 285 m in height (Fig. 1). A strip of tall buildings ½ km wide extends several kilometres to the north. Surrounding these areas are several kilometres of relatively low buildings. Lake Ontario lies 1 km to the south.

The proximity model used in the NAE 9 m x 9 m wind tunnel was mounted on a 7.3 m diam turntable 18.3 m downstream of spires used to develop the boundary layer (Fig. 2). A power law profile of 0.33 was used for approaches from west, north and east, while a profile exponent of 0.15 was used for the lake exposure to the south. The strip of tall buildings to the north was simulated by adding larger blocks to the general urban roughness between the spires at the inlet and the turntable.

Plan dimensions of the building are shown in Fig. 3, which also gives the layout of surface pressure taps on each of four levels. The column layout of the 57-storey steel frame structure is given in Fig. 4. Floors are concrete and cladding and partitions were designed to allow movement with respect to the frame, which deflects mainly by shearing of adjacent floors in parallel planes with no significant column shortening. The centre of mass tends to be near the geometric centre, but the elastic axis is generally about 5 m to the south on the E-W centreline.

3. INSTRUMENTATION OF BUILDING

Reference wind speed and direction were measured by a 3-cup anemometer and vane mounted on a radio transmitting antenna mast at a height of 286 m above grade (47 m above the roof). Differential pressure transducers were located along eight vertical lines at four levels to measure surface pressures against the pressure in a pneumatic line connecting the reference ports of all 32 transducers to the ambient pressure in a central recording location on the 33rd floor. Ambient absolute pressure in the recording room was also recorded.

Strain gauges were mounted near the ends of spandrel beams and on two columns at the 11th floor where a change in column section was expected to result in higher stresses under wind loading. Although more than 50 pairs of gauges remained serviceable throughout the project, only four to six pairs were monitored on a regular basis.

Three accelerometers were installed at the 202 m level in March 1975, two to measure E-W motion at the north and south walls, and one to measure N-S motion. A displacement tracking device to measure horizontal X-Y displacements near the geometric centre of the building and two more accelerometers were installed in July 1977. The displacement tracker uses two pairs of photocells to sense the position of a laser beam directed vertically from foundation level (at 15 m below grade) to the 234 m level.

Figure 1. Commerce Court Tower (arrow) in downtown Toronto viewed from 330 deg (0 = building west)

Figure 2. Commerce Court model in NRC 9 m × 9 m wind tunnel viewed from 10 deg (0 = model west)

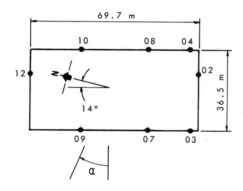

LEVEL	ELEV. m
ANEMOMETER	285
ROOF	239
3	202
4	166
5	138
7	102

e.g., TAP 304

3 - LEVEL

04 - TAP NUMBER

Figure 3. Location of pressure taps on building (a sub-set of those on the model)

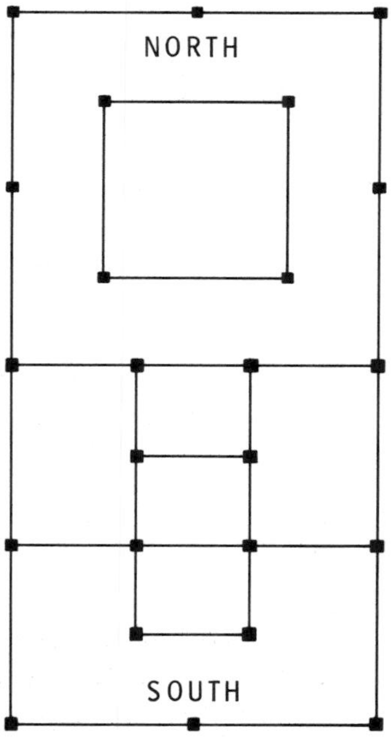

Figure 4. Framing of 57-storey building

Outputs from the photocells control two servo motors that move the carriage on which the photocells ride to compensate for movements of the top of the building relative to the laser beam.

4. FIELD DATA COLLECTION

A mini-computer controlled digital sampling and storage of data from 48 channels continuously from early in 1973 until the middle of 1980 (except when interrupted by power failure or malfunctions). Although some of the details varied, such as sampling rates, choice of sensors for recording, threshold wind speeds and so on, the most common format was to record two sets of summary data per hour regardless of wind speed. As well, whenever the wind speed exceeded 18 m/s or some other selected value, complete time histories of all 48 channels were recorded for 35 min.

Summary data provided the arithmetic mean, standard deviation, minimum and maximum for each of the 48 sensors for the 5-min interval at the end of the hour and for that 5-min interval with the highest wind speed. The sampling rate was 20 s^{-1}, but in general the time histories consisted only of every tenth sample for an effective rate of 2 s^{-1}. A special mode shape survey was run in August 1980 in which several additional accelerometers were temporarily installed on the roof and at the five instrumentation levels. Wind speeds encountered on the mode shape survey were 7 and 14 m/s.

The strongest winds recorded came mainly from one of two sectors, E-N-E and S-W (relative to building). There were also strong winds from the N-W, but after completion (in 1976) of the 285 m high First Canadian Place building very few records were triggered by the anemometer directly downwind at approximately the same height. Very few strong winds were experienced from the S-E and S, so that these field measurements cannot provide useful information for comparison with model data.

The highest 5-min mean reference speed of 33 m/s was recorded during an exceptionally long-lasting storm on 26 January 1978. The largest peak pressure difference measured was −640 N/m^2. Largest displacements measured at the 234 m level at the geometric centre of the building were about 220 mm, and peak accelerations were 10 to 15 x 10^{-3} times the acceleration due to gravity (10 to 15 millig).

5. REVIEW OF PRESSURE COEFFICIENT COMPARISONS

The dynamic reference pressure, by which measured surface pressures were divided to form non-dimensional coefficients, was calculated from the anemometer speed at 286 m, outdoor air temperature from a nearby meteorological station, and barometric pressure in the instrument room on the 33rd floor. Measured surface pressures were, in fact, not absolute readings but differences from the barometric pressure in the instrument room; unfortunately, there appears to be no easy method of relating this so-called static reference pressure of field measurements to something comparable in the wind tunnel, short of modelling the effective porosity of the building envelope.

5.1 Stack Effect

Another difficulty is posed by buoyancy of heated indoor air; in cold weather, pressure differences across the walls of a 57-storey building as a result of stack effect are of the same order as those from moderately strong winds that make up the bulk of useful observations in field measurements. Reference dynamic pressure and the difference between the inverses of indoor and outdoor temperatures are treated as two independent variables whose effects on surface pressure differentials can be separated by multiple linear regression. A third source of pressure differentials, also not found in wind tunnel testing, is the mechanical system for distributing air throughout the building.

The removal of stack effect is helped by the strong correlation with temperature (partial correlation coefficient is typically greater than 0.85), but some uncertainty is introduced. A more difficult problem is the variable success found in correlating the wind-induced component with the dynamic reference pressure; here the partial correlation coefficient usually ranges from a high of 0.85 to as low as 0.15. The main factor seems to be the availability of sufficiently strong winds to provide a good range of the independent variable. The process whereby field measurements are transformed to pressure coefficients is illustrated in Fig. 5 for two sensor locations, 308 and 312, for two different wind directions.

5.2 Internal Pressure Coefficient

The slopes in the diagrams on the left of Fig. 5 represent a combined external and internal mean pressure coefficient; those on the right are coefficients for the standard deviation of the pressure difference; though they are not significantly affected by either stack effect or the air-handling system, the fluctuating pressure coefficients may still contain an internal component not present in wind tunnel experiments. Correlations are generally better for the fluctuating coefficients (on the right) than for the mean coefficients because there is no need for multiple regression to discriminate between wind and some other effect.

The pressure coefficients obtained by determining the slopes of best-fit regression lines in the least square sense (Fig. 5) are compared with those measured directly in the wind tunnel, with the following proviso: that an attempt is made to extract the mean internal pressure coefficient by subtracting the average of the 32 differences between matching sensor locations in model and building for each wind direction evaluated. This averaged difference is interpreted as the difference between the common internal reference used in the building, and the free-stream static reference used in the wind tunnel. For the most part the difference would not be considered unreasonable as an internal pressure coefficient, usually about -0.15 in the Commerce Court. On the other hand, where insufficient high winds were available to give strong correlations with pressure differences attributed to wind effect an offset of as much as 0.65 might result.

The two sensors treated in Fig. 5 are also used as examples of the agreement between model and full-scale pressure coefficients as a function of wind direction, with mean external coefficients to the left and root-mean-square about the mean coefficients (standard deviation) to the right (Fig. 6).

580

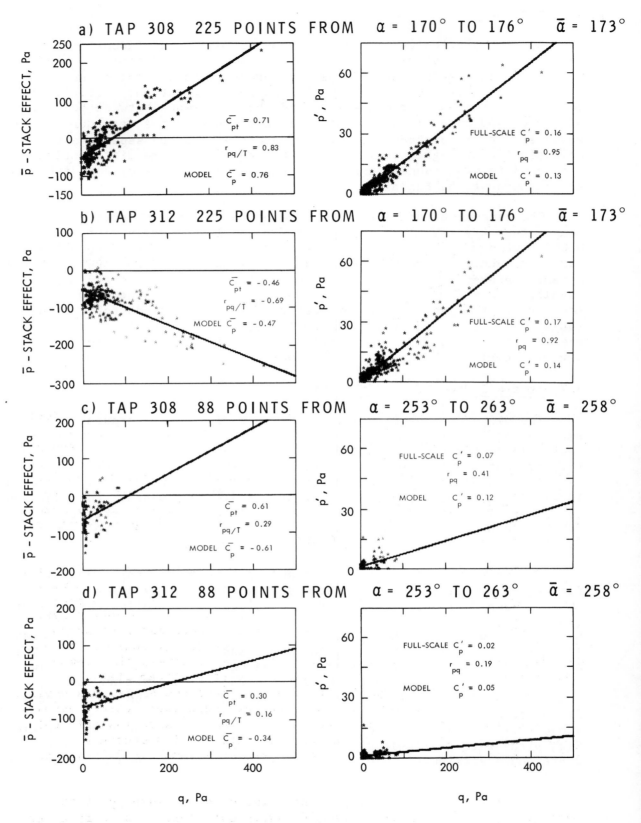

Figure 5. Measured full-scale surface pressures versus dynamic reference pressure at 286 m

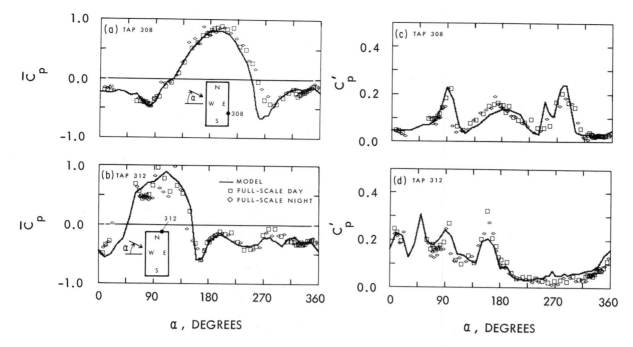

Figure 6. Comparison of model and full-scale
pressure coefficients versus wind direction
(0=building west) for pressure taps 308 and 312.

5.3 Peak Pressure Coefficient

Peak pressure coefficients, usually the end product sought for
cladding design, are essentially "location" parameters for extreme value
distributions. Sometimes they are estimated from measurements of the mean
and standard deviation coefficients, but the trend in wind tunnel practice
is now to measure the peaks and record them directly as the wind effect
needed in design. Clearly, some assurance is required that the extreme
values discovered in the wind tunnel bear as close a resemblance as
possible to those observable on real buildings. Close attention must be
paid to time scaling as well as to geometric scaling, and sampling rates in
the field and the wind tunnel should be high enough to capture the shortest
transient peaks.

A fairly simple experimental approach was used to process the field
data from Commerce Court. A negative exponential distribution was fitted
to "parent" populations of spikes of either pressure or suction in time
series records from various sensors, grouped according to the general
character of the distributions. This gave rise to perhaps the simplest
possible analytical expressions for extreme values, and these expressions
provided exponential curves with acceptable fit to histograms of extreme
values, both from the building and from the wind tunnel (Fig. 7). An added
advantage of fitting extreme value histograms with analytical expressions
comes when there is a need to predict not only the extreme peak for a
typical design storm but also several lesser spikes, as in the estimation
of cumulative damage effects on window glass[7].

6. STRUCTURAL RESPONSE TO WIND LOADING

Instrumentation to monitor structural response was introduced in

582

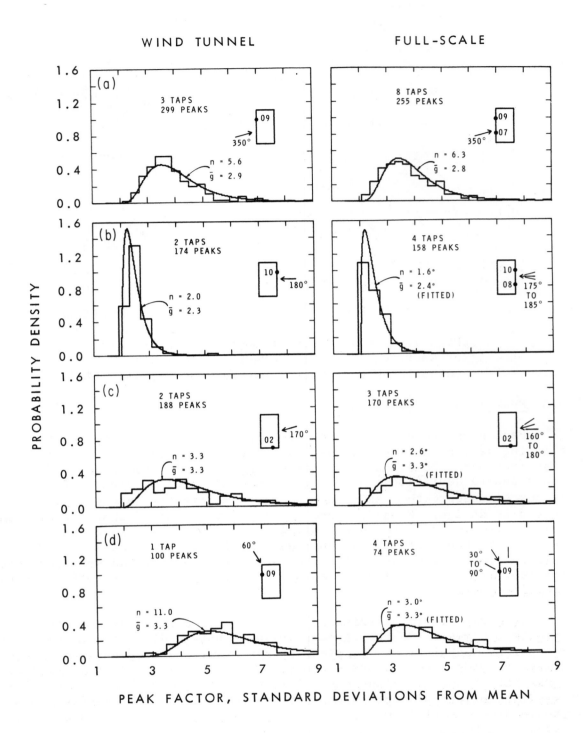

Figure 7. Histograms of peak pressures measured in the wind tunnel (left) and on the full-scale building (right)

stages, beginning with strain gauges at the 11th floor, followed in 1975 by accelerometers, then displacement tracking equipment and more accelerometers in 1977. One of the first interesting discoveries was the remarkable similarity in the "dynamic signature" of the strain gauge readings during a strong wind and the corresponding displacement traces. Another was the distinct shift in natural frequencies (in all observed modes) downwards with increasing amplitude of motion[3], a phenomenon noted by others observing motion of tall buildings[8].

Early attempts by NRCC to relate amplitude of motion to dynamic reference pressure in the same manner as for deriving pressure coefficients were moderately successful for mean along-wind displacements. Predictions based on either the detailed method of the National Building Code of Canada[9] or wind tunnel tests at the University of Western Ontario[1] appeared to be representative of the mean and standard deviations of both along-wind and cross-wind movements in an easterly wind. It is important to note that agreement depended on using measured building frequencies in the calculations.

7. NAE AEROELASTIC MODEL

Indications of torsional behaviour sparked interest in aeroelastic tests using a rather more elaborate model than would normally be considered necessary for providing design information. The N-S modes are translational, with no torsion and tend to have the same, or slightly lower, frequencies when compared with corresponding E-W translational modes. The first E-W mode of vibration is predominantly translation, increasing linearly with height, and is coupled with a predominantly torsional mode at a frequency only 30 per cent above the first. The two modes combine to make the accelerations of the north wall in the E-W direction markedly more severe than those of either the geometric centre or the south wall of the building.

The aeroelastic model stiffness was based on several practical considerations and resulted in a frequency scale of 1:53 at a full scale reference wind speed of 14 m/s, increasing to about 1:56 (E-W) or 1:58 (N/S) at 28 m/s. The increase results from the decrease in frequencies of the building as wind speed increases. The shift is about the same for all modes, and amounts to 10 to 15 per cent in moving from reference wind speeds of 7 m/s to those of about 30 m/s. The choice of length and frequency scaling fixes time scaling and velocity scaling as far as dynamic behaviour of the aeroelastic model is concerned. The higher modes relate to the fundamental mode approximately in the ratios 1:3:5, as expected for a cantilever deforming in shear rather than bending. Seven mass levels were chosen by placing one at each of the five instrumented levels in the building and then dividing each of the bottom two modules in two to make all the modules approximately the same height (Fig. 8). The action of the coupled E-W modes incorporating both translation and rotation is diagrammed in terms of tip displacements in Fig. 9. Mode shapes are compared for the building and the model in Fig. 10 on the basis of accelerometer readings taken during the special survey of the building just before the removal of recording equipment in October 1980.

Figure 8. Frame of aeroelastic model with seven lumped mass levels

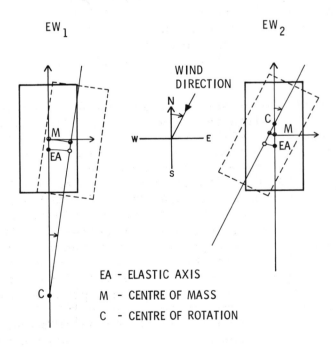

Figure 9. Tip modal deflections for first two coupled modes
(positive deflection east and north; positive rotation clockwise)

8. NEW DISPLACEMENT AND ACCELERATION COMPARISONS

Figure 11 gives a fairly representative picture of the correlation for mean tip displacements measured on the 21 degree-of-freedom model and the real building. The model measurements were made at a reference speed of 24 m/s in full-scale terms (velocity scaling 1:3.55). Building measurements at speeds from 10 to 30 m/s were first reduced to coefficient form in mm/(N/m^2) by linear regression; then "best estimates" were calculated for a reference dynamic pressure of 365 N/m^2. The horizontal bar of each cross marks the range of wind directions included in each of 20 estimates of E-W displacement (east positive, 0 = wind from the west). The vertical bar gives the 95 per cent confidence band for each estimate, and the crossing point is the best estimate of the value. The numbers of readings used to establish the estimates range from 17 to 198. Note that the building displacements are essentially actual measurements taken at the geometric centre of the building at a reference speed of 24 m/s, since a large percentage of the observations were made at speeds of 18 to 25 m/s. A similar graph for comparison with model measurements made at 45 m/s (not observed on the real building) gave much the same correlations.

Figure 12 makes the same sort of comparison for N-S displacement, north positive, along the E-W or Y axis. The vertical scale showing displacement in mm for a reference pressure of 365 N/m^2 is half that of Fig. 11. As stiffness is approximately the same in the two directions, the deflection should be roughly in proportion to the area presented to the wind, in this case half as much for the N-S face as for the E-W face.

The correlation between dynamic reference pressure and displacements on the building is much poorer in the N-S direction, shown by the lengths of the vertical bars of the 95 per cent confidence bands. One reason for this is probably the smaller response to wind force resulting from the narrower face in the N-S direction. In comparing the building displacements with the model results it is tempting to speculate that some of the differences, for example from about 300 to 330 deg, might be due to differences in local wind direction. Although this is a possibility, it would be difficult to verify, given the problems of siting instruments to map the incident wind directions in the field.

8.1 Acceleration Spectra

Detailed studies have yet to be made of the correlations between model and building accelerations, but it is possible to illustrate the relative contributions of the first and higher modes to acceleration and to compare building and model measurements. Power spectra in Fig. 13 show sharper peaks and a greater fall-off of contribution by the second mode in the model than in the building. The agreement between building and model appears closer in the E-W coupled modes (Figs 14, 15). The predominantly translational lower mode is illustrated in Fig. 14 by taking the sum of E-W accelerations of the north and south walls, and the predominantly rotational mode in Fig. 15 by taking the difference.

Damping estimated by the half-power-bandwidth method is about 2 to 3 per cent for the model and 3 to 4 per cent for the building. The difference between structural damping alone (1 per cent) and total damping for the model could be partially attributed to aerodynamic damping (less than 1 per cent). For the building, the variation of frequency with amplitude would contribute to an apparent increase in damping.

586

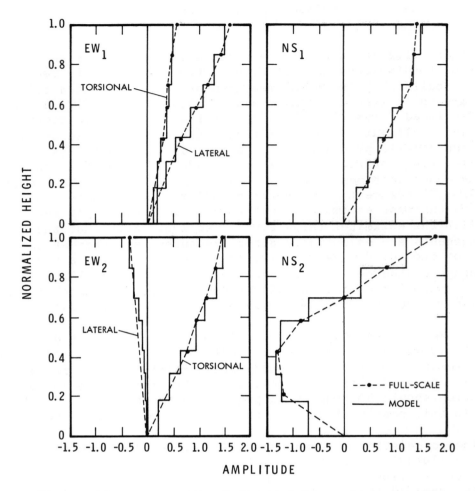

Figure 10. Measured model and full-scale mode shapes

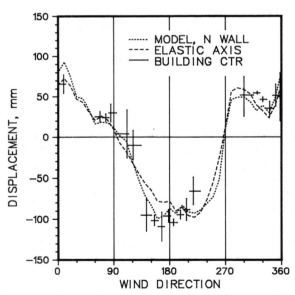

Figure 11. Comparison of model and full-scale east-west mean tip displacements versus wind direction (0 = building west) for dynamic reference pressure of 365 N/m^2 (24 m/s)

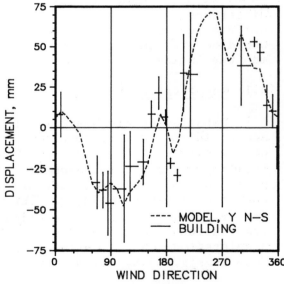

Figure 12. Comparison of model and full-scale north-south mean tip displacements versus wind direction (0 = building west) for dynamic reference pressure of 365 N/m^2 (24 m/s)

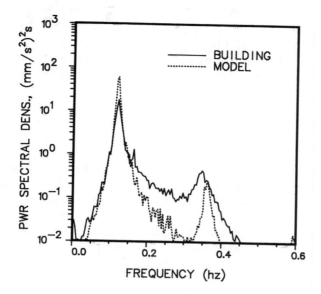

Figure 13. Comparison of model and full-scale north-south acceleration power spectra – translational modes

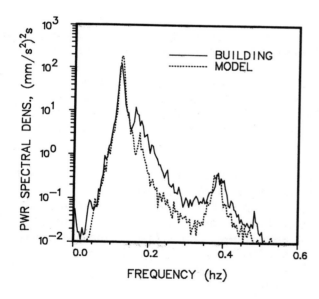

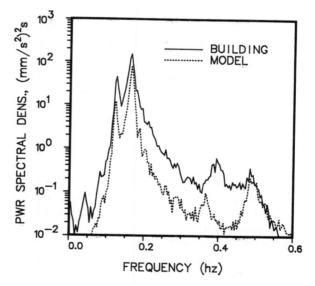

Figure 14. Comparison of model and full-scale east-west acceleration power spectra – translational modes emphasized

Figure 15. Comparison of model and full-scale east-west acceleration power spectra – rotational modes emphasized

588

9. RELATING WIND TUNNEL RESULTS TO BUILDINGS

A basic requirement of acceptable prediction of wind effects and building response is reliable input data appropriate to the building and the site. Field measurements are invaluable in providing after-the-fact check points on specific buildings, but it is important to acknowledge their obvious limitations. In addition to being time-consuming and expensive, full-scale experiments are inherently less deterministic and accurate than the wind tunnel experiments they are supposed to validate.

Where appropriate data are available the Commerce Court experience confirms that model testing, according to the methods followed in the two laboratories used in the comparisons, will yield results applicable to full scale, within experimental error in some cases. The discrepancies that have been discovered are indications that further investigation is needed; at this stage it would be premature to conclude that either the modelling technique or the result in question requires correction. Comparisons to date suggest that because field observations are subject to so many uncontrolled variables there may be little hope of predicting full-scale behaviour to within better than about 10 to 15 per cent, even when the input data on which the model is based are accurate.

Much remains to be done in fully utilizing the data collected at Commerce Court and on the models tested in the NAE 9 m x 9 m wind tunnel, particularly in the matter of aeroelastic response. It is already clear that torsion makes important contributions to the accelerations experienced at the north wall of Commerce Court in comparison with those of the central core.

The measurement of peak pressures for cladding design gives rise to two important concerns: sampling rate and internal pressures. Sampling rates on buildings have to be at least 20 per second to capture the peak values of interest, and 30 per second would probably be even better. Sufficiently high rates at model scale may be difficult to achieve. At least ten observations of individual peaks will generally be needed to establish a representative peak coefficient for any given surface pressure tap location.

Internal pressures and internal pressure coefficients to describe wind effects are also essential for accurate assessment of net loads on cladding. There are, as well, other applications for such information, for example, heat and moisture transfer calculations and the design of details to ensure rain-tightness of the building envelope. Both field observations and development of modelling techniques to simulate porous walls could provide valuable assistance to several members of the design team; further development in this area is highly desirable.

ACKNOWLEDGEMENTS

The cooperation of a great many organizations and individuals was needed at various stages of the Commerce Court project. It is a pleasure to acknowledge once again the unfailing assistance of the Canadian Imperial Bank of Commerce, Carruthers & Wallace, Structural Consultants, the University of Western Ontario Boundary-Layer Wind Tunnel Laboratory, and

the National Aeronautical Establishment; and, in particular, of
J.T. Templin, K.R. Cooper, J.H. Rainer and F.K. Hummel. This paper is a
contribution from the Division of Building Research, National Research
Council of Canada, and is published with the approval of the Director of
the Division.

REFERENCES

1 A.G. Davenport, M. Hogan and N. Isyumov, A study of wind effects on the
   Commerce Court Tower Part 1, University of Western Ontario, London,
   Canada, Engineering Science Research Report, BLWT-7-69, (1979).

2 W.A. Dalgliesh, Comparison of model/full-scale wind pressures on a high-
   rise building, Journal of Industrial Aerodynamics, 1 (1975) 55-66.

3 W.A. Dalgliesh and J.H. Rainer, Measurements of wind induced
   displacements and accelerations of a 57-storey building in Toronto,
   Canada, Proc., 3rd Colloquium on Industrial Aerodynamics, Buildings
   Aerodynamics, Part 2, Aachen, Germany, 14-16 June 1978, pp. 67-78.

4 W.A. Dalgliesh, J.T. Templin and K.R. Cooper, Comparisons of wind tunnel
   and full-scale building surface pressures with emphasis on peaks, Wind
   Engineering Proc., 5th International Conference on Wind Engineering,
   July 1979, (Ed. by J.E. Cermak) 1 (1980), pp. 553-565.

5 J.T. Templin and K.R. Cooper, Design and performance of a multi-degree-
   of-freedom aeroelastic building model, Journal of Wind Engineering and
   Industrial Aerodynamics, 8 (1981) pp. 157-175.

6 J.T. Templin and K.R. Cooper, Torsional effects on the wind-induced
   response of a high-rise building, Presented 4th U.S. International
   Conference on Wind Engineering Research, Seattle, Washington, 26-29 July
   1981.

7 W.A. Dalgliesh, Assessment of wind loads for glazing design, Proc.,
   IAHR/IUTIAM Symposium on Practical Experience with Flow Induced
   Vibrations, Karlsruhe, Germany, 3-6 September, 1979, Springer Verlag,
   Berlin/Heidelberg (1980) pp. 696-708.

8 L.C.H. Lam and R.P. Lam, An experimental study of the dynamic behaviour
   of a multistorey steel-framed building, Proc., Institution of Civil
   Engineers, Part 2, September 1979, pp. 707-720.

9 Supplement to the National Building Code of Canada, National Research
   Council of Canada, Associate Committee on the National Bulding Code,
   NRCC 17724, 1980.

MODEL AND FULL SCALE TESTS OF THE
ARTS TOWER AT SHEFFIELD UNIVERSITY

Brian E. Lee, Department of Building Science, Sheffield University, Sheffield, U.K.

1.   INTRODUCTION

The purpose of this paper is to present a comparison of the dynamic wind forces measured on the Arts Tower at Sheffield University with those determined from wind tunnel model studies carried out in the University's Department of Building Science. The full scale studies of the structural behaviour of the Arts Tower and of its response to the wind have been carried out in conjunction with the Structural Design Division of the U.K. Building Research Establishment who have also been instrumental in funding the wind tunnel modelling programme. The wind tunnel project to be described in this paper has embodied a high frequency, rigid body modelling technique capable of reproducing fundamental translational, straight line mode shapes and is used to determine modal forces, rather than as a full aeroelastic model to measure the response for a particular building. It is considered that this approach to modal force determination may have a useful potential for future design guide application.

2.   FULL SCALE MEASUREMENT OF DYNAMIC WIND LOADS

The Arts Tower, Figure 1, is situated within the University site near the centre of the City of Sheffield. It stands with its north-south axis, minor axis, at an angle of -20° to true north on a sloping site.

The building is 78m high, 36m wide and 20m deep. Structurally it consists of a cast in situ reinforced concrete core with deep floor slabs spanning between the core and external reinforced concrete columns. Above the first floor there are 88 columns evenly spaced around the periphery of the building, each 203 x 406mm. Below the first floor level the number of columns is reduced to 16, each 965 x 965mm. Non-structural blockwork partitions, which vary in arrangement, are located on each floor and the building is clad in both glass and lightweight panelling. The building has a link bridge to an adjacent low rise building on the west side at mezzanine level between ground and first floors. There is a deep basement and the building is founded on piles driven into shale.

In order to determine the dynamic properties of the building and to calibrate it in preparation for the wind response measurements, two series of forced vibration tests have been carried out. The first of these tests, performed in April 1976, used a single eccentric mass vibrator supplied by CEBTP and was located on the 19th floor of the building. The second series of forced vibration tests was performed using the BRE multiple unit vibrator set in which 4 vibrators were located on the 18th floor of the building, each one, on the periphery at the centre of a face. These tests were carried out in February 1979. The details of each of these series of tests has been previously reported, the first by Jeary, Lee and Sparks (1979) and the second by Ellis, Evans, Jeary and Lee (1979).

The details of the structural characteristics of the building are given in Table 1 and the mode shapes for the three fundamental modes of vibration are given in Figure 2. An inspection of the plan mode shapes of the building during the first series of forced vibration tests has shown that the shear centre of the building does not coincide with the geometrical centre but that it lies to the north on the north-south axis. Thus the E-W mode shape contains a torsion component and the torsional mode shape contains a translation component. This coupling effect should then account for the difference between the E-W modal mass and the N-S modal mass, Table 1, whose values ought to be identical since they have the same mode shapes, Figure 2. The difference in modal masses suggests that the E-W mode torsional component amounts to approximately 9% of the apparent translation of the centre.

The method by which the modal wind force measurements on the Arts Tower have been determined from recordings made during wind storms has also been described by Jeary, Lee and Sparks (1979) and will not be repeated here. The full scale data given in previous publications i.e. Jeary, Lee and Sparks (1979) and Evans and Lee (1981, a), undergoes a continuing process of change as new wind records become available to join the ensemble averaging process by which the modal responses are evaluated. Hence Table 2 which presents the modal forces as equivalent r.m.s. values contains different data from the previous versions of the corresponding table. The equivalent r.m.s. value of the modal force can alternatively be expressed as a power spectral density function modal force value as follows:

$$\sigma_{r_{r.m.s.}} = \left(\frac{Sr \cdot \pi \cdot fr}{4 \ \zeta r}\right)^{\frac{1}{2}} \quad \text{Hurty \& Rubenstein (1964) Equ. 11.58}$$

or as an r.m.s. modal acceleration value by

$$\ddot{x}_{r_{r.m.s.}} = \left(\frac{\pi \ fr \ Sr}{4 \ \zeta r}\right)^{\frac{1}{2}} \times \frac{\phi_r}{Mr} \quad \text{ESDU 76001(1976) Equ. 27.}$$

where for the r th mode

σ_{rms} = equivalent rms modal force

$\ddot{x}_{rms}$ = rms modal acceleration

S   = power spectral density function modal force

f   = modal frequency

ζ = modal damping

ϕ = mode shape function

M   = modal mass

In Table 2 each pair of values of the rms modal loads for the N-S and E-W modes are presented for a set of values of wind speed and wind direction. The wind speeds referred to are those measured by the building anemometer mounted 6m above roof level, 84m above street level, and have been corrected for the proximity effects of the flow over the building roof, Evans and Lee (1981, b). The wind directions are referred to true north.

3.   THE INCIDENT WIND FLOW

3.1 Description of Site Conditions

The Arts Tower stands on sloping ground in the university area about a mile from the city centre. The ground rises to the north-west on a slope of about 1:25 reaching the crest of a hill at a height equivalent to the roof of the Arts Tower about 1½ miles away. In the opposite direction the ground continues to slope away from the building towards the city centre. The ground terrain in the sector from west through north to east consists mainly of low rise, two and three storey high density housing units, mainly in long terraces. The general terrain features through the other sector, south facing, is of rather taller buildings spaced further apart in the height range 4 to 10 storeys, with older low rise industrial buildings between them. In the terms of the U.K. wind loading code of practice, BS I (1972), the northern terrain sector would probably be classified as terrain category 3 and the southern sector as terrain category 4.

Wind measurements on, and adjacent to, the site of the Arts Tower have been made over varying periods of time by three local anemometers. An analysis of the relationships between the outputs of these three anemometers, each of which stands on a different building and is at a different height from the others, has been made by Evans and Lee (1981, b). The general conclusion drawn from this study is that the anemometer on the tall building is significantly influenced by the flow over the building roof and that both other anemometers on low buildings are significantly influenced by the proximity of adjacent buildings. All three effects are directionally dependent and whilst it is possible to apply directional correction factors to the indicated speed based on new wind tunnel studies of the local flow patterns, it has not been possible to derive a general relationship for the variation of wind speed with height from the three instruments. Thus the full scale modal force data from the Arts Tower has been referenced to the anemometer located on the building roof, corrected for directional effects, and not to an equivalent wind speed at a standard reference height, normally 10m, thought to be representative of local wind flow conditions.

3.2 Modelling the Wind Flow

The following simplifications have been made to the real flow situation in order to accommodate the modelling problem within the limitations of our resources. First the surrounding ground has been assumed to be flat and secondly the roughness of the surroundings has been assumed to be that of Category 3 terrain. This second assumption is considered to be most reasonable in view of the fact that all the full scale data is from wind directions for which this terrain category is appropriate. Since no measurements of the turbulence characteristics of the incident flow have been made in full scale it has been assumed that the general description for flow over this terrain category given by Counihan (1975) is appropriate.

The aim of the atmospheric boundary layer flow modelling process has been to reproduce a **range** of mean flow and turbulence characteristics based on the surveys by Counih,an (1975) and ESDU 72026 (1972). Briefly these are as follows

(a) Mean Velocity Profile, Power Law Index, $\alpha = 0.30$

(b) Roughness length, $zo = 1.0m$

(c) Reynold's stresses near ground

$$0.002 \leqslant \frac{-uv}{U^2_G} \leqslant 0.003$$

(d) Turbulence Intensities near ground

$$\frac{u'}{u*} = 2.6, \frac{v'}{u*} = 1.3, \frac{w'}{u*} = 2.0$$

U-component based on local mean velocity = 20 to 30%

(e) Turbulence Spectra. Shape to conform to equation by Harris (1968), where position determines integral scales of turbulence.

The method employed for modelling the flow in the Sheffield University 1.2 x 1.2m Boundary Layer Wind Tunnel has been fully described by Lee (1977) and is based on an intermediate working section length, accelerated roughness flow, technique. This techniques uses a castellated fence, a row of triangular section tapering spires and an array of floor roughness blocks whose total fetch length is 4.55m.

The simulated atmospheric boundary layer flow produced by this modelling technique may be briefly described as follows

(a) Boundary layer height = 850mm

(b) Mean velocity profile, power law index, $\alpha = 0.29$

(c) Zero plane displacement, d = 10mm

(d) Roughness length, zo = 3mm

(e) Reynolds stresses near ground

$$0.003 \leq \frac{-\overline{uv}}{U_o^2} \leq 0.004$$

(f) Turbulence intensities near ground

$$\frac{u'}{u^*} = 2.58, \quad \frac{v'}{u^*} = 1.23, \quad \frac{w'}{u^*} = 1.84$$

u-component based on local mean velocity = 20 - 40%

(g) Turbulence Spectra. The turbulence spectra has been measured at Y/H values of 0.125, 0.31 and 0.5, the central measurement shown in Fig. 3. In matching the modelling data to Harris's equation the reduced frequency ordinate yields a unique value of the function Lx/U for the experiment. These values of Lx are as follows:

Y/H = 0.125   Lx = 0.255m

Y/H = 0.3125   Lx = 0.296m

Y/H = 0.50   Lx = 0.314m

The values of Lx are used in determining the length scale ratio of the modelled flow to the full scale flow.

In general the comparison of the modelled flow characteristics with the corresponding full scale design values is considered satisfactory.

3.3 The Boundary Layer Length Scale Factor

It is now generally accepted that two factors are required to determine the average length scale factor of a modelled atmospheric boundary layer, one based on the mean velocity characteristic and another based on the turbulence characteristic of the flow.

The roughness length, zo, is the most suitable length parameter associated with the mean flow structure on which to base a scale ratio. A comparison of the design value of zo for the terrain category in question, 1.0m, with the corresponding value measured in the simulated flow, 3mm gives a model to full scale length factor of 333.

A useful method for determining a length scale factor based on the integral scale of turbulence has been explained by Cook (1976). Using the values of Lx, given in the previous section, with a graphical iterative technique a scale factor can be simply determined. The length scale factors determined by this process were:

Y/H = 0.125   Scale factor = 320

Y/H = 0.3125   Scale factor = 370

Y/H = 0.50   Scale factor = 415

On the basis of these factors, together with that determined from the roughness length, a value of 350:1 has been used as the scale factor most representative of the lower layers of the atmospheric flow simulation.

4. THE MODEL/BALANCE FORCE TRANSDUCER

A rigid body modelling technique has been evolved by which modal loads may be determined. The system has been designed to operate at specific model building resonances and consists of a series of strain gauged balance transducers, each one of which, when fitted with the appropriate model building, resonates at a predetermined frequency. Additionally the system contains an integral variable damping mechanism, thus enabling both model frequency and model damping to be varied independently. Figure 4 illustrates the model/balance transducer system and Figure 5, its location in relation to the wind tunnel turntable. The response characteristics of the modelled Arts Tower building have been investigated at specific frequencies within the range 60-125 Hz and for damping values in the range $0.005 < \zeta < 0.05$ in the atmospheric boundary layer flow described in Section 3.

The fundamental translational modes have been examined for both N-S and E-W modes and both resonant and non-resonant dynamic response components have been evaluated. The idealization of the full scale mode shapes shown in Figure 2 by straight line modes in the modelling system is considered acceptable. A typical total system response is shown in Figure 6 where the different contributions from the background and resonant dynamic response components can be seen. It should be noted that the resonant response, used to evaluate the modal loads, is always made with high pass and low pass filters set at 60 Hz and 149 Hz respectively, in operation. This approximation may result in an underestimate in the modal forces of the order of 5%.

The model of the Arts Tower Building, built to a linear scale of 350:1, is constructed from 3mm thick carbon fibre reinforced plastic side panels with dural end plates. The model is filled with polyurethane foam to inhibit panel vibration.

No attempt has been made in the modelling technique to reproduce the modal coupling known to exist at full scale between the E-W translational mode and the fundamental torsional mode.

5. WIND TUNNEL MODEL RESULTS

In order that a direct comparison can be made between equivalent r.m.s. forces measured on both model and full scale buildings the values of damping chosen for the model and building have been nominally identical in this case.

However, this equality of damping ratios is not always necessary. The full procedure by which model forces measured under a variety of conditions of frequency, damping, bandwidth and wind speed can be normalized to represent those acting on a geometrically similar full scale building having arbitrary dynamic characteristics are currently under investigation. Such a normalizing procedure will be necessary in order to present dynamic wind forces in the form of design guide information.

In order to determine appropriate values for the frequencies of the model/balance transducer configuration in both N-S and E-W bending modes the model to full scale length ratio has been assumed to be 1:350 and the corresponding velocity ratio has been assumed to be 1:3. For the model results presented, this scaling ratio when applied to the wind tunnel speed of 9.4 m/s results in a corresponding full scale wind speed of 28.2 m/s at the equivalent full scale height of 84m, since the model and full scale modal frequencies are related by the relationship:

$$f_m = f_f \times \frac{U_m}{U_f} \times \frac{D_f}{D_m}$$

where f_m is the model modal frequency

f_f is the full scale modal frequency

$\dfrac{U_m}{U_f}$ is the model to full scale velocity ratio

$\dfrac{D_m}{D_f}$ is the model to full scale length ratio

The joint choice of these scaling ratios implies that the N-S mode whose frequency is 0.68 Hz in full scale can be modelled using an 80 Hz balance transducer and the E-W mode, 0.86 Hz full scale, modelled by a 100 Hz balance transducer.

The model results for the isolated building in the simulated atmospheric boundary layer are shown in Figure 7 which depicts the variation of r.m.s. force with wind direction for the E-W mode and the N-S mode. The interference effects caused by the surrounding buildings can be seen in Figure 8 where the variation with wind direction of the r.m.s. force for the N-S mode is shown both with and without a site model. The values of r.m.s. force presented in these figures are those measured directly by models. In order to convert these measurements to the appropriate full scale values the following relationship may be used, Melbourne (1974),

$$\frac{\text{Model Force}}{\text{Full Scale Force}} = \frac{(U_m)^2}{(U_f)} \times \frac{(D_m)^2}{(D_f)}$$

i.e. Full Scale Force $= (3)^2 \times (350)^2 \times$ Model Force

$= 1.1 \times 10^6 \times$ Model Force.

At present the full scale data available for comparison is sparse and does not adequately cover the full range of wind speeds and directions, though it is hoped that more data will become available in due course. From the wind tunnel results presented in Figure 7 it can be seen that the r.m.s. force values for the N-S mode, vary from 0.0025 N at 360° through a minimum of 0.0015 N at 310° to 0.0020 N at 270°. After scaling up the equivalent full scale values are 2.75 KN, 1.65 KN and 2.2 KN respectively for an 84m wind speed of 28.2 m/s. The corresponding values for the E-W mode are 1.3 KN at both 360° and 270° and a minimum of 1.0 KN at 310°.

The layout of the university site is given by Evans and Lee (1980, b) who show the Arts Tower lies to the north east of the site surrounded by numerous low-rise buildings with two seven storey buildings to the south at distances of $\sim$ 100m and $\sim$ 200m. The effects of aerodynamic interference from the surrounding buildings may be seen in Figure 8. The more noticeable interference effects occur when the wind is to the south and west of the Arts Tower, where the results indicate an increase in the r.m.s. force thought to be caused by the increased buffeting on the Arts Tower from the wakes of the adjacent low rise buildings. Since this buffeting is restricted to the lower levels of the Arts Tower the large moment arm necessary to produce a more significant influence on the dynamic load is avoided.

The model data which corresponds most nearly in wind direction terms to the available full scale data are given in Table 3, where the N-S modal loads includes the effect of the university site and the E-W modal loads do not. The comparison of this data with that in Table 2 for the full scale loads is considered most encouraging, despite the scatter and statistical uncertainties which surround some of the full scale data, particularly those derived from small numbers of wind storm recordings.

ACKNOWLEDGEMENT

This work has been performed under contract from the U.K. Building Research Establishment and is reproduced by their kind permission. The contribution of Dr. R.H. Evans who undertook the experimental work, is gratefully acknowledged.

REFERENCES

1. British Standards Institute (1972) CP3, Chapter 5, Part II. Code of practice for determination of wind loads.

2. Cook, N.J. (1976) Data for the wind tunnel simulation of the adiabatic atmospheric boundary layer, Part C. B.R.E. Note N 135/76.

3. Counihan, J.C. (1975) Adiabatic atmospheric boundary layers: A review and analysis of data from the period 1880-1972. Atmos. Envir., Vol. 9, pp. 871-905.

4. Ellis, B., Evans, R., Jeary, A. and Lee, B.E. (1979). The
    wind induced vibration of a tall building. Paper G13.
    IAHR/IUTAM Symposium, Karlsruhe. Sept.

5. ESDU (1972) Characteristics of wind speed in the lower
    layers of the atmosphere near the ground. Strong winds
    (Neutral Atmosphere) Engng. Sci. Data Unit, Item 72026.

6. ESDU (1976) The response of flexible structures to atmospheric
    turbulence. Engng. Sci. Data Unit, Item 76001.

7. Evans, R.A. and Lee, B.E. (1981, a) The assessment of dynamic
    wind loads on a tall building. 4th U.S. National Conf. Wind
    Engineering Research, Seattle, USA.

8. Evans, R.A. and Lee, B.E. (1981, b) The problems of anemometer
    exposure in urban areas : A wind tunnel study. Meteorological
    Magazine, Vol. 110, pp. 188-199.

9. Harris, R.I. (1968) Measurements of wind structure at heights
    up to 598 ft. above ground level. Symp. Wind Effects on Buildings
    and Structures, Loughborough, University.

10. Hurty, W.C. and Rubinstein, M.F. (1964) Dynamics of structures.
    Prentice Hall, New Jersey, USA.

11. Jeary, A., Lee, B.E. and Sparks, P.R. (1979). The determination
    of modal wind loads from full scale building response measure-
    ments. Proc. 5th Int. Conf. Wind Engineering. Colorado State
    University. Publ. by Pergamon Press. (1980).

12. Lee, B.E. (1977). The simulation of atmospheric boundary layers
    in the Sheffield University 1.2 x 1.2m boundary layer wind
    tunnel. Department of Building Science Report No. BS 38.
    Sheffield University.

13. Melbourne, W.H. (1974). Response of slender towers to wind
    action. Proc. 5th Australasian Hydraulics and Fluid Mechanics
    Conference, Canterbury, New Zealand.

TABLE 1

Arts Tower - Structural Properties

	N-S mode	E-W mode	θ mode
Fundamental Natural Frequency, Hz	0.68	0.86	0.79
Modal Mass, Kg	6.35×10^6	5.84×10^6	-
Modal Inertia, Kg/m^2	-	-	8.14×10^8
Modal Lateral Stiffness, N/m	1.15×10^8	1.70×10^8	-
Modal Torsional Stiffness, Nm/rad.	-	-	1.99×10^{10}
Damping, % critical	0.86	0.86	0.90

TABLE 2

Full Scale - Modal Wind Loads

Mean Wind Direction 0	Mean Wind Speed (84m) m/s	Equivalent r.m.s. Modal Force $\sigma_{r_{r.m.s.}}$ KN	
		N-S mode	E-W mode
254	20	2.71	2.03
254	24.1	3.13	1.59
305	18.3	2.01	1.68
305	25	4.25	2.86
305	29.7	5.23	3.40
305	34.6	5.50	3.50
338	22.9	1.95	0.81
338	27.2	2.77	1.47
338	30.4	3.12	1.86

TABLE 3

Model Data - Modal Wind Loads (Scaled Up)

Mean Wind Direction	Mean Wind Speed (84m) m/s	Equivalent r.m.s. Modal Force $\sigma_{r_{r.m.s.}}$ KN	
		N-S mode (site model)	E-W mode (no site model)
254	28.2	2.07	1.21
305	28.2	1.68	1.00
338	28.2	2.12	1.26

Figure 1   The Arts Tower

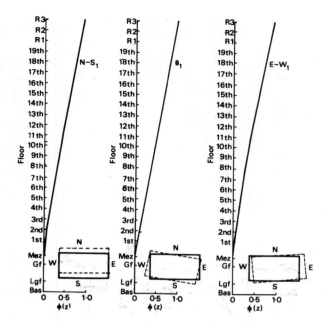

Figure 2   Arts Tower   fundamental   mode   shapes

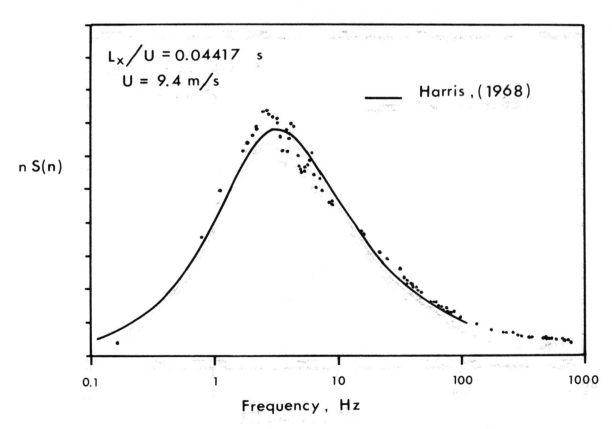

Figure 3   Spectrum   of   u - component   of   turbulence , $Y/H = 0.3125$

Figure 4

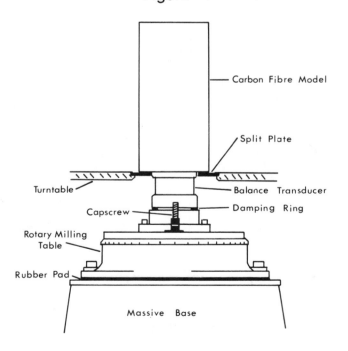

Carbon Fibre Model

Split Plate

Turntable

Balance Transducer

Capscrew

Damping Ring

Rotary Milling
Table

Rubber Pad

Massive Base

Figure 5   Experimental   Arrangement

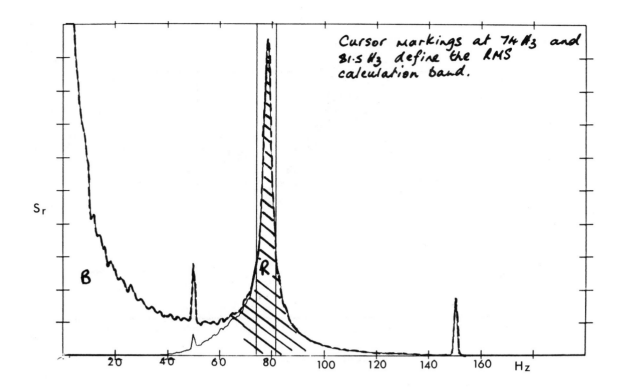

Cursor markings at 74 Hz and 81.5 Hz define the RMS calculation band.

Figure 6   R M S   spectrum   of   80 Hz   balance   output

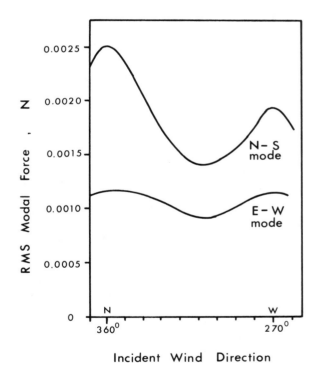

Figure 7   Variation   of   rms   modal   force   with   wind   direction

No   site   model

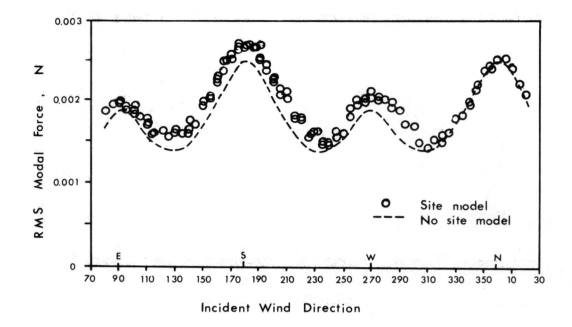

Figure 8   Effect of site model on variation of rms modal force

N-S mode .

COMPARISON OF MODEL AND FULL-SCALE TESTS
OF THE AYLESBURY HOUSE

J D Holmes, CSIRO Division of Building Research (Australia)

1. INTRODUCTION

It is now over ten years since the Building Research Establishment (B.R.E.) of the United Kingdom commenced the full-scale measurement program on the Aylesbury site and about eight years since the experimental building was dismantled. Interest in the Aylesbury house continues not only because of the uniqueness and quantity of the full-scale data on wind pressures on a low-rise building that were obtained, but also because the building has become, in recent years, a test case for wind tunnel modelling techniques for low-rise buildings.

Using the full-scale data base (Eaton and Mayne[1], Eaton, Mayne and Cook[2]) for comparison, detailed wind tunnel testing of the Aylesbury experimental building has been carried out at the Universities of Oxford[3,4,5](U.O.), Western Ontario[6,7](U.W.O.), James Cook[8,9](J.C.U.) and at the Virginia Polytechnic Institute and State University[10,11](V.P.I), at geometric scaling ratios ranging from 1/500 to 1/25. Following these studies, an International Aylesbury Comparative Study, using 1/100 scale models of the building supplied by the Building Research Establishment is about to commence.

In this paper, a short review of the previous full-scale and model studies is given followed by an introduction to the international study.

2. THE FULL-SCALE MEASUREMENTS

A large amount of data was collected by B.R.E. at the Aylesbury site between 1972 and 1974, including pressure measurements on the terraced estate houses, and load cell measurements from the experimental house. However, most of the interest by wind tunnel modellers has been in the point pressures on the experimental Aylesbury house, described in detail in readily available publications[1,2], and only these measurements will be discussed here.

2.1 Experimental Building

Figure 1 shows the experimental building which had plan dimensions of 7 m x 13.3 m, and a height to the eaves of 5 m. An adjustable roof allowed for the pitch to be set at any angle between 5° and 45°. Figure 2 shows the positions on the building where pressure transducers could be installed, although not all of these were used in any single run.

Figure 1   Aylesbury experimental building with 22.5° pitch roof

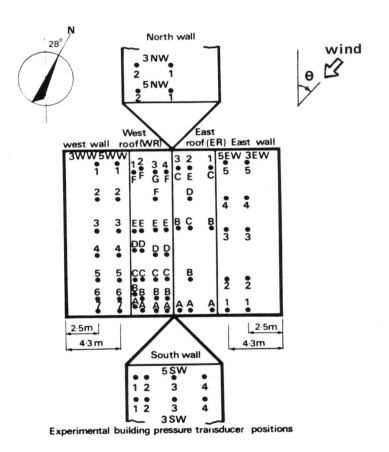

Experimental building pressure transducer positions

Figure 2   Pressure transducer positions

2.2 Test Site

The nearby housing estate on the eastern side of the experimental building is shown in Figure 3. The 10 m high anemometer mast, located about 25 m from the south wall of the experimental building is also visible in Figure 3. Most of the winds in the test runs came from the South to West quadrant, and the terrain for these directions is shown in Figure 4 (reproduced from Apperley et al.[6]). The shortest distance from the experimental building to the first 5 m hedge was estimated to be 135 m (Tieleman et al.[11]). The direct shielding at the distance of 27 hedge heights or greater is not significant, but the hedges and trees shown in Figure 4 clearly did contribute to the general upwind roughness for winds from the West and South-West directions.

Figure 3   Aylesbury site looking East

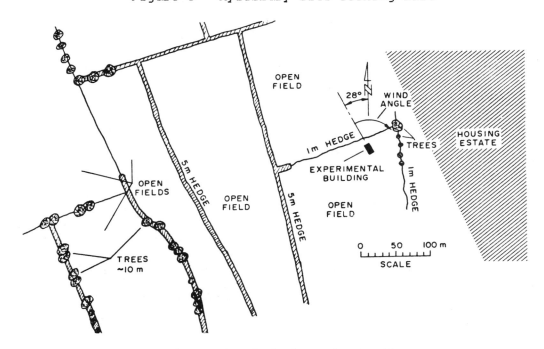

Figure 4   Aylesbury test site

2.3 Wind Characteristics

Characteristics of the test runs are shown in Table I, following. Velocity measurements were available from three heights: 10 m, 5 m and 3 m, for most of the runs. The earlier measurements showed non-logarithmic velocity profiles, but for some reason, the later measurements followed the logarithmic law fairly well. The roughness length appeared to vary from about 50 mm, for southerly wind directions for which the hedges were parallel to the flow, (Runs A31B[1] and A35F-J[2]), to around 150 mm for south-westerly and westerly directions, for which the hedges and trees upwind had a significant effect (Runs A32[1] and A38A-G[2])*.

The turbulence intensities also reflected the effect of the hedges, with an average value of about 27% at the 5 m height for the south-westerly winds to an average of 22% for southerly winds.

Table I   Characteristics of the full scale Aylesbury runs

(a) Results given in Reference 1.

Record no	Date	Time GMT	$\bar{\beta}$	$\bar{\theta}$	$\bar{u}_{10}$ (m/s)	$\hat{u}_{10}$ (m/s)	Roof pitch of experimental building
A1	12. 3.72	13.59	40°	70°	7.6	15.0	*
A5	25. 5.72	23.24	200°	230°	6.1	13.2	22.5°
A7	26. 5.72	12.43	235°	265°	11.9	18.6	22.5°
A11	1.12.72	12.06	205°	235°	9.4	20.9	22.5°
A12A	1.12.72	13.38	205°	235°	8.4	15.6	10.0°
A12B	1.12.72	14.07	205°	235°	8.4	14.7	15.0°
A25B	12.11.73	15.48	250°	280°	8.4	12.9	22.5°
A29	14.12.73	16.04	330°	0°	8.6	13.2	22.5°
A31B	8. 1.74	14.35	150°	180°	13.3	21.2	22.5°
A32	8. 1.74	15.55	235°	265°	14.3	24.4	22.5°

$\bar{\beta}$ Mean wind direction relative to true north.

$\bar{\theta}$ Mean wind direction relative to N-S axis of experimental building.

$\bar{u}_{10}$ Mean wind speed averaged over the length of the record.

$\hat{u}_{10}$ Maximum 2 second speed during record.

* Results refer to houses in estate.

---

* Note that the roughness lengths for Runs A38B to A38G quoted by Tieleman et al.[11], were too small by a factor of ten.

(b) Results given in Reference 2.

Record no	Date	Time GMT	$\bar{\beta}$	$\bar{\theta}$	$\bar{u}_{10}$ (m/s)	$\hat{u}_{10}$ (m/s)	Roof pitch of experimental building
A35F	27.1.74	23.16	180°	208°	12.4	18.9	22.5^{o}
A35G	27.1.74	23.41	180°	208°	11.2	18.8	15.0^{o}
A35H	28.1.74	00.04	180°	208°	11.1	17.0	10.0^{o}
A35J	28.1.74	00.26	180°	208°	10.5	14.0	5.0^{o}
A38A	10.2.74	11.49	205°	233°	12.4	21.1	15.0^{o}
A38B	10.2.74	12.16	205°	233°	11.3	20.4	10.0^{o}
A38C	10.2.74	12.40	205°	233°	11.1	18.4	5.0^{o}
A38D	10.2.74	13.12	205°	233°	11.5	19.9	27.5^{o}
A38E	10.2.74	13.38	205°	233°	11.4	18.3	35.0^{o}
A38F	10.2.74	14.08	210°	238°	10.2	18.8	45.0^{o}
A38G	10.2.74	14.48	220°	248°	8.3	14.9	22.5^{o}

2.4 Pressure Measurements

For the 20 runs in Table I for which pressure measurements were obtained on the experimental house, mean, root-mean square fluctuating, and 2 second and 1/32 second peak pressures were recorded for record lengths of 1024 seconds duration. Spectra were also measured for a number of locations on roof and walls. The backs of the pressure transducers were vented to a manhole in the field on the easterly side of the experimental building. The later runs (Table I(b)) had a 0.7 m eaves overhang on the middle third of the western side of the building.

A comparison of Runs A7 and A32 for which the building had a 22.5° roof pitch, and a wind direction of 265° (relative to the N-S building axis - see Fig. 2); shows considerable scatter (e.g. Fig.9 of Apperley et al.[6]). For example, a linear regression of the mean pressure coefficients from Run A7 on those from A32 gave a relationship: $(\bar{C}_p)_{A7} = 0.55 (\bar{C}_p)_{A32} - .05$, with a correlation coefficient of 0.61 (Holmes and Best[8]). This variability, whether due to the natural variability in the wind, or to measurement errors, tends to set a limit on what can be expected when comparing full-scale and wind tunnel test results.

3. THE WIND TUNNEL TESTS

3.1 Characteristics of the Tests

The characteristics of the wind tunnel tests carried out at U.O., U.W.O., J.C.U. and V.P.I. are summarised in Table II. Other tests carried out at the C.S.T.B., France (Barnaud and Gandemer[12]) and at the B.R.E. itself[2], were of a preliminary nature, or less extensive than those in Table II. Bray[13], at the University of Bristol, carried out tests only for the estate houses.

Table II   Model studies of the Aylesbury house

Laboratory	Geometric Scaling Ratio	Approx. h/z_0*	Turbulence Intensity at z = 5 m (for W/S.W winds)	Roof Pitches	Wind Directions, θ (see Fig.2)	Hedges Modelled?
U.Oxford[3,4,5]	1/75	75	0.19[+]	5°,10°, 15°,22½°	233°,248°,265°	Yes
U.Western Ontario[6] (Exposure 2)	1/500	3200[x]	0.20	5°,22½°	180°,208°,230°,233° 235°,248°,265°,270°	Yes
U.Western Ontario[7] (Exposure 4)	1/100	120	0.28	22½°	265°	Yes
James Cook U.[8]	1/50	150	0.20	10°,22½°	180°,235°,265°	No
James Cook U.[9] "	1/50 (1/100)	150 (100)	0.23 (0.18)	10°,22½° (22½°)	180°,235°,265° (180°,235°,265°)	Yes (No)
Virginia P.I.&S.U.[10]	1/50	-	0.27	22½°	0°-360° at 30° incr.	No
Virginia P.I.&S.U.[11]	1/25,1/50, 1/100	-	0.23,0.27,0.29	22½°	0°-360° at 30° incr.	No.

* Approximate full-scale values: 33 (W/S.W. winds); 100 (S.winds)

+ Measured at centre of turntable

x This value has been queried by Holmes[14]

A range of geometric scaling ratios between 1/500 and 1/25 was used, but not all roof pitches were reproduced at each laboratory. The 22.5° pitch was common to all tests, however. Except for the V.P.I. tests, the wind directions were chosen to reproduce the full-scale conditions. The upwind hedges were not reproduced for the V.P.I. tests and for the early J.C.U. tests. Non-logarithmic velocity profiles were obtained for the V.P.I. tests, and probably for the 1/500 U.W.O. tests, and the turbulence intensities at 5 m height for westerly and south-westerly wind directions were low in the U.O., U.W.O. (1/500), J.C.U. and V.P.I. (1/25) tests. Although not shown in Table II, the ratio of building dimension to turbulence length scale (this can be represented as h/λ_u, where λ_u is the peak wavelength of the normalised longitudinal turbulence spectrum) was probably close to the full-scale value for the U.W.O. 1/500 tests, about twice the full-scale value for the U.O. and J.C.U. tests, and about three to four times of the full-scale value for the U.W.O. 1/100, and V.P.I. tests.

The U.W.O. tests at 1/500 scale were carried out in a naturally grown boundary layer over uniform surface roughness (nylon cloth), with the addition of the local hedges and trees (Exposure 2). In all other tests, fences, grids or spires were added to accelerate the boundary layer growth and/or to produce large turbulence scales to match the large geometric scaling ratios of the model buildings. Alternative boundary layers were used in the U.W.O. tests (both scales) and to a lesser extent in the U.O. tests. However, most attention in comparison studies has been directed to the best simulation in each case; the characteristics of these are shown in Table II.

The pressure measuring systems used were of a sufficient frequency response to allow comparison with the shortest duration full-scale peaks except for the U.W.O. 1/500 scale tests, in which one second full-scale peaks were the shortest that could be reproduced. In the U.O. measurements no 'restrictor' system was used in the pressure tubing so that a resonant peak occurred in the frequency response curve; this would cause some overestimation of r.m.s. and peak pressures. The same was probably true of the V.P.I. test results although details of the tubing response were not provided, in this case.

Figure 5 shows one of the 1/50 scale J.C.U. models containing a 'Scanivalve' with pressure transducer, and pressure tubes with restrictors.

Figure 5   1/50 scale wind tunnel model (10° pitch)

3.2 Comparative Studies

The wind tunnel data have been compared both with the full-scale results and with other wind tunnel data in different ways. Greenway and Wood[3,4] compared the U.O. model results with the full-scale data, and with the U.W.O. 1/500 scale data, graphically, and by linear regression analysis. Holmes and Best[9] and Vickery (the younger)[7] using the same techniques, concentrated on the 22.5° pitch, 265° wind direction data (Runs A7/A32 in full-scale) for comparison of data from different wind tunnels. In the above cases, the pressure coefficients used for comparison were based on the 10 m mean wind speed, as in the full-scale tests. Tieleman et al.[10,11], however, compared pressure coefficients based on the local mean wind speed at the height of the pressure measurement positions, in an attempt to remove the effect of variations in mean velocity profile. These variations were quite large in both individual full-scale runs, and between wind tunnel tests (see values of h/z_0 in Table II and Figure 3 of Tieleman et al.[11]).

This comparison technique used by V.P.I. is more valid for windward side pressures than for the wake regions, but is better than using the 10 m velocity in the present case. Of course, if all profiles were the same, the same relationships between the pressure coefficients from different tests would apply, whichever reference velocity was used.

Figure 6 shows a comparison of mean pressure coefficients (based on 10 m mean velocity) from the J.C.U. 1/50 scale tests (with hedges), and the full scale data from four different runs. This is fairly representative of the sort of comparison obtained between wind tunnel results, and the combined full-scale data, although better agreement has been obtained with certain individual full-scale runs.

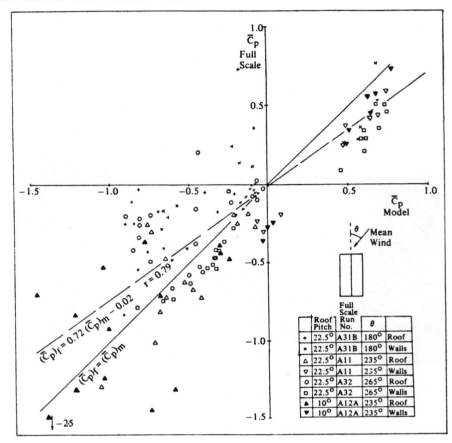

Figure 6   Full scale and model (1/50 J.C.U.) comparison for
mean pressure coefficients

The results in Figure 6 indicated higher values on average for the wind tunnel data compared with the full-scale values, as indicated by the slope of 0.72 for the linear regression line. This can partly be explained by the differences in mean velocity profile, indicated by the low value of roughness length z_0 used in the model tests (i.e. high value of h/z_0 – Table II). However there are many individual cases in which the full-scale values exceed the model C_p's in absolute value. Higher suctions were observed at the windward edge of the roof in all wind tunnel tests for the case of wind nearly normal to the ridge ($\theta = 265°$) and normal to the gable end ($\theta = 180°$), but higher values were found in the full-scale case for oblique winds ($\theta = 235°$).

Correlation coefficients between full-scale and model mean pressure coefficients were generally found to be between 0.75 and 0.9, with lower values when r.m.s. pressure coefficients were compared, and slightly higher values for the peak pressure coefficients.

The regression analyses also indicated, by the intercepts, small but significant differences between the static reference pressures used in the various tests. This should be a cause for some concern requiring closer attention in future. Variations in the mean velocity profiles, as already discussed, were clearly apparent in the slopes of the regression lines, when the 10 m mean velocity was used for the dynamic pressure reference.

Figures 7 and 8 show 'local' pressure coefficients based on the upwind mean wind speed at the height of the pressure tappings for centre and edge wall tappings (taken from Tieleman et al.[11]). Mean C_p's are shown in Figure 7 and r.m.s. values in Figure 8. The relatively narrow range of the wind tunnel results and the agreement with full-scale in Figure 7 is encouraging, considering the variations in model scaling ratio and boundary layer simulation procedures used. In the case of the r.m.s. pressures, the range of wind tunnel results and agreement with full-scale is not so good. However, low r.m.s. C_p's can generally be associated with tests where the turbulence intensities are low (Table II). This observation is in agreement with some supplementary measurements by Greenway and Wood[5] in which the turbulence intensity was varied above and below the optimum, with other boundary layer characteristics being kept the same. Corresponding changes to the r.m.s. pressure coefficients (and to a lesser degree in the peak C_p's) were found.

The ratio of building dimension to turbulence length scale appears to be a parameter of lesser importance at least for the point pressures under consideration here. Figures 7 and 8 contain results from the larger scale V.P.I. tests, in which the turbulence scales were only about one-quarter to one-third of 'optimum' and from the U.W.O. 1/500 scale tests, in which good matching of turbulence length scales was claimed.

4. INTERNATIONAL COMPARATIVE STUDY

The wind tunnel studies described previously have been useful in setting the limits to which model tests can reproduce the vagaries of the real wind loading process on an actual building. It is likely that closer attention to velocity profiles, turbulence intensities, static reference pressures and perhaps to the detailed geometry at the separation points on the models, will produce some improved comparisons between full-scale and model Aylesbury building results. However, the inherent variability in the full-scale data is such that greatly improved results are unlikely to occur.

On the other hand, there is less excuse for the differences that are presently occurring between wind tunnel test results as a result of differing modelling and testing procedures. In an attempt to resolve these differences, an International Aylesbury Comparative Study has been proposed. A model of the Aylesbury building at a scale of 1/100, supplied by the Building Research Establishment, will be tested by any laboratory which wishes to participate. Three identical models will be circulated in Europe/Africa, North/South America and Asia/Australasia respectively.

It has been agreed that two flow conditions will be specified:

(a) the laboratory's best simulation of the Aylesbury site allowing for the variation of roughness with wind direction if necessary;

614

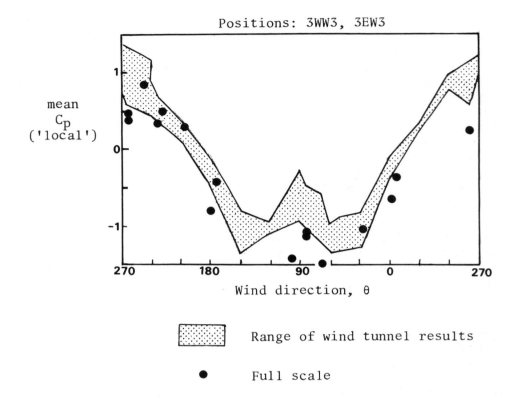

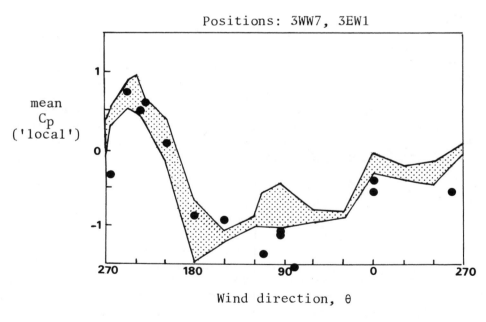

Figure 7   Comparison of 'local' mean pressure coefficients
(derived from Tieleman et al.[11])

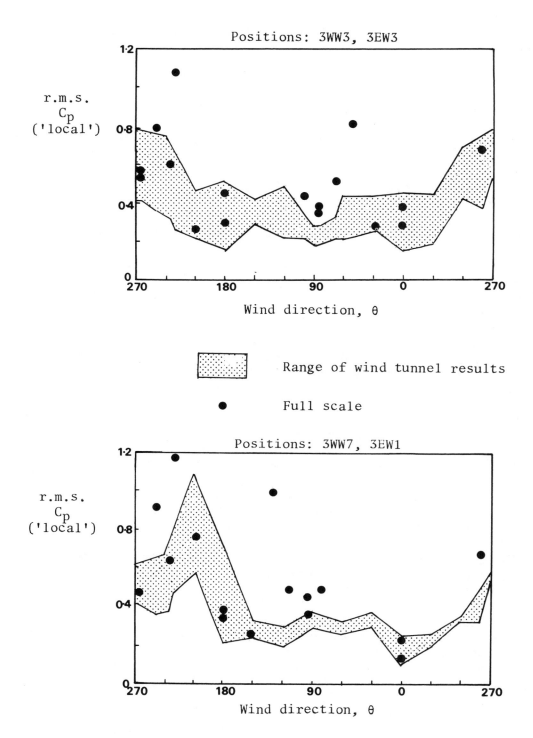

Figure 8   Comparison of 'local' r.m.s. pressure coefficients
(derived from Tieleman et al.[11])

(b) a 'standard' rural boundary layer, which will have closely specified roughness and turbulence parameters.

Different levels of measurement complexity will be specified with increasingly detailed statistical analysis required at the higher levels. This will allow laboratories with less sophisticated facilities and instrumentation to participate.

It is hoped that the study will provide a 'benchmark' data base which may be used by newcomers to the wind engineering field for calibration of wind tunnel modelling techniques. In many respects, it is similar to the C.A.A.R.C. model study conducted a few years ago for tall buildings (Melbourne[15]), although this was not open to all wind tunnels, and no full-scale data was available for comparison, in that case. A preliminary comparison of data from the Aylesbury study will be made at the Sixth International Conference on Wind Engineering (Gold Coast, Australia, March 1983).

A detailed schedule for the study is available from the three coordinators (Dr N J Cook, Building Research Establishment, U.K. (collator); Dr A G Davenport, University of Western Ontario, Canada; and the writer).

5. CONCLUSIONS

i) The full-scale experiments carried out by the Building Research Establishment at Aylesbury, U.K., have provided the best data base on wind pressures on low-rise buildings available to date. However, there are some inherent variabilities in the data which will limit the agreement that can be achieved using wind tunnel tests. How much of this variability can be attributed to measurement errors and how much to natural variations is unknown at this stage.

ii) The wind tunnel testing programs that have been carried out to date on models of the Aylesbury building at range of scaling ratios, have generally shown favourable agreement with the full-scale data and with each other, when differences in mean velocity profile and static pressure reference are taken into account.

iii) The turbulence intensity, and if a reference position well above the building is used, the mean velocity profile, are important parameters to be scaled correctly in a wind tunnel test.

iv) Turbulence length scale similarity, which is difficult to achieve when large geometric scales are used, does not seem to be a parameter of the greatest importance, although the limits of distortion have yet to be defined clearly (Vickery[7] had a contradictory conclusion to this however).

v) The International Aylesbury Comparative Study will, it is hoped, narrow existing differences between wind tunnel test results, and provide a standard data base, which future wind tunnel modellers concerned with low-rise buildings can use.

6. ACKNOWLEDGEMENTS

The author wishes to thank the participants in previous Aylesbury studies for freely providing reports and information in the past. In particular, Messrs Eaton, Mayne and Cook for the full-scale studies, and Wood, Surry, Apperley and Tieleman for the wind tunnel experiments, are mentioned.

The contributions of his fellow coordinators, Nick Cook and Alan Davenport, in setting up the International Comparative Study, are also gratefully acknowledged by the author.

7. REFERENCES

1. K.J. Eaton and J.R. Mayne, The measurement of wind pressures on two-storey houses at Aylesbury, Building Research Establishment, Current Paper CP 70/74, July 1974 (also J. Ind. Aerodyn., 1 (1975) 67-109).

2. K.J. Eaton, J.R. Mayne and N.J. Cook, Wind loads on low-rise buildings - effects of roof geometry, Building Research Establishment, Current Paper CP 1/76, January 1976 (also Proc. 4th International Conference on Wind Effects on Buildings and Structures, Lodon, 1975, Cambridge University Press, 1977, pp.95-110.

3. M.E. Greenway and C.J. Wood, Wind tunnel pressure measurements on the Aylesbury low-rise housing estate, Part I, Simulation design and mean pressures, University of Oxford, Report OUEL 1213/77, 1977.

4. M.E. Greenway and C.J. Wood, Wind tunnel pressure measurements on the Aylesbury low-rise housing estate, Part II, Mean, r.m.s. and extreme pressures with frequency spectra, University of Oxford, Report OUEL 1271/78, 1978.

5. M.E. Greenway and C.J. Wood, Wind tunnel pressure measurements on the Aylesbury low-rise housing estate, Part III, Additional experiments, University of Oxford, Report OUEL 1272/78, 1978.

6. L. Apperley, D. Surry, T. Stathopoulos and A.G. Davenport, Comparative measurements of wind pressure on a model of the full-scale experimental house at Aylesbury, England, J. Ind. Aerodyn., 4(1979), 207-228 (also unpublished University of Western Ontario report and Proc. 3rd Colloquium on Industrial Aerodynamics, Aachen, 1978, pp.195-209).

7. P.J. Vickery, A wind tunnel study on a model of the full-scale experimental house at Aylesbury, England, University of Western Ontario, B.E.Sc. thesis, April 1981 (also, D. Surry and P.J. Vickery, The Aylesbury experiments revised - further wind tunnel tests and comparisons, Proc. 5th Colloquium on Industrial Aerodynamics, Aachen, 1982).

618

8. J.D. Holmes and R.J. Best, Wind tunnel measurements of mean pressures on house models and comparison with fulll-scale data, James Cook University of North Queensland, Wind Engineering Report 3/77, July 1977 (also Proc. 7th Australasian Hydraulics and Fluid Mechanics Conference, Adelaide, 1977, Institution of Engineers, Australia, 1977, pp.26-30).

9. J.D. Holmes and R.J. Best, Further measurements on wind tunnel models of the Aylesbury experimental house, James Cook University of North Queensland, Wind Engineering Report 4/78, October 1978 (also Wind Engg., 2 (1978) 203-220).

10. H.W. Tieleman, R.E. Akins and P.R. Sparks, A comparison of wind-tunnel and full-scale wind pressure measurements on low-rise structures, J. Wind Engg. and Ind. Aerodyn., 8 (1981) 3-19 (also Proc. 4th Colloquium on Industrial Aerodynamics, Aachen, 1980, pp.1-16).

11. H.W. Tieleman, R.E. Akins and P.R. Sparks, Model/model and full-scale/model comparison of wind pressures on low-rise structures, Proc. Colloque 'Construire avec le Vent', Nantes, 1981, pp.IV-5-1 - IV-5-21.

12. G. Barnaud and J Gandemer, Determination en soufflerie simulant le vent naturel des coefficients de pression sur les structures basses, C.S.T.B. (France), Report ADYM-12-74, 1974.

13. C.G. Bray, Wind tunnel modelling of the wind loading on low-rise houses in a housing estate, U. Bristol Thesis, Nov. 1976.

14. J.D. Holmes, Discussion of Reference [6], J. Wind Engg. and Ind. Aerodyn., 6 (1980) 181-182.

15. W.H. Melbourne, Comparison of measurements on the C.A.A.R.C. standard tall building model in simulated model wind flows, J. Wind Engg. and Ind. Aerodyn., 6 (1980) 73-88.

COMPARISON OF MODEL AND FULL SCALE TESTS ON BRIDGES

A. G. Davenport, Professor and Director, Boundary Layer Wind Tunnel Laboratory, University of Western Ontario, Faculty of Engineering Science, London, Ontario, Canada   N6A 5B9

1.   INTRODUCTION

The action of turbulence on the aerodynamic behaviour of suspension bridges has been a subject of considerable interest for more than a decade. Studies at the Boundary Layer Wind Tunnel Laboratory in 1968 provided the first insights into its importance. Subsequently this theme has been taken up by Wardlaw and Irwin, Scanlan, Walshe, Okauchi, Melbourne, Holmes and several others.

The investigations carried out at the BLWT Lab. commenced with a comparative study of the behaviour of both section and full bridge models of the Murray McKay suspension bridge across Halifax Harbour, Nova Scotia. In this, the full bridge models were carried out in both uniform and turbulent boundary layer flow.

The study was undertaken with the initial objective of obtaining information for the design and erection of this bridge which was complete in 1969. The main span of this bridge is 1400 ft. with symmetrical sidespans of 513 ft. The shallow trusses, only 11' 0" in depth, gave this bridge an extremely slender silhouette; the light weight orthotropic deck system made the bridge potentially more sensitive to wind than the significantly heavier bridges of more conventional design.

On the complete bridge (as distinct from the construction stages, also studied) the following tests were run:

1/40 scale section model tests; 1/320 scale section model tests; 1/320 scale full bridge model tests in uniform flow; and 1/320 scale full bridge model tests in turbulent boundary layer flow.

Typical measured dynamic responses of these models in wind normal to the bridge axis are given in Fig. 1. Three observations were made:

1)   The section models exhibited at a critical wind speed a coupled vertical torsional oscillatory instability typical of truss-stiffened section models.

2)   The full bridge model in uniform flow exhibited a divergent instablility at a speed well in excess of the critical speed for the sectional model.

3) The full bridge model in turbulent flow showed only random vertical oscillations increasing steadily in amplitude with increased wind speed and also turbulent intensity; no instability or significant torsional motion was observed.

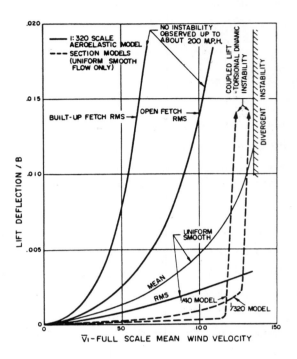

Fig. 1   **Summary of Predictions of Bridge Response From Section Models and Full Bridge Models Tested in Uniform Flow and Flow Simulating Natural Wind (completed bridge, $a = 0$)**

The responses of these models of the same bridge could hardly have been more dissimilar! The possibility that differences in behaviour could be explained by the relatively small scale of the full bridge model was eliminated by the reasonably good agreement in the behaviour of the two section models having scales of 1/40 and 1/320. The explanation therefore seemed to relate to:

● inherent differences in the behaviour of full bridge and sectional models; and

● the influence of turbulence on the general behaviour and in particular the instability characteristics.

The influence of turbulence in exciting random, buffeting movements was to be expected and has been observed in many situations. The dramatic modification of the instability characteristics was perhaps more surprising but have been observed with prismoidal shapes by Novak. Most disconcerting however was the dissimilarity between the section model and full bridge model in uniform flow.

This last result was disconcerting for two reasons: First, because of the satisfactory qualitative agreement which Farquharson and Vincent and Frazer and Scruton had obtained much earlier between section models and full bridge models; these tests in fact formed the basis for the validity of section model testing procedures. Second, section model tests had satisfactorily predicted - at least qualitatively - the unstable behaviour of several full scale bridges including the Tacoma Narrows, Golden Gate and Deer Isle bridges.

The failure of the Murray McKay full bridge model to follow these earlier trends posed a question fundamental to the aerodynamic testing of suspension bridges. Are section models capable of predicting the full scale bridge behaviour? If they are, then clearly the Murray McKay aeroelastic model had failed to perform "correctly"; if not then the technique of section model testing could not always be relied on.

Considering these questions, one or two explanations for the incompatible behaviour suggested themselves. In the first instance the Halifax Narrows Bridge was relatively light compared to other bridges and it seemed possible that aerodynamic and mechanical influences generally neglected in section model testing — in particular the participation of the towers and cables in responding to wind — could be exerting a stronger control. Secondly, being intrinsically a stable bridge deck configuration, the wind speeds associated with the instability (of the section model) were high and consequently the static deformations (principally twists) at those speeds began to be noticeable. While these deformations varied significantly across the span of the full bridge they did not obviously on the section model. The well known sensitivity of instability to angle of attack suggested that the static deformations could have interfered with the instability mechanism.

Confronted by the lack of agreement between full bridge and section model experiments and only strong suspicions as to the explanation for this, the experiments carried out in the early stages of suspension bridge aerodynamics were re-examined.

Two bridges were selected for study, the Tacoma Narrows H-section and the Golden Gate Bridge. In both instances the full scale bridges had demonstrated instability and section model tests were available: the Tacoma Narrows Bridge was also tested as a full bridge model in uniform flow. To these bridges was added the flat plate for which theoretical predictions of aeroelastic instability could be made.

The objective was to test these sections in a configuration as close as possible to the full bridge model in uniform flow and in turbulent boundary layer flow. To do this a rig was developed in which sections of the deck, whose mass distribution and inertia were correctly scaled, were suspended on pairs of parallel, taut piano wires running at the level of the shear centre. The decks thus vibrated as taut strings primarily in a fundamental half-wave mode shape. By adjusting the tension and wire spacing the correct ratio between torsion and vertical motion could be obtained. These models have been termed "taut strip" models. The lengths of the models were taken the same as the full bridge main span and ran practically the full width of the boundary layer wind tunnel (i.e. approximately 6-8 ft. long).

622

Tests were carried out in uniform and turbulent flows of different turbulence intensities and boundary layer roughnesses. The latter were generally representative of atmospheric turbulent boundary layers (4).

For the Tacoma Narrows H-section, the measured dynamic vertical and torsional motions are shown in Fig. 2. From this it can be seen that;

a)   in uniform flow the response is characterized by one degree of freedom torsional instability at a reduced velocity of approximately 2.5 and both torsional and vertical oscillation peaks of limited amplitude at lower reduced velocities;

b)   moderate turbulent flow (9%) causes little change in the behaviour;

c)   strong turbulent flow slightly raises the critical reduced velocity, eliminates the peak torsional and vertical movements at lower reduced velocities but induces a random vertical respsonse monotonically increasing with wind speed.

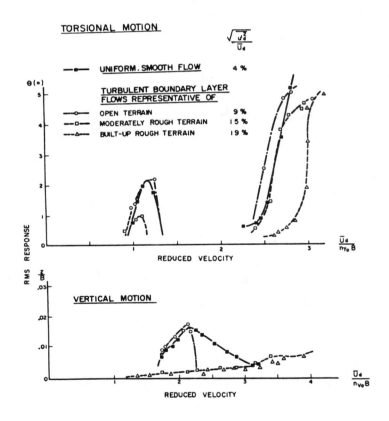

Fig. 2   Dynamic Response of "Tacoma" H-Section Taut Strip Model in Different Flow Regimes

Fig. 3 shows typical oscillograph traces obtained in different flow conditions. From this it appeared that with allowances for differences in damping, the taut strip model was in reasonable agreement with earlier experiments with both section models and full bridge models. Full scale observations indicated that until the date of failure, only random vertical motions in several different modes were observed. Only on the day of failure was significant torsional motion observed. This occurred for a brief period in the fundamental symmetric torsional made (having a frequency of about 10 cpm), then developed the catastrophic single noded asymmetric mode having a frequency of about 12 cpm. The reduced velocity $V/n_T B$ at failure cannot be accurately determined but seems to be less than about 6.0 and the single amplitude about 35° (RMS $\simeq 25^O$).

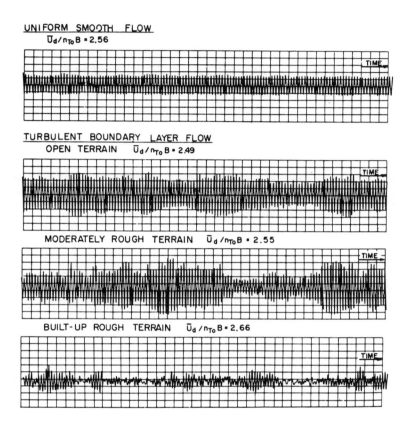

Fig. 3   Traces of Dynamic Torsional Response of "Tacoma Narrows" H-Section Taut Strip Model in Different Flow Regimes ($\zeta_T \simeq 0.25\%$)

These studies drew the final conclusion that turbulence could have a profound effect on the behaviour and that testing procedures capable of absorbing this new feature needed to be studied. Similar conclusions were reached by Irwin and Wardlaw in their detailed studies for the Lion's Gate Bridge in Vancouver. This very large model nearly 25 ft. in length also helped remove any lingering doubt associated with the use of the smaller scale models in the earlier BLWT studies. More recently Walshe has demonstrated other effects and Okauchi has described the 1/10th scale bridge testing facility in Japan.

Essential to the improvement of bridge testing procedures is confirmation of the results against full scale data. The purpose of this paper is therefore to provide two instances of full scale measurement for comparison. The first concerns an up-date of comparisons with the

Golden Gate Bridge. The second concerns the Bronx Whitestone Bridge.

2. THE GOLDEN GATE BRIDGE

During the initial years, both the Golden Gate bridge and, shortly afterwards, the Bronx Whitestone Bridge showed signs of the same (but much milder) susceptibility to dynamic response to wind which later caused the destruction of the Tacoma Narrows Bridge. This, and the fact that the motions of the Golden Gate Bridge were carefully monitored during the critical period, are by themselves important reasons for studying the wind response of this bridge. The measurements of the motion represent a reference standard against which to test any procedure intended to predict the response of a bridge section to wind including any tendency to instability.

The general layout of the bridge is shown in Fig 4. The measurements concerned the period prior to the installation of the bottom lateral system to its stiffening truss. The lower modes of vibration of the bridge, as calculated by Vincent, are shown in Fig. 5. The lowest frequency is the first asymmetric mode. The unusually sinuous shape of the first symmetric mode should also be noted.

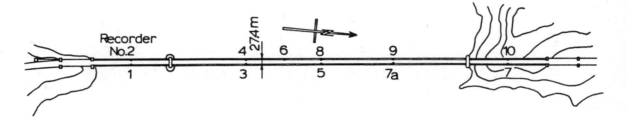

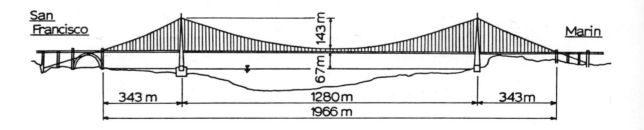

Fig. 4   General Layout of the Golden Gate Bridge

The field observations of the bridge, reported by Vincent, were made over a period of fifteen years using suspended mass accelerometers located at a number of locations shown in the general layout of the bridge in Fig. 4.

Figure 6 shows a reconstruction of four of the accelerometer traces. These were achieved by digitizing the original records and replotting the results. Time marks provided every five minutes assisted in the difficult task of synchronizing the records. The records depicted in Fig. 6 were made during a violent storm on June 6, 1950 with a westerly wind of approximately

38 mph (17 m/s) mean speed from a direction perpendicular to the bridge axis. The maximum double amplitude at the centre of the main span was roughly 45 inches (114 cm).

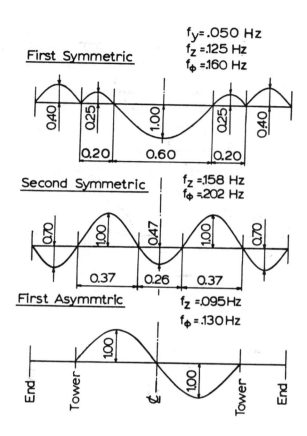

Fig. 5   First Three Vibration Modes of Golden Gate Bridge (After Vincent)

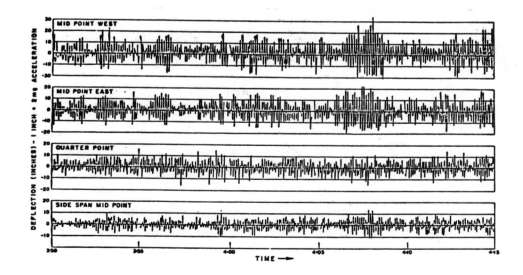

Fig. 6   Oscillograph Recprds of Vertical Acceleration of Golden Gate Bridge During Storm of June 6, 1950

Vincent's observations on the bridge motion and more detailed analysis of records such as those in Fig. 6 suggests the following. The motion of the bridge consists primarily of the symmetric and the first anti-symmetric modes of vibration. The former dominates the motion at the mid span and the latter, the motion at the quarter points. Most of the time vertical motion was more significant than the torsional motion although the west side movements were generally larger than the east side suggesting some degree of coupling. The accelerations due to the first anti-symmetric mode at the quarter point were significantly less than the accelerations in the first symmetric mode at the mid span. Both of these features which had held true during almost the entire life of the bridge were reversed during the storm of December 1, 1951. On this occasion the wind speed rose to nearly 45 mph (20 m/s) more or less perpendicular to the bridge and the bridge motion was dominated by anti-symmetric torsional motion reaching a double amplitude at the quarter point of approximately 145 inches (368 cm). Superficial damage to the bridge led to the addition of a lower lateral system to increase the torsional stiffness.

The transition from dominantly symmetric vertical motion to anti-symmetric torsional motion was clearly a dramatic feature of the aerodynamic behaviour. Vincent attributed it in part to the protection afforded by hills to the north of the bridge which might produce strongly asymmetric flow on the bridge. A more important contributing factor was perhaps the fact that the asymmetric frequencies were the lowest and hence the first to become unstable. Vincent also pointed out that because of the very long span the possible opportunities for "variation in the wind velocity, direction and behaviour" would be greater than over a short span bridge.

3. COMPARISON OF TEST RESULTS WITH BRIDGE RESPONSES

The taut strip model was examined both by Miyata and later by Tanaka who made a number of corrections to the earlier experiments. Figs. 7 and 8 show comparisons of model response spectra for different levels of turbulence with those of the Golden Gate Bridge. All amplitude scaling is converted to full scale using model properties and calculated bridge frequencies: the frequency scale has not been converted. The reason for the latter is that the matching of the observed and calculated frequenices is not exact. The suggestion is that in the first symmetric modes, the observed frequencies are slightly (approximately 5 percent) higher.

A comparison is also made of response amplitudes in Fig. 9. Fig. 9 (a) summarizes the vertical oscillations observed on the bridge on a number of occasions. Fig 9 (b) corresponds to the incident of December 1, 1951 with the asymmetric torsional motion.

It should be noted again that the structural damping of the bridge is still an uncertain factor. Especially when the bridge shows aerodynamic instability as is seen in Fig. 9 (b), the critical velocity is sensitive to the structural damping as well as turbulence level. Therefore, the exact knowledge of damping characteristics is vitally important in estimating the full scale bridge behaviour from the wind tunnel test results.

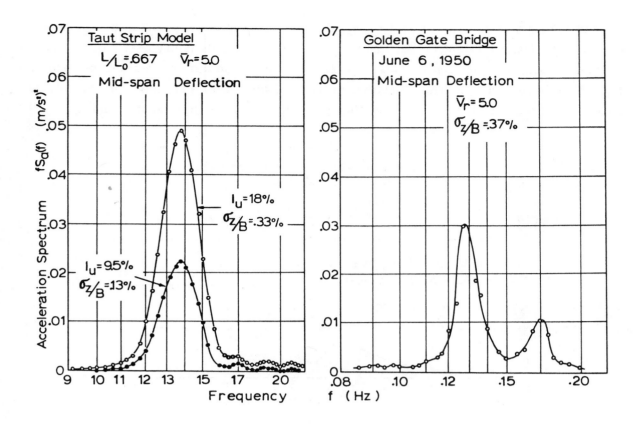

Fig. 7 Comparison of Vertical Response Spectra

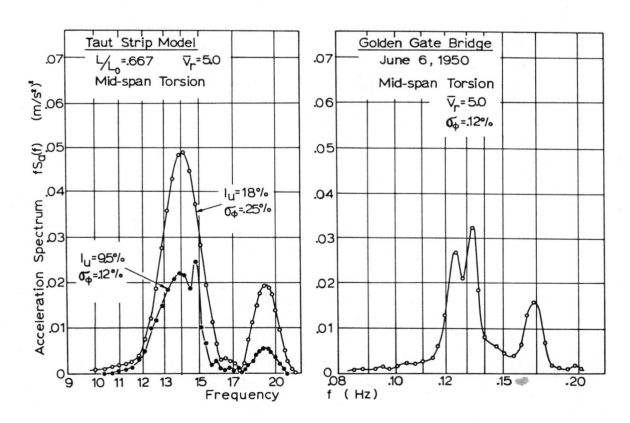

Fig. 8 Comparison of Torsional Response Spectra

For comparison given in Fig. 9 (b), the torsional damping is left at the measured value ($\zeta_T = 0.7\%$). The agreement in Fig. 19 (a) seems to be quite satisfactory. Most of the measured amplitudes are embraced by two response curves based on different turbulence conditions. It should be noted that the wind directions in the field observations are rather imprecisely defined, and off normal to the bridge axis, they generally reduce or suppress instability.

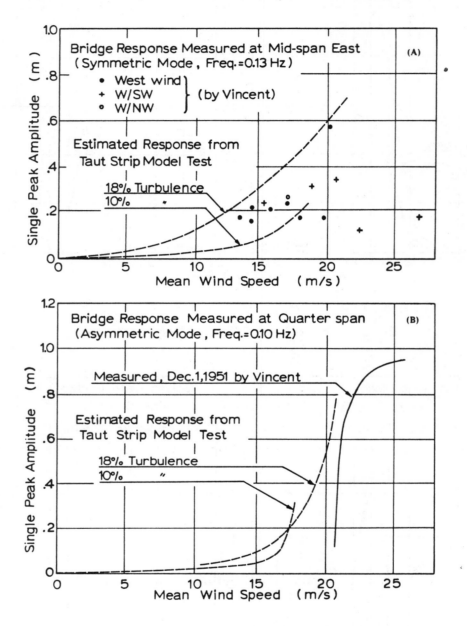

Figs. 9 (a) & (b)   **Comparison of Response Amplitudes of Symmetric and Asymmetric Modes**

The vertical damping of the model was approximately 0.3% which is undoubtedly somewhat less than in reality. However, as the aerodynamic damping in vertical motion is relatively high, the effect of structural damping on response amplitude is expected to be small. In Fig. 9 (a) the peak amplitude is assumed to be 3.6 times rms amplitude.

Although the comparisons are seen to be reasonably good there is still room for further nailing down of certain key parameters such as the structural and aerodynamic damping.

4. BRONX WHITESTONE BRIDGE

This section summarizes a study of another particularly interesting bridge in which both different wind tunnel testing procedures and full scale observations were compared. This bridge has a 700 m center span suspension bridge across the East River in New York City. The siting and overall dimensions are shown in Fig. 10. The bridge was built as a girder stiffened suspended structure carrying four roadway lanes and two sidewalks. Because the bridge vibrated in moderatley strong winds during construction, modifications were made to the design before it was opened to traffic. These comprised the addition of inclined stay cables from tower tops to stiffening girders and the installation of a frictional device in the tower. From its opening in 1939 until its traffic capacity was expanded in 1946, the bridge functioned satisfactorily, although it underwent occasional episodes of minor oscillation. The oscillations which occurred with winds between 18 and 27 m/sec lasted as long as the winds blew strongly and ranged in amplitude up to about 40 cm.

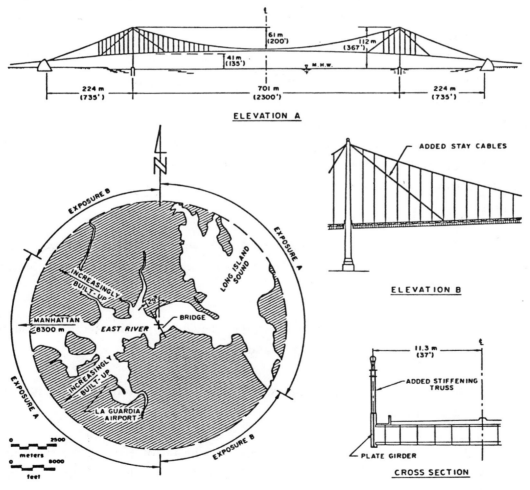

Fig. 10   Location and Details of Bridge

During this period, which followed the failure of the first Tacoma Narrows Bridge, the bridge was kept under close surveillance by its designer, O. H. Ammann, who was also a member of the board investigating the collapse of the Tacoma Narrows Bridge. In view of the similarity of the sections of these two bridges, section model wind tunnel tests were made at the University of Washington. Typical test results indicated that the bridge could not survive in winds over about 20 m/s, whereas the bridge in fact had often withstood far greater winds with minor vibration and no damage. As a result, when it became necessary to modify the structure in order to increase its traffic capacity, Ammann gave little weight to these wind tunnel tests. He instead based the alterations on a Stiffness Index which he developed and which correlated well with observed bridge performance.

In modifying the bridge, the sidewalks were removed, the roadway was widened to six lanes and a stiffening truss was built on top of the existing girders, see cross-section in Fig. 10. This is the present configuration of the bridge. The bridge has functioned well for the 32 years since it was modified, although it still has had episodes of minor vibration during strong winds.

5. FULL SCALE MEASUREMENTS

5.1 General

Full scale observations were made of the vertical acceleration at the mid and quarter points of the main span and of the wind speed and direction near the middle of the main span. At both locations the accelerometers are positioned at each of the two stiffening trusses. Their output is passed through integrating circuits and the resulting vertical displacements are record-ed on a strip chart recorder. Summing and differencing the total displacment at the upstream and downstream stiffening trusses provides measurements of the vertical and torsion bridge motions.

Continuous strip chart records for six periods with significant wind speeds and wind direct-ions approximately normal to the bridge centreline were selected for detailed analysis. The objectives were,

i)   to provide comparisons of full scale behaviour with the findings of the wind tunnel studies for this bridge; and

ii)   to fine tune existing estimates of the dynamic properties, in particular the structural damping.

Wind speeds are given as 10 min. average at a height of 10 m above the bridge deck. A summary of the selected records is given in Figure 11. The wind direction was near normal to the bridge for every cases. Records of the bridge motion at the mid-point of the main span for these storms were digitized at a sampling rate of 2 Hz. For the first symmetric vertical and torsional modes this corresponds to approximately 11 and 7 equally spaced samples per cycle of oscillation respectively. Typical traces of the bridge motion, reconstructed from these digitized records, are presented in Figure 11. The subsequent analysis of the digitized records included evaluations of various statistical properties of the response, calculations of auto-correlation functions and special analysis.

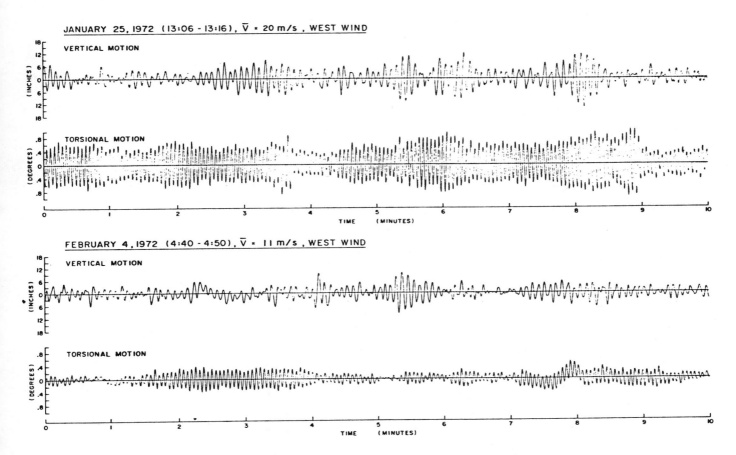

Fig. 11 Typical Time Records of Wind Induced Motion at Mid-Span of Bridge

5.2 Response Estimates

The dynamic amplitudes of the bridge motion during the various storms, summarized in Figure 12, were computed using both 5 and 10 minute averaging times. Also shown are the intervals of the records during which the mean wind speed remained approximately constant. The variation of the RMS response approximately follows changes in recorded wind speed and the RMS estimates based on 5 and 10 minute averaging times are in most cases similar. This suggests that the response over most parts of the records can be treated as locally stationary over a 10 minute period.

Probability distributions of the dynamic response for all storms are presented in Figure 13. The parent distributions of both vertical and torsional motions are seen to be approximately Gaussian. Also shown in Figure 13 are the distributions of observed 10 minute extreme values.

632

The average values of these extreme value distributions are approximately 3.4 and 3.1 standard deviations above the parent mean for vertical and torsional motions respectively. These observed average peak factors agree well with theoretical predictions. As shown elsewhere by Davenport, the average factor of extreme values over a sampling interval T for a narrow-band Gaussian distribution is,

$$\bar{g} = \mu + \frac{\gamma \mu}{\mu^2 - 1}, \quad \text{where} \quad \mu = \sqrt{2} \ln \nu T + \frac{\ln \sqrt{2} \ln \nu T}{\sqrt{2} \ln \nu T}$$

is the mode of extreme values, ν is the effective cycling rate and $\gamma = .5772$.

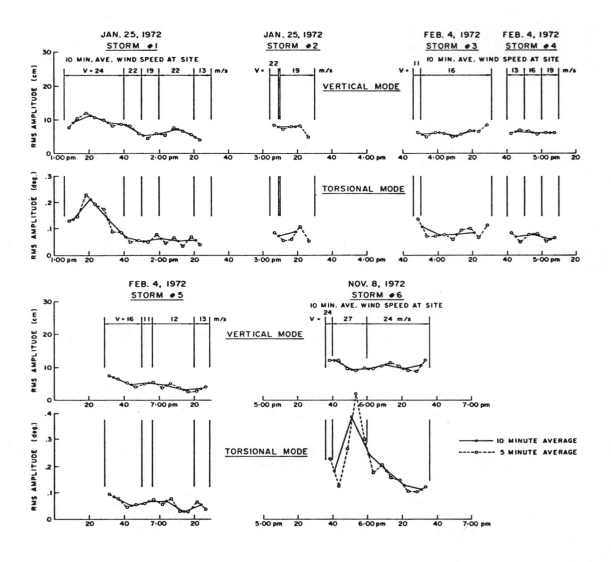

Fig. 12   Summary of Storm Data

Taking $\nu \approx n_o \sqrt{8\zeta}$, where n_o is the natural frequency and ζ is the total damping.

theoretical estimates of the g for $T = 10$ minutes are

$$\overline{g} = (vertical)$$

$$\overline{g} = (torsional)$$

This relatively close agreement with observed values suggests that both the vertical and torsional response of the bridge over the observed range of wind speed remain narrow-band Gaussian. It is also noteworthy that the observed full scale value of g are consistent with those observed in the taut strip model study.

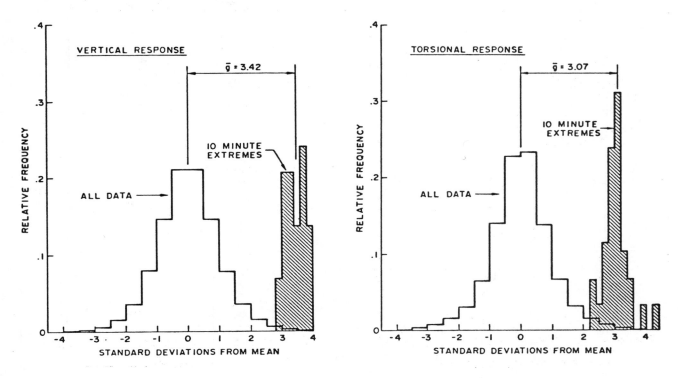

Fig. 13   Histograms of Full Scale Motions at Bridge Centreline

Typical power spectra of the full scale motion are presented in Figure 14. Most of the energy is concentrated at the 1st symmetric vertical and torsional modes of vibration with negligible contributions from higher harmonics. The natural frequencies estimated for the 1st symmetric vertical and torsional modes agree very well with those obtained in the earlier ambient vibration measurements and assumed in the wind tunnel study. The agreement is also excellent for the higher modes determined from these spectral estimates. For example, the 2nd symmetric torsional and the 2nd and 4th symmetric vertical frequencies obtained in the ambient vibration study were 0.422, 0.300 and 0.558 Hz respectively.

6.   COMPARISON WITH WIND TUNNEL FINDINGS

Figure 15 compares the results of the full scale measurements with the taut strip and section data for several configurations of damping and turbulence intensity. The full scale results are the peak values for the various records listed in Table 3. The model results effectively bracket the damping and turbulence levels anticipated for the bridge, with probable values being:

| Damping | Vertical | 1.2 – 2.0% | Intensity | West | 15 – 20% |
| | Torsional | 1.0 – 1.6% | | East | 10 – 15% |

These values of damping are those suggested earlier in the study; the turbulence levels are suggested by the upstream exposures for the two directions.

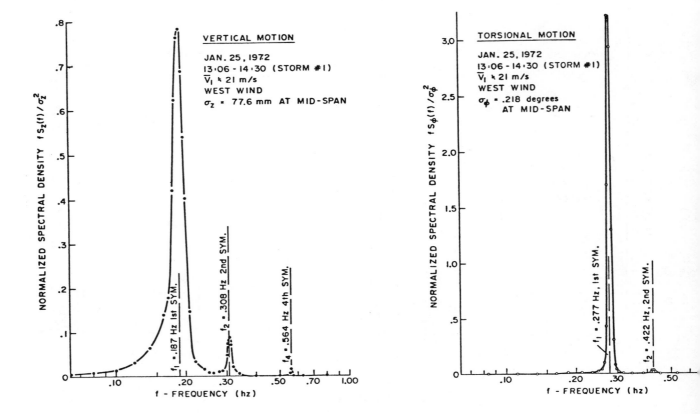

Fig. 14   Typical Full Scale Response Spectra at Mid-Span

Some comments on the factors which contribute to the variability of the full scale data are in order. Some of the scatter undoubtedly results due to the intrinsic variability of the structure of strong winds which can only be approximated in the wind tunnel. Other factors affecting the scatter, however, are inherent to the actual observations. These include,

a)   the fact that the time intervals available for analysis were not identical;

b)   wind directions were nominal and some records selected may have been for slightly yawed winds;

c) the wind speeds in some cases were not known precisely; and

d) the estimates are point estimates and therefore contain intrinsic statistical variability in estimating the peak values.

Despite the scatter there is a strong indication that these wind tunnel tests in turbulent flow are consistent with full scale. This is not the case for section model tests conducted in smooth flow.

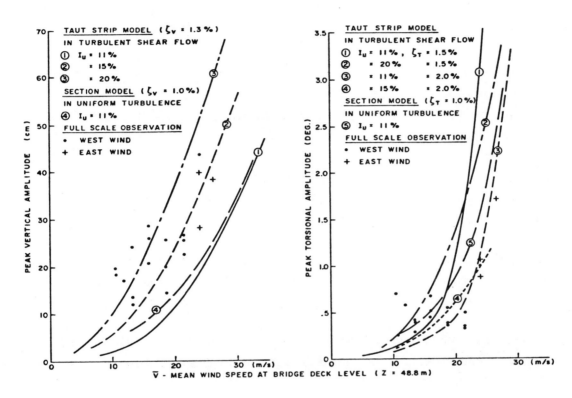

Fig. 15   **Summary of Wind Tunnel and Full Scale Bridge Response, Wind Normal to Bridge Centreline**

ACKNOWLEDGEMENTS

The writer acknowledges the assistance of the Golden Gate Bridge Authority, the Triborough Bridge Authority and H. Rothman of Paul Weidlinger & Associates in making available the records on which the full scale performance is based. The writer also acknowledges the work put into these studies by numerous colleagues past and present in particular Dr. N. Isyumov, Dr. H. Tanaka, Dr. T. Miyata, Dr. I. Hirai and Dr. M. Hogan.

REFERENCES

Davenport, A. G., Isyumov, N., Fader, D. J. and Bowen, C. F. P., "A Study of Wind Action on a Suspension Bridge During Erection and on Completion, U.W.O., Eng. Sc. Res. Rept., BLWT–3–1969, 1969.

Davenport, A. G., Isyumov, N. and Miyata, T., "The Experimental Determination of the Response of Suspension Bridges to Turbulent Flow, Proc. 3rd Int. Conf. on Wind Effects on Bldgs. and Struct., Tokyo, 1971.

Davenport, A. G., "The use of Taut Strip Models in the Prediction of the Response of Long Span Bridges to Turbulent Wind, Proc. Symp. on Flow Induced Struct. Vibrations, Karlsruhe, 1972.

Davenport, A. G. and Tanaka, H., "The Aerodynamic Stability of the Golden Gate Bridge: Comparison of the Full Scale Bridge Motion with Taut Strip Wind Tunnel Model", presented at Symp. of Full Scale Measurements of Wind Effects on Tall Bldgs. and Other Structures, London (Ont.), 1974.

Davenport, A. G., Isyumov, N., Rothman, H. and Tanaka, H., "Wind Induced Response of Suspension Bridges – Wind Tunnel Model and Full Scale Observations", Proc. 5th Int. Conf. on Wind Engineering, Fort Collins, Colorado, 1979.

Farquharson, F. B and Vincent, G. S., "Aerodynamic Stability of Suspension Bridges with Special Reference to the Tacoma Narrows Bridge, Part V", Univ. Washington, Eng. Expt. Station, Bulletin 116, 1954.

Holmes, J. D., "Prediction of the Response of a Cable-Stayed Bridge to Turbulence", Proc. 4th Int. Conf. on Wind Effects on Buildings and Structures, Heathrow, U.K., September 1975, Cambridge University Press, 187–197, 1977.

Irwin, H. P. and Schuyler, G. D., "Experiments of the Full Aeroelastic Model of Lion's Gate Bridge in Smooth and Turbulent Flow", N.R.C. Report LTR–LA–206, 1977.

Melbourne, W. H., "Model and Full Scale Response to Wind Action of the Cable Stayed Box Girder West Gate Bridge", Practical Experience With Flow-Induced Vibrations, Symp. Karlsruhe/ Germany, pp. 625–632, Sept. 3-6, 1979.

Okauchi, I., Tayima, J. and Akiyama, H., "Response of Large Scale Bridge Model to Natural Wind", Proc. 5th Int. Conf. on Wind Engineering, Fort Collins, Colorado, U.S.A., July 1979.

Scanlan, R. H., and Gade, R. H., "Motion of Suspended Bridge Spans Under Gusty Wind", J. Struct. Div. ASCE Vol. 103, No. ST9, 1867–1883, (1977).

Tanaka, H. and Davenport, A. G., "Response of Taut Strip Models to Turbulent Wind", J. of Eng. Mech. Div. ASCE, February 1982.

COMPARISON OF MODEL AND FULL SCALE TESTS OF A BRIDGE AND A CHIMNEY STACK

Professor W.H. Melbourne, Department of Mechanical Engineering, Monash University, Australia

1. INTRODUCTION

To the uninitiated it would seem a relatively straightforward exercise to measure the response of a structure to wind action and to then compare these measurements with those made at model scale. Those who attended the meeting on "Full scale measurements of wind effects on tall buildings and other structures", University of Western Ontario, June 1974, and who subsequently counted the actual number of programmes which resulted in producing readily comparable generalised results can testify that there must be a few hidden problems. Not the least of these problems is that of obtaining accurate estimates of the freestream wind conditions. However, the purpose of this paper is not really to lament on the difficulties but is rather to highlight the need for full scale validation of physical and mathematical models and to present a preview of two such studies. The model and full scale measurement programmes on the cable-stayed, box girder, West Gate Bridge and on the 265 m reinforced concrete chimney stack at Mt. Isa are virtually complete. One will serve to illustrate the situation where failing to model excitation mechanisms which occur in full scale and vice versa can lead to lack of confidence in the model scale tests, and the other will reinforce our knowledge that we do not yet know everything about Reynolds number and turbulence effects.

2. WHY THE NEED FOR FULL SCALE TESTING

The obvious is simply that virtually all models require simplifying assumptions to be made about what forces and mechanisms are important and these assumptions need to be validated before confidence in the model can be established. The real situation in the wind engineering field is somewhat more complex because the old fluid mechanics bogey of being unable to adequately model the ratio of viscous to inertia forces both in respect of turbulence characteristics and shear (boundary) layer stability. This scaling inequality leads to the situation where we know there are severe Reynolds number problems associated with flow around curved surfaces and to the lingering doubt that even when separation is fixed by a sharp corner that there might be some unknown, but significant, shear layer behaviour which is Reynolds number dependent.

It would be fair to say that the great majority of scaling problems in the wind engineering field finally get back to Reynolds number scaling problems in one form or another, and which most of us cannot avoid. Hence, full scale testing in some of the wind engineering areas is not needed just for validating purposes, but to supply actual parameter values which cannot be adequately modelled.

3. WEST GATE BRIDGE

Wind tunnel tests were conducted on a scale model of the full West Gate Bridge and later full scale response measurements were made, these

have been reported in part by Melbourne (1, 2). Some section model
measurements were made in smooth flow by Vickery (3) and in turbulent flow
by Walker (4).

As shown in Figure 1, the full scale West Gate Bridge deck is 846 m
long with a centre span of 336 m. The box, including cantilevered
roadways, has a chord of 37.4 m and an effective aerodynamic depth of 4.1
m. The first vertical bending mode frequency, from the model tests, was
estimated to be 0.33 Hz.

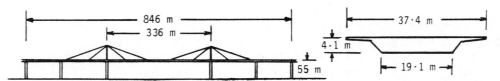

Figure 1. Aerodynamic outline dimensions of West Gate Bridge

The vertical displacement response at centre deck is given in Figure 2
for the full model and a sectional model tested in the turbulent wind model
and for a sectional model tested in steady smooth flow. Both the full and
sectional models tested in turbulent flow showed a steady increase of
response with mean wind speed, oscillating in the first vertical bending
mode. Response in the first torsional mode was very small and in the full
model it was difficult to detect any response other than the first vertical
bending mode. There were no indications of any critical velocity effects.
The sectional model in smooth flow showed little response until the deck

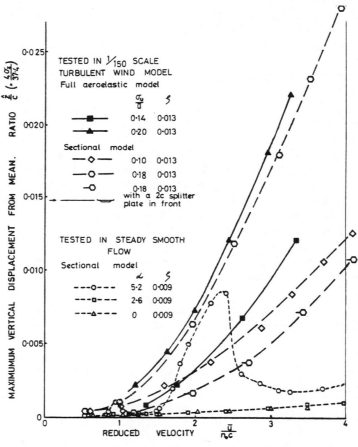

Figure 2. West Gate Bridge vertical displacement at centre deck from full
and sectional model tests in a turbulent wind model and in smooth flow.

was set at an angle of attack of 5.2°, at which, response to strong vortex shedding wake excitation was very evident, (flow was separating from the top leading edge without reattachment). It was concluded that this response in smooth flow was not related in any way to that shown by the models in turbulent flow or what was expected in full scale.

Although there have been several examples of bridge decks being excited by vortex excitation in low turbulence, relatively low velocity, wind flows, (notably the Long's Creek Bridge, Wardlaw 5, and the Wye Bridge, Smith 6), any attempt to predict maximum response of long span bridges for design purposes, by testing either full or sectional models in other than correctly modelled turbulent flows, is to ignore the very mechanisms likely to cause the maximum response.

From the model tests in turbulent flow in two levels of turbulent intensity and theoretical estimates by Holmes (7) it was concluded that most of the cross-wind excitation for this slender deck could be attributed to incident turbulence with some lesser contribution from broad band wake excitation. However, one aspect which did not tally with the theoretical estimates was the effect of turbulence on the deck response for which the theory predicts linear dependence of response with turbulence intensity. The full bridge model data was fitted by the following equation showing the functional dependence of the vertical response, at centre deck, $\hat{z}$, normalised by deck width, c, on reduced velocity, turbulence intensity, damping and wind angle to the deck normal.

$$\frac{\hat{z}}{c} = 0.0176 \left| \frac{\overline{u}}{n_o c} \right|^{2.5} \left| \frac{\sigma_u}{\overline{u}} \right|^{2.2} \left| \frac{1}{\zeta} \right|^{0.2} \cos \beta \qquad (1)$$

Approximately half of the full scale data have now been reduced. Whilst it is too early to draw very precise conclusions, the trends are relatively clear. Generally the full scale response is a little greater than predicted by the model studies. Above a turbulence intensity of 10% the response is increasing with more than a linear dependence on turbulence intensity. The full scale data is quite variable and even when all the data are analysed it may not be easy to draw precise conclusions.

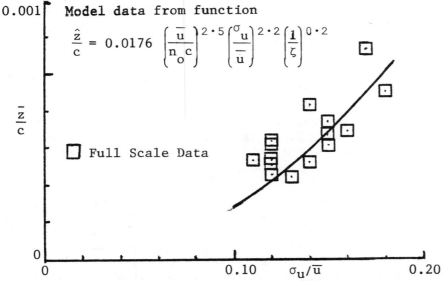

Figure 3. Full scale average hourly maximum centre deck vertical displacement as a function of the turbulence intensity of the incident wind flow, at $\overline{u}/n_o c = 0.1$, $\zeta = 0.01$, $\beta = 0^0$

Here then is a case where the physical model did predict quite different response characteristics to that predicted by the theoretical models. A number of possible explanations for the underprediction of response in more turbulent flows by the theoretical model have been followed up by the author. The explanation appears to relate to the quasi-steady assumption that the lateral correlation of the fluctuating lift forces is directly related to the lateral correlation of the incident flow longitudinal or vertical components. There is now ample evidence of the mechanism by which negative pressures are created under reattaching shear layers (Melbourne 8) and the lateral extent (correlation) of these pressure fields is very little related to lateral velocity correlations. Evidence for the difference in these lateral correlations is currently being put together by the author as characterised in Figure 4.

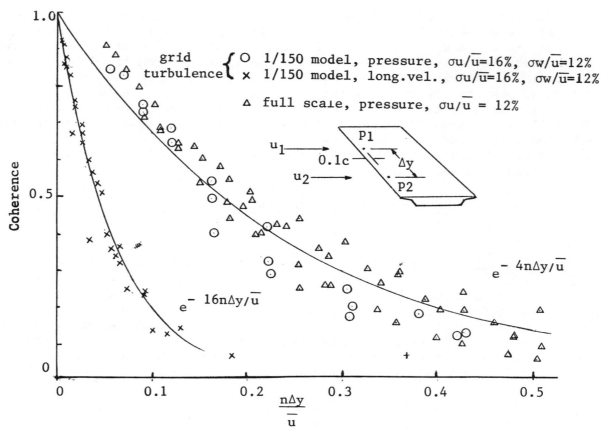

Figure 4   Lateral coherence of deck pressures and longitudinal velocity component in model and full scale.

The conclusions to be made from the comparison of model (physical and theoretical) and full scale studies is that one must be very careful to ensure that the model permits the full scale excitation mechanisms to exist at model scale. Whilst this may appear trivial or axiomatic, let us reflect on the many tens of thousands of dollars spent on tests in smooth flow in quite recent years. In the case of bridge decks, it might be said, in defence, that they produced conservative estimates, whilst in the case of determining surface pressure on tall buildings, the disastrous consequences of under-predictions of design loads may well be heard through the annals of litigation for many years to come. There are currently texts and many wind loading codes still on sale which purport to contain design information, but which, in fact, have pressure coefficients based on measurements in smooth flow.

4.   MT. ISA CHIMNEY STACK

In 1978 a 265 m high reinforced concrete stack was built at Mt. Isa
for the Lead Smelter Plant of Mt. Isa Mines Ltd. Tileman (A/sia) Pty. Ltd.
were the contracted designers and builders. In 1977 aeroelastic model
studies of the proposed stack were undertaken at Monash University to
assist in the design. Following completion of the stack Mt. Isa Mines Ltd.
agreed to support a programme to measure the response of the stack to wind
action. Mt. Isa Mines Ltd. staff installed accelerometers, pressure
transducers and strain gauges in 1979 and monitored their outputs through
six storms from 24th October 1979 to 24th February 1980. These data and
meteorological data were analysed at Monash University with assistance from
Tileman (A/sia) Pty. Ltd. during 1980, including topographical modelling of
the site to enable reduction of wind data over the height of the stack from
anemometer data measured at several lower level sites.

The objective of the full scale measurements was to obtain data which
could be used to compare with physical and mathematical model simulations
of the response of large chimney stacks, in particular those being undertaken at Monash University. The full results of these studies (Melbourne,
Cheung & Goddard) are as yet unpublished. However, it is possible to
present a limited amount of the data here to illustrate the problems of
scale in that complex region where curved surfaces are concerned.

4.1   The Model Studies

It was necessary for practical, structural reasons to build an aeroelastic model of the stack to as large a scale as possible. However, it is
difficult to develop a large scale wind model with accurately scaled
turbulent length scales through the full boundary layer depth. In this
case a compromise was reached by using a model to full scale length ratio
of $L_r = 1/200$ and because model wind speeds in excess of 15 m/s were
required, the 2 x 2 m working section had to be used.

The characteristics of the wind model used are given in Figure 5.
Because of the effect turbulence has on flow around a circular cylinder,
that parameter dominated the final selection of the wind model used. As
can be seen in Figure 5, the turbulence intensity over the top half of the
stack is close to that required for open country terrain and over the lower
half approaches that for built-up terrain, which are the conditions for the
strong wind directions at Mt. Isa.

There was thought to be a slight possibility that turbulence intensities over the top half might be less than those described above and hence
a few check measurements were made in a lower turbulence latter wind model,
the characteristics of which are given also in Figure 5. This latter wind
model was called the Low Turbulence Wind Model to differentiate it from the
previously described wind model used for most of the test programme and
which will be called the High Turbulence Wind Model.

The full scale configuration of the stack as modelled was a little
different to that subsequently built. The modelled stack tapered almost
linearly from 21.8 m diameter at the base to 10.7 m diameter at the top
(265 m), whereas the stack as built tapered from 21.8 m diameter at the
base to 12.3 m at a height of 205 m and was then parallel to the top (265
m). The reinforced concrete wall thickness varied from about 650 mm at the
base to 230 mm at the top. The Youngs modulus for the reinforced concrete

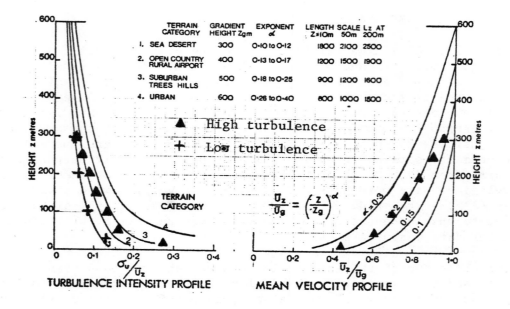

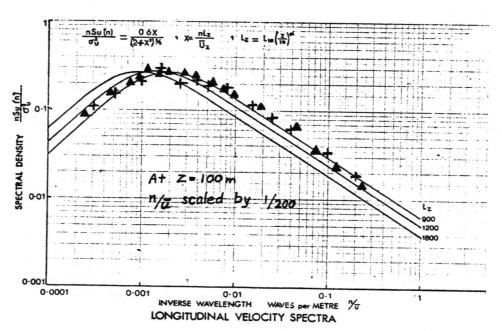

Figure 5   1/200 scale natural wind models used for the Mt. Isa Stack tests

wall was taken at 2.6×10^{10} N/m^2 and the density as 2400 kg/m^3. The full scale frequency of the modelled stack was calculated to be 0.27 Hz.

The aeroelastic model of the stack was defined by the need to keep the ratio of inertia to stiffness forces constant between model and full scale. This requirement can be expressed as

$$\frac{\rho \overline{u}^2}{EI/L^4} = \text{Const}$$

or

$$\frac{\rho_r \overline{u}_r^2 L_r^4}{E_r I_r} = 1$$

where
ρ = density
$\bar{u}$ = mean wind velocity
E = Youngs modulus of elasticity
I = 2nd moment of area
L = length

subscript r refers to a ratio of model value over full scale value.

To ensure that the fluid and structure inertia forces are kept in the same ratio, it was necessary that ρ_{air}/ρ_{stack} be kept constant. Because the density of the air is the same in model and full scale, i.e. $\rho_r = 1$, the density of the model stack must be kept the same as in full scale. Sometimes this density requirement can be kept by using an equivalent lumped mass system in the model. However, in the case of modelling reinforced concrete structures, there is an epoxy base material, used for repair work, marketed under the trade name Devcon B, which has a similar density to reinforced concrete (2400 kg/m^3) and which is machinable. The model of the 265 m stack wind shield was made by plastering Devcon B onto a mandrel to the scaled inside dimensions of the stack and then machining externally to give the required wall thickness and outer diameter.

The full scale and model properties used and resultant scaling ratios were as follows:

$$E_{Full\ scale} = 2.6 \times 10^{10} \text{ N/m}^2$$

$$E_{model} = 5 \times 10^9 \text{ N/m}^2$$
(measured on test sample)

$$\therefore E_r = \frac{5 \times 10^9}{2.6 \times 10^{10}} = \frac{1}{5}$$

for
$$\rho_r = 1$$

$$L_r = 1/200$$

$$I_r = L_r^4 \quad (\text{because } \rho_r = 1)$$

$$E_r = 1/5$$

from equation 1 the velocity scale ratio is given by

$$\bar{u}_r^2 = \frac{E_r I_r}{\rho_r L_r^4} = 1/5$$

$$\therefore \quad \bar{u}_r = 1/\sqrt{5} = 0.45$$

also the time scale ratio is given by

$$T_r = L_r/\bar{u}_r = 0.0112$$

and the moment scale ratio is given by

$$M_r = \rho_r L_r^3 \bar{u}_r^2 = 25 \times 10^{-9}$$

Strain gauges were glued onto the outside of the shell near to the base of the model stack and the output from two four gauge bridges gave base overturning moment in along-wind and cross-wind directions. A photograph of the model is shown in Figure 6, which also shows a simplified linear mode, model of the existing 76 m steel stack, which was included to check the interference effects of building the new stack close by.

Figure 6   1/200 scale model of the 265 m Mt. Isa Mines Lead Smelter Stack

4.1.1 Model Measurements and Results

The general model testing procedure consisted of measuring along-wind and cross-wind base overturning moments on the model stacks for various values of freestream mean wind velocity at the height of the top of the new stack wind shield (265 m) and various values of the structural damping. In each case the mean and standard deviation of the moments were measured, and from an on-line computer analysis the probability distributions and evaluation of the peak factors were obtained to facilitate an estimate of the peak moments (maximum per hour on average).

The base overturning moments for the 265 m windshield for the A2 Stack as a function of wind belocity and damping in the High Turbulence Wind Model is given in Figure 7. The cross-wind response shows a rapid rise to

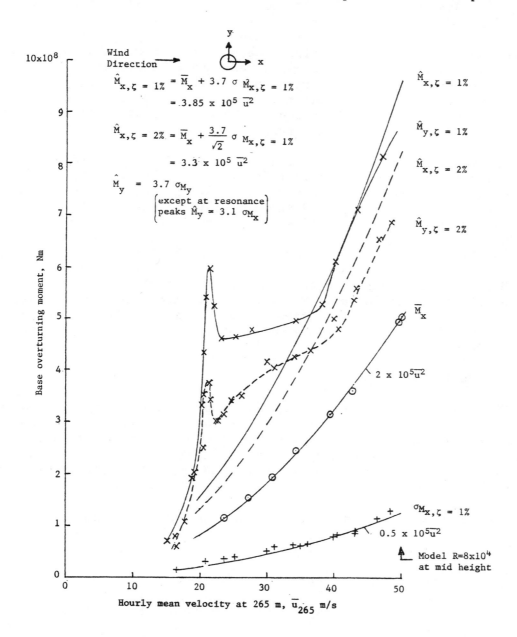

Figure 7 Mt. Isa 265 m (A2) Stack – Base overturing moments as a function of wind velocity and damping, n_o=0.28 Hz high turbulence wind model

a resonance peak (due to wake excitation) at a mean reduced frequency of,

$$\frac{n_o \bar{D}}{\bar{u}} = \frac{0.28 \cdot 15.5}{20.6} = 0.21$$

The more relevant diameter over the top half of the stack is about 13 m, which gives a reduced frequency of 0.18. The along-wind moments rise proportional to the square of the velocity. At the design criterion of $\bar{u}_{265}$ = 36 m/s, and ζ = 1%, the along-wind and cross-wind peak moments are effectively the same for design purposes.

The cross-wind moments were measured for critical damping ratios of ζ = 1% and 2% (using hanging chain dampers inside the model stack). The along-wind moments were measured for ζ = 1% only and for limit state calculation a moment curve for ζ = 2% has been estimated on the assumption that the standard deviation of response is inversely proportional to the square root of the damping ratio. The peak factor was about 3.7 for all conditions except for the cross-wind response at and near the resonant peaks where the lowest values were about 3.1. Whilst the lower values of peak factor indicated the usual move towards dependence of variables in the wake excitation process, there was no suggestion that for a damping ratio of 1% that a critical lock-in situation was close at hand.

In Figure 8 the mean base pressure as a function of stack height has been given in pressure coefficient form. It can be seen that at R = 4 x 10^4 in the Low Turbulence Wind Model the base pressure is much more negative than would be expected in full scale at R = 10^6 to 10^7, but not as much as for classical low turbulence sub-critical flow. However, the base pressure in the High Turbulence Wind Model, even though at nominally sub-critical Reynolds numbers, is very nearly in the full scale range. This leads to the conclusion that the flow conditions with the High Turbulence Wind Model could be expected to be close to that which are likely to occur in the full scale situation. The mean along-wind base overturning moment would still be slightly overestimated; but comment with respect to the scale effect on the cross-wind moments will be made after discussion of the full scale data.

4.2   The Full Scale Studies

The Lead Smelter Stack at Mt. Isa Mines is a 265 m reinforced concrete wind shield with provision for two flues, although only one was installed initially. The outside diameter tapers from 21.85 m at the base to 12.27 m at a height of 205 m and then remains constant over the top 60 m. Measurements during six storms were made on the full scale stack from the following instrumentation:

(i)   2 accelerometers, NS and EW mounted at level 258 m.

(ii)   4 strain gauges at level 43 m.

(iii)   4 pressure transducers at level 219 m.

(iv)   7 anemometers and wind direction vanes, the most important being a propeller vane anemometer mounted 79.3 m above ground level on a mine headframe, a distance of 270 m west of the stack.

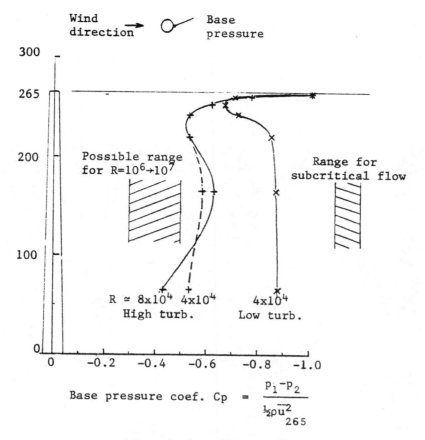

$$Cp = \frac{P_1 - P_2}{\frac{1}{2}\rho \overline{u}^2_{265}}$$

Figure 8 Mt. Isa 265 m(A2) Stack – Base pressures measured
in the high and low turbulence wind models

A seven channel tape recorder and ten channel UV recorder were used to
record outputs from the accelerometers and in varying combinations the
strain gauges and pressure transducers during each storm. As the duration
of each storm was of the order of an hour it was possible to get outputs
for relatively low wind speeds, and calibration levels were included on
every record. The accelerometers were calibrated after the measurement
series at the first mode frequency. The strain gauges could only be
calibrated by reference to the accelerometers. The pressure transducers
were calibrated before and after the measurement series and showed no
significant change. The anemometers were part of a most extensive
meteorological monitoring network at the Mt. Isa Mines and mean and
standard deviation wind speeds averaged over 5 minute periods were directly
available for all stations.

Without any doubt the most difficult data to establish at heights
between 200 and 300 m are the wind speed data. Because of the availability
of excellent anemometer systems at Mt. Isa it was decided not to install
any other anemometers on the stack, which even with extensive (and
expensive) support systems would still have required position error
calibrations. To use the existing anemometers it had to be assumed that
wind speeds up to 270 m could be extrapolated on the basis of profiles
dominated by surface roughness conditions. This assumption has been made,
with the implicit inaccuracies associated with non-stationarity and thermal
effects. However, considerable care has been taken to use data records
from storms of over thirty minute duration which built up steadily in wind
speed, and for which there was sensible correlation from at least one of
the lower level anemometers with the main Headframe Anemomter at 79 m near
the stack.

648

The actual profiles of wind speed and turbulence intensity were determined from a topograhical model study carried out in the 4 x 3 m working section of the Monash University wind tunnel. A photograph of the 1/500 scale model is shown in Figure 9.

Figure 9   1/500 topograhcal model of Mt. Isa Mines, used to obtain wind speed and turbulence intensity profile data

4.2.1 Full Scale Measurements

Only .the results of the top displacement measurements as obtained from the accelerometers will be discussed here. Examples of the one minute x-y traces from the accelerometers which were used to evaluate the cross-wind and along-wind response and the wind direction are given in Figure 10. These records showed that the stack response was almost totally in the first mode with a frequency of n=0.263 Hz; higher modes are just apparent. The displacements at the top of the 265 m windshield, which hereafter will be referred to as top displacements, were calculated, using the assumption that motion was simple harmonic in the first mode giving

$$\ddot{x} = -\omega^2 x \qquad\qquad (2)$$

where x is displacement

             $\ddot{x}$ is acceleration

             ω is circular frequency $(= 2\pi n)$

Corrections of 2.5% and 4.9% were made to account for accelerometer tilt and for the accelerometers being 7 m below the top, respectively.

During the storm of the 5th December 1979 we have the only evidence of the wind speed clearly exceeding the critical wind speed for cross-wind response. With reference to the traces in Figure 10 and the peak displacements given in Figure 11, the events during that storm can be described as follows:

Time
(24 hour clock)

20.55 to 21.00	Wind speed rising; small cross-wind and along-wind displacements.
21.00 to 21.05	5 minute mean wind speed rising to 20 m/s; response now dominated by cross-wind displacement with maximum mean to peak displacement of 116 mm compared with along-wind maximum peak to peak displacement of 18 mm.
21.05 to 21.10	Wind speed continuing to rise, to the maximum 5 minute mean wind speed of 23 m/s; peak cross-wind displacements beginning to fall to about 70 mm whilst along-wind displacements continuing to rise to a maximum mean to peak displacement of 44 mm.
21.10 to 21.20	Wind speed dropping relatively quickly to 5 minute mean of less than 10 m/s; both cross-wind and along-wind response falling quickly.
21.20 to 21.50	5 minute mean wind speed falls to 2 m/s; cross-wind and along-wind displacements fall to between 2 and 10 mm.
21.50 to 21.55	5 minute mean wind speed rises again nearly to 15 m/s; significant cross-wind response begins but dies again as 5 minute mean wind speed drops below 10 m/s.

The traces in Figure 10 show quite well how the cross-wind response dominated through the critical wind speed range and then as wind speeds increase beyond the critical range the cross-wind response fell as the along-wind response continued to rise until the maximum wind speeds were reached.

650

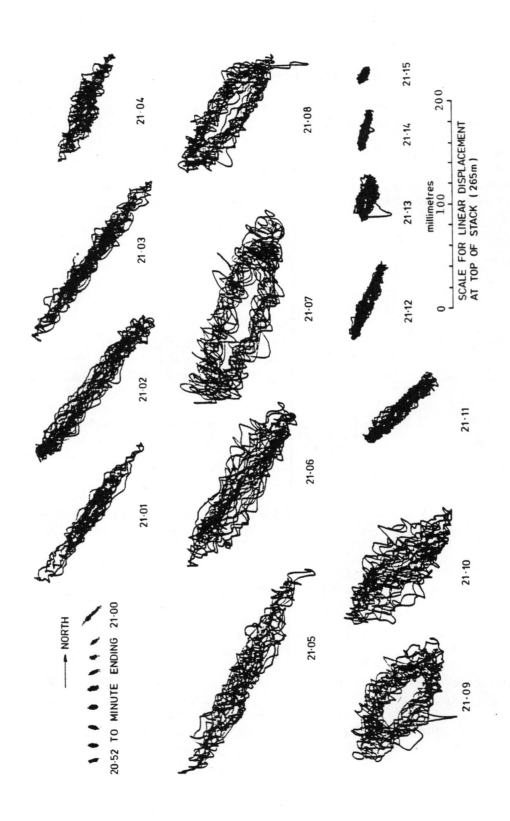

Figure 10   Traces from top of stack NS and EW accelerometer records for each minute during the storm of 5th December 1979 (times shown are the ending for each one minute record).

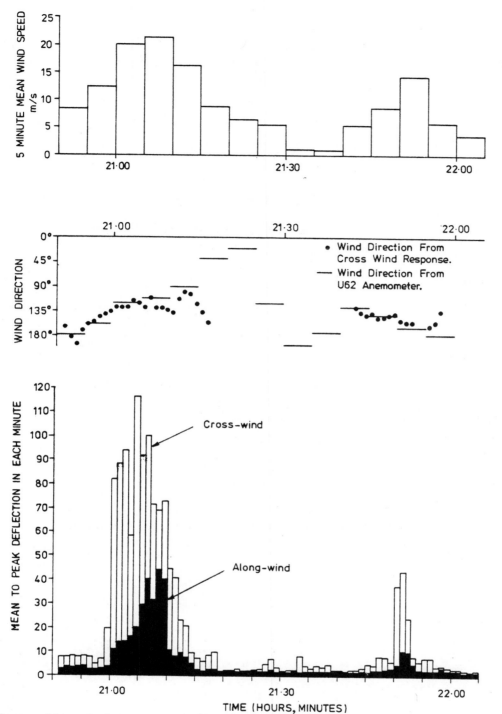

Figure 11   Wind speed, wind direction and top of stack peak to peak displacment as a function of time during the storm of 5th December 1979

4.3   Comparison between Model and Full Scale

To make a comparison between the model base overturning moment measurements and the full scale top displacements , it is necessary to transfer one of these sets of data. In this case the full scale base overturning moments have been calculated from the top displacements through a modal analysis of the stack. The modal analysis was tuned to give the same first mode frequency as that measured in full scale by adjusting the value of Youngs modulus, and which gave the linear relationship between base overturning moment and top displacement as moment (Nm) = 1.15×10^9 top displacement (m).

The base overturning moments from model and full scale in coefficient form are compared in Table 1. The moment coefficients are defined as follows:

$$C_m = M \,/\, \tfrac{1}{2}\, \rho \, \overline{u}_h^2 \, dh^2$$

where   M   is base overturning moment   (mean $\overline{M}$, peak $\hat{M}$)

           ρ is air density

           $\overline{u}_h$ is mean wind speed at the top of the windshield

           d   is outer diameter averaged over the top half of the windshield

           h   is height of windshield

The full scale stack damping was estimated to be $\zeta = 0.0065$ and the model data has been corrected to refer to this value of damping. The mean base overturning moment for the full scale stack was determined from records during the storm of the 5th December 1979. The peak moments of the fluctuating component from the model were determined using peak factors of 3.7 and 3.1 for along-wind and cross-wind respectively, as noted in Figure 7. The full scale peak moments, which are the maximum in a 5 minute period, have been multiplied by a factor of 1.25 to correct them to the maximum likely to occur (on average) during a period of 1 hour at the same mean wind speed.

TABLE 1

Comparison of Full Scale and Model Base Overturning Moment Coefficients for damping $\zeta = 0.0065$. Peak moments are the maximum on average per hour.

	Full Scale	Model High Turbulence	Model Low Turbulence
Mean, $C_{\overline{M}_x}$	0.26	0.36	0.54
Fluctuating along-wind peak, $C_{\hat{M}_x}$	0.22	0.33	0.33
Fluctuating cross-wind peak (at critical velocity), $C_{\hat{M}_y}$	0.78	2.95	3.15
Reynolds number	1.8×10^7	3×10^4	3×10^4

The difference in mean moments can be simply explained in terms of expected Reynolds number effects, refer to Figure 8. The equivalent quasi-steady drag coefficients to give the above moments are 0.6, 0.76 and 1.14 for full scale, model high and low turbulence respectively. The model along-wind fluctuating components are nearly double that measured in full scale. The gust factor for the full scale stack is 1.8 and seemingly in the range which might be expected from a gust factor approach, considering that the turbulence intensity over the top part of the stack is estimated to be about 7%. The model cross-wind fluctuating components are about four times that measured in full scale. Recent unpublished work by Cheung at Monash University has shown that fluctuating lift coefficient values for smooth two-dimensional circular cylinders in a 5% to 10% turbulent flow field vary from about 0.4 at Rn = 10^4 to less than 0.1 at Rn = 10^6, and this goes a long way towards explaining the scale effect in this case.

The general conclusion from this example is that Reynolds number scale effects can be very large indeed. Whilst much of it can be explained, as in this example, there will be many cases of structures involving turbulent flow over curved surfaces where results at model scale will bear little relation to that likely to exist at full scale Reynolds numbers.

5. ACKNOWLEDGEMENTS

In any such studies, as have been previewed in this paper, there have been many people and organisations involved. I would like to acknowledge assistance from the Australian Research Grants Commission, The West Gate Bridge Authority, Mt. Isa Mines Ltd. and Tileman (A/Asia) Ltd, the many people in those organisations who have helped in these studies and the team at Monash University over the past ten years who have been actively involved in both the model and full scale studies.

6. REFERENCES

1 W.H. Melbourne, Cross-wind response of structures to wind action, 4th Int. Conf. on Wind Effects on Buildings and Structures, London,(1975), Cambridge Press, 343-358.

2 W.H. Melbourne, Model and full scale response to wind action of the cable stayed box girder West Gate Bridge, Proc. Symp. Practical Experiences with flow-induced vibrations, Karlsruhe, (1979), Springer-Verlag, 625-632.

3 B.J. Vickery, Unpublished University of Sydney Investigation Reports,(1973).

4 D. Walker, The effects of turbulence on the aerodynamics of a bridge deck, M.Eng.Sc. Thesis, Monash University (1975).

5 R.L. Wardlaw, Some approaches for improving the aerodynamic stability of bridge road decks, Proc. Int. Conf. Wind Effects on Buildings and Structures, Tokyo,(1971) 931-940.

6 I.J. Smith, The wind induced dynamic response of Wye Bridge, Proc. Conf. on Environmental Forces on Engineering Structures, Imperial College, London, (1979) 91-105.

7 J.D. Holmes, Prediction of the response of a cable stayed bridge to turbulence, Proc. 4th Int. Conf. Wind Effects on Buildings and Structures, London, (1975), Cambridge Press, 187-197.

A COMPARISON OF MODEL AND FULL-SCALE BEHAVIOUR IN WIND OF TOWERS AND CHIMNEYS

R. I. Basu and B. J. Vickery, Boundary Layer Wind Tunnel Laboratory, University of Western Ontario, Faculty of Engineering Science, London, Ontario, Canada, N6A 5B9

1. INTRODUCTION

In the literature search for this paper the authors could find no reference to completed studies in which both model and full-scale observations were available. The CN Tower (Toronto, Canada) was the subject of extensive wind tunnel testing and has been instrumented. There is little doubt that eventually this project will yield valuable results but at this point in time there is little to report.

The Sydney Tower [5] was also the subject of an extensive wind tunnel test program and is now being instrumented but it will be some time before there is sufficient data gathered to make meaningful comparisons. A large reinforced concrete chimney examined in a wind tunnel test program (Ref. 8 of [5]) was instrumented and although little data has been gathered this is discussed later in this paper.

Although data from model and full-scale is limited there have been numerous full-scale observations of chimneys and towers of circular cross-section. To compare these results with model data, even if it were available, would be of little value because of the differences in Reynolds Number but a comparison with theoretical predictions based upon theoretical models developed from wind tunnel studies is of great significance.

The content of this paper is therefore concerned with a comparison of full-scale observations with predictions using a theoretical model outlined in reference [5] and described in detail in references 9 to 12, in [5]. The theoretical or mathematical model employed was derived following observations at model scale and has been tested against wind tunnel results from the aeroelastic testing of a wide range of models at sub-critical values of Reynolds Number.

2. SELECTION OF INPUT PARAMETERS FOR THEORETICAL MODEL

The input parameters can be conveniently considered in three categories, namely parameters defining the wind field, those which describe the structural properties and finally the

the aerodynamic parameters which determine the relationship between the fluid flow and the forces induced.

For neutral atmospheres the magnitude and variation with height of the mean wind and turbulence characteristics are usually assumed to be determined soley by the mean wind speed at a specified height (usually 10 m) and z_0, the surface roughness length. The model of wind employed in this study was developed by ESDU [1,2,3]. The principal virtues of the ESDU model are that it is founded on a broad data base, and that it gives a comprehensive description of the wind field. For the computation of response a description of both the longitudinal and lateral components are required. Although numerous models of spectra, coherence, turbulence intensity etc. have been proposed few studies have developed a model as complete as that of ESDU.

The dynamic properties of the structure of concern are the natural frequencies and mode shapes, and the structural damping. In the case studies that follow the natural frequencies and mode shapes were calculated using the well known Stodola method. The structural damping cannot be predicted by theoretically based approaches. Estimates of structural damping are made on the basis of measurements made of this parameter on structures of similar construction. This is a parameter which is difficult to predict with precision. An analysis of measurements of structural damping made on chimneys, towers, etc. by Basu and Vickery [4] suggest the following ranges of structural damping as a fraction of critical:

Unlined steel stacks and similar structures such as distillation columns	*0.002 − 0.01*
Lined steel stacks	*0.004 − 0.016*
Reinforced concrete chimneys and towers	*0.004 − 0.02*

For the computation of response, using the mathematical model briefly outlined in [5], of full-scale reinforced concrete structures in the sections that follow a value of 1% is assumed. In cases where a value of structural damping is quoted then of course the quoted value is used in the computations.

The mathematical model employed for predicting response is briefly outlined, and referenced, in a paper [5] presented earlier at this Workshop. Four aerodynamic parameters define the relationship between the wind and the mean and fluctuating force. They are C_D, the mean drag coefficient; $\widetilde{C}_L$, the rms lift coefficient; S, the Strouhal Number and K_{a_o} the dimensionless negative aerodynamic damping. At the very least all four parameters are functions of Reynolds Number, aspect ratio and the surface roughness of the cylinder. In varying degrees C_D, $\widetilde{C}_L$ and K_{a_o} are sensitive to turbulence. In the context of the present study, which is concerned with the prediction of response of full-scale structures, it is fair to restrict our attention to transcritical Reynolds Numbers (transition precedes separation) and to scales of turbulence many times larger than the diameter of structures under consideration. This restriction makes the selection of the values for parameters less problematical although, as is discussed below, substantial uncertainty remains with some of the parameters. Within the constraints just described C_D, $\widetilde{C}_L$ and S are regarded as dependent only on surface roughness and aspect ratio. The parameter K_{a_o} is dependent on the level of turbulence and on aspect ratio, and also perhaps on other influences but the dearth of measurements of this parameter preclude any firm statements on their nature. It is clear, from the remarks made in the paragraph, that the behaviour of these parameters are very complex. It is not possible to discuss this subject fully here; an analysis of the measurements of the parameters has been made by Basu and Vickery [4]. An abbreviated treatment of each parameter, in two-dimensional conditions with a view to suggesting values appropriate to the computations of response of full-scale structures in the next section, follows. A few comments regarding aspect ratio effects concludes this section.

The mean drag coefficient is the best established of the four parameters under consideration. A large number of experiments have been conducted measuring the influence of surface roughness on the mean drag coefficient. The data falls in two categories; the first contains measurements made on nominally smooth cylinders $(k_s/D \lesssim 10^{-5}$, where k_s is an equivalent sand roughness height), and the second contains measurements made on artificially roughened cylinders where the range $k_s/D = 3 \times 10^{-4} - 3 \times 10^{-2}$ encompasses most of the data. Relative roughnesses typical of steel stacks and reinforced concrete chimneys and towers range from about 10^{-5} to 10^{-3}. For full-scale applications, therefore, it becomes necessary to interpolate between the two sets of data. An analysis of experimental data by Basu and Vickery [4] was undertaken with a view to providing a basis from which to select an appropriate value of C_D. The analysis suggests that the transcritical value of C_D increases from 0.55 for nominally smooth cylinders and approaches 1.2 for the limit of the data, i.e. $k_s/D = 3 \times 10^{-2}$.

The data available on the transcritical rms lift coefficient, $\widetilde{C}_L$, amounts to a small fraction of that available on the mean drag coefficient. The same division between smooth and rough cylinder results that exists with data on C_D also exists in the case of $\widetilde{C}_L$. Furthermore the scatter is greater. Smooth cylinder values for $\widetilde{C}_L$ range from about 0.09 to 0.15. Measurements of $\widetilde{C}_L$ on rough cylinders are few but seem to suggest a value of about 0.28 for a maximum roughness of about 10^{-2}. Because of the limited data the interpolation between the smooth and rough cylinder $\widetilde{C}_L$'s is of a speculative nature.

The analysis referred to above also includes consideration of full-scale measurements of the aerodynamic parameters. In the case of C_D and $\widetilde{C}_L$ the full-scale results and laboratory results were found to be broadly consistent. For the Strouhal No., S, this was not the case. Laboratory based values of S were found to be consistently higher than the, addimittedly few, full-scale values. A decreasing value of S with increasing roughness is suggested by the data. The final values of S recommended by the analysis were adjusted to reduce the gap between the full-scale and model scale results. For a range of roughness similar to that associated with data for C_D and $\widetilde{C}_L$ the Strouhal No. varies from about 0.24 for smooth cylinders down to about 0.21 for rough cylinders.

Measurements of K_{a_o}, the negative aerodynamic damping coefficient, are rare. However its value can often be inferred from response measurements. The general form of the variation of K_{a_o} with Reynolds is similar to the variation of C_L. Maximum values of K_{a_o} are attained at Reynolds Nos. of about 10^4 to 10^5, followed by a steep descent at critical Reynolds Nos. and then a partial recovery at transcritical Reynolds Nos. The limited data suggests that in the subcritical regime K_{a_o} attains a maximum of about 2.2, and in the transcritical the value 1.1 can be regarded as a representative figure. The latter value is assumed in the computations described in the next section. A more complete discussion of this parameter is contained in [4]. The values of K_{a_o} quoted above refer to smooth flow conditions; the presence of large scale turbulence with intensities of the order of 10% reduce these values by a factor of about 2.

The values for parameters discussed in the paragraphs above pertain to two-dimensional conditions. In general full-scale structures such as towers, chimneys etc. are of proportions which cause the value of aerodynamic parameter to be modified significantly. In common with much of the data discussed in this section, consideration of finite aspect ratio effects is hampered by the dearth of data. This is particularly the case with transcritical Reynolds Nos. This subject is discussed in full in [4] .

3. COMPARISON OF COMPUTED RESPONSE AND OBSERVED RESPONSE

The across-wind response of five structures is considered in this section. The mathematical model briefly described in a previous paper [5] is used, employing the input data discussed in the previous section, to yield estimates of response. These estimates are compared with the measured response.

The overall dimensions of the five structures are given in Fig. 1; also indicated are the dynamic properties of each structure. For the purposes of assessing the performance of a mathematical model the ideal situation would be if measured values of the various input parameters were available. Full-scale measurements are an expensive undertaking and are, consequently, rare. In the five cases under examination data on the input parameters required to compute is, in general, sketchy. The approach used in selecting values for input parameters has been as consistent as possible. Measurements of relative roughness of the surface of structures are very few. Since C_D, C_L and S are, according to the analysis briefly described in the previous section, functions of relative roughness it becomes necessary to assume a value for k_s. A value for k_s of 1mm is assumed for concrete surfaces and 0.5 mm is assumed for steel surfaces. Where a parameter value has been measured, or quoted, in the study concerned then the recommendations of the previous section are ignored in preference for the measured or quoted value.

A summary of the input parameters selected on the basis of the above remarks is given in Table 1.

658

3.1 Case Study 1 — 194 m Reinforced Concrete Chimney

This power station chimney is located on the lakeshore, and for the wind direction of interest, has an uninterrupted fetch of open water of some 200 km. Response measurements were made in the form of strains near the base of the chimney. Measurements of wind were made on a tower on the lakeshore. This is assumed to be at 10 m height. The sample length of strain measurement was 15–30 seconds. A simultaneous measurement of wind speed range was made.

A number of assumptions are required in order to transform the results to a usable form. A mean hourly wind speed range is calculated assuming a roughness length, z_o, of 0.001 m. A conversion factor of 1.25 is calculated which, when applied to the 15–30 sec. wind speed sample of 22.2–24.4 m/s, yields a wind speed range of 17.8–19.6 m/s. This is consistent with measurements of mean hourly wind speeds of 19.4 and 19.2 m/s made at nearby meteorological stations noting that the winds were the result of a large-scale weather system.

The main features of the structure are given in Fig. 1a. It is not obvious what the appropriate peak factor should be. Clearly assuming a sample length of 15–30 sec. is not correct. The fact that these measurements were made during a large-scale storm, and remarks made in a report on the measurements, suggest that the response was being monitored over some hours. For the purposes of computing peak response a peak factor based on an hour sample length is used.

The computed and measured response is shown in Fig. 2. The main source of uncertainty in the measured response is with the mean hourly wind speed which is based on a number of assumptions. It is interesting to note that if the measurements of mean hourly wind speed made by the meteorological authorities are appropriate to the site of the chimney then the agreement between observed and computed response is very better.

Also shown in the figure is the computed response assuming structural damping, as a fraction of critical, of 0.02. There is strong evidence that the structural damping of reinforced concrete structures is a function of amplitude, or more accurately the strain. A structural damping of 0.02 is more appropriate at design stress levels than the figure of 0.01.

3.2 Case Study 2 — 180 m Reinforced Concrete Chimney

Measurements of response and the associated wind field were made and reported by Muller and Nieser [6]. The fetch is described as 'flat terrain' and for the nearest 400 m the chimney is surrounded by refinery structures up to 30 m in height. The turbulence characteristics of the 'flat terrain' is assumed to apply since the internal boundary formed by the change of roughness associated with the refinery structures is probably less than a third of the height of the chimney. A roughness length of 0.1 m is assumed.

Initial computations of response yielded values grossly below the estimated response. A closer examination of the results presented suggests a numerical error by a factor of about four in the plotting of response; the results reported by Muller and Nieser are reproduced here unaltered as Fig. 3. Also reproduced, as Fig. 4, is their plot of the power spectra of the along-wind tip displacement. The displacement was deduced from measurements of acceleration. The spectrum corresponds to a mean wind speed (at 180 m height) of 31.1 m/s. At this wind speed the along-wind rms displacement is shown, in Fig. 3, to be 11 cm. This is not consistent with the spectrum of displacement shown in Fig. 4. Making the usual assumption of constant spectral density of force in the vicinity of the natural frequency the mean square response, $\tilde{y}^2$, can be shown to be

$$\tilde{y}^2 = \frac{\pi f_o \, S_F(f_o)}{4 \, \eta \, K^2} \tag{1}$$

where $S_F(f_o)$ = spectral density of force at natural frequency, k = stiffness; η = damping as a fraction of critical. The peak of the displacement spectrum is

$$S_y(f_o) = \frac{S_F(f_o)}{4 \, \eta^2 \, K^2} \tag{2}$$

Substituting eqn. (2) in (1) yields the following expression for mean square displacement

$$\tilde{y}^2 = \pi f_o \, \eta \, S_y(f_o)$$

The damping logarithmic decrement quoted is 0.045; the substitution of this and $f_o = 0.26$ Hz. and $S_y(f_o) = 352.6/0.26$ yields a rms displacement of 2.78 cm which differs from the rms displacement of 11 cm shown in Fig. 3 by a factor of about 4.0.

For the computation of response the following values for the input parameters are assumed, $C_D = 0.74$; $\tilde{C}_L = 0.18$; $S = 0.22$ and $\eta_s = 0.007$. It is assumed that the factor of 4.0 applies to the across-wind response. The computed rms response, multiplied by a factor of 4, is shown in Fig. 5. The computed response compares fairly well for moderate wind speeds except that the peak is underpredicted by about 30%. At high wind speeds, where the across-wind response is almost entirely due to lateral buffeting, response is overpredicted by a substantial amount; this is probably due to the use of a drag coefficient of 0.74 in the calculation of the across-wind buffetting by the lateral component of turbulence.

3.3 Case Study 3 — 50 m. Observation Tower

The measured response, and a description of relevant parameters is given in Vickery [7]. The overall dimensions of the structure and some of its dynamic properties are shown in Fig. 1c. The immediate fetch for the observation tower is approximately 1 km of open water, beyond which the terrain is described as urban to suburban. It seems likely that the wind field at heights of interest will be influenced by both terrains. A roughness length of 0.4 m is selected as a compromise.

The response due to excitation of the 4.88 m diameter stem and of the turret are computed separately. The value of input parameters assumed for the excitation of the stem were $C_D = 0.75$; $C_L = 0.18$ and $S = 0.23$. The higher value of S is justified by the presence of the turret which can reasonably be expected to alleviate to some extent the three-dimensional effects of flows associated with free ends of structures. For excitation of the turret the values assumed were $C_D = 0.4$; $\tilde{C}_L = 0.1$ and $S = 0.22$. The reducing values of C_D and $\tilde{C}_L$ are to account, in a very approximate way, for the low aspect ratio of the turret. Knowledge of fluctuating forces on cylinders of very low aspect ratio is limited. However for the case under consideration this is of little consequence since, as can be seen from Fig. 6, the response at the only wind speed at which response was measured is due mostly to excitation of the stem.

The comparison between measured and computed response is particularly good in this case.

3.4 Case Study 4 — 80 m. Distillation Column

The overall dimensions of a distillation column are shown in Fig. 1d. Over a period of some months at certain wind speeds large amplitude vibrations were observed; peak amplitudes of 0.38 m and 0.69 m were measured. As part of the study the structural damping was determined from the ambient vibrations. A frequency of 0.53 Hz., was measured and the structural damping in terms of the logarithmic decrement was estimated to be

$$\delta_s = 0.020 \pm 0.004$$

The values for the aerodynamic parameters assumed were $C_D = 0.72$; $\tilde{C}_L = 0.17$ and $S = 0.21$.

The magnitude of the vibration amplitudes strongly suggest the presence of instability which in this context means that motion-induced forces are considerably in excess of the forces arising from "stationary cyclinder" vortex shedding. This prompted the presentation of computed response in the form shown in Fig. 7 where the maximum response is shown as a function of structural damping. The measured amplitudes were peak values and the computed response is presented in terms of its root-mean-square value. The peak factor is a function of the ratio of structural damping to negative aerodynamic damping. For stable structures where this ratio is of the order of 2 and greater the response is essentially narrow-band Gaussian and peak factors calculated using the usual formulae are applicable. Where the ratio is of the order of 1 the peak factor is reduced, and for very low structural damping when response approaches a Sinusoidal distribution the peak factor tends to a value of $\sqrt{2}$. This aspect of large amplitude response is discussed in full in Vickery and Basu [8], and on the basis of results presented in the aforementioned paper a peak factor of approximately 3 is appropriate here.

The surrounding terrain was quite non-uniform in character. Therefore response was computed for two roughness lengths which sould enclose the true roughness length. Response in this high-amplitude unstable regime is very sensitive to the level of structural damping. The form of the response illustrates the difficulty in predicting response in this regime. It is difficult, if not impossible, to estimate the structural damping of a full-scale structure with the precision necessary to yield meaningful predictions of response.

3.5 Case Study 5 — 265 m. Reinforced Concrete Chimney

The full-scale measurements of wind and response for this structure are reported in a paper by W. H. Melbourne in these proceedings. Five-minute mean wind speeds and one minute peak along-wind and across-wind displacements are presented.

The aerodynamic parameters used in the response calculations are given in Table 1. A value for z_o, the roughness length, of 0.13 m is assumed on the basis of a description of the terrain. The structural damping ratio assumed is 1%.

Large statistical variability is associated with short averaging times of the kind used in the measurement of response in this case. The measured response is presented in Fig. 8 and has not been amended. Also shown is the computed response. Statistical variability from two sources contribute to the scatter; first, the variance of the displacement has a variability associated with it, and second, the peak factor itself has some variability when the number of cycles is small. In this case the variability from the first source will dominate. The coefficient of variation of the root-mean-square response is given approximately by the following expression:

$$cov = \frac{1}{2\sqrt{B_e T}}$$

where B_e = is an equivalent bandwidth for the process; T = averaging time. In this case B_e is assumed to be $2 f_o \eta_s$, where η_s is the structural damping ratio. For $\eta_s = 0.01$, $f_o = 0.27$ Hz and $T = 300$ secs, $cov = 40\%$. This should be regarded as an upper bound; the equivalent bandwidth is likely to be broader in reality.

In Fig. 8 a line has been drawn through the observed data representing the likely trend of the variation of response with wind speed. The predicted peak response is about 50% greater than suggested by the data. It is difficult to be certain of causes of the differences; apart from short-comings of the mathematical model the most obvious of possible sources of error are in the statistical variability of the observed data (response and wind) and in the selection of input data.

The latter seems a likely source of uncertainty in view of the systematic nature of the difference between observed and predicted response.

4.   CONCLUSIONS

While the theoretical model employed herein is capable of describing the response to shedding over a wide amplitude range there remain difficulties in the specification of the basic parameters at prototype scale. In view of the doubts surrounding the magnitude of these parameters it would be unreasonable to expect agreement in any one study to better than about $\pm$ 30% in the stable range and, because of the extreme sensitivity of response to the difference between aerodynamic and structural damping, a substantially greater margin in the unstable range. With the above comments in mind it can be stated that, for the five studies presented, there was encouraging agreement between predicted and observed results but that further progress in strongly dependent upon the establishment of the relevant aerodynamic parameters at transcritical values of Reynolds Number and accurate determination of structural damping.

ACKNOWLEGEMENTS

The work described herein was supported by a grant from the National Science and Engineering Research Council of Canada.

REFERENCES

1.   E. S. D. U., (1972), Characteristics of Wind Speed in the Lower Layers of the Atmosphere Near the Ground: Stong Winds (Neutral Atmosphere), Engineering Sciences Data Unit, London, England, ESDU Item No. 72026.

2.   E. S. D. U., (1974), Characteristics of Atmospheric Turbulence Near the Ground, Part II, Engineering Sciences Data Unit, London, England, ESDU Item No. 74031.

3.   E. S. D. U., (1975), Characteristics of Atmospheric Turbulence Near the Ground, Part III, Engineering Sciences Data Unit, London, England, ESDU Item No. 75001.

4.   Basu, R. I. and Vickery, B. J., (1981), "Across-Wind Response of Tall Slender Structures of Circular Cross-Section to Atmospheric Turbulence; Part II: Input Parameters, Faculty of Engineering Science, Report No. BLWT–2–81, University of Western Ontario, London, Ontario, Canada.

5.   Vickery, B. J., (1982), "Aeroelastic Modelling of Towers and Chimneys", Int. Workshop on Wind Tunnel Modelling Criteria and Techniques for Civil Engineering Applications, Natural Bureau of Standards, Gaithersburg, MD., U.S.A.

6. Muller, F. P. and Nieser, H., (1976), "Measurements of Wind-Induced Vibrations on a Concrete Chimney", J. Indus. Aerodyn., 1, 3, 239–248.

7. Vickery, B. J., (1974), "Three Studies Involving a Comparison of Full-scale Behaviour With That Determined Theoretically and/or Experimentally", Symp. Full Scale Measurements of Wind Effects on Tall Buildings and Other Structures, University of Western Ontario, London, Ontario, Canada, June 1974.

8. Vickery, B. J. and Basu, R. I., (1982), "Across-Wind Vibrations of Structures of Circular Cross-Section, Part 1: Development of a Model for Two-Dimensional Conditions", submitted for publication in J. Ind. Aerody. Wind Eng.

TABLE 1

SUMMARY OF VALUES ASSUMED FOR AERODYNAMIC PARAMETERS

Structure No.	Aspect Ratio	Relative Roughness	C_D	$\tilde{C}_L$	S
1	11.1	0.57×10^{-4}	0.58	0.14	0.19
2	32*	1.8×10^{-4}	0.74	0.18	0.22
3 (Stem)	>13.3	2.0×10^{-4}	0.75	0.18	0.23
4	20	1.3×10^{-4}	0.72	0.17	0.21
5	22*	0.8×10^{-4}	0.69	0.16	0.21

* based on diameter at 5/6 of overall height

664

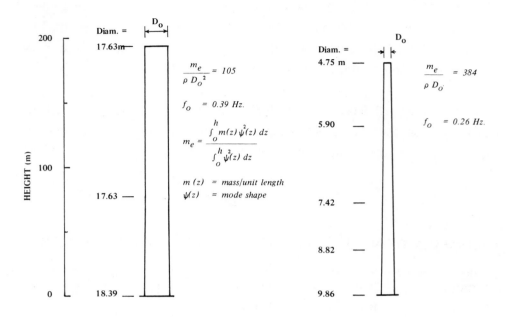

(a) 194 m. R. C. CHIMNEY

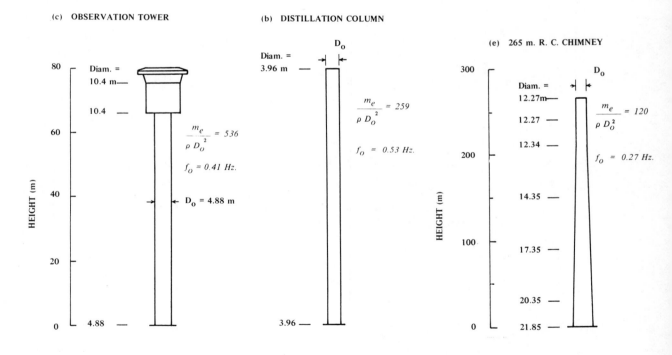

FIG. 1   ELEVATION AND SELECTED DYNAMIC PROPERTIES OF CASE-STUDY

STRUCTURES

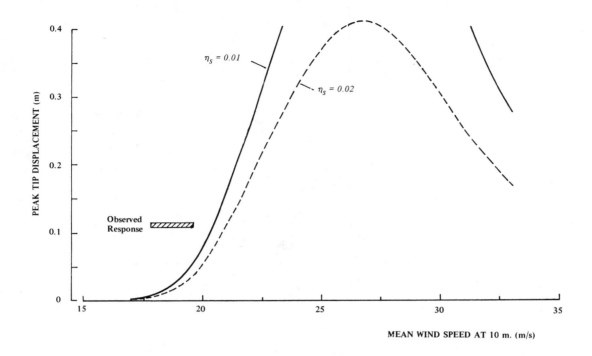

FIG. 2   OBSERVED AND COMPUTED RESPONSE OF A 194 m. R. C. CHIMNEY

(Case Study No. 1)

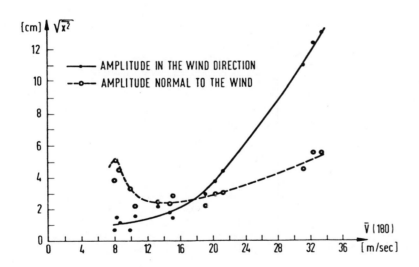

FIG. 3   MEASURED RESPONSE OF CASE STUDY NO. 2 STRUCTURE (After Muller and Nieser [6])

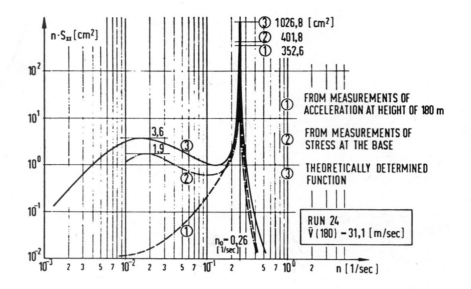

FIG. 4 SPECTRUM OF ALONG-WIND TIP DISPLACEMENT OF CASE STUDY NO. 2 STRUCTURE (After Muller and Nieser [6])

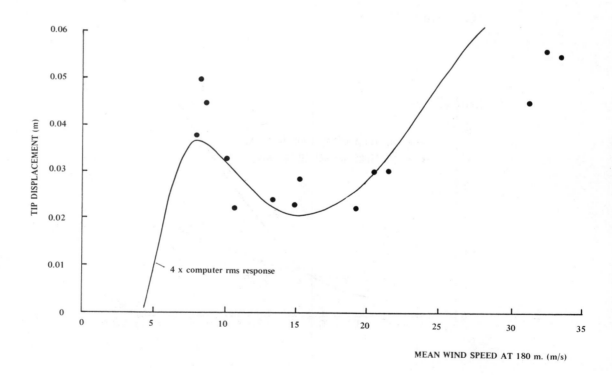

MEAN WIND SPEED AT 180 m. (m/s)

FIG. 5 OBSERVED AND COMPUTED RESPONSE OF 180 m. R. C. CHIMNEY (Case Study No. 2)

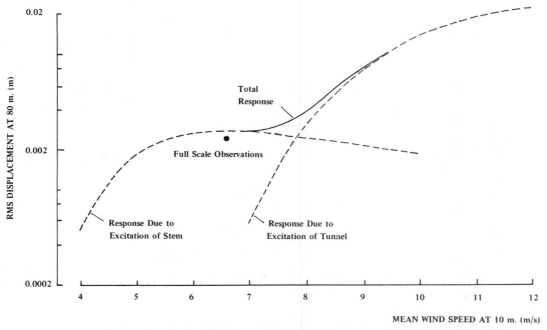

FIG. 6   OBSERVED AND COMPUTED RESPONSE OF OBSERVATION TOWER (Case Study No. 3)

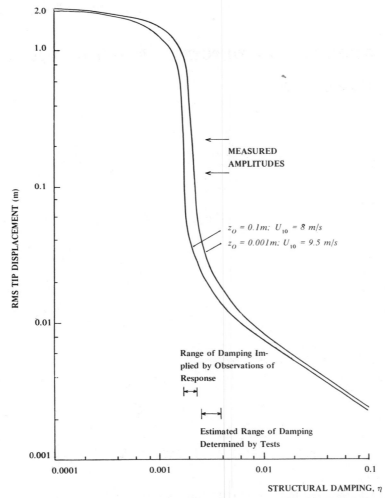

FIG. 7   OBSERVED AND COMPUTED MAXIMUM RMS RESPONSE OF A DISTILLATION COLUMN (Case Study No. 4)

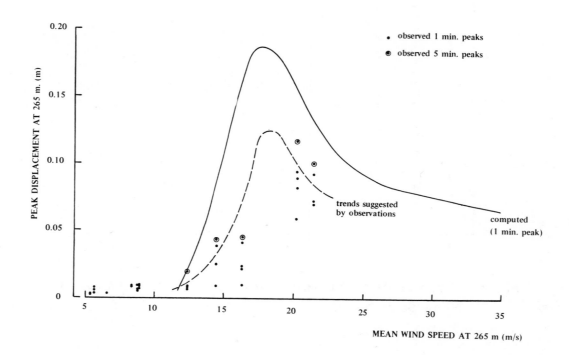

FIG. 8 **OBSERVED AND COMPUTED RESPONSE OF A 265 m. R. C. CHIMNEY**

(Case Study No. 5)

COMPARISON OF MODEL AND FULL SCALE
NATURAL VENTILATION STUDIES

S. Chandra, A. A. Kerestecioglu, P. W. Fairey, III, W. Cromer
Florida Solar Energy Center
Cape Canaveral, FL 32920

1. INTRODUCTION

Historically, man has achieved human comfort in summertime by opening building windows and ventilating. Design guidelines on window size, shape, placement in buildings have been suggested after investigating scale model buildings in uniform speed tunnels in the fifties and early sixties (Refs. 1, 2, 3) and later on by model testing in boundary layer wind tunnels (Refs. 4, 5, 6). To our knowledge, except for the qualitative flow visualization comparisons performed by the Texas researchers (Ref. 7), the literature is devoid of any quantitative comparison of model and full scale naturally ventilated buildings. This is why, as part of our investigation into quantitative understanding of natural ventilation, we decided to undertake this comparison of full scale and model scale internal velocities of naturally ventilated rooms. The full scale studies were performed at the Florida Solar Energy Center in Cape Canaveral, Florida. The model scale studies were performed by the Colorado State University Fluid Dynamics and Diffusion laboratory personnel under the guidance of Dr. Jack E. Cermak.

2. FULL SCALE AND MODEL BUILDINGS

The FSEC site is located within a mile of the Atlantic Ocean as shown in Figure 2.1. The FSEC Passive Cooling Lab (PCL), an experimental building with a fixed roof supported by columns whose floor plan and ceilings are reconfigurable, is the building used in this study. Figure 2.2 is a photograph of the PCL showing its south and east facades. The east wall shows the two 120° overhangs above the openings. This southeast room was the room where the ventilation experiments were carried out. Figure 2.3 shows a closer view of the southeast room with the wingwalls in place. The purpose of the wingwalls are to hopefully increase the ventilation in the room for south, southeast, northeast and northerly winds. The room is not otherwise cross ventilated. Figure 2.4 shows the surrounding buildings -- this photo was taken from a point approximately 400 feet southeast of the PCL.

For testing in the CSU wind tunnel, they built a 1:25 scale model faithfully reproducing the PCL with its movable walls and ceilings. The model was constructed from 1/16" (for roof) and 3/16" thick (for walls) acrylic with steel and aluminum framing as needed. Figure 2.5 shows the floor plan of the PCL. The southeast room is the test room. The velocities and pressure taps for this room are shown in the drawing. All measurements were at mid room height (i.e., 4'0" full scale) above PCL floor in the model. Table 1 compares the model and full scale dimensions as tested.

Table 1.   Model and Full Scale Dimensions as Tested
Nominal Model Scale is 1:25

Dimension	Model		Full Scale	Scale
Interior room length	217	mm	17.60 ft	1:24.7
Interior room width	141	mm	11.70 ft	1:25.3
Interior room height	95	mm	8.05 ft	1:25.8
Window width	48.5	mm	3.88 ft	1:24.4
Window Height	36.5	mm	3.00 ft	1:25.0

3.   EXPERIMENTAL PROCEDURES

The full scale tests were conducted during evening and early night hours so that atmospheric ΔT was nearly zero corresponding to a thermally neutral atmosphere like that in the wind tunnel.   The atmospheric measurements consists of six instruments on a mast located about 150 feet away from the building.   The mast had instrumentation at two levels, approximately 10 m and 1 m above ground.   Each level had a precision cup anemometer (threshold 0.5 mph, full accuracy 1.5 mph), wind vane and a thermocouple in a naturally aspirated radiation shield.   The velocities inside the room were measured by a TSI model 1620 omnidirectional air speed probe.   These appear to have an accuracy of ±10% over a range of 0.1 to 5 mph.

The probes have a small (2 mm) sphere with hot wires as the sensing element and are temperature compensated.   They have a nonlinear output which was curve fitted and read by the data processing software.   The response time constant of the field probels are given in Table 2.

Table 2.   Response Time for Field Probes

Probe	Time Constant
omniprobe	< 2 seconds
cup anemometer	< 2 seconds
wind vane	< 10 seconds

Generally only one omniprobe was present in the room at location #1 or #2.   As noted later, most of the time, the height above floor was 4'5" rather than 4'0".   Some data was also obtained by bringing the lower cup anemometer from outside to location #1 inside.

All data channels were simultaneously recorded every 10 seconds, the fastest scan rate available.   Typically, data were averaged over five minute intervals.

The CSU, Fluid Dynamics and Diffusion Lab meteorological wind tunnel (MWT) was used for all model studies (Figure 3.1).   Neighbouring buildings were modeled from styrofoam or masonite and installed upwind as appropriate.   An atmospheric surface layer approximately 30 m deep was simulated by following techniques recommended by Cook (Ref. 8).   Six-ft wood spires were positioned across the MWT at the test section entrance.   These were followed by a 7-inch trip and varying degrees of surface roughness. The four spires were located at 17-inch intervals, while the trip was continuous across the tunnel. In addition, 7.87-inch roughness cubes were positioned near the MWT entrance to further enhance development of the desired boundary layer. The floor of the test section was covered with 28 feet of one-inch roughness cubes, followed by 48 ft of one-half inch roughness and terminated with 8 ft of one-quarter inch smooth masonite upon which the PCF model rested. Graphic illustration of the MWT configuration (complete with pertinent dimensions) is included _ Figure 3.2 and 3.3.

The velocities were measured by quick response hot film anemometers with the sensor axis vertical. TSI 1211-10 and 1210-20 cylindrical hot film sensors were used. CSU experience has been that mean velocities measured are accurate to ±10 percent, similar to expectations of full scale accuracy.

4. RESULTS

The full scale tests were conducted during February and March of 1982. Four out of five of the full scale data sets reported here correspond to a thermally neutral atmosphere with a ΔT <0.1°C and were collected during the evening and early night hours with ambient temperatures between 19°C and 21.5°C. Wind speeds ranged between 4 and 16 mph encompassing the speeds prevalent during ventilation conditions (4-7 mph or less) and stronger winds. Wind directions were generally between east and south -- summertime directions coming from the ocean.

Figures 4.1 and 4.2 compare the full scale and model scale internal to external velocity ratios for several wind directions. In 4.1 one sees the comparisons for the configuration without the wingwalls for internal probe location #1 (see Figure 2.5) Note that the field data overlapped for two different days (3/13 and 3/14) for WD = 135 and were about 30 percent lower than the model data. The data set for WD = 90, collected daytime under bright sun and high ΔT and fairly low winds (4-6 mph) appear to be $\cong$ 30 percent higher than the model data.

All three sets of data in Figure 4.2 are for the configuration with the wingwalls in place for internal probe location #2 (see Figure 2.5). All data are for nearly neutral atmospheres. Unfortunately the field and model internal probe locations were not identical in the z coordinate as noted in the graph. As can be seen, for WDs of 135 and 180 (SE and S) the full scale ratios exceed the model scale by about 30 to 40 percent and for WD = 90° by about 200 percent.

In order to understand this discrepancy, Figures 4.3 through 4.6 were drawn from the data of 2/2, 2/15 and 2/16. (Figure 4.6 contains some additional points.) Figure 4.3 compares the atmospheric velocity ratio at 1.1 m and 9.7 m above ground. It can be seen that the atmospheric mean speed variation with height is well modeled in the tunnel and so is the turbulence at the lower level as evidenced by Figure 4.4. However, Figure 4.5 shows that the atmospheric turbulence at the 9.7 m level is about twice as great as the tunnel. This is perhaps explained by the fluctuations in the atmospheric wind direction shown in Figure 4.6. The wind tunnel fluctuations in wind direction are less than perhaps two degrees.

5. DISCUSSIONS

It may be useful to list and discuss the differences during the full scale and model tests.

i) <u>Small dimensional differences</u> -- See Table 1. Due to the likely predominantly 2-D nature of the flow at the internal velocity probe locations, this is unlikely to explain the large observed differences.

ii) <u>Differences in the probe types</u> -- The external wind measurements were done by sensitive cup anemometers which should measure the same mean horizontal vector wind speed as the vertical hot film probe in the model scale. One could argue that since the wind tunnel internal probe measured only the horizontal mean airspeed and the full scale internal probe was omni-directional, the model mean velocities would be lower if the mean flow was three dimensional. To check this, full scale tests were conducted where the omni-directional probe and a cup anemometer were placed as close together as possible at measurement location #1, with the small (2 mm) omni probe upstream of the 6-inch cup to minimize interference. Figure 4.7 compares the readings from the two probes. The higher airspeeds were obtained with the wingwalls with the probes 4'5" above the floor and the lower airspeeds (< 0.5 mph) for probes located 4'0" above the PCL floor and without the wingwalls. Both sets of data were for southeasterly winds.

It can be seen that below about 1.5 mph the cup reads lower than the omniprobe, and above 1.5 mph the readings are essentially identical. It could be that at lower airspeeds the airflow is more three dimensional but the more likely reason for this behavior is the fact that the cup threshold is 0.5 mph and full accuracy of the cup is not reached until 1.5 mph. With this information, we proceeded to plot V_1 (omniprobe) and U_1 (cup) as shown in Figure 4.8 for the criterion of $U_1 > 1.5$ mph. A cup at 10 meters was not available and so the U_1 and V_1 were nondimensionalized with an outside atmospheric cup anemometer located level with the internal probes or at 1.57 m above ground. The other difficulty with this data set was due to an instrumentation problem with the WD vane. Lot of data on WD had to be discarded and as a result only 1 minute (6 point) averages could be plotted on figure 4.8. (All the other graphs have been plotted with 5 minute averages). In spite of the data scatter it does appear that there is closer agreement between full scale and model data.

iii) <u>Height Differences</u> -- The PCL floor is 8½" above zero ground ($\equiv$ CSU tunnel floor) and that is the way it was built at CSU. However, the real ground is wavy and the field mast base is actually 9½" below zero ground. There was some confusion in sensor heights. The model scale results were taken with internal probes located at midheight of the room and external measurement reported with tunnel floor as zero height. As noted earlier, prior to March 12, 1982 all field internal probes were located at slightly above room midheight (4'5" above PCL floor rather than @ 4'0"). The outside cup locations also varied with the 10 m cup at 9.7 meters before 3/12/82 and at 9.9 m after. The lower cup was at 1.1 m before 3/1/82 and at 1.57 m during 3/4 and 3/5/82. These should cause only minor differences in the comparisons.

iv) Reynolds No. differences -- The model test were conducted at a Re of 1.26e+05 corresponding to a tunnel speed of 21.5 mph (9.6 m/sec), a kinematic viscosity of 1.7e-05 sq.m/sec and a model aerodynamic radius of 0.224m. The full scale Re equals 1.68e+05 per mph of wind speed for a kinematic viscosity of 1.5e-05 sq.m/sec. So full scale Re for the 4-16 mph range would vary between 6.7 and 26.9e+05 -- 5 to 20 times greater than model scale. However, a plot of internal airspeed vs. Re for NE winds showed it to be independent of Re for Re > 0.7e+05 although the internal turbulence intensity kept on varying up to Re=1.26e+05. This would seem to imply Re would have little effect on mean internal airspeeds.

v)   Time Scale and Wind Direction -- If a time scale is defined as the aerodynamic radius divided by the wind speed, the tunnel time scale was 0.01 seconds and the field time scale varied between 0.78 and 3.13 seconds corresponding to 16 to 4 mph winds. This in conjunction with the differences in wind direction variations and sensor time constants may explain part of the differences between field and model scale turbulence intensities.

vi) Nature of Flow -- The mean pressure differences at the apertures (closed) without the wingwalls are very small. Table 3 lists the Cp's for the two pressure taps #24 and #26 (see figure 2.5) as measured in the model.

TABLE 3. Mean Pressure Coefficients (Cp's) at
taps #24 and #26

WD	#24	#26	ΔCp
90 E	+0.517	+0.472	0.045
135 SE	-0.096	-0.134	0.038

Indeed the forcing mean pressures are about an order of magnitude smaller than that prevalent for cross ventilated rooms. So the flow will definitely be affected by turbulence characteristics. So for these cases the disagreements as seen in figures 4.1 and 4.2 are perhaps not surprising in view of figure 4.5 which shows that the turbulence at the 10 m level was not exactly modeled. The large 200 percent discrepancy for easterly winds with wingwalls can be due to the highly unsteady nature of the flow. Full scale smoke pictures taken show that the apertures were alternating as inlets and outlets. Moreover there are not sufficient number of data points near WD = 90.

However, comparing and extrapolating the V_1's for the southeastern winds one notes that full scale $V_1/U_{9.9}$ goes from about 0.07 for no wingwall

(Figure 4.1) to about 0.4 with the wingwall (figure 4.8, assuming $U_{1.57}$/$U_{9.9} \sim 0.7$) -- a 500% increase! This seems to indicate that now the flow is predominantly mean pressure driven. And indeed the discrepancy between model and full scale appears to decrease (Figure 4.8).

6. CONCLUSIONS

To the authors' knowledge, this is the first time quantitative comparisons have been performed for model and full scale naturally ventilated structures. The results show small disagreements between full scale and model internal velocities for flows not driven by mean pressures. This is partly explained by the inexact modelling of the atmospheric turbulence of speed and direction. For flows which are mean pressure driven, e.g., SE winds with wingwalls, the agreement between model and full scale internal velocities are quite good.

These data lead us to believe that boundary layer wind tunnel testing is indeed a good method for performing tradeoff studies as has been done by Sobin, Aynsley and Vickery (Refs. 4, 5, 6) -- especially since in good natural ventilation designs, airflows are produced mainly by mean pressure differences.

Future research should be aimed at turbulence modelling (vibrating the model in the tunnel may be an interesting approach). Also some model to full scale comparisons should be conducted with uniform speed wind tunnels to determine the validity of existing natural ventilation data collected in uniform speed tunnels.

7. NOMENCLATURE

TI - Turbulence intensity defined as standard deviation divided by the average of the instantaneous airspeed indications.

U - Mean airspeed in the horizontal plane i.e. all model velocities and all full scale velocities measured by the cup anemometers. Subscripts 1 and 2 i.e., U_1, U_2, refer to locations #1 and #2 in Figure 2.5. Two or three digit subscripts like $U_{1.57}$ or U_{10} refer to free stream velocity at 1.57 or 10 m (full scale) above ground.

V_1, V_2 - Mean airspeed measured in full scale by the TSI omniprobes (model 1620)

ΔT - Atmospheric temperature difference in the field between that measured at 0.71 m and 9.46 above zero ground.

WD - Wind Direction with respect to the PCL measured at a height of 9.7 or 9.9 m above ground.

8. ACKNOWLEDGEMENTS

The authors appreciate the cooperativeness of Dr. J.E. Cermak and his colleagues at the Colorado State University, Fluid Dynamics and Diffusion Laboratory, in conducting the model tests. This work is funded by U.S. Department of Energy, San Francisco Operations office, contract No. DE-AC03-SF11510 under the Passive and Hybrid Solar Cooling program; Ms. J. Neville, program manager.

9. REFERENCES

1.   Texas A&M research reports #22, 33, 36, 45, 50 now out of print. For a summary by one of the original authors see Evans, B.H., "Energy Conservation with Natural Airflow Through Windows," ASHRAE Transactions, vol. 85, part 2, 1979, pp 641-650.

2.   Van Straaten, J.F., Thermal Performance of Buildings, Elsevier Architectural Science Series, Elsevier 1967, ch. 14.

3.   Givoni, B., Man, Climate and Architecture, 2nd Ed. Applied Science, London, 1976.

4.   Sobin, H., Window Design for Passive Ventilative Cooling: An Experimental Model Scale Study, Proceedings, AS/ISES International Passive and Hybrid Cooling Conference, Maimi Beach, 1981, pp. 191-195.

5.   Aynsley, R.M., Tropical Housing Comfort by Natural Airflow, Building Research and Practice, U.K. July/August 1980, pp 242-252. Also see Proceedings of the Fifth International Wind Engineering Conference, Fort Collins, CO, 1979, Pergamon Press.

6.   Vickery, B.J., The Use of Wind Tunnel in the Analysis of Naturally Ventilated Structures. Proceedings - AS/ISES International Passive and Hybrid Cooling Conference, Miami Beach, 1981, pp. 728-742.

7.   Smith, E.G., The Feasibility of Using Models for Predetermining Natural Ventilation. Research Report #26, Texas Engineering Experiment station, Texas A&M University, June 1951.

8.   Cook, N.J., "Determination of the Model Scale Factor in Wind Tunnel Simulations of Adiabatic Atmospheric Boundary Layer," J. Industrial Aerodynamics, vol. 2, No. 4, pp 321-331.

676

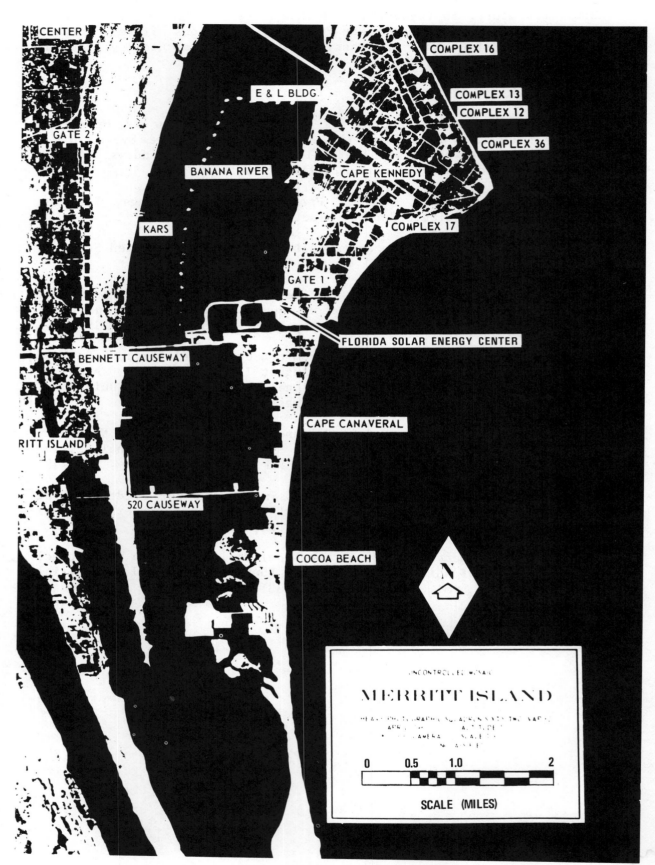

FIGURE 2.1   FSEC Site and Surrounding Area

Figure 2.2   The FSEC PCL, SE view
without wingwalls

Figure 2.3   Closeup of the test room
exterior showing the
removable wingwalls in place

Figure 2.4   Buildings near the PCL (SE view from a point
about 400 ft. from PCL)

678

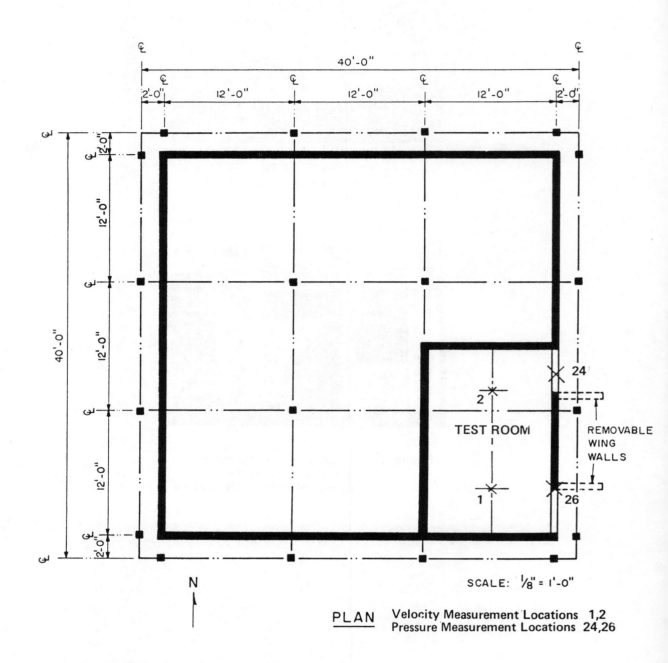

FIGURE 2.5. Plan View of the PCL Model
(full-scale dimensions showing velocity and pressure locations)

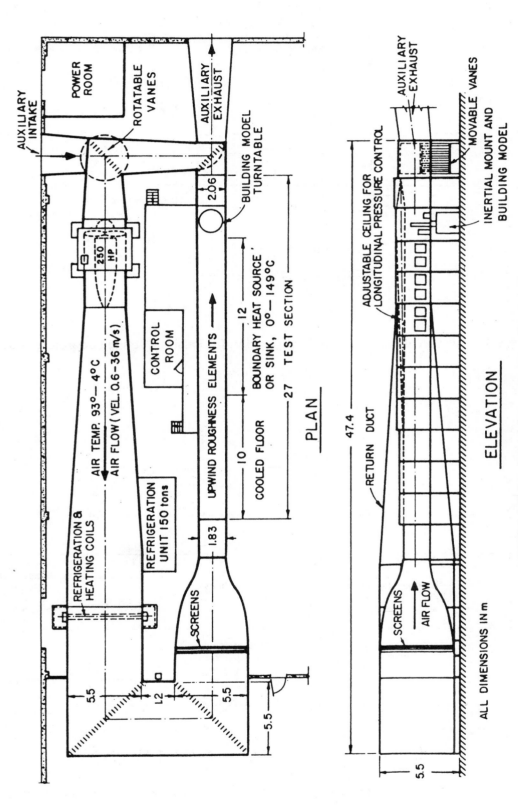

PLAN

ELEVATION

ALL DIMENSIONS IN m

FIGURE 3.1. METEOROLOGICAL WIND TUNNEL
FLUID DYNAMICS & DIFFUSION LABORATORY
COLORADO STATE UNIVERSITY

680

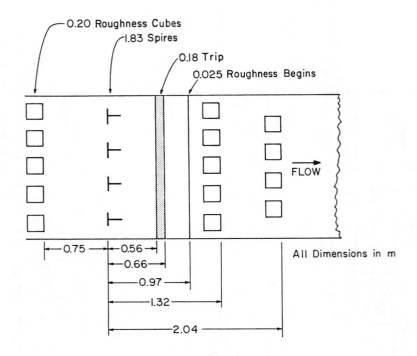

FIGURE 3.2. Wind-Tunnel Entrance Configuration for Passive Cooling Laboratory Study

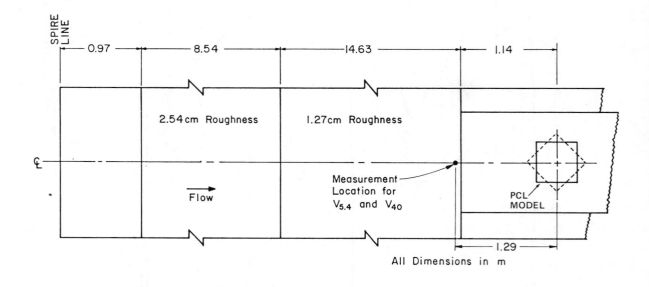

FIGURE 3.3. Wind-Tunnel Test Selection Configuration for Passive Cooling Laboratory Study

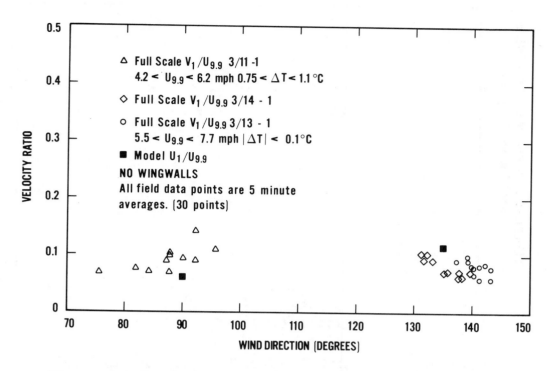

Figure 4.1 Comparison of full scale and model internal velocities without wingwalls

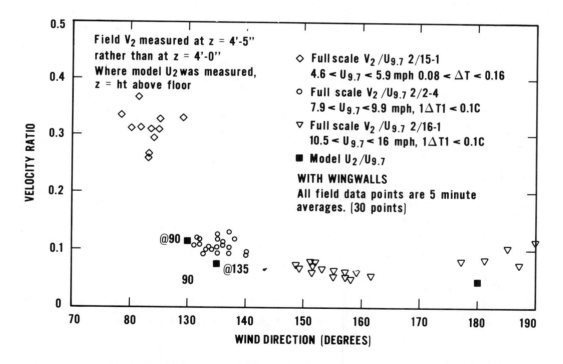

Figure 4.2 Comparison of full scale and model internal velocities with wingwalls

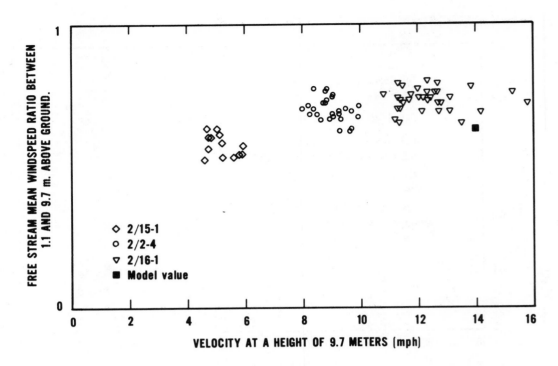

Figure 4.3 Comparison of mean windspeed variation with height

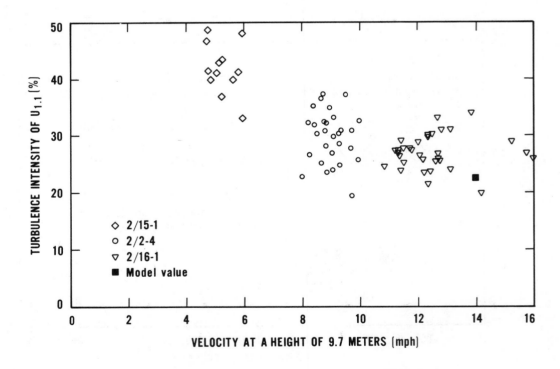

Figure 4.4 Comparison of full scale and model free stream turbulence intensities at 1.1 m. (full scale) above ground.

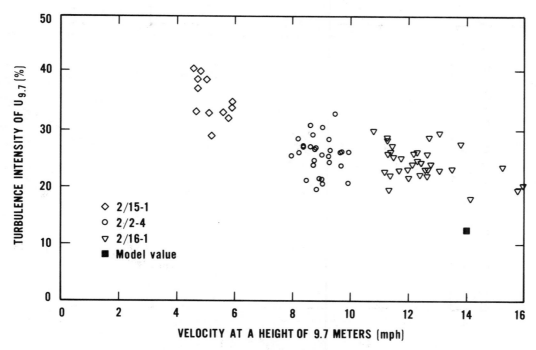

Figure 4.5 Comparison of full scale and model free stream turbulence intensities at 9.7 m. (full scale) above ground.

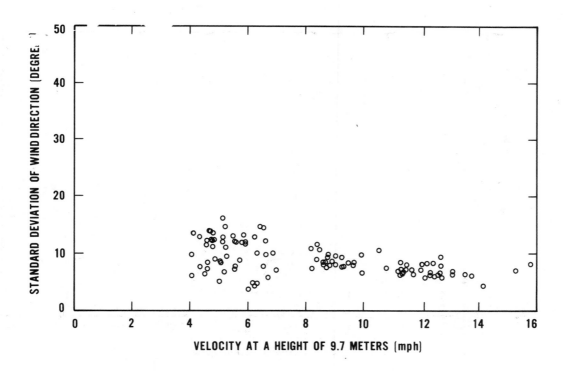

Figure 4.6 Full scale fluctuations of wind direction

684

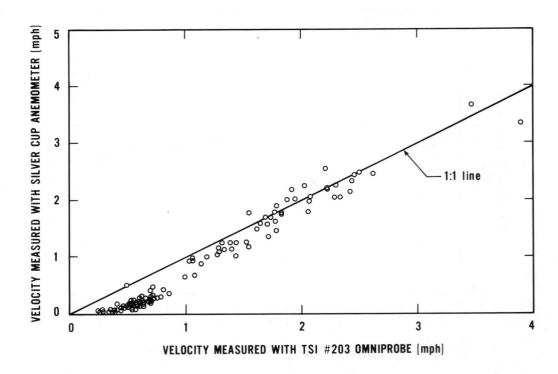

Figure 4.7 Comparison of omnidirectional probe and
cup anomometer both located at position 1

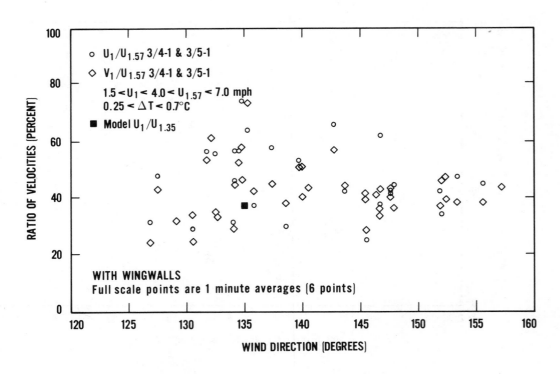

Figure 4.8 Comparison of full scale and model internal
velocities at location 1

A PREPARED CRITIQUE OF PAPERS ON THE VALIDATION
OF WIND TUNNEL TESTING

P. R. Sparks, Virginia Polytechnic Institute and State University, U.S.A.

In preparing this critique I have tried to take the position of an Engineer searching for information to prepare a guide for fellow Engineers about the validity of design information provided by wind-tunnel tests. I have done this mainly on the basis of the papers presented in this session but I do have to admit to inside information on some of the tests reported, having been connected to some extent with the full-scale and model tests reported both by Holmes and by Lee.

As Melbourne points out it is surprising how few full-scale tests have produced results in a form that can be used for the validation of wind-tunnel tests. I believe that this is because many tests are over ambitious in their scope and end up being essentially instrumentation projects. One quickly realizes that the highest correlation that one is likely to encounter is that between storm conditions and malfunction of measuring or recording equipment. A long term committment is also required for these tests and useful results may take many years to obtain. The pressure for rapid publication and the availability of only short term funding has caused many research workers to shy away from this type of work.

The 6 papers presented at this session cite some of the best examples of full-scale testing but one would be hard pressed to find sufficient information to write another 6 papers of this type. This is a sad state of affairs in a field which is still essentially empirical.

The cynical Engineer may look at the papers and conclude that provided one knows what wind pressures, deformations and dynamic characteristics exist for a particular structure, a wind-tunnel model can be created to simulate them reasonably well. The obvious counter to this is to ask the Engineer who is designing the structure, if he is able to predict, at the time of the initial wind-tunnel tests, the structural characteristics of his structure, how the loads are to be carried or even what shape it is eventually going to be. There is a tendency towards complacency by the Structural Engineer in these days of large computer programs for structural analysis. Recent work by the author, as yet unpublished, suggests that elements not usually considered in design or incorrect assumptions of behaviour may have dramatic effects on the actual characteristics of buildings, not always resulting in conservative structures. Natural frequencies may be doubled or even halved compared to the initial design predictions. Even when the frequencies are correctly predicted there may be significant differences between the actual way in which the wind loads are distributed to the components and that assumed in the design. The level of structural damping in a structure is still to a large extent an inspired guess. Yet many papers show that these parameters are of fundamental importance in setting up wind tunnel tests and interpreting the results, particularly with regard to long-span bridge testing and any aeroelastic studies of buildings, chimneys and towers.

685

The interaction between a bridge's movement and the air flow past it generally dictates the use of some form of dynamic model. These models have become very sophisticated and Davenport's paper shows a very intelligent use of existing full-scale data to explain and correct deficiencies in modelling techniques.

In most buildings the movement of the structure has very little influence on the air flow past it. In view of the inherent deficiencies in the prediction of dynamic characteristics of buildings, mentioned earlier, I believe that elaborate aeroelastic models may be of limited use for initial wind-tunnel studies. In reference 3 of Dalgleish's paper he reports fairly good agreement between the original aeroelastic model results and the observed deflection, after the original aeroelastic data had been modified to take into account errors in the original assumed natural frequencies. No mention is made of the uncorrected data upon which the original structural design would have been based. The original model was also only 2 dimensional and, as Dalgleish points out, torsion plays a significant role in the response of that building.

Dalgleish's use of an aeroelastic model based on observed full-scale characteristics, provides a valuable tool for checking the validity of the modelling of the approach flow but it is an extremely elaborate model if indeed the interaction between the building movement and the flow can be neglected. I believe there is considerable potential in the use of rigid models which provide information about modal force spectra which are independent of the dynamic characteristics of the building. These can then be applied easily to a wide range of structural characteristics which the designer may consider likely to exist in his building. Session 3 contained papers by Kareem and Tschanz which described methods for making measurements of these spectra.

In this session Lee reports a technique which makes use of the analogue which exists between the base bending moment and the modal force for linear mode shapes in translation. Unfortunately no similar analogue exists between the base torque and the modal torque. Despite this deficiency this technique has an advantage that the integration of pressures is done by the model just as occurs at full-scale. An important feature of his work is the active role taken by the full-scale structure. The structure was actually calibrated and used as a transducer. Although a great deal of data has been collected, most has yet to be analysed, and the applicability of the technique has yet to be fully tested. What is made clear by Lee and strongly emphasized by Holmes, is the difficulty in defining the reference conditions under which the full-scale measurements took place. Lee shows that the anemometer on the top of the Arts Tower was strongly influenced by the flow over the roof and by the wind direction. He was able to make use of wind-tunnel tests to correct observed values. Since a validation of wind tunnel techniques will generally be based on some normalization procedure, it is extremely important that conditions are defined correctly. Measurements at one point are unlikely to be sufficient if one is trying to pin-point reasons be differences between forces measured in the wind tunnel and at full-scale.

Problems associated with different approach conditions are highlighted in Holmes' paper. The approaching flow conditions were fairly well documented at Aylesbury but not closely modelled in the wind-tunnel tests quoted. When one is dealing with low rise buildings in the lower few meters of the boundary layer, velocity profiles and turbulence intensities become very important. Provided the turbulence intensity is matched well and local pressure coefficients are used, it appears that fair agreement can be obtained between model and full-scale results for mean pressures. The author has observed at other sites much greater variation in wind direction at full-scale than that which is modelled in the wind tunnel. In areas where mean pressure coefficients are very sensitive to wind direction, such directional variations manifest themselves as large r.m.s. coefficients. This may be the case at Aylesbury where, for some wind directions, the r.m.s. coefficients are badly underestimated in the wind-tunnel tests.

Another problem associated with reference conditions is pointed out by Dalgleish. Pressure measurements are always made as differential pressures. In a wind tunnel the static side of a pitot-static tube is often used. No similar absolute reference can be found in full-scale measurements. Dalgleish also points out that in cold weather the stack effect in a tall building may create pressures differences as large as a moderate wind. Clearly one has to be careful what results are used to validate wind-tunnel tests. Conversely one must also be aware that even in a precise wind-tunnel model the pressures measured on this type of building are only part of the total pressure acting on the building skin.

In the case of chimneys and towers of circular crossection, I often wonder why one does not make as many full-scale measurements as possible, develop a set of standard design curves and forget about wind-tunnel testing. The Reynolds Number problems seem so difficult to overcome that very little faith can be placed on the results. Melbourne is particularly candid about this in his remarks.

Basu and Vickery have made a valiant attempt to develop a theoretical technique, based on a series of wind-tunnel measurements, to predict the response of circular section towers and masts. They met with some success but in the end conclude that structural parameters must be specified far more accurately than can be done at present, if meaningful predictions of behavior are to be made.

So, are wind tunnel techniques validated? Not yet, not until many more full-scale measurements have been made. What comparisons have been made, have indicated shortcomings and as a result improved techniques have been developed. International comparisons of the type described by Holmes may help greatly in establishing acceptable standards for wind-tunnel testing, but knowing the shortcomings of the Aylesbury data, I would like to see more full-scale tests used in the comparison. This may necessitate more full-scale projects but they need not be as elaborate as the Aylesbury project. A wood shell would be sufficient and if mounted on a turntable would provide a much broader base of data. Such a building has been constructed very cheaply at Virginia Polytechnic Institute for the study of wind loads on solar collectors.

What should one advise an Engineer comissioning a wind tunnel test or using the results of such a test? On the whole he should not expect the results to be any more accurate than any of his other design parameters which result from a combination of random variables. Unfortunately he may be unaware of the shortcomings in the accuracy of those parameters.

There have been relatively few wind-tunnel tests carred out on large-span bridges but several of those tested have been instrumented. Model tests are usually very elaborate and carried out extremely carefully because of the nature of the instability problem. An Engineer can probably expect fairly reliable data from such a test. He can also be fairly confident of pressure measurements on a high rise building especially when used as a check on pressures defined by codes of practice. He should be more skeptical of response predictions, realizing that they depend to a large extent on his predictions of the dynamic properties of the structure.

An Engineer is unlikely to commission a wind tunnel test on a low-rise building but he may use a code of practice based on wind-tunnel results. At this stage he should realize that there is a great deal of uncertainty in these design pressures and should bear that in mind in his design.

Finally, if he is to commission a test on a circular chimney he should ask the wind tunnel operator how confident he is about being able to provide accurate results. If the operator is completely confident, the Engineer should go elsewhere.